Linear Programming and Economic Analysis

Linear Programming and Economic Analysis

ROBERT DORFMAN
Professor of Economics
Harvard University

PAUL A. SAMUELSON
Professor of Economics
Massachusetts Institute of Technology

ROBERT M. SOLOW
Professor of Economics
Massachusetts Institute of Technology

DOVER PUBLICATIONS, INC.
New York

Published in Canada by General Publishing Company, Ltd., 30 Lesmill Road, Don Mills, Toronto, Ontario.
Published in the United Kingdom by Constable and Company, Ltd., 10 Orange Street, London WC2H 7EG.

This Dover edition, first published in 1987, is an unabridged and unaltered republication of the work first published by The McGraw-Hill Book Company, New York, in 1958. It is reprinted by special arrangement with The RAND Corporation, 1700 Main St., Santa Monica, California 90406-2138.

Manufactured in the United States of America
Dover Publications, Inc., 31 East 2nd Street, Mineola, N.Y. 11501

Library of Congress Cataloging-in-Publication Data

Dorfman, Robert.
Linear programming and economic analysis.

Reprint. Originally published: New York : McGraw-Hill, 1958.
Bibliography: p.
Includes index.
1. Economics—Mathematical models. 2. Linear programming.
I. Samuelson, Paul Anthony, 1915– II. Solow, Robert M.
III. Title.
HB135.D67 1987 330'.0724 87-24382
ISBN 0-486-65491-5 (pbk.)

Foreword

Is linear programming something new or just a new name for old methods? Does it help in analyzing economic and business problems? Can it solve practical problems? The RAND Corporation, a private research corporation, whose primary contract is with the United States Air Force, has found linear programming expedient in practical problems and fruitful in analytic procedure. In part this is simply the result of the fact that much of standard economic analysis is linear programming.

In 1951, RAND and the Cowles Foundation for Research in Economics jointly sponsored *Activity Analysis of Production and Allocation*, a book that dealt with the mathematical and computational features of linear programming. Now RAND presents a book emphasizing the economic interpretation of linear programming.

While this book is intended not as a text but as a general exposition of the relationship of linear programming to standard economic analysis, it has been successfully used for graduate classes in economics. It is hoped that it will satisfy the curiosity of all economists, from first-year graduate students learning the "old" classical principles to their teachers who want to know what is "new."

Preface

Linear programming has been one of the most important postwar developments in economic theory. Its growth has been particularly rapid, thanks to the joint efforts of mathematicians, business and defense administrators, statisticians, and economists. Yet the economist who wants to learn how linear programming is related to traditional economic theory can nowhere find a comprehensive treatment of its many facets. The present book hopes to give the economist, who knows existing economic theory but who does not pretend to be an accomplished mathematician, a broad introduction to the theory of linear programming, or, as it is sometimes called, activity analysis. It hopes also to be useful to the practitioner of managerial economics, and possibly to provide the growing body of mathematicians interested in programming problems with insights into the vast body of modern economic theory.

When asked by The RAND Corporation to undertake the book, we agreed to avoid higher mathematics. We planned to stress the economic aspects of the problem, paying attention to practical problems of computation and giving important concrete applications but laying no stress on them. So vast has the theory become that we have had to be selective, reluctantly deciding to omit many interesting topics and applications. Thus, we have not dealt with the important role of linear-programming concepts in statistical decision theory.

On the other hand, we have gone into the extensive interrelations between the celebrated von Neumann theory of games and linear programming, particularly since every economist will want to know the interrelations between game theory and traditional economic theories of duopoly and bilateral monopoly. And modern economists will be interested in the interrelations between linear programming and modern welfare economics and the insights that linear programming gives into the determinateness of Walrasian equilibrium—as perfected by the recent works of K. Arrow, G. Debreu, L. W. McKenzie, and others.

This book can also serve as an expository introduction to the student interested in the Leontief theory of input-output, which has played so important a role in the last twenty years. Similarly, we have treated extensively problems of dynamic linear programming, not only because of their intrinsic interest but also because of their vital connections with the economist's theory of capital—that most difficult field of modern economic theory. Had we more space and time at our disposal we might have added some material summarizing the related "dynamic programming" methods of Richard Bellman, also developed at RAND. This new theory is of considerable interest to economists but mathematically more difficult than what we have attempted here. Fortunately, Bellman has just published a full exposition of his own.

Our task took more time than we had expected, primarily because we found ourselves in somewhat the same situation as the friend of Dr. Samuel Johnson who explained that he had hoped to become a philosopher but "cheerfulness kept breaking in." Our task of quickly providing an explanation has been frustrated because originality kept breaking in—as gaps were discovered in the existing theory or as whole new fields for analysis suggested themselves. The RAND Corporation has been extraordinarily patient in putting up with our explorations and extraordinarily generous in providing interested scholars with our research memoranda for a period of more years than we dare recall. However, in a field characterized by such intimate cooperation among numerous individuals from diverse disciplines, there is no need to stake out claims for new results. And needless to say, the book is the joint work of the three authors, with each taking responsibility for all.

Our acknowledgments can be brief, since footnotes within the text and a selective annotated bibliography at the end will relate our work to the literature. Yet we cannot fail to mention the names of George B. Dantzig, A. W. Tucker, H. W. Kuhn, David Gale, Tjalling C. Koopmans, A. Charnes, A. Wald, and John von Neumann, who laid the foundations of the theory of linear programming.

And within The RAND Corporation itself we must give our thanks to many people. First, to Professor Armen Alchian of UCLA whose many suggestions in theoretical interpretation have improved the work. Second, to Charles J. Hitch, the head of the RAND Economics Division, who had the original idea for such a work. Third, to Joseph A. Kershaw. Fourth, to Melvin Dresher, Reuben Kessel, and Russell Nichols, who read and improved parts of the manuscript. Finally, to a number of others at RAND for countless favors over a long period of time.

We alone take responsibility for all flaws, but we dare to hope that this group operation may be a minor exception to the view of those who, forgetting that the King James Bible was the work of a committee, categorically deny value to any work not produced by a single, isolated individual.

Robert Dorfman
Paul A. Samuelson
Robert M. Solow

Contents

Foreword v

Preface vii

1. Introduction 1
2. Basic Concepts of Linear Programming 8
3. The Valuation Problem; Market Solutions 39
4. The Algebra of Linear Programming 64
5. The Transportation Problem 106
6. Linear-programming Analysis of the Firm 130
7. Application to the Firm; Valuation and Duality 166
8. Nonlinear Programming 186
9. The Statical Leontief System 204
10. The Statical Leontief System (Continued) 230
11. Dynamic Aspects of Linear Models 265
12. Efficient Programs of Capital Accumulation 309
13. Linear Programming and the Theory of General Equilibrium 346
14. Linear Programming and Welfare Economics 390
15. Elements of Game Theory 417
16. Interrelations between Linear Programming and Game Theory 446

APPENDIX A: Chance, Utility, and Game Theory 465
APPENDIX B: The Algebra of Matrices 470

Bibliography 507

Index 513

Linear Programming and Economic Analysis

1

Introduction

1-1. HISTORICAL SKETCH

At any time, an economy has at its disposal given quantities of various factors of production and a number of tasks to which those factors can be devoted. These factors of production can be allocated to the different tasks, generally, in a large number of different ways, and the results will vary. There is no more frequent problem in economic analysis than the inquiry into the characteristics of the "best" allocation in situations of this kind.

We have just outlined a rudimentary problem in welfare economics or in the theory of production. It is also a problem in *linear* economics, the word "linear" being introduced to call attention to the fact that the basic restrictions in the problem take the form of the simplest of all mathematical functions. In this case the restrictions state that the total amount of any factor devoted to all tasks must not exceed the total amount available; mathematically each restriction is a simple sum.

This illustration suggests that many familiar problems in economics fall within the scope of linear economics. Like Molière's M. Jourdain and his prose, economists have been doing linear economics for more than forty years without being conscious of it. Why, then, a book on the subject at this date? Because until recently economists have passed over the linear aspects of their problems as obvious, trivial, and uninteresting. But in the last decade the stone which the builders rejected has become the headstone of the corner. New methods of analysis have been developed that depend heavily on the linear characteristics of economic problems and, indeed, accentuate them. The most flourishing of these methods are linear programming, input-output analysis, and game theory.

These three branches of linear economics originated separately and only gradually grew together. The first to be developed was game theory, the central theorem of which was announced by John von Neumann[1] in 1928. The main impact of game theory on economics was delayed, however,

[1] "Zur Theorie der Gesellschaftsspiele," *Mathematische Annalen*, **100**:295–320 (1928).

until the publication of *Theory of Games and Economic Behavior*[1] in 1944. Briefly stated, the theory of games rests on the notion that there is a close analogy between parlor games of skill, on the one hand, and conflict situations in economic, political, and military life, on the other. In any of these situations there are a number of participants with incompatible objectives, and the extent to which each participant attains his objective depends upon what all the participants do. The problem faced by each participant is to lay his plans so as to take account of the actions of his opponents, each of whom, of course, is laying his own plans so as to take account of the first participant's actions. Thus each participant must surmise what each of his opponents will expect him to do and how these opponents will react to these expectations.

It was von Neumann's remarkable achievement to demonstrate that something definite can be said about such a welter of cross-purposes and psychological interactions. He showed that under certain assumptions, which we shall have to examine, each participant can act so as to be guaranteed at least a certain minimum gain (or maximum loss). When each participant acts so as to secure his minimum guaranteed return, then he prevents his opponents from attaining any more than their minimum guaranteeable gains. Thus the minimum gains become the actual gains, and the actions and returns for all participants are determinate.

The implications of this theory for economics are evident. It holds out the hope of banishing oligopolistic indeterminacy from economic situations in which von Neumann's assumptions are satisfied. The military implications are also evident. And, it turns out, there are important implications for statistical theory as well. Since 1944 the development of these three fields of application of game theory has gone forward actively.

Input-output analysis was the second of the three branches of linear economics to appear. Leontief published the first clear statement of the method in 1936[2] and a full exposition in 1941.[3] Input-output analysis is based on the idea that a very considerable proportion of the effort of a modern economy is devoted to the production of intermediate goods, and the output of intermediate goods is closely linked to the output of final products. A change in the output of any final product (say automobiles) implies changes in the outputs of the intermediate goods (copper, glass, steel, etc., including automobiles) used in producing that final product

[1] John von Neumann and Oskar Morgenstern, Princeton University Press, Princeton, N.J., 1944. Third edition, 1953.

[2] W. W. Leontief, "Quantitative Input and Output Relations in the Economic System of the United States," *Review of Economic Statistics*, **18**:105–125 (August, 1936).

[3] W. W. Leontief, *The Structure of American Economy, 1919–1929*, Harvard University Press, Cambridge, Mass., 1941. Second edition, Oxford University Press, New York, 1951.

and, indeed, in producing goods used in producing those intermediate goods, and so on.

In its original version, input-output analysis dealt with an entirely closed economic system—one in which all goods were intermediate goods, consumables being regarded as the intermediate goods needed in the production of personal services. Equilibrium in such a system exists when the outputs of the various products are in balance in the sense that just enough of each is produced to meet the input requirements of all the others. The specification of this balance and its pricing implications was Leontief's first objective.

Beginning with World War II, interest has shifted to a different view of Leontief's model. In this view final demand is regarded as being exogenously determined, and input-output analysis is used to find levels of activity in the various sectors of the economy consistent with the specified pattern of final demand. For example, Cornfield, Evans, and Hoffenberg have computed employment levels in the various sectors and, hence, total employment consequent upon a presumed pattern of final demand,[1] and Leontief has estimated the extent to which fluctuations in foreign trade influenced activity in various domestic sectors.[2] The input-output model, obviously, lends itself well to mobilization planning and planning for economic development.[3]

The last of the three branches of linear economics to originate was linear programming. Linear programming was developed by George B. Dantzig in 1947 as a technique for planning the diversified activities of the U.S. Air Force.[4] The problem solved by Dantzig has important similarities to the one studied by Leontief. In any operating period the Air Force has certain goals to achieve, and its various activities of procurement, recruitment, maintenance, training, etc., are intended to serve those goals. The relationship between goals and activities in an Air Force plan is analogous to the relationship between final products and industrial-sector outputs in Leontief's model; in each case there is an end-means connection. The novelty in Dantzig's problem arises from the fact that in Leontief's scheme there is only a single set of sector output levels that is consistent with a specified pattern of final products, while in Air Force planning, or

[1] J. Cornfield, W. D. Evans, and M. Hoffenberg, "Full Employment Patterns, 1950," *Monthly Labor Review*, **64**:163–190 (February, 1947), 420–432 (March, 1947).

[2] W. W. Leontief, "Exports, Imports, Domestic Output, and Employment," *Quarterly Journal of Economics*, **60**:171–193 (February, 1946).

[3] See, for example, H. B. Chenery and K. S. Kretschmer, "Resource Allocation for Economic Development," *Econometrica*, **24**:365–399 (October, 1956).

[4] The fundamental paper was circulated privately for several years and published as G. B. Dantzig, "Maximization of a Linear Function of Variables Subject to Linear Inequalities," in T. C. Koopmans (ed.), *Activity Analysis of Production and Allocation*, pp. 339–347, John Wiley & Sons, Inc., New York, 1951.

in planning for any similar organization, there are generally found to be several different plans that fulfill the goals. Thus a criterion is needed for deciding which of these satisfactory plans is best, and a procedure is needed for actually finding the best plan.

This problem is an instance of the kind of optimizing that has long been familiar to economics. Traditionally it is solved by setting up a production function and determining that arrangement of production which yields the desired outputs at lowest cost or which conforms to some other criterion of superiority. This approach cannot be applied to the Air Force, or to any other organization made up of numerous components, because it is impossible to write down a global production function relating the final products to the original inputs.[1] Instead it is necessary to consider a number (perhaps large) of interconnected partial production functions, one for each type of activity in the organization. The technique of linear programming is designed to handle this type of problem.

The solution of the linear-programming problem for the Air Force stimulated two lines of development. First was the application of the technique to managerial planning in other contexts. A group at the Carnegie Institute of Technology took the lead in this direction.[2] Second, a number of economists, with T. C. Koopmans perhaps in the forefront, began exploring the implications of the new approach for economic theory generally.[3] The present volume belongs to this general direction of effort. We shall regard linear programming as a flexible and powerful tool of economic analysis and hope that the applications to be presented below will justify our position.

These are the three major branches of linear economics. The relationship between input-output analysis and linear programming is evident. Input-output analysis may be thought of as a special case of linear programming in which there is no scope for choice once the desired pattern of final outputs has been determined.

The connection of these two with game theory is more obscure. Indeed,

[1] This statement is a little too strong. A global production function *can* be constructed, but its construction presupposes that the relationships among the levels of operation of the different parts of the organization have already been determined, i.e., that the hardest part of the problem has been solved. In other words, the heart of the problem is the construction of the over-all production function with which the usual economic analysis starts.

[2] For a typical application, see A. Charnes, W. W. Cooper, and B. Mellon, "Blending Aviation Gasolines: A Study in Programming Interdependent Activities in an Integrated Oil Company," *Econometrica*, **20**:135–159 (April, 1952).

[3] For work in this spirit see the symposium volume: T. C. Koopmans (ed.), *Activity Analysis of Production and Allocation*, John Wiley and Sons, Inc., New York, 1951, particularly chap. 3 by Koopmans, chap. 7 by Samuelson, and chap. 10 by Georgescu-Roegen.

after the sketches we have given of the problems handled by the three techniques, it may seem surprising that there is any relationship, and, as a matter of history, the connection was not perceived for some time after the three individual problems and their solutions were well known. The connection resides in the fact that the mathematical structures of linear programming and of game theory are practically identical. Is this a pure coincidence?[1] Probably it does not pay to search for an economic interpretation. It may make the connection seem less mysterious if we put it this way: Both game theory and linear programming are applications of the same branch of mathematics—the analysis of linear inequalities—a branch which has many other applications as well, both in and out of economics. The connection is analogous with the connection between the growth of investments at compound interest and Malthusian population theory.

1-2. OUTLINE OF THE BOOK

Linear programming is the core of linear economics, and we take it up first. Chapter 2 sets forth the basic concepts and assumptions of linear programming and illustrates them by two examples, one from home economics and one from the theory of international trade. The truism that the problem of allocation and the problem of valuation are inseparable applies as well to linear programming as to other modes of economic analysis. The valuation aspect of linear programming is explored in Chap. 3.

Chapters 2 and 3 together take up the leading ideas of linear programming; Chap. 4 goes on to the mathematical properties of linear-programming problems and practical methods of solution. This latter chapter is somewhat technical and may be omitted since it adds no new economic concepts. Readers who are interested in actual solutions will find it indispensable, however.

Chapter 5 presents a particularly simple and important application of linear programming. It deals with this problem: Suppose that a homogeneous commodity is produced at a number of places and consumed at a number of places, and suppose also that the total demand at each point of consumption and total supply at each point of production are known. How much should each consuming point purchase from each producing

[1] Applied mathematics abounds in such coincidences. To take an example from physics: the well-known "parallelogram of forces" is used both (1) in mechanics to find the resultant of a number of forces, and (2) in electricity to find the current and phase (i.e., timing) of an alternating current affected by resistances, inductances, etc. This is just coincidence; these two problems have no physical connection in spite of their mathematical identity.

point so that all demands are satisfied and total costs of transportation are kept as small as possible? This "transportation" or "assignment" problem is interesting not only for its own sake but because it has useful generalizations.

In Chap. 6 the linear-programming approach is applied to the theory of the competitive firm. The conclusions are consistent with those of the marginalist theory of production. But, as we noted earlier, the marginalist theory invokes the concept of a global production function comprehending all the activities of the firm, while in a multiproduct or multistage firm it may be more convenient to work with a number of partial production functions. Chapter 7 covers the imputation of values to the resources used by a competitive firm.

Chapters 6 and 7 were restricted to competitive firms because of one of the linearity assumptions. In a competitive firm, gross revenue is a linear function of the physical volume of sales, namely, the sum over all the kinds of commodity sold by the firm of price times quantity sold. In a firm not in perfect competition the relationship between revenue and physical sales volume is more complicated; it is, in fact, nonlinear. Chapter 8 discusses the analysis of such firms and the problem of relaxing some of the linearity assumptions in linear programming.

Input-output analysis is taken up next. The basic input-output system is set forth, illustrated, and discussed in Chap. 9. Chapter 10 is a more technical discussion of the system and may be omitted by readers who wish to avoid the more mathematical aspects of the subject. It deals with more difficult questions of interpretation than does Chap. 9, including an examination of Leontief's strongest assumption—that there is a unique combination of factor and material inputs for the product of each economic sector.

Chapters 11 and 12 extend the input-output model dynamically, i.e., to a sequence of time periods, and link it up with the theory of capital. In this pair of chapters, again, the earlier chapter is primarily conceptual and the later is devoted to the more difficult and technical problems. Here, almost uniquely in this volume, our presentation takes issue with previously published results. We have mentioned above that in Leontief's static system there is only one set of levels of sector outputs that will produce a specified pattern of final products. There is therefore no room for choice once the pattern of final output has been determined. Leontief has extended his system dynamically in a way that preserves this fully determined character. Our position is that the possibility of holding intermediate and final products in inventory makes choices inevitable, so that Leontief's analysis ignores an important aspect of economic dynamics. But we cannot pursue the issues here; the reader will have to wait until Chap. 11. These chapters also arrive at some new criteria for economic efficiency in a dynamic context and some new conclusions concerning the operation of competitive markets in a dynamic context.

Rather surprisingly, linear programming has turned out to be the most powerful method available for resolving the problems of general equilibrium left unsolved by Walras and his immediate successors. Under what conditions will there exist an equilibrium position for an economy in which all prices and all outputs are nonnegative? Under what conditions is this equilibrium unique? The techniques at Walras' disposal did not permit him to reach satisfactory answers to these questions. Solutions by means of linear programming are given in Chap. 13. Linear programming has also proved to be an easy and powerful method for deriving the basic theorems of welfare economics and is used for this purpose in Chap. 14.

The final two chapters deal with game theory. Chapter 15 deals with the basic concepts of game theory as applied to economic problems and discusses some methods of practical solution of game situations. Chapter 16 explores thoroughly the mathematical connections between game theory and linear programming.

The crucial dependence of game theory on the measurability of utility warrants some discussion, particularly in view of the old issue of the relevance of the measurability of utility for economics. Appendix A is devoted to this issue.

The reader will shortly become aware that linear economics makes liberal use of the results of matrix algebra. The text is nearly, but not completely, free of matrices. Nevertheless, to help readers who wish to gain some insight into matrix methods we have added Appendix B on matrix algebra, which, it is hoped, despite being called an appendix, will not be a useless appendage.

2

Basic Concepts of Linear Programming

2-1. INTRODUCTION

Since at least the time of Adam Smith and Cournot, economic theory has been concerned with maximum and minimum problems. Modern "neoclassical marginalism" represents the culmination of this interest.

In comparatively recent times mathematicians concerned with the complex problems of internal planning in the U.S. Air Force and other large organizations have developed a set of theories and procedures closely related to the maximization problems of economic theory. Since these procedures deal explicitly with the problem of planning the activities of large organizations, they are known as "linear programming." The mathematical definition of linear programming is simple. It is the analysis of problems in which a linear function of a number of variables is to be maximized (or minimized) when those variables are subject to a number of restraints in the form of linear inequalities. That definition is a bit arid, to be sure, but there is nothing difficult about it.

The difficulties begin to enter when we raise the question of applying methods derived from linear programming to economic problems. Notice that the word "linear" occurred twice in stating the mathematical definition of linear programming. Can economic problems be cast in this strict format without doing them mortal violence? On the surface it may not seem so. The U-shaped cost curves, the gently curving isoquants, the nests of indifference lines on which so much of economic theorizing depends seem to stand in the way of expressing meaningful economic problems in terms of strictly linear relationships.

Yet it can be done, and with advantage. That is the theme of this and the following five chapters. We shall develop, in some detail, the way in which economic problems have to be reformulated in order to be amenable to the methods of linear programming. The gain from this reformulation will be seen to be twofold. First, we shall be able to bring to bear on economic problems the powerful computational and solution methods developed for handling linear-programming problems. Second, by looking at familiar problems from an unfamiliar point of view we shall gain some new insights of economic importance.

A word of caution before we embark. The linear-programming models we shall develop will, of course, not be strictly accurate representations of the economic situations with which they deal. Strict descriptive faithfulness is an unreasonable demand to make of any conceptualization. The most completely accepted of economic concepts—the production function, the demand curve, or whatnot—would fail if held up to that standard. What we have a right to ask of a conceptual model is that it seize on the strategic relationships that control the phenomenon it describes and that it thereby permit us to manipulate, i.e., think about, the situation.

In the present chapter we shall illustrate the application of linear programming to economic problems by discussing two examples. The first of these—the so-called diet problem—was brought into prominence in recent years by mathematical linear programmers, who used it as a kind of trial run for their new methods. The second example—the theory of comparative advantage—was devised a long time ago by economists, who had no thought of linear programming in mind. Both bring out important aspects of the concepts and uses of linear programming.

2-2. THE DIET PROBLEM

The diet problem is famous in the literature of linear programming because it is the first economic problem ever solved by the explicit use of this method.[1] It was originally intended merely to serve as an illustration and test of the use of the method, but, like so many toy models, it has turned out to have unexpected but important practical applications. The essential issue in this problem is that a diet to be acceptable must meet certain quality specifications; e.g., it must contain so many calories, so many units of riboflavin, etc. Moreover, the quality of a diet in terms of these specifications is the mathematical sum of the qualities of its component parts, i.e., of the foods that comprise it. These characteristics—attention to quality specifications derived by addition from the qualities of components—are the structural elements on which the solution to the problem depends.

Do problems with this structure have any important place in economics? They do. They occur in such industries as livestock feeding, gasoline and textile blending, and ice-cream manufacturing, to name a few. Thus they enter into many significant business decisions and play

[1] History of the problem: 1941—independently formulated and approximately solved by Jerome Cornfield in an unpublished memorandum. 1945—solved by G. J. Stigler, not using linear programming; published in *Journal of Farm Economics*, **27**:303–314 (May, 1945). 1947—solved by G. B. Dantzig and J. Laderman by use of linear programming; not published.

a role in determining the shapes of supply and demand curves in many industries.

Now consider a hyperscientific and hard-pressed housewife who desires to provide an adequate diet for her family at the minimum possible cost. What foods shall she buy, and how much of each? To answer this question she must take into account the data we now outline.

2-2-1. Health Standards. The National Research Council (NRC) has published a table purporting to show, on the basis of present scientific knowledge, the minimum (annual) amounts of different nutritional elements—calories, niacin, vitamin D, etc.—that a typical adult should have. Opinions change rapidly in this field, and no claim can be made for great accuracy in such a specification. Moreover, the penalties for having less than these amounts are known only for extreme cases of unbalanced diet; and there is the further point that too much of some elements, such as calories, may be as harmful as too little. But for our purposes we may take the table as definitive and write it symbolically as shown in Table 2-1.

TABLE 2-1. MINIMUM STANDARDS OF NUTRITIONAL ELEMENTS

Nutritional elements	*Minimum standards*
1	C_1
2	C_2
3	C_3
.	.
.	.
.	.
i	C_i
.	.
.	.
.	.
m	C_m

Each of the requirements $C_1, \ldots, C_m$ is, naturally, positive.

2-2-2. Nutritional Composition of Foods. Our second bit of information comes from biologists and chemists. It analyzes the nutritional content of a large number of common foods (cooked in some agreed-upon way). We may call these foods, measured in their appropriate units, $X_1, X_2, \ldots, X_n$. We shall make the (somewhat doubtful) assumption that there is a constant amount of each nutritional element in *each* unit of any given food; so that if 10 units of X_1 gives us 100 calories, 20 units will give us 200, and 100 units will give us 1,000 calories—all this independently of the other X's that are being simultaneously consumed. This "constant-return-to-scale" and "independence" assumption helps to keep the problem within the simpler realms of linear-programming theory. It also permits us to summarize our second type of information in one rectangular table (Table 2-2).

TABLE 2-2. NUTRITIONAL CONTENT OF UNITS OF VARIOUS FOODS

Nutritional element	Food							Minimum standards
	X_1	X_2	X_3	. . .	X_k	. . .	X_n	
Element 1	a_{11}	a_{12}	a_{13}	. . .	a_{1k}	. . .	a_{1n}	C_1
Element 2	a_{21}	a_{22}	a_{23}	. . .	a_{2k}	. . .	a_{2n}	C_2
....	. . .	. . .	. . .	. . .	. . .	. . .	. . .	. . .
Element i	a_{i1}	a_{i2}	a_{i3}	. . .	a_{ik}	. . .	a_{in}	C_i
....	. . .	. . .	. . .	. . .	. . .	. . .	. . .	. . .
Element m	a_{m1}	a_{m2}	a_{m3}	. . .	a_{mk}	. . .	a_{mn}	C_m

In words, the amount of the third nutritional element contained in the seventh food is a_{37}. If we think of one slice of toast as having 50 calories, we could say $a_{\text{calories, toast}} = 50$ (calories per slice), etc.

Usually the number of foods will be much greater than the known number of nutritional elements, so that $n > m$. (But this need not be the case; indeed it would not be the case on a desert island or for a community subject to many taboos.) So long as each prescribed element is actually present in at least one food, it is clear that the given standard of nutrition can somehow be reached. (This means that we must not have *all* the a's zero in any row.) Ordinarily, the prescribed standard of nutrition $(C_1, C_2, \ldots, C_i, \ldots, C_m)$ can be reached and surpassed in a variety of different ways or diets; but the different diets will not all be equally tasty or cheap.

How do we test whether a given diet, say

$$(x_1, x_2, \ldots, x_k, \ldots, x_n) = (100, 550, \ldots, 3.5, \ldots, 25{,}000)$$

is adequate? Here x_k denotes the quantity of X_k. We must test each nutritional element in turn. Since each unit of the first food contains a_{11} units of the first element, we get altogether $a_{11}x_1$ of such an element from the first food. Similarly we get $a_{12}x_2$ of this first element from the second food. We must compare the sum of this element *from all foods in the diet* with the prescribed minimum C_1 to make sure that

$$a_{11}x_1 + a_{12}x_2 + \cdots + a_{1k}x_k + \cdots + a_{1n}x_n \geq C_1$$

and similarly for the second element, we must have

$$a_{21}x_1 + a_{22}x_2 + \cdots + a_{2k}x_k + \cdots + a_{2n}x_n \geq C_2$$

and so forth, for the ith or mth element.

We have not yet introduced the cost of food into the picture, but when we do it will become apparent that it is desirable not to have to pay for any excess consumption of food. In the above equations we should like, if possible, to have the equality signs hold rather than the inequalities, to avoid paying for excess nutrition. But this will not always be possible, as an ambitious dietician might discover after trying to find a diet that *exactly* reaches the prescribed standard in every respect. And even where it is in fact possible, she will discover that it is an exceedingly difficult arithmetical feat to find such an exact diet. Moreover, and this may surprise her still more, it may turn out to be most economical *not* to follow such an exact diet, since there will often turn out to be a cheaper diet that overshoots the mark in some respect.[1]

2-2-3. Economic-price Data. Thus far no mention has been made of the economic costs, in terms of dollars, of the various diets. In theory we can hope to get from the Bureau of Labor Statistics (BLS) data on the prices of the different foods, such as might be indicated in Table 2-3.

TABLE 2-3. PRICE (PER UNIT) OF DIFFERENT FOODS

Number or name of food	X_1	X_2	. . .	X_k	. . .	X_n
Price (per unit)	p_1	p_2	. . .	p_k	. . .	p_n

For any given diet, $x_1, x_2, \cdots, x_n$, the total cost would be easily calculated as the sum of the costs of each of the n foods (it being understood that in most relevant diets only a few of the possible foods would appear, the rest having zero weight). Mathematically, the total dollar cost of a diet would be

$$Z = p_1x_1 + p_2x_2 + \cdots + p_kx_k + \cdots + p_nx_n$$

We may state the full problem as that of minimizing this last sum subject to the m basic inequalities which guarantee that the minimum of each nutritional element is in fact secured. That is,

$$Z = p_1x_1 + \cdots + p_nx_n$$

is to be a minimum subject to

$$\begin{aligned} a_{11}x_1 + \cdots + a_{1n}x_n &\geq C_1 \\ a_{21}x_1 + \cdots + a_{2n}x_n &\geq C_2 \\ \cdots\cdots\cdots\cdots&\cdots\cdots \\ a_{m1}x_1 + \cdots + a_{mn}x_n &\geq C_m \end{aligned} \qquad (2\text{-}1)$$

and

$$x_1 \geq 0,\ x_2 \geq 0,\ \ldots,\ x_n \geq 0$$

It is clear that if the set (2-1) of dietary conditions can be met at all, then there is some minimum cost at which it can be met, and this cost

[1] To show that an exact diet may be impossible, consider the case in which every food contains more than twice as much of the first element as of the second; and suppose that the NRC asks for equal amounts of the two elements. Obviously, the guinea pig must end up consuming an excess of the first element if he is to have enough of the second element.

will correspond to one or more least-cost diets, which we shall refer to as optimal diets.[1]

2-3. A NUMERICAL EXAMPLE

A simple hypothetical example will illustrate the nature of the problem. Assume only two nutritional elements 1 and 2, or "calories" and "vitamins," with $(C_1, C_2) = (700, 400)$. Assume that there are five foods. Let the first, X_1, contain only calories and be measured in units that result in the coefficient $a_{11} = 1$, with a_{21} being zero; let the second, X_2, contain only vitamins as indicated by a given $a_{22} = 1$, with a_{12} being zero; let the third food be like the first in that it contains only calories so that $a_{13} = 1$ and $a_{23} = 0$; let the fourth food contain equal amounts of both elements, and let us define a unit of the fourth food as being the quantity such that a_{14} and a_{24} are equal to each other and to unity; finally let a unit of the fifth food possess twice as many calories as the fourth food, so that $a_{15} = 2$ and $a_{25} = 1$. Finally, we must assume some prices to make the problem complete. Let $(p_1, p_2, p_3, p_4, p_5) = (2, 20, 3, 11, 12)$, where all prices are in dollars per unit. Our numerical data can be summarized in Table 2-4. Our problem is to find a best diet $(x_1, x_2, \ldots, x_5)$ and the least cost Z as indicated by the question mark.

TABLE 2-4. SYMBOLS AND DATA FOR NUMERICAL DIET PROBLEM

	Symbols						Numerical data					
	Nutrients per unit of food					Standard	Nutrients per unit of food					Standard
	1	2	3	4	5		1	2	3	4	5	
Element:												
Calories	a_{11}	a_{12}	a_{13}	a_{14}	a_{15}	C_1	1	0	1	1	2	700
Vitamins	a_{21}	a_{22}	a_{23}	a_{24}	a_{25}	C_2	0	1	0	1	1	400
Prices	p_1	p_2	p_3	p_4	p_5	Z	2	20	3	11	12	(?)

If one tackled this problem by trial and error, by luck, or good judgment, one would finally find that (1) the cheapest Z is 4,700. It happens that (2) this can be reached in only one way: by the diet

$$(x_1, x_2, x_3, x_4, x_5) = (0, 0, 0, 100, 300)$$

[1] It is important to remember that the conditions imposed may be such that no solution to the problem exists. Suppose, for example, that the diet is intended for a patient suffering from digestive disturbances. If he requires large amounts of fats but, at the same time, has a circulatory disorder that cannot tolerate fats, there may still be found a starchy diet that will meet the situation. However, if a disorder of the pancreas is also present, there may be no known resolution of the incompatible biological demands. In general, it is quite possible for an innocent-looking problem to contain mutually contradictory conditions which render it insolvable in principle.

with nothing of the first three foods being bought. Note that (3) there are only as many foods bought as there are nutritional elements, the rest being consumed at zero level. Finally, it happens that (4) this "best diet" is also an "exact" one, yielding no surplus of either element.

How do we arrive at these answers? For the moment such questions may be deferred. Let us first ascertain how general these results are. Is our first conclusion, of a single best Z, universally true in linear programming? The answer is "yes" for all well-behaved programs. There cannot be two different best Z's, for in any pair of unequal Z's one will be better than the other. Moreover, in a meaningful linear-programming problem there will be a limitation on how good (i.e., how large or how small, as the case may be) Z can be made, and linear-programming problems are almost invariably set up so that the most extreme permissible values of Z will actually be assumed.[1] Thus there will be a greatest (or a least) possible Z, and this will be the optimum.

But our second conclusion—that the x's are unique—is not universally true. Often the best Z will be reached by a number of alternative x's, and if so by an infinite number of such. For example, suppose that the first three foods had been given extremely cheap prices compared with the last two. Obviously, the best diet would be found among the first three foods. But suppose Food X_1 and Food X_3, which are exactly alike, were given equally low prices. Then the best way of getting our calories could involve any one of an infinite number of combinations of X_1 and X_3, providing only that their sum added up to 700.

In this last case our final diet could involve three foods instead of only two. But even in this case there could be no harm in setting either x_1 or x_3 equal to zero and achieving the best Z with *as few foods as there are nutritional elements.* This suggests a general proposition in the field of linear programming:

THEOREM. *In a linear minimum or maximum problem involving n variables (i.e., x's) and m inequalities* [*e.g., Eq.* (2-1)], *the number of nonzero x's will never have to be greater than m.*

This general statement of conclusion 3 above will have to be proved later.[2] Note that the theorem would not help much if m were greater than n, as can happen in many problems. Note too that we could some-

[1] There is a nice mathematical distinction between requiring that the diet include (1) at least 700 calories a day and (2) more than 700 calories a day. In case 2 there is no definite minimum to the daily caloric intake (the skeptical reader may try to name the smallest permissible intake; 700.0001 calories is too high, 700.0 calories is too low) or, consequently, to Z. Linear-programming problems almost invariably conform to case 1, as we stated in the text.

[2] See Chap. 4 for further discussion of this important theorem, originally due to G. B. Dantzig. The value of the theorem is that it narrows considerably the variety of diets that have to be considered as candidates for being the cheapest diet.

times have more than $n = m$ zeros. A simple example will demonstrate this possibility. Suppose the price of X_4 were extremely low compared with all other prices; then it stands to reason that all our required calories and vitamins could be bought most cheaply by purchasing 700 units of X_4, and buying a single food would be the best way to get two elements.[1]

The last example shows that our cheapest diets will not always be exact, so that conclusion 4 above, that an "exact" optimal diet exists, is *not* generally true. Often some of the m side conditions or constraints will turn out *not* to be binding; however, they cannot be thrown away, because for other prices or standards they may become binding.

It is intuitively clear that changing any C_i, for example, increasing the calorie requirements, will cause a definite change in the best Z; but it is also clear that changing a nonbinding C will have zero effect on Z, until it begins to be binding. The rates of change of the form dZ/dC_i are in the nature of *marginal costs* and will be found to have an important economic and mathematical significance, related to "shadow prices" and so-called Lagrangean multipliers.

2-4. SOLUTION BY ELIMINATION

There are two factors that make the diet problem and, as we shall see, all linear programming problems hard to solve. They are so closely related that it takes some subtlety to distinguish them. Yet, for the sake of clarity, we shall try.

The first complicating feature is the sheer numerousness of the conditions to be met, represented by Eqs. (2-1). If there were only one restriction, say the restriction of providing adequate calories, the problem would be trivial. You would simply find the food that provided most calories per dollar and procure only that one.[2] Incidentally, this case of trying to minimize or maximize something subject to a single requirement is the standard case of economic analysis. The firm in production theory is

[1] The reader may be tempted to stretch the theorem to cover this case. He may argue that for this example we can forget about vitamins, since C_2 is no bottleneck. In effect, then, we have only one rather than two effective constraints, and a new $m = 1$ can be substituted for $m = 2$. But there are many problems involved in defining such a new m that are glossed over in this discussion. It should be emphasized that m is the number of inequalities or possible equalities, not the actual number of equalities, that turn out to be satisfied.

[2] This extreme example highlights one of the artificialities of the diet problem. A diet consisting of a single food would undoubtedly meet a given requirement for calories at minimum cost and would undoubtedly be unpalatable. There is nothing in the construction of the diet problem to ensure palatability and little reason to expect that even with a dozen nutritional requirements a minimum-cost diet would be edible. One solution to the Dantzig-Laderman problem consisted principally of corn meal and evaporated milk; no steak or ice cream all year long.

assumed to maximize profits subject to a single production function or to minimize costs for a given volume of output. The consumer is conceived of as maximizing satisfaction subject to a single budgetary restraint.

Add just one more restraint, to make a total of two, and the problem cannot even be expressed without invoking some special concepts that would interrupt the thread of the exposition if introduced at this point. It is inevitable, then, that the problem for a general number of restrictions will require novel methods of expression and solution designed to meet this novel situation.

The second complicating feature is the fact, noticed several times already, that the requirements do not have to be fulfilled exactly. This is no complication, to be sure, when there is only one requirement. If we seek the minimum-cost diet that provides at least 3,000 calories a day, we shall surely not find it to be one that provides 3,005 calories a day, so we can treat our inequality as if it were an equality condition. Not so when there is a single additional restraint. The cheapest diet that provides at least 3,000 calories and 1,500 milligrams of thiamin a day may turn out to yield 3,010 calories or 1,550 milligrams of thiamin (but not both). Thus the inequality signs have to be respected.

This complication, the fact that we cannot tell in advance which restraints will be satisfied with equal signs in the best solution and which (if any) with inequalities, is not a fine point. It is really central. Suppose that some occult source could provide us with a helpful hint such as that there will be an excess of thiamin in the optimal solution. Then we could forget about thiamin and solve for the cheapest diet that provided adequate calories, as if we had a one-restriction problem. Or suppose that this useful source of clues told us that both requirements would be met exactly in the end. Then the problem would be a bit more complicated, but still, as we shall see almost immediately, this advance information would be extremely valuable. In actual practice, though, we have to solve problems in initial ignorance of which restrictions are really going to be binding and which we can ignore safely. This makes life harder.

At this stage we cannot present generally adequate methods of solution but we can present two elementary approaches that suffice for the diet problem with only two restraints. These will be taken up in the present and the following sections. The first of these elementary methods is solution by elimination.

Let us see how elimination works by applying it to the vitamin-calorie problem. Suppose, first, that someone told us that the vitamin restraint would turn out to be not binding, i.e., that when we have found the cheapest diet with adequate calories it will turn out to have at least enough vitamins. Then we could forget about the vitamin restriction and look for the cheapest diet that satisfied the calorie side condition

$$a_{11}x_1 + a_{12}x_2 + a_{13}x_3 + a_{14}x_4 + a_{15}x_5 = C_1$$

or

$$1x_1 + 0x_2 + 1x_3 + 1x_4 + 2x_5 = 700$$

We can use this equation to solve for any variable, except x_2 which has a zero coefficient. Suppose we decide to eliminate x_5 as redundant; then we solve for it and substitute into our cost data to get

$$\begin{aligned} Z &= p_1x_1 + p_2x_2 + \cdots + p_5x_5 = 2x_1 + 20x_2 + 3x_3 + 11x_4 + 12x_5 \\ &= p_1x_1 + p_2x_2 + \cdots + p_5 \frac{C_1 - a_{11}x_1 - a_{12}x_2 - \cdots - a_{14}x_4}{a_{15}} \\ &= \left(p_1 - \frac{a_{11}}{a_{15}}p_5\right)x_1 + \left(p_2 - \frac{a_{12}}{a_{15}}p_5\right)x_2 + \cdots \\ &\qquad + \left(p_4 - \frac{a_{14}}{a_{15}}p_5\right)x_4 + \text{constant} \\ &= (2 - 6)x_1 + (20 - 0)x_2 + (3 - 6)x_3 + (11 - 6)x_4 + \text{constant} \\ &= -4x_1 + 20x_2 - 3x_3 + 5x_4 + \text{constant} \end{aligned}$$

Our cost function has now been expressed in terms of one less variable than we had originally. But presumably these four variables are now free to move independently over all nonnegative values.

How should we optimally adjust x_2? Because it has a positive coefficient, it is clear that $\partial Z/\partial x_2 = 20 > 0$ and every increase in x_2 sends up costs. Therefore we go into reverse and reduce x_2 in order to effect savings. This we continue until we reach the limit $x_2 = 0$. The exact same can be said for x_4, which also has a positive coefficient.

So far so good. But applying the same reasoning to the other x's leads to a perplexing situation: x_1 and x_3 have negative coefficients, and it would seem that increasing them indefinitely would be in order. This is surely absurd. Or is it? Will increasing x_1 and/or x_3 cause the total of calories to become excessive, and therefore be a foolish procedure? No, it will not; it will not because x_5 is always being reduced so as to keep total calories $= 700 = C_1$. That is the meaning of our earlier substitution.

In disposing of one objection we encounter another. If x_1 and x_3 are increased enough, x_5 will ultimately become a negative number, which it is not permitted to do since it is nonsense to consume a negative amount of any food. This means that x_1 and x_3 can be increased only until x_5 is zero; from that point on, if we increase x_1, we must decrease x_3, and vice versa. This means that, with x_2 and x_4 already set equal to zero, and with x_1 and x_3 made so large as to set $x_5 = 0$, we are finally left with the following choices for x_1 and x_3, as can be seen from the first equation of this section:

$$-\frac{1}{2}x_1 - \frac{1}{2}x_3 + \frac{700}{2} = x_5 = 0 \qquad \text{that is, } x_1 = -x_3 + 700$$

We substitute into Z to get

$$\begin{aligned} Z &= p_1x_1 + p_3x_3 = 2x_1 + 3x_3 \\ &= p_1(-x_3 + 700) + p_3x_3 = (p_3 - p_1)x_3 + \text{constant} \\ &= 1x_3 + \text{constant} \end{aligned}$$

Since x_3 is now our remaining free variable and since it has a positive coefficient, we shall realize economies by making it as small as possible. When x_3 is set equal to zero, then obviously—from the above relation between x_1 and x_3, or from the original calorie constraint—we must have $x_1 = 700$.

At long last we have our optimal diet:

$$(x_1, x_2, x_3, x_4, x_5) = (700, 0, 0, 0, 0)$$

As we had reason to expect earlier, where there is only one effective constraint, there need be only one nonzero variable.

An intuitive economist might have arrived at this result almost immediately. He is used to working with the concept "marginal utility of the (last) dollar spent on each commodity." In this problem, he would replace utility by calories and look for the most calories per dollar, or for the maximum of

$$\frac{a_{11}}{p_1} = \frac{1}{2} \qquad \frac{a_{12}}{p_2} = 0 \qquad \frac{a_{13}}{p_3} = \frac{1}{3} \qquad \frac{a_{14}}{p_4} = \frac{1}{11} \qquad \frac{a_{15}}{p_5} = \frac{2}{12} \qquad (2\text{-}2)$$

Clearly x_1 is the cheapest way of getting calories. It is too bad that this simple device will not get the solution to more complicated problems.

As a matter of fact, the more tedious method of substitution outlined above can follow many paths. With good luck we might have picked a path which would have gotten us our solution in almost a single step. Suppose we had used our calorie relations to solve for x_1 rather than x_5. Then our cost would have turned out to be

$$\begin{aligned} Z &= \left(p_2 - \frac{a_{12}}{a_{11}}p_1\right)x_2 + \left(p_3 - \frac{a_{13}}{a_{11}}p_1\right)x_3 + \left(p_4 - \frac{a_{14}}{a_{11}}p_1\right)x_4 \\ &\qquad + \left(p_5 - \frac{a_{15}}{a_{11}}p_1\right)x_5 + \text{constant} \\ &= 20x_2 + 1x_3 + 9x_4 + 8x_5 + \text{constant} \end{aligned}$$

All the coefficients of the variables are positive; each variable is best set to zero; from our constraint we find that $x_1 = 700$. Hindsight always helps.

We have labored hard to get the best solution. The only trouble with our solution is that *it is wrong*. We have already been informed that the best diet for our original problem is

$$(x_1, x_2, x_3, x_4, x_5) = (0, 0, 0, 100, 300)$$

What is lacking about (700, 0, 0, 0, 0)? Clearly it yields the correct calories, but it fails to yield the specified amount of vitamins. It is definitely the cheapest diet under the assumption made that only the calorie constraint would be binding. But this was a gratuitous assumption that we had no right to make, as can be determined by seeing whether the full conditions of the problem are satisfied.

Nevertheless, our work has not been entirely in vain. We have the answer to the problem in which vitamins are of no importance. (Or alternatively if the first food, x_1, contained a great quantity of vitamins so that a_{21} were very large instead of being zero and so that we could be sure that the vitamin requirements would be more than satisfied, then our solution would be a correct one.) This is clearly a *lower bound* to the best obtainable cost: If calories alone cost at least the amount

$$700 \times 2 = 1{,}400$$

then a diet adequate in every respect must cost that much or more.[1]

The main purpose of this discussion was, however, expository. When only one constraint is binding, the problem of elimination by substitution is at its simplest and the logic of the process is revealed most clearly.

We may recapitulate just what was done in this process:

1. We found an expression for one of the variables by using our constraint.

2. We substituted this expression into our Z sum wherever the dependent variable appeared, thus *eliminating* the dependent variable from our Z sum.

3. The remaining variables were *not* perfectly free to move as they pleased. When one became zero, it hit an inflexible stop. Worse than that, when a movement of the independent variables caused the eliminated dependent variables to hit zero, we again ran into an inflexible wall and could at best move along that wall.

4. But our minimizing procedure, within these constraints, was logically simple. We kept repeating firmly to ourselves: "Every day in every way, we must be getting better and better. We just keep moving, as long as we are moving *down* the cost trail." (Specifically, when we had chosen to eliminate x_5, we then moved to $x_2 = 0$ because the positive coefficient of x_2 meant that this would be a downward direction; then we moved

[1] Thus our optimal Z must be at least as great as the cheapest way of buying calories alone, or $Z \geq \min_k (p_k C_1/a_{1k})$. This must equally be true with respect to vitamins or any other element, or $Z \geq \min_k (p_k C_i/a_{ik})$, where $i = 1, 2, \ldots, m$. The best of these lower bounds is given then by $Z \geq \max_i \min_k (p_k C_i/a_{ik})$. In these expressions $\max_i$ means "maximum with respect to i" and similarly for $\min_k$. Note finally that if some of the a's in our problem could be negative, e.g., if some caloric food destroyed vitamins contained in other foods, this line of reasoning would fail.

further downward by setting $x_4 = 0$. Since x_1 and x_3 had negative coefficients, our next downward move involved increasing one or both of them; this went on until we hit the "geometrical plane or wall" represented by

$$x_2 = x_4 = x_5 = 0 = \frac{C_1}{a_{15}} - \frac{a_{11}}{a_{15}} x_1 - \frac{a_{13}}{a_{15}} x_3$$

We proceeded to edge our way along this wall in a downward direction by decreasing x_3 which had a positive coefficient in the expression for Z defined in terms of x_3 along this final wall. If x_3's coefficient had been negative instead of positive, we would have increased it at the expense of x_1, up to the $C_1 = 700$ limit. If its coefficient had been zero, any point on the wall would have been indifferently good.)

So much for the process of elimination of dependent variables when there is only one constraint. If there are two or more constraints, the logic of the process is unchanged; but the numerical steps are considerably more tedious. Let us illustrate by examining briefly our simple calorie-vitamin problem, where we have been told that both constraints are in fact to be binding. Here we have two effective constraints, and so we can eliminate two variables. Actually, in this case, except for x_1 and x_3 we can eliminate *any* two variables from the numerical relations

$$\begin{aligned} 1x_1 + 0x_2 + 1x_3 + 1x_4 + 2x_5 &= 700 \\ 0x_1 + 1x_2 + 0x_3 + 1x_4 + 1x_5 &= 400 \end{aligned}$$

Applying the methods of high-school or more advanced algebra, we shall soon find that it is much easier to eliminate the "pure" variables, x_1 and x_2 or x_3 and x_2, than any other pair. In fact, to express x_4 and x_5 in terms of the remaining variables involves solving two simultaneous equations (or as a mathematician would say, involves "inverting a matrix"). This is logically easy to do but tedious in practice.

Let us therefore agree to eliminate x_1 and x_2, to get

$$\begin{aligned} x_1 &= 700 - 1x_3 - 1x_4 - 2x_5 \\ x_2 &= 400 - 0x_3 - 1x_4 - 1x_5 \end{aligned}$$

We now substitute these into our cost expression

$$\begin{aligned} Z = \sum_1^5 p_i x_i &= 2x_1 + 20x_2 + 3x_3 + 11x_4 + 12x_5 \\ &= 2(700 - 1x_3 - 1x_4 - 2x_5) + 20(400 - 1x_4 - 1x_5) \\ &\qquad\qquad + 3x_3 + 11x_4 + 12x_5 \\ &= (-2 + 3)x_3 + (-2 - 20 + 11)x_4 + (-4 - 20 + 12)x_5 + 9{,}400 \\ &= 1x_3 - 11x_4 - 12x_5 + 9{,}400 \end{aligned}$$

Because x_3 has a positive coefficient (or "*net* cost"), we must obviously reduce it to zero. Just as clearly the negative coefficients of x_4 and x_5 mean that we must increase them *at the expense of* x_1 *and* x_2. But x_1 and x_2 can never be reduced below zero. When they both reach zero, x_4 and x_5 take on the values (100, 300), which we earlier said were the best values.[1]

In the general case in which we know that there are r ($\leq m$ and $\leq n$) independent and consistent binding constraints, we can always eliminate r variables and substitute for them in the Z expression. The resulting expression for Z will be defined in terms of the remaining $(n - r)$ quasi-free variables; and depending on their coefficients, we can proceed to find some $(n - r)$ variables that can be set equal to zero. The final values of the nonzero variables can be found by solving our r effective equations. If, and only if, we have selected the right set of effective constraints, will the whole process be consistent.

The picturization of this process in terms of higher geometry is conceptually very helpful, but will be reserved for a later section.

2-5. GRAPHIC SOLUTION

The elimination method of solution is not practically useful because it depends on knowing in advance which restraints can be used for performing eliminations, i.e., which restraints will be binding. We discussed it chiefly in order to show how the levels of the various choice variables (e.g., the food quantities) are interrelated. The graphic method, to be discussed now, is practical when there are as few as two restraints and serves to clarify some additional aspects of programming problems. This method can handle any number of choice variables, so long as only two constraints are present. In Sec. 3-3, in which we study the so-called "dual" to the diet problem, we shall be led to an alternative graphical method which can handle any number of constraints so long as there are only two variables.

The graphic method depends on considering the quantity of each food element—in the case of our example, vitamins and calories—that can be purchased for a fixed sum, say \$1,000, spent on each food. These quantities, for our numerical example, are given in Table 2-5 and shown graphically in Fig. 2-1.

In Fig. 2-1 a point, indicated by a ringed numeral, is plotted for each food, showing the quantity of calories and vitamins provided by \$1,000 worth of it. A glance at the figure (or at Table 2-5, for that matter) shows that there are two foods which cannot, at the quoted prices, enter

[1] To be rigorous, we must verify that *both* x_1 and x_2 should be forced to zero levels. If we had eliminated x_4 and x_5, the resulting coefficients of x_1, x_2, x_3 would all be positive, providing round-about verification of what can be directly shown.

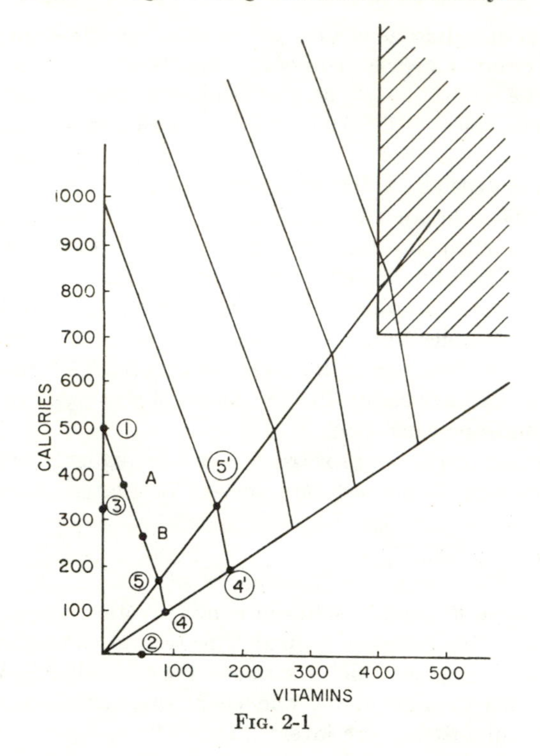

Fig. 2-1

into an economical budget. Food 3 provides fewer calories per dollar than Food 1 and no more vitamins, and so will never be purchased. Similarly, Food 2 is inferior to either Food 4 or Food 5 with respect to both nutritive elements. We therefore drop Foods 2 and 3 out of consideration forthwith and attend only to the other three.

Now, suppose we split $1,000 between Foods 1 and 5, spending, say, $500 on each. From Table 2-5 we compute that the resulting diet will

Table 2-5. Quantity of Calories and Vitamins per $1,000 Spent on Each of Five Foods*

Element	Food					Requirement
	1	2	3	4	5	
Calories..........	500	0	333.33	91.91	166.67	700
Vitamins.........	0	50	0	91.91	83.33	400

* Computed from the hypothetical data of Table 2-4. Thus, Food 1 provides one calorie per unit and costs $2 per unit. Hence $1,000 worth of Food 1 provides $1{,}000 \times \frac{1}{2} = 500$ calories.

provide 333.33 calories [½(500 + 166.67)] and 41.67 vitamins. This is indicated by point A in Fig. 2-1. Similarly, $250 worth of Food 1 and $750 worth of Food 5 will yield 250 calories and 62.5 vitamins, shown by point B in the figure. Note that points A and B both lie on the straight line connecting the points for Foods 1 and 5, and it is easy to see that all combinations of $1,000 spent on those two foods will correspond to points on that line. The line connecting Foods 4 and 5, similarly, represents the results of splitting $1,000 between them. Now, what about splitting $1,000 between Foods 1 and 4? The yields of such divisions would lie along the straight line connecting those two foods, but we haven't drawn the line because for any point on that line there will be some point on one of the previous two lines which represents a diet that provides more of both elements without costing any more. The two lines which we have drawn, then, represent the locus of efficient diets.

Of course, this locus represents just the efficient ways of spending $1,000. There will be analogous loci for all other levels of expenditure. A sum of $2,000 spent on Food 4 will provide twice as much of both food elements as $1,000 and so will be represented by a point twice as far out along the radius vector through point 4. We have designated this point by 4′. There is a similar point, 5′, for the results of spending $2,000 on Food 5, and similar points could be graphed for all other foods. Using those points we can construct a locus of efficient expenditures for $2,000 and, similarly, for any other budget. We have drawn the loci for $1,000, $2,000, . . . , $5,000.

Now consider the shaded area in Fig. 2-1. Its lower left-hand corner is at 400 vitamins, 700 calories. Thus, this is the region of adequate diets, and our problem is to determine the least expensive diet in this region. Obviously it will be the diet in the region which lies on the lowest possible efficient diet locus, and, equally obviously, the point in question is the lower left-hand corner itself. We see at once that this point lies on a line segment connecting Foods 4 and 5. In other words, the minimum-cost diet will consist of Foods 4 and 5 in some amounts yet to be determined and will exclude all other foods. Furthermore, the minimum-cost diet will not provide an overage of either food element; i.e., both restrictions are binding. This is indicated by the fact that the minimum-cost diet, observed diagrammatically, provides just the stated amounts of the two food elements.

The rest is elementary algebra. We have two unknowns, the amounts of Foods 4 and 5, and two restraints, one for each food element. Thus we have exactly the correct number of the equations to determine the two unknowns and to write them

$$1x_4 + 2x_5 = 700$$
$$1x_4 + 1x_5 = 400$$

The solution is, as we found previously, $x_4 = 100$, $x_5 = 300$.

Obviously, any number of foods can be handled in this way, provided there are only two nutritional requirements. Each of the latter requires an axis, so if there were a third nutrient to be accounted for, this method would need three dimensions.

2-6. COMPARISON WITH THE THEORY OF CONSUMPTION

The diet problem can be stated (but not solved!) by means of the familiar concepts of indifference maps, budget lines, etc., used in the theory of consumption. We shall state the problem in these terms in order to see what in this problem is new and different and requires a new and different approach.

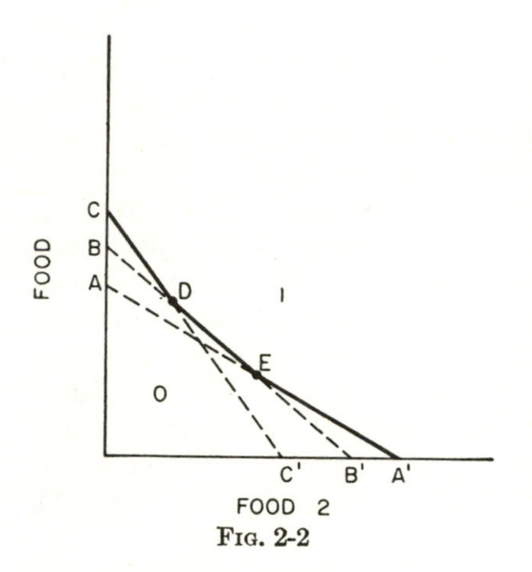

FIG. 2-2

In order to picture the situation graphically let us suppose that the problem involves two foods and three nutritive elements. In this special case we have to choose the quantities x_1, x_2 of the two foods so as to make the total cost

$$Z = p_1x_1 + p_2x_2$$

a minimum subject to the three nutritive restraints

$$\begin{aligned} a_{11}x_1 + a_{12}x_2 &\geq C_1 \\ a_{21}x_1 + a_{22}x_2 &\geq C_2 \\ a_{31}x_1 + a_{32}x_2 &\geq C_3 \end{aligned}$$

The x's, of course, cannot be negative.

These data are depicted in Fig. 2-2, where the quantities of the two foods are measured along the two axes and each of the lines AA', BB', and CC' represents those combinations of the two foods that exactly meet (i.e., with no excess) one of the three requirements.

This is not an indifference map, though its axes are the same as the axes of an indifference map. In order to construct an indifference map we must express the preference scale implicit in the problem. It is a very simple scale. All diets that meet the three requirements are satisfactory and, as far as the problem goes, are equally satisfactory. Let us assign them a utility index of 1. All other diets are unsatisfactory. Let us give them a utility index of 0. In the diagram, then, all diets on or above the broken-line boundary $CDEA'$ have a utility index of 1 and all those below it an index of 0. If we add a third dimension, for utility, we obtain the picture shown in Fig. 2-3. This is the utility map for the diet problem. To solve the problem, all that is needed is to find the lowest budget line that touches the raised portion of the diagram in at least one point. We have sketched in one budget line—one that obviously is not the lowest.

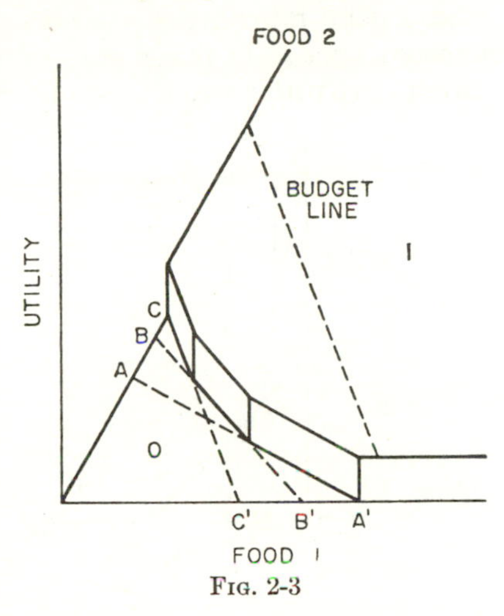

Fig. 2-3

The peculiarities of this indifference map are evident at once. Instead of a utility hill we have a mesa with a precipitous face; instead of a family of indifference curves we have two indifference regions; instead of continuously curved contours we have some broken lines. The concept of a marginal rate of substitution, the basic concept of the theory of consumers' choice, doesn't apply to this diagram except for points on the boundary of the two regions. This diagram does not lead to any of the helpful principles that indifference maps usually teach us.

The trouble is that this diagram and the utility scale that it represents are artifices, and ignoble ones at that. The problem as it came to us involved three restraints, one for each nutritive element. Our procedure was to construct a utility scale that incorporated all the restraints and thereby to reduce the problem to the following apparently standard form: Find the least expensive diet with a utility index of unity. But, alas, this kind of trickery can only sweep difficulties under the rug; it cannot dispose of them. In this case all we have accomplished is to substitute one mathematically intricate condition for three simple ones.[1]

Thus the presence of several conditions to be met distinguishes the diet problem fundamentally from the familiar problems of the theory of con

[1] The condition that the budget line touch the raised portion of the diagram may not seem intricate, because the eye can solve it easily. But remember that in dealing with problems that can be visualized the eye will put the most gigantic electronic computer to shame.

The condition in question can be expressed mathematically in a number of ways. The following is perhaps the easiest. Let $I(u)$ be the step function defined by $I(u) = 0$ if $u < 0$, $I(u) = 1$ if $u \geq 0$. Then $I(a_{i1}x_1 + a_{i2}x_2 - C_i)$ equals 1 if the ith nutritive requirement is satisfied and equals zero otherwise. The utility index of any diet—x_1, x_2—can then be written

$$U(x_1, x_2) = \min_i I(a_{i1}x_1 + a_{i2}x_2 - C_i) \qquad i = 1, 2, 3$$

This is clearly not a very nice function to have to work with.

sumption. An interesting intermediate case that points up the complications attendant on numerous restrictions is the case of consumer choice when a point-rationing scheme is in effect. Then any good will have both a money price and a ration price associated with it, and the consumer will have to live within two limitations, one on the total value of consumption

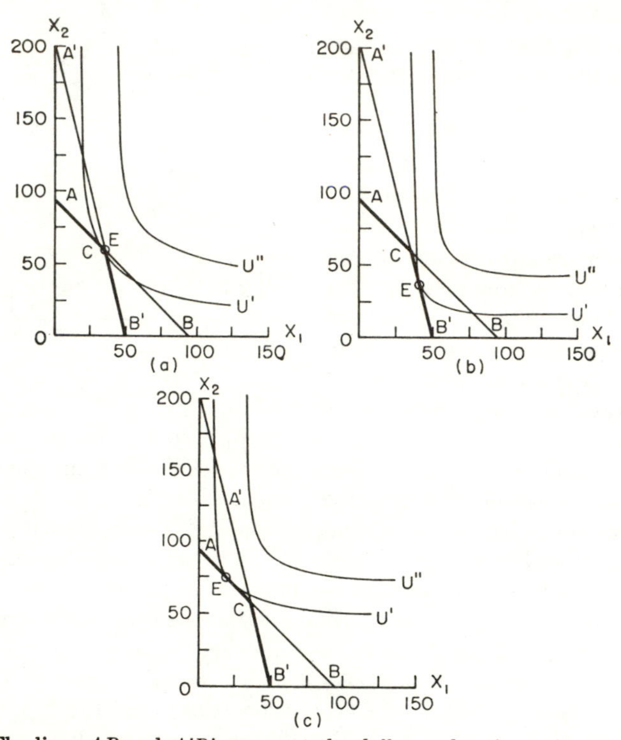

FIG. 2-4. The lines AB and $A'B'$ represent the dollar and ration-point budget equations, respectively. The heavy locus ACB' represents the locus available to the consumer, since the "scarcest currency" is always the bottleneck. In *a*, this locus touches but doesn't cross the highest indifference curve at C. In *b*, this phenomenon occurs along CB', where dollars are redundant; in *c*, the ration points are redundant. When there are only two x's and when both constraints are known to be binding, there is no room left for maximizing behavior; only when there are more goods does the problem become interesting.

and one on its ration-point value. The situation in which there are two commodities, which shows the essential issues, is illustrated in Fig. 2-4. Each panel of that figure shows two indifference curves U' and U'', a conventional price line AB, and a ration-point budget line $A'B'$. Let us look first at Fig. 2-4*b*. There is only one pattern of consumption in which the consumer uses both all his dollar budget and all his ration points, namely, the pattern represented by point C. It is evident from this figure that point C does not represent an optimal budget; point E corresponds to a superior attainable budget even though it involves leaving some dollar purchasing power unused. The three panels together show three possible situations: in Fig. 2-4*a*, all dollars and points are used; in Fig. 2-4*b*, dollars are redundant; and in Fig. 2-4*c*, points are redundant.

Thus as soon as two restraints are imposed on the consumer we can no longer assume that the budget equality will be satisfied and must express our restraints by means of "not-greater-than" inequalities. But inequalities are much less useful in analysis than equalities, essentially because equalities can be used to eliminate variables and inequalities cannot. The appearance of the inequalities, as in Fig. 2-4*a* and *b*, disqualifies the usual analysis via the calculus and its graphic analogs. For example, it is not true in either Fig. 2-4*a* or *b* that at the optimal consumption point the marginal rates of substitution of the commodities are proportional to their money prices. The problem would be much simplified if only we knew in advance which limitation would actually be binding. But discovering this is part of the solution.

This simple point-rationing example is our first glimpse of nonlinear programming. The point-rationing case is one in which some nonlinear function of variables of choice is to be maximized but where the scope for choice is constricted by linear restraints. The same type of problem, as we shall see, arises in connection with imperfect competition.

Thus we see that new approaches are needed when we have to analyze problems of economic choice subject to several distinct limitations. It seems worthwhile to point out another contrast between the linear-programming approach—as exemplified by the diet problem—and the theory of consumption, this being a contrast of emphasis rather than of content. The familiar theory of consumer choice, whether expressed in terms of curves of diminishing marginal utility or in terms of convex indifference curves,[1] was designed to explain how a limited budget is allocated among diverse commodities. This theory is consequently much handier for explaining how much is consumed of each commodity purchased, and how those quantities respond to changes in underlying data, than it is for explaining which commodities are purchased and which are not, and how this selection responds to changes in the data. This some-or-none decision is often the important one, as the diet problem suggests. The standard two-commodity indifference map shows that the range of relative prices at which positive amounts of both commodities will be purchased depends on the amount of curvature in the indifference contours; if the two commodities are close substitutes, the range of price ratios at which both are purchased may be very narrow.

Whenever the optimal allocation of a budget entirely excludes one or

[1] A curve is called "convex" if all its chords lie entirely on or above it.

more of the commodities considered, the familiar marginal equalities are no longer strictly valid and should be replaced by inequalities. These conclusions follow straightforwardly from the conventional theory, and we mention them only because they are so frequently glossed over and because the conventional analysis becomes decidedly awkward when the zero-consumption case arises.[1]

The linear-programming approach lays its major emphasis on the problem of which commodities are to be included in the budget. The possibility that one or more of the available commodities may not be consumed at all therefore presents no particular difficulties or complications. This added generality, naturally, has to be paid for, and the price is that the linear-programming approach is consistently more complicated than the marginal one.

2-7. SOME CONCEPTS AND GENERALIZATIONS

We have now seen an illustrative linear-programming problem and two elementary methods of solution. The problem selected was typical in form but certainly not in substance. The reason for this last reservation is that linear-programming problems arise in many varied contexts. Let us therefore turn to the formal characteristics of the problem.

In the first place, we had to minimize something. All linear-programming problems are concerned with either minimizing or maximizing something. In general, the quantity to be minimized or maximized is called the "objective function"; in the diet problem the objective function is the aggregate cost of the food purchased. There is some temptation to classify linear-programming problems into two types according as the objective function is to be minimized or maximized, but this is pointless because there is no real difference between the two operations. Every time a quantity is maximized, some other quantity—for example, its negative—is minimized. Thus we do not distinguish between minimizing and maximizing problems and discuss each aspect of the method in terms of whichever type of extremum seems most convenient at the time.

In the second place, there were certain conditions which prevented us from making the objective function infinitely small. The equations

[1] Even Hicks' fundamental exposition in *Value and Capital* falls into error when it comes to this point. See John R. Hicks, *Value and Capital*, p. 20, Oxford University Press, New York, 1939. Walras was aware of the problem and struggled with it unsuccessfully. See Leon Walras, *Elements of Pure Economics* (William Jaffé, transl.), pp. 166ff, George Allen & Unwin, Ltd., London, 1954. Walras' difficulties illustrate how awkward the marginal approach is when applied to situations, which clearly are very prevalent, in which some available commodities are excluded entirely from the budget.

expressing these conditions are called, in general terminology, the "constraints," or "restraints," of the problem. The restraints are of two sorts, both generally and in the diet problem. There are the restraints expressing the fact that certain variables (the food quantities in the diet problem) cannot be negative. This type of restraint, identical in every respect to the form in which we have encountered it, appears almost universally in linear-programming problems. Also, there are the restraints which express the special conditions of the problem. The minimum nutritional requirements in the diet problem were of this kind. The nature and number of special restraints vary, of course, from problem to problem.

In the third place there are "choice variables," the numbers which are to be chosen so as to minimize (or maximize) the objective function and to satisfy all the restraints, e.g., the food quantities in the diet problem. The objective function is, in general, a linear function of the choice variables, and the restraints are either equalities or inequalities involving linear functions of the choice variables.

Quite frequently, though not invariably, each choice variable indicates the extent to which something is to be done. In the diet problem, for instance, x_1 indicated the extent to which Food 1 was to be purchased. For this reason it is convenient (and usually sufficiently accurate) to think of each choice variable as indicating the level of some operation, called an "activity," or a "process." Intuitively, as we have just suggested, a process is some physical operation, and it may be almost any physical operation, e.g., consuming something, storing something, selling something, throwing something away, as well as manufacturing something in a particular manner. In the diet problem, each process consisted in purchasing something (and presumably consuming it). But it is often convenient to think of fictitious processes which do not correspond to any physical operation. For instance, in the diet problem we may think of the process of purchasing excess calories.[1] We shall see below that in many important problems the choice variables represent prices, in which case the processes have no physical significance whatsoever.

[1] The advantage of such fictitious processes is that they enable us to get rid of the inequality signs which, as we already know, are troublemakers. For instance, by using the process of buying excess calories we can write

Calories from Food 1 + calories from Food 2 + · · · + calories from Food n
− excess calories = minimum caloric requirement

instead of

Calories from Food 1 + calories from Food 2 + · · ·
+ calories from Food n ≥ minimum caloric requirement

These two restrictions are obviously equivalent, since the number of excess calories, technically called the "slack variable," is inherently nonnegative.

The essential characteristic about a process from the point of view of linear programming is not its physical nature (the process may be entirely fictitious and have no nature) but the way it enters into the objective function and the restraints. This characteristic is contained entirely in the coefficients by which the choice variable corresponding to the process is multiplied in the objective function and in the restraints. The coefficient by which the choice variable is multiplied in the objective function is known as the "value" of the process. The list of coefficients by which that variable is multiplied in the restraints is known as the "process vector."[1] Some of these coefficients may, of course, be zero. For example, using the data in Table 2-4 for the two-element diet problem, Process 5 consists in buying Food 5, its value is 12, and its process vector is (2,1). Since the process vector includes all the information about the process that is relevant to linear programming, the terminology is frequently shortened, so that when we talk about the process we mean simply the list of its coefficients. This terminology is not only shorter but, in a sense, more exact since a process always possesses coefficients (or else it could not enter the problem) and need not possess anything else.

Just as each choice variable specifies the level at which some process is to be used, so a list of levels of all the choice variables that occur in a problem specifies the levels at which all processes considered are to be used. Such a list is therefore called a "program." For example, the list (0, 0, 0, 100, 300) is a program. If the list satisfies all the restraints, it is called a "solution," or a "feasible program." If a solution maximizes (or minimizes, as the case may be) the objective function, it is known as an "optimal solution," or an "optimal feasible program." Linear programming is concerned with finding optimal solutions and studying their characteristics.

This whole array of definitions is on an abstract level. But as soon as they are applied to an economic problem, another substantive aspect arises. Consider, for instance, the objective function. We have defined it simply as whatever is to be maximized. In an economic context the objective function is usually the measure of social valuation adopted by whatever social unit controls the values of the choice variables. To invest the objective function with meaning then involves locating the social unit which has effective control (which is not always easy) and ascertaining its objectives (which is almost never easy). Similarly the whole structure, social, economic, and technical, in which the deciding unit is embedded is involved in the practical identification of the choice variables, the processes, and the restraints, which specify the field of choice open to the unit of decision. These are the significant and difficult

[1] These coefficients come from a *column* of Table 2-4, but for convenience we write them in a row.

problems from the point of view of the economist. We shall deal with some of them in later chapters. At present we concentrate on presenting linear programming as a conceptual apparatus which can justify itself in practical analysis.

2-8. ILLUSTRATION FROM THE THEORY OF COMPARATIVE ADVANTAGE

The classical theory of international trade illustrates the concepts of linear programming in an entirely different context. More than a century ago the English economist David Ricardo outlined a simple theory to explain the pattern of international trade and to proclaim its benefits to all participating countries. A traditional numerical example (somewhat simplified) is of the following form: Portugal can divert resources from food to clothing production and in effect convert one unit of food into one unit of clothing; England, on the other hand, can convert one unit of food into two units of clothing.

Almost certainly Portugal will specialize completely in food, England completely in clothing. England will export clothing in exchange for food imports, the "terms of trade" (or barter-price ratio) being almost certainly somewhere between 1 food:1 clothing and 1 food:2 clothing. Both countries will be better off than if they do not specialize. World production will be optimal.

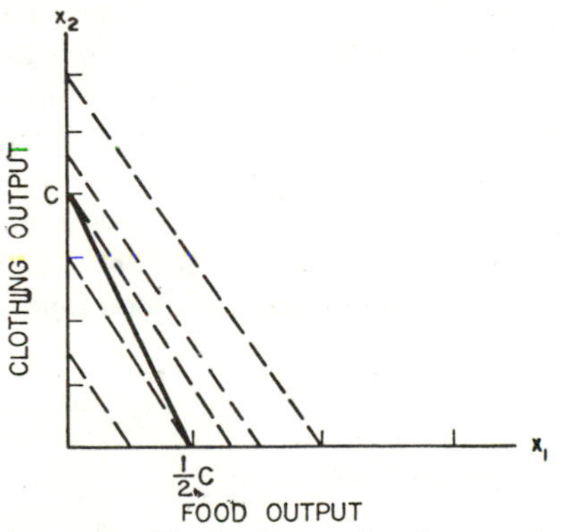

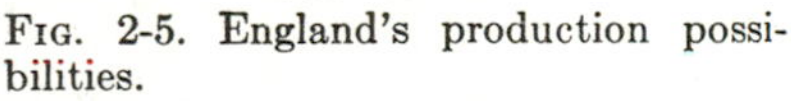
Fig. 2-5. England's production possibilities.

These four standard economic conclusions can be derived from the linear-programming framework. Mathematically, we may consider any one country—say England—to be subject to a linear relation ("production-possibility" curve) of the type shown in Fig. 2-5.

If there exists an international price ratio p_1/p_2, somewhere between 1 and 2, at which England can exchange food for clothing, the real value of its National Product (expressed in clothing units) may be written as

$$Z = \frac{p_1}{p_2} x_1 + x_2 \tag{2-3}$$

or, say,

$$Z = 1.5x_1 + x_2$$

The problem is to maximize National Product (NP) subject to the technical production-possibility curve of England. Or mathematically, to maximize a linear sum of the form (2-3) subject to a linear inequality of the form $2x_1 + x_2 \leq C$. In this inequality C is the maximum output of clothing when no food is produced. According to our assumption, clothing output must be cut back two units for every unit of food produced. Thus the maximum output of clothing when x_1 units of food is produced is $x_2 = C - 2x_1$, and this is equivalent to the inequality stated. This is a typical problem of linear programming.

Graphical or numerical experimentation will soon convince one that the highest NP will be reached only when $x_1 = 0$, $x_2 = C$, and $Z = NP = C$. This may be seen by considering in Fig. 2-5 the *contour lines of equal NP*. These are a family of parallel lines, all with a slope of p_1/p_2; these parallel lines will be less steep than the single production-possibility line of Fig. 2-5, because the price ratio is less than 2. We want to get up to the highest NP line, which we can only do by climbing northwest along the production-possibility line until we reach the X_2 intercept, where food production is zero. (If negative numbers were allowed, we would want to continue moving northwest indefinitely. This is what happens in an arbitrage situation in which two powerful and rich agencies try to maintain different relative price ratios between the same pair of commodities.)[1] Any other solution will lead to a lower NP. Thus, if unemployment is allowed to develop so that England is *inside* its production-possibility curve, NP will suffer. Or if it should specialize on the wrong commodity, food, then NP will turn out to be only $0.75C = 1.5C/2$.

The reader may easily verify that for Portugal the best production pattern, when p_1/p_2 is greater than 1, involves zero clothing production and complete specialization on food. Only in that way will she maximize her NP:[2]

$$Z' = 1.5x_1' + x_2' = \frac{p_1}{p_2} x_1' + x_2'$$

subject to

$$x_1' + x_2' \leq C'$$

We have expressed the international problem of resource allocation as two separate problems, one for each country. The reader should translate the formulations we have arrived at into the terminology introduced in Sec. 2-7.

[1] In this case, little chaps named Gresham who are on their toes can make a lot of money, buying cheap and selling dear and repeating the process indefinitely—or rather until one of the agencies runs out of one of the goods or changes its mind about its price pegging. Buying and selling introduces negative as well as positive numbers. But in linear programming, all variables must usually be positive.

[2] The justifications for Portugal's national-product formula and her production-possibility inequality are analogous to those just given for England.

2-9. THE EFFICIENCY FRONTIER

Throughout the foregoing argument we have assumed that we were given a food-clothing price ratio p_1/p_2 greater than 1 and less than 2. We found that England would then concentrate on clothing production and Portugal on food. Furthermore, since each unit of English resources produces most "value" if devoted to clothing production and each unit of Portuguese resources produces most "value" if devoted to food, there is no need for a national or international planning authority to bring about the efficient allocation. Under atomistic competition it will come about automatically.

However, fingers were made before forks, and we can imagine this situation as it might appear to a naïve scientist from Mars who had never heard about prices and competitive private enterprise. He might still ask the noncommercial question: What is the "optimal" pattern of world production of food and clothing between England and Portugal? If he were acute, the Martian scientist would be troubled by his own question, particularly by the word in quotation marks—optimal. Optimal in what sense? Certainly not—as we have already agreed—in the sense of money value, since this is a precommercial Martian. The scientist might be tempted to consider evaluating food and clothing by their "intrinsic worth"; but unless he had been contaminated by a course in heavy German philosophy, he would soon realize that this is an undefinable concept for food and clothing in a world where some people are more like peacocks and others more like gluttons.

A sagacious Martian would soon settle for a more modest definition of the optimum. He would say, "I don't know how the final choice between food and clothing is to be made, whether English millionaires will have the greatest say, or the United Nations Commission on Living Standards. But it is my job as a production expert *to give the world the best menu from which to choose.* Or, in other words, for each specified amount of any one good—say food—I must make sure that the production of the other good is as large as possible."

In saying this, the scientist has unwittingly defined a definite class of problems in linear programming. Prices as such have nothing to do with the problem, although—like Voltaire's God—it may be desirable to invent them if they do not exist! To the economist, at least, it will seem natural to introduce a system of "shadow" or accounting prices or some sort of system of numerical points.[1]

But first let us write down the mathematical problem that the scientist

[1] If such shadow prices are really useful, then it follows that many problems of linear programming may benefit on the computational side from a process of imitating the market mechanism.

has posed for himself. He wants (1) to maximize the total of clothing production in Portugal and England, subject to (2) a prescribed total of food, and subject to (3) the two linear-production-possibility constraints of the two countries, and where (4) all quantities must by their nature not be negative numbers. This can be stated mathematically using the notation introduced already. Recall that x_1 and x_2 denote the output of food and clothing, respectively, in England, and let C denote the total of resources available for food and clothing production in England. Similarly, x'_1, x'_2, and C' denote food output, clothing output, and resources in Portugal. Finally X_1 and X_2 denote the total outputs of food and clothing in the two countries. Then the mathematical problem is

$$Z = X_2 = x_2 + x'_2$$

is to be a maximum subject to

$$x_1 + x'_1 = X_1 \qquad \text{a prescribed constant}$$
$$2x_1 + x_2 \leq C$$
$$x'_1 + x'_2 \leq C'$$

and $$x_1, x_2, x'_1, x'_2 \qquad \text{all nonnegative}$$

This is a typical problem in linear programming, which is defined as the problem of maximizing one linear relation subject to a number of linear inequalities. It is important to realize that every linear programming problem has the same formal structure. To drive this home, let us give systematic new names, z_1, z_2, z_3, z_4, to the four unknowns of this problem and rewrite it as follows:

Maximize

$$Z = 0z_1 + 1z_2 + 0z_3 + 1z_4$$

subject to

$$2z_1 + 1z_2 + 0z_3 + 0z_4 \leq C_1$$
$$0z_1 + 0z_2 + 1z_3 + 1z_4 \leq C_2$$
$$-1z_1 - 0z_2 - 1z_3 - 0z_4 \leq C_3$$

and $$z_1 \geq 0,\ z_2 \geq 0,\ z_3 \geq 0,\ z_4 \geq 0$$

Here, of course, $x_1 = z_1$, $x_2 = z_2$, $x'_1 = z_3$, $x'_2 = z_4$, $C = C_1$, $C' = C_2$, and $-X_1 = C_3$. Of the three main inequalities above, only the last requires explanation: it corresponds to the equality $x_1 + x'_1 = X_1$ in the earlier formulation. We have taken the liberty of writing this equality as an inequality in order to keep our problem within the standard form so far discussed. This is legitimate provided there is no harm in overshooting the prescribed amount of X_1 (e.g., provided X_1 is not poisonously radioactive or, if it is, provided we can dispose of any excess of it at zero cost). Of course, in most economic problems we would not want to ignore the valuable original information that the last inequality will in fact turn out to be a binding equality.

Note that with C and C' being given, there will be a different best Z (or X_2) for each prescribed X_1. Actually, our economic intuition—if pushed far enough—tells us that the resulting "menu," or world production-possibility curve, looks like Fig. 2-6. The broken-line curve ARB is known as the "efficiency frontier" because it gives the maximum possible world clothing output for each postulated level of food output and, conversely, the maximum possible food output for any preassigned output of clothing.

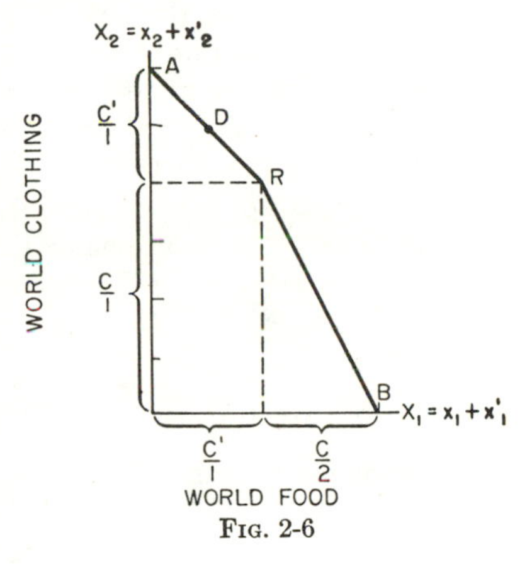

FIG. 2-6

The flatter of the two line segments of the efficiency frontier has a slope of -1, equal to Portugal's food-clothing technical ratio; the steeper segment has a slope of -2 corresponding to the similar technical ratio for England. The absolute maximum of X_2 (corresponding to zero X_1) occurs at A, where all resources in both countries are used to produce x_2 alone, giving us (from the technical relations) $C/1 + C'/1$ of clothing. The maximum of food comes at B, where all resources in both countries are going to food and yielding, in all, $C/2 + C'/1$ of food.

The critical corner point on the X_2X_1 curve occurs at R, where England is specializing completely on clothing and Portugal on food. It might be called the "Ricardo point," since it is there that the classical theory of comparative advantage tells us we shall *almost* certainly end up.[1]

There is another remarkable economic feature of our final so-called world opportunity-cost relation between food and clothing. The curve is a *concave* one: In economic terminology, the "marginal opportunity cost" of converting one good into the other is increasing as we want more of any one good. Or more accurately, it is *nondecreasing*, since along any one of the line segments it is constant, neither increasing nor decreasing. But between line segments it is strongly increasing, a

[1] The classical economists recognized that the word "almost" is needed in order to take account of the possibility that the final market-price ratio might be at the *limit* of the range of differing comparative costs instead of *within* the range. This case was considered especially likely if one country was big compared with the other. In such cases, where the final price ratio settles at the cost ratio of one of the countries, production within that country would be indeterminate. The actual result would have to be dictated by the final pattern of international demand.

result that may at first seem surprising in view of the so-called constant-cost assumptions of the classical theory of comparative advantage.

Economic and mathematical importance attaches to the concept of marginal cost as defined by $-dX_2/dX_1$, the absolute slope of our curve. To the left of the Ricardo point this is exactly equal to one of the technical coefficients of the problem, that of Portugal. To the right, it is equal to the corresponding coefficient of England. At the critical point R it is, strictly speaking, undefined, since the right-hand and left-hand limits that define a mathematical derivative are different. We may adopt the convention whereby marginal cost at such a point is defined as any number between the limiting right- and left-hand slopes. A similar problem arises in defining marginal cost (MC) at the limiting intercept points where the curve hits the axes. It is natural and convenient to define MC at the X_2 intercept as any number equal to or less than the absolute right-hand slope at that point; that is, MC is 1 or any number *less* than 1. Similarly, at the X_1 intercept the MC, or slope, is any number equal to or *greater* than 2.[1]

2-10. ECONOMIC CONSIDERATIONS

An economic theorist who is used to thinking of problems in terms of market situations will immediately be struck by a rather remarkable fact: The purely technical concept of MC, or slope, which could be arrived at from the pure logic of the maximum problem, without reference to prices or markets, does behave remarkably like a market-price ratio. His economic intuition tells him as follows:

1. When $1 < p_1/p_2 < 2$, each country will specialize on its best product and the world will in fact be at the Ricardo point where $1 \leq \text{MC} \leq 2$.
2. When $p_1/p_2 < 1$ (or >2), both countries will specialize completely on the same product and the world will actually be at one (or the other) intercept with $\text{MC} \leq 1$ (or ≥ 2).
3. When $p_1/p_2 = 1$ (or $= 2$), we shall actually be anywhere on one (or the other) of the two line segments with MC, if uniquely defined, being $= 1$ (or $= 2$).

Not only does a market-price ratio have the properties of world MC, but by creating a shadow-price ratio even where none existed and playing the game of competition, we could end up at any specific point of the

[1] We define *MC* at any point on the curve as the numerical slope of any straight line that "touches" but does not "cut" the curve at that point, i.e., the slope of a line that never lies inside the curve but does touch it at the point in question. This corresponds to the general notion of tangency, but carries over to the case of curves with corners, where there may be infinitely many "tangents."

final optimal locus.[1] Moreover, in the special case of this problem in which the production conditions in the two countries are independent, the problem of decision making can be in a certain sense decentralized and partitioned into separate parts.

Thus, if we invent a parameter λ, which is to play the role of a shadow price p_1/p_2, we can split our original maximum problem into two quasi-separate ones. Instead of maximizing

I $$Z = x_2 + x_2'$$

subject to

$$\begin{aligned} x_1 + x_1' &= X_1 \\ 2x_1 + x_2 &\leq C \\ x_1' + x_2' &\leq C' \end{aligned}$$

let us separately maximize

II $$z = \lambda x_1 + x_2$$

subject to

$$2x_1 + x_2 \leq C$$

and II′ $$z' = \lambda x_1' + x_2'$$

subject to

$$x_1' + x_2' \leq C'$$

There is some choice of λ for which separate optimal solutions to the latter problems, II and II′, do add up to the optimal solution to I for any X_1. Also it is clear that solving both II and II′ is equivalent to solving the world problem of maximizing

I′ $$Z = z + z' = \lambda(x_1 + x_1') + (x_2 + x_2')$$

subject to

$$\begin{aligned} 2x_1 + x_2 &\leq C \\ x_1' + x_2' &\leq C' \end{aligned}$$

To arrive at an optimal point (on our X_2X_1 locus) of each of the following types:

1. At the upper, or X_2, intercept, A
2. On the upper flat-line segment, AR
3. At the Ricardo point, R

[1] In order to produce at point D on Fig. 2-6 we should have to have a price ratio $p_1/p_2 = 1$. But with that price ratio, any point on the line segment AR corresponds to an equally large total value of product, $p_1x_1 + p_2x_2$. Thus some supplementary guidance would be needed to arrive at a specified output point like D. This problem will be discussed in the next chapter.

4. On the lower steep-line segment, RB
5. At the lower, or X_1, intercept, B

we must have

1. $\lambda \leq 1$
2. $\lambda = 1$
3. $1 \leq \lambda \leq 2$
4. $\lambda = 2$
5. $2 \leq \lambda$

As λ grows from small to large, X_1 goes from nothing to the maximum, always at the expense of X_2. The correspondence between λ and the separate solutions is shown in Fig. 2-7.

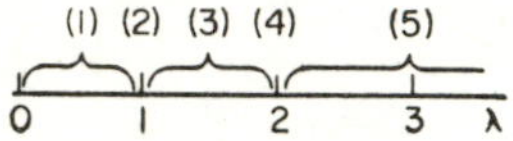

Fig. 2-7

The international-trade example has taken us a little further than the minimum-cost diet example. In the diet problem we regarded food prices as given and came out with a unique optimal diet. In the international-trade example we saw that to each given structure of prices there corresponds an optimal output, and, further, we saw how the optimal output changed in response to changes in the price structure. In the diet problem we found a single optimal diet; in the international-trade problem we traced out an entire family of optimal outputs by permitting price ratios to change. The international-trade problem illustrates "parametric programming," i.e., programming in which one of the parameters is permitted to vary. These contrasts are not inherent in the problems but result only from our choice of considerations to be discussed in each context.

3

The Valuation Problem; Market Solutions

We have already noted that linear programming is based on a mathematical problem. It happens that mathematical linear-programming problems come in pairs; every mathematical linear-programming problem is intimately related to another problem called its "dual." This statement would be no more than an interesting mathematical curiosity if it were not for the fact that if an economic problem can be formulated as a linear-programming problem, then there will generally be a related economic problem that corresponds to the dual. We shall soon see some examples.

These facts are not intuitively evident, and, indeed, it took a while after linear programming had been discovered for the dualism feature to be recognized and appreciated. But they should not be surprising to an economist who, after all, is familiar with the fact that resource allocation and pricing are two aspects of the same problem. An economist would expect that since linear programming solves the allocation problem, it would solve the pricing problem also, and this, in essence, is what the dualism property consists in. In this chapter we shall develop the duals of the two problems discussed in Chap. 2 and shall investigate their economic meaning and implications.

3-1. THE MATHEMATICAL DUAL

Let us take things up in historical order. In the development of linear programming some dual problems were encountered before their usefulness was appreciated. To follow this sequence we shall set up some duals first and subsequently consider their interpretation.

A typical linear-programming problem can be expressed in the following way: It is required to find the n numbers $x_1, x_2, \ldots, x_n$ which make the expression

$$z = c_1x_1 + c_2x_2 + \cdots + c_nx_n \tag{3-1}$$

as great as possible, where $c_1, \ldots, c_n$ are given constants, subject to the restrictions that no x shall be negative and that the x's shall satisfy the m inequalities:

$$\begin{aligned} a_{11}x_1 + a_{12}x_2 + \cdots + a_{1n}x_n &\le b_1 \\ a_{21}x_1 + a_{22}x_2 + \cdots + a_{2n}x_n &\le b_2 \\ \cdots\cdots\cdots\cdots\cdots\cdots& \\ a_{m1}x_1 + a_{m2}x_2 + \cdots + a_{mn}x_n &\le b_m \end{aligned} \tag{3-2}$$

The diet and international-trade examples of the last chapter had precisely this form.

We now construct a different problem by rearranging the data of the problem just stated. Conditions (3-2) consist of m inequalities. First we introduce a variable to correspond to each of the inequalities and call these new variables $u_1, u_2, \ldots, u_m$. Then we form the sum of the cross products of these new variables with the constants on the right-hand side of the inequalities; i.e., we form

$$z' = b_1u_1 + b_2u_2 + \cdots + b_mu_m \tag{3-3}$$

Next we form an inequality involving the u's to correspond to each of the variables x, in the original problem, using for that purpose the coefficients of the x's in the original problem. For example, by using the coefficients of x_1 in the original problem we derive the inequality

$$a_{11}u_1 + a_{21}u_2 + \cdots + a_{m1}u_m \ge c_1$$

Note that we have reversed the sense of the inequality sign. Next we pick up the coefficients of x_2 in the original problem and cross-multiply them with the u's to obtain the second dual constraint. Continuing in this way we derive the whole set of dual inequalities:

$$\begin{aligned} a_{11}u_1 + a_{21}u_2 + \cdots + a_{m1}u_m &\ge c_1 \\ a_{12}u_1 + a_{22}u_2 + \cdots + a_{m2}u_m &\ge c_2 \\ \cdots\cdots\cdots\cdots\cdots\cdots& \\ a_{1n}u_1 + a_{2n}u_2 + \cdots + a_{mn}u_m &\ge c_n \end{aligned} \tag{3-4}$$

We then consider the problem of finding a nonnegative set of values $u_1, u_2, \ldots, u_m$ which make expression (3-3) as *small* as possible while satisfying the inequalities (3-4). This is another linear-programming problem, called the "dual" of the problem involving the x's.

The relationship between a problem and its dual may be summarized as follows:

1. The dual has one variable for each constraint in the original problem.
2. The dual has as many constraints as there are variables in the original problem.

3. The dual of a maximizing problem is a minimizing problem, and vice versa.

4. The coefficients of the objective function of the original problem appear as the constant terms of the constraints of the dual, and the constant terms of the original constraints are the coefficients of the objective function of the dual.

5. The coefficients of a single variable in the original constraints become the coefficients of a single constraint in the dual. Stated visually, each column of coefficients in the constraints of the original problem becomes a row of coefficients in the dual.

6. The sense of the inequalities in the dual is the reverse of the sense of the inequalities in the original problem, except that the inequalities restricting the variables to be nonnegative have the same sense in the direct problem and the dual.

All this is no more than a peculiarly intricate formal manipulation. To make sense of it we shall form and interpret the duals of the two examples discussed in Chap. 2, taking up the international-trade example first because it is the easier of the two.

3-2. THE DUAL OF THE INTERNATIONAL-TRADE EXAMPLE

The international-trade example concerned the allocation of production of food and clothing between two mythical countries called England and Portugal. England was supposed to have a comparative advantage in clothing production. We found that the optimal allocation of effort depended on the terms of exchange between food and clothing, but that at no terms of exchange would England produce the food and Portugal the clothing. We now ask a different question: Suppose that the terms of exchange (or relative prices) of the two commodities are given. What will be the value of English resources relative to Portuguese resources?

To answer this question let us define a unit of resources in either country as sufficient resources to produce a unit of clothing.[1] Then, recalling the numerical data assumed, one unit of English resources can produce one-half unit of food and one unit of Portuguese resources can produce one unit of food. Let a unit of food be worth p_1 and a unit of clothing be worth p_2. A unit of English resources can then be used to

[1] Notice that we are adopting an economic rather than a physical unit so that, for example, a unit of English labor need not consist of the same number of man-hours as a unit of Portuguese labor. We are assuming also that the resources within each country are completely substitutable so that there is only a single resource limitation in each country. Further, we are making the assumption of constant returns to scale in each country.

produce either $\frac{1}{2} p_1$ of food or p_2 of clothing and, of course, will be devoted to whichever of these has the greater value. Similarly a unit of Portuguese resources can be used to produce either p_1 of food or p_2 of clothing and, again, will be used to produce the greater value.

If we assume, as we did in Chap. 2, that $p_2 < p_1 < 2p_2$, English resources will be used to produce clothing, the price of which is p_2, and Portuguese resources will be used to produce food, the price of which is p_1. Thus we arrive again at the optimum found in Chap. 2. In addition, we arrive at a rule for valuing the services of the resources themselves. On standard marginal-productivity principles, the value of the services of a resource equals the value of its output in its most profitable employment. Thus if u and u' denote the values of English and Portuguese resources, respectively, $u = p_2$ and $u' = p_1$.

This answers the question with which we started this section, but what has it to do with the relationship between a linear-programming problem and its dual? Just this: our analysis was simply a thinly disguised solution to the dual of the comparative-advantage example. To see this we restate the original problem, this time bringing in resources explicitly. Denote the number of units of English resources devoted to food and clothing production by y_1, y_2, respectively, and similarly the allocation of Portuguese resources by y'_1, y'_2. Let Q, Q' be the number of units of resources in the two countries. Then the restrictions on resource allocation are:

For England:

$$y_1 + y_2 \leq Q$$

For Portugal:

$$y'_1 + y'_2 \leq Q'$$

The total value of output, which is to be maximized subject to these restrictions, is

$$z = p_1(\tfrac{1}{2} y_1 + y'_1) + p_2(y_2 + y'_2)$$

Next we construct the dual of this problem by following the directions given in Sec. 3-1. The dual problem will have two variables, one for each of the two constraints of the direct problem—call them u and u'. We need not be concerned with their significance yet. The objective function of the dual uses the constants of the restraints of the original problem as coefficients of the dual variables. Thus it is

$$z' = Qu + Q'u'$$

The dual has four constraints, one corresponding to each variable in the original problem. The first of these constraints, for example, is that u times the coefficient of y_1 in the first constraint of the direct problem plus u' times the coefficient of y_1 in the second constraint of the original problem[1] must be no smaller than the coefficient of y_1 in the objective function of the original problem. Thus, the dual of the international-trade example is as follows:

Minimize

$$z' = Qu + Q'u'$$

subject to

$$\begin{aligned} u &\geq \tfrac{1}{2} p_1 \\ u &\geq p_2 \\ u' &\geq p_1 \\ u' &\geq p_2 \end{aligned}$$

These relations, plus a little detective work, will tell us what the new variables u and u' must signify. If the constraints are to make sense economically, u and u' must be prices because in each case u and u' are compared in magnitude with p_1 and p_2, which are prices. The objective function shows that u must be the price of English resources because it is multiplied by Q, the quantity of English resources, and no other interpretation would give the product Qu an intelligible meaning. Similarly u' must be the price of Portuguese resources. Thus z', the quantity to be minimized, is the total valuation of English and Portuguese resources. We can now interpret the restraining inequalities of the dual problem. They require that English and Portuguese resources be given high enough values so that whether they are used to produce food or clothing, the value of the resources used will be at least as great as the value of the goods produced.

This dual problem can be solved at a glance. If $p_2 < p_1 < 2p_2$, as we have been assuming, the solution is obviously $u = p_2$, $u' = p_1$. Comparing this result with the previous solution we see that u, u' are just the values of output of the two types of resources when devoted to their most profitable uses. The dual problem, then, amounts to finding the smallest total valuation of resources z' such that the value of output is completely imputed to the resources.

Things stand as follows: We assumed that the prices of food and clothing were known and started with the problem of finding the optimal allocation of food and clothing production between England and Portugal. That turned out to be a linear-programming problem. Then we formed the dual of this allocation problem and interpreted it. The dual was the problem of assigning values to English and Portuguese resources in such a way as to minimize the total resource valuation without giving rise to unimputed profits in any possible use of the resources. This system of valuation is reminiscent of the operation of a competitive market in which resource users are forced by competition to offer to resource owners the

[1] This coefficient happens to be zero, so that this term does not appear explicitly in this example.

full value to which their resources give rise, while competition among the resource owners drives down resource prices to the minimum consistent with this limitation.

This analysis illustrates (but of course does not prove) some important general properties of the relationship between the two members of a pair of duals. These are, indeed, the features that make the dualism property important both mathematically and economically. Let us state them in general terms and then apply them to the international-trade example.

1. The maximum value of the objective function of a maximizing problem equals the minimum value of the objective function of its dual. Similarly, the minimum of the objective function of a minimizing problem equals the maximum of the objective function of its dual.

2. When a linear-programming problem has been solved, some of its choice variables will be positive and others may be equal to zero. The dual problem has one constraint corresponding to each variable in the original problem. If a variable is positive in the solution to the direct problem, then the corresponding constraint will be satisfied with exact equality in the solution to the dual. If a variable takes the value zero in the solution to the direct problem, then the corresponding constraint in the solution to the dual will usually be satisfied with an inequality.

Similarly, the dual problem has one variable for each constraint in the original problem. If any constraint is satisfied by an inequality in the solution to the direct problem, the corresponding variable in the solution to the dual will equal zero. And if a constraint is satisfied with exact equality in the solution to the original problem, the corresponding dual variable will usually have a positive value in the dual solution.

These are the crucial rules which tell us which of the dual variables have to be solved for and which of the dual constraints will have equal signs and thus be available for use in an elimination process. By virtue of these rules, once a linear-programming problem has been solved, its dual reduces to a straightforward system of linear equations.

3. The dual of the dual problem is the original problem itself. The reader can verify this readily by constructing the dual of the problem expressed in Eqs. (3-3) and (3-4).[1]

To test these properties by applying them to the international-trade example, first recall that in solving the resource-allocation problem we found that the maximum attainable total value of output was

$$z = p_1Q' + p_2Q$$

In solving the resource-valuation problem we found that the minimum permissible total imputed value of resources was $z' = p_2Q + p_1Q'$, or the same. Thus the first general property is confirmed in this case.

To test the second general property note that in the resource-allocation

[1] Proofs of these assertions will be given in Chap. 4.

problem there were four choice variables, y_1, y_2, y'_1, y'_2, of which two, y_2 (English resources devoted to clothing) and y'_1 (Portuguese resources devoted to food) were positive in the optimal solution. The assertion was that the dual constraints corresponding to these two variables, the second and third constraints as written, should be satisfied with equal signs, while the remaining constraints are satisfied with inequality signs. This is just what happened. Furthermore, since both constraints in the original problem were satisfied with exact equality, both dual variables should have positive values in the solution to the dual, and this occurred also. The reader can check easily to assure himself that the direct international-trade example is the dual of its dual, in accordance with the third assertion.

3-3. THE DUAL OF THE DIET PROBLEM

In the preceding section we constructed the dual of the international-trade example and interpreted it, finding that the dual was simply the problem of imputing values to the services of the factors of production. The reader can test how well he is catching on to this kind of manipulation by thinking back to the diet problem and trying to surmise the significance of its dual.

In the diet problem[1] we have to minimize the total cost of a diet chosen from among five foods and still provide given amounts of calories and vitamins. Symbolically, we have to minimize

$$z = 2x_1 + 20x_2 + 3x_3 + 11x_4 + 12x_5$$

subject to

$$\begin{aligned} x_1 \qquad + x_3 + x_4 + 2x_5 &\geq 700 \\ x_2 \qquad + x_4 + x_5 &\geq 400 \end{aligned}$$

Following the dualism rules, the dual of the minimizing problem in five variables and two restraints is the following maximizing problem in two variables and five restraints:

To maximize

$$z' = 700u_1 + 400u_2$$

subject to

$$\begin{aligned} u_1 \qquad &\leq 2 \\ u_2 &\leq 20 \\ u_1 \qquad &\leq 3 \\ u_1 + u_2 &\leq 11 \\ 2u_1 + u_2 &\leq 12 \\ u_1 \qquad &\geq 0 \\ u_2 &\geq 0 \end{aligned}$$

[1] The data are given in Chap. 2, Table 2-4.

Each of the numbers on the right-hand side of these restraints is a price, the price of some food. Thus if the restraints are to make sense, the left-hand sides must be prices too. In the objective function 700 is a number of calories and 400 a number of vitamins. So z' will have the dimensions of a value if we interpret u_1 as the price of calories and u_2 as the price of vitamins.

On this interpretation the function to be maximized, z', is the imputed value of an adequate diet, i.e., the imputed value of 700 calories plus 400 vitamins. The five inequalities, one for each food, state that in every case the price of a food must be at least as great as the imputed value of the calories and vitamins that it provides. For example, suppose $u_1 = \$1$ and $u_2 = \$10$. Then we can compute a table, shown as Table 3-1, comparing the cost of each food with the value of the nutritive elements that it contains. To illustrate the calculation, consider Food 5. It contains 2 calories (each worth \$1) and 1 vitamin (worth \$10), giving a total value of \$12. The two prices we have assumed satisfy the inequality conditions because the value of the nutrients is in no case greater than the price of the food.

TABLE 3-1. MINIMUM-COST-DIET EXAMPLE: FOOD PRICE AND VALUES OF NUTRITIVE CONTENT

Food	Value of nutrients (1)	Price of food (2)	Excess of (1) over (2)
1	1	2	−1
2	10	20	−10
3	1	3	−2
4	11	11	0
5	12	12	0

The problem dual to the minimum-cost-diet problem can then be stated: To assign nonnegative values to the two food elements in such a way that the total "nutritive value" assigned to a food never exceeds its market price and, consistent with this limitation, so that the total nutritive value of the minimal acceptable diet is as great as possible. This amounts to imputing the market values of the foods to the nutritive elements, which are what make the foods worth purchasing.

Perhaps this doesn't sound like a very sensible way to impute values to the nutritive elements. We intend to justify it in the next few sections. But first, look at the useful properties of the values we have assumed, \$1 per calorie and \$10 per vitamin. They tell us that buying Food 1 is a bit like paying \$2 for a table d'hôte dinner when the prices of the individual dishes add up to \$1. We may assume that the economical housewife will never buy a food unless the value of its nutrients is at least as great as its price. Thus, if she lets herself be guided by the prices we have been assuming, the wise housewife will eschew the first three foods and buy only the fourth and fifth, and this, as we already know from Chap. 2, is the way to get an adequate diet at minimum cost. So these prices are sensible guiding prices for the diet problem. They are also, we shall see below, the solution to the dual problem.

Granted, then, that the dual problem yields sensible prices such that if a person follows the principle of never buying a food if its price is greater than the value of its nutritive content, that person will be guided to a minimum-cost, adequate diet. Still the question remains: Why does the dual problem have this property? The answer to this question is not transparent, and we approach it slowly.

3-4. THE SIMPLE CASE OF "PURE" FOODS

Suppose we tackle a rather simple problem first—that in which all foods are "pure" foods, each containing something of one nutritional element and nothing of all others. Thus, X_1 may have only calories, X_2 only vitamins, X_3 perhaps calories only (but not necessarily with the same number as X_1 or with the same market price), etc. In this pure-foods case our problem obviously breaks up into m different, independent, simple problems. Among all the calorie foods we select that one which most cheaply gives us our calorie requirements. Similarly, we select among the pure foods containing only vitamins the cheapest way of buying vitamins. Our final cost of an optimal diet is the sum of the cost of getting calories *and each* of the other elements.

Let us consider calories alone. For this purpose we might as well number all the pure calorie foods $X_1, X_2, X_3, \ldots$, with market prices $p_1, p_2, p_3, \ldots$, and with unit calorie contents $a_{11}, a_{12}, a_{13}, \ldots$. We must minimize

$$z = p_1x_1 + p_2x_2 + \cdots$$

subject to

$$a_{11}x_1 + a_{12}x_2 + \cdots \geq c_1$$
$$x_1 \geq 0, \quad x_2 \geq 0, \quad \ldots$$

We could get our answer by simply picking the greatest of a_{11}/p_1, a_{12}/p_2, etc., and concentrating on the corresponding food.

It will turn out to be a little more convenient to define a shadow imputed price for calories. Let this be called u_1. As yet we don't know its value (in dollars per calorie). The "unit profitability" of any food can be thought of as dollars of calorie value it contains minus its cash cost per unit, or as

$$\pi_1 = a_{11}u_1 - p_1$$
$$\pi_2 = a_{12}u_1 - p_2$$
$$\cdots\cdots\cdots\cdots$$
$$\pi_i = a_{1i}u_1 - p_i$$

These are definite numbers once u_1 is given a definite value. At first glance one might suppose that we should pick the largest of these profitability numbers and get our calories from the corresponding food. But a second thought will convince us that this is not valid. If one food cost twice as much as another and had twice as many calories, its profitability would be twice as great; but there would be no advantage whatever in choosing that food over the other.

Our profitabilities have to be put on a per-dollar basis if they are to be comparable. We could work with

$$\frac{\pi_1}{p_1} = \frac{a_{11}}{p_1} u_1 - 1$$
$$\cdots\cdots\cdots$$
$$\frac{\pi_i}{p_i} = \frac{a_{1i}}{p_i} u_1 - 1$$

The greatest of these would give us our cheapest source of calories. For any positive u_1, or calorie shadow price, this would be the same thing as picking the greatest a_{1i}/p_i, or number of calories per dollar.

Let us suppose that the kth food is the best one, so that

$$\frac{a_{1k}}{p_k} \geq \frac{a_{1i}}{p_i} \qquad i \neq k$$

Our best calorie diet is $(x_1, x_2, \ldots, x_k, \ldots, x_i, \ldots) = (0, 0, \ldots, c_1/a_{1k}, \ldots, 0, \ldots)$, and its cost is

$$z = 0 + 0 + \cdots + p_k \frac{c_1}{a_{1k}} + \cdots + 0 + \cdots$$

Obviously the (cheapest) extra or marginal cost of increasing our calorie requirements from c_1 to $c_1 + 1$ units would be

$$MC_1 = \frac{p_k}{a_{1k}}(c_1 + 1) - \frac{p_k}{a_{1k}} c_1 = \frac{p_k}{a_{1k}} = \frac{\partial z}{\partial c_1}$$

The *cost* of calories—and note that we do *not* say the *worth* of calories—would seem to be given by p_k/a_{1k}. This is the shadow price of calories, or

$$u_1 = \min_i \frac{p_i}{a_{1i}} = \frac{\partial z}{\partial c_1} = MC_1$$

With our shadow price now determined, we can go back and look at our original profit figures $\pi_1, \pi_2, \ldots,$ etc. Our objection to them—that they are not on a per-dollar basis—now disappears. In every case the profits are negative except in the case of our very cheapest calorie source X_k. Thus

$$\pi_1 = a_{11}u_1 - p_1 \leq 0 \qquad \text{because } u_1 = \frac{p_k}{a_{1k}} \leq \frac{p_1}{a_{11}}$$

$$\pi_2 = a_{12}u_1 - p_2 \leq 0$$

$$\cdots\cdots\cdots\cdots$$

$$\pi_k = a_{1k}u_1 - p_k = 0$$

$$\cdots\cdots\cdots\cdots$$

$$\pi_i = a_{1i}u_1 - p_i \leq 0$$

If the unusual should happen and some other food also had exactly zero profit, then it would be a matter of indifference as to how we divided our calorie expenditure between the cheap food and X_k.

In effect we have determined the highest imputed shadow price u_1 that is possible for calories; or more accurately, the highest possible calorie price compatible with some food in the system "breaking even." All other nonoptimal foods will show a loss. In order to solve a minimum problem (lowest cost from selecting best x's), we have in effect chosen to solve a maximum problem (highest price for calories). Thus our correct u_1 is the solution of

$$z = c_1u_1$$

to be a maximum subject to

$$\begin{aligned} a_{11}u_1 &\leq p_1 \\ a_{12}u_1 &\leq p_2 \qquad u_1 \geq 0 \\ &\cdots\cdots \\ a_{1i}u_1 &\leq p_i \end{aligned}$$

All that we have established in the pure-calorie-foods case also holds for the pure-vitamin-foods case and for any other pure foods. Let us recapitulate what we have established.

1. We can define the profitability of any pure food in terms of dollars per unit, i.e., the $\pi_1, \pi_2, \ldots$.
2. To do this we must first know the imputed (shadow) price of the nutritional elements, i.e., the u_1 per calorie, the u_2 per vitamin, $\ldots, u_m$, etc.
3. The shadow prices of the nutrients have a number of properties:
   a. They are in dollars per unit of each nutritional element.
   b. They reflect marginal costs, i.e., the least cost of an extra prescribed unit of the element in question.
   c. They must be such as to make profits negative for a category of foods, namely, those foods that are not bought at all.
   d. They must be such as to make profits zero for any food that is bought in positive quantities.
   e. They are, in economic jargon, "derived demands," and these values depend on the prices of the foods ($p_1, \ldots, p_n$), on the

food contents (a_{ij}), and (possibly) on the specified requirements ($c_1, \ldots, c_m$).

f. The shadow prices ($u_1, \ldots, u_m$) are the solution of the "dual," or "transpose," problem to our original problem. They give us the maximum amount of what might be called "economic rent" that can be imputed to the nutritional elements; and the sum total of optimal rents on nutrients must add up to exactly the same value as the total optimal cost of foods.[1]

3-5. GENERAL CASE OF MIXED FOODS

3-5-1. Marginal Costs and Imputed Values. The above conclusions were developed for the almost trivial case of pure foods, where no food had more than one nutritional element. It so happens that they remain substantially unchanged if we consider the more realistic and complicated case in which foods are mixtures of many elements. A slightly more complicated argument is necessary to demonstrate this; at the same time the actual task of finding a best diet is also slightly more complex.

The argument depends on an analysis of the marginal cost of the nutritive elements and, in particular, on the fact that each nutritive element has a determinable marginal cost even when the nutrient cannot be purchased individually but has to be bought in mixed bundles called foods.

Take an example. Suppose the foods available, their prices, and the nutritive requirements are given in Table 2-4. Take as a point of departure from which to compute marginal costs the diet consisting of 25 units of Food 1, 5 units of Food 2, 25 units of Food 3, 50 units of Food 4, and 300 units of Food 5. It is easy to verify that this diet meets the nutritive standard (i.e., is feasible), but we know from our previous solution in Chap. 2 that it is not the cheapest diet that does so.

The marginal cost of a calorie, say, can be defined as the increase in the cost of the diet that would result from increasing the number of calories required by unity. In the present instance there are three ways

[1] The interested reader can verify all this by working out our earlier arithmetical example of calories and vitamins on the assumption that only the goods X_1, X_2, X_3 can be bought. The data and solution are, respectively,

a	C	1	0	1	700
		0	1	0	400
P	Z	2	20	3	?

$$(x_1, x_2, x_3) = (700, 400, 0)$$

$$(u_1, u_2) = (2, 20)$$

$$2(700) + 20(400) = 9{,}400 = Z = Z' = 700(2) + 400(20)$$

in which a unit increase in caloric requirement could be met: (1) purchase one more unit of Food 1, at a cost of \$2; (2) purchase one more unit of Food 3, at a cost of \$3; (3) increase consumption of Food 5 by one unit and simultaneously reduce consumption of Food 4 by one unit, at a net cost of $\$12 - \$11 = \$1$. The third method is obviously the cheapest, thus determining the marginal cost of calories at \$1.[1]

This example suggests several generalizations. First, the pure foods 1, 2, and 3, being relatively expensive, did not affect the calculation. We would have found the same marginal costs even if those foods had been omitted from the list of possibilities. The essential requirement for establishing marginal costs for the separate nutritive elements is the possibility of substituting foods of different nutritive composition for each other, as in method 3.

Second, the marginal costs *of the nutrients* depend not only on the prices *of the foods* and their nutritive contents, but also on the diet used as a point of departure. If, for example, the diet considered above had not included at least one unit of Food 4, method 3 for increasing its caloric content would not have been available and the marginal cost of calories would have been \$2, derived from method 1. Another illustration of the dependence of marginal costs on the diet used as a point of departure is given by the diet consisting of 400 units of Food 5. This meets the minimum requirement—indeed it provides 100 calories to spare. As a result the marginal cost of calories with reference to this diet is zero, since the same diet would suffice if the caloric requirement were increased to 701. In general, each feasible solution to a programming problem—in this example each feasible diet—implies a set of marginal costs. There may be numerous feasible diets, however, corresponding to a single set of marginal costs.

Third, this set of marginal costs would inform us at once, if we had not known it in advance, that the suggested diet is needlessly expensive. This is shown by the fact that the prices of three of the foods, namely, the first three, are greater than the values of their constituents valued at marginal cost. Thus, Food 3 provides one calorie at a cost of \$3, while a calorie can alternatively be obtained by method 3 for a net cost of \$1. Clearly money can be saved without changing the nutritional content of the diet by substituting 25 units of Food 5 for 25 units each of Food 3 and 25 units of Food 4, i.e., by eliminating Food 3 from the diet and restoring its nutritive content by method 3.[2] Since the nutritive elements can in

[1] *Exercise.* In what ways can the consumption of vitamins be increased without affecting the consumption of calories? What is the marginal cost of vitamins?

[2] It will be instructive for the reader to see if the other two wasteful foods can be eliminated from the diet (they can) and to compare the resulting "economical diet" with the optimum obtained in Chap. 2.

effect be purchased individually, by artful substitution if they are not offered in pure form, the table d'hôte analogy suggested earlier applies. No food can be included in the economical diet if its price is greater than the marginal cost of its nutritive content.

By similar reasoning a diet cannot be optimal if there is any available food, whether included in the diet or not, whose price is less than the sum of the marginal costs of its constituent nutrients. In the first place it would be a mathematical contradiction for such a food to be included in the diet. The reason is that the marginal costs are the lowest prices at which the various nutrients can be obtained. For any food in the diet to be priced below the marginal-cost value of its constituent nutrients would mean that a package of nutrients could be acquired at less than its marginal-cost value. This would mean that at least one nutrient could be purchased at less than its lowest cost. This deduction may be more transparent if it is remembered that the marginal costs of the nutrients reflect the prices of the foods in the diet; change those and you change the marginal costs.

In the second place if any food not in the diet were priced below the sum of the marginal imputed costs of its nutritive content, then that diet could not be optimal. The reason is that if this food were substituted for a combination of foods in the diet with the same nutritive content, then money would be saved.

The upshot of this argument is that (1) the price of every food must be at least as great as its marginal-cost value computed on the basis of an optimal diet, and (2) the price of any food included in the optimal diet must be exactly equal to the marginal-cost sum of its nutritive contents. These statements, it should be noted, do not characterize the prices of the foods, which are regarded as given data in this discussion. Rather, they characterize the marginal costs of the nutrients, which are to be imputed from the food prices. We have been calling these marginal costs of nutritive elements, chosen so as to satisfy these requirements, "shadow prices."

3-5-2. Dual Problem. We have not mentioned the dual problem of linear programming so far, but it is emerging all the same. The requirement we have just arrived at, namely, that the sum of the marginal costs of nutrients associated with each food in an efficient diet must be not less than the price of that food, is the same as the constraining inequalities of the dual linear-programming problem. Written algebraically, it is

$$\pi_i = a_{1i}u_1 + a_{2i}u_2 + \cdots + a_{mi}u_m - p_i \leq 0 \qquad i = 1, 2, \ldots, n$$

i.e., the excess of the nutritional worth of the ith food over its price cannot be positive. Moreover, the equality (as distinct from the inequality) must hold for all foods included in the optimal diet.

3-5-3. Five-food Example. Let us revert to the five-food example given in Table 3-2.

TABLE 3-2. NUTRIENTS AND PRICES IN FIVE-FOOD EXAMPLE

	Calories	Vitamins	Prices	Best diet
Foods:				
X_1	1	0	2	(?)
X_2	0	1	20	(?)
X_3	1	0	3	(?)
X_4	1	1	11	(?)
X_5	2	1	12	(?)
NRC requirements.........	700	400	$Z = ?$	
Nutrient shadow prices.....	(?)	(?)		

Our unknowns are the best diet and its cost; also the correct shadow prices for vitamins and calories. We happen to have been told that the best diet is (0, 0, 0, 100, 300). It can also be revealed that the correct shadow prices for calories and vitamins are $(u_1, u_2) = (1, 10)$. How can we find this out for ourselves? And what good is this last information once we have acquired it?

To find it out for ourselves, we might set down the rule:

Find shadow prices that permit of no positive profits anywhere in the system and that permit zero profits somewhere in the system.

This rule certainly will help us to rule out a number of price configurations. Thus we could never have the vitamin price greater than 20 because that would make

$$\pi_2 = 1u_2 - 20 > 0$$

Likewise, we could never have the calorie price greater than 2 since that would make $\pi_1 > 0$. We might simply examine all our profit figures and try all combinations of the u's until we end up with a pattern of profits:

$$\pi_1 = a_{11}u_1 + \cdots + a_{m1}u_m - p_1 \leq 0$$
$$\cdots\cdots\cdots\cdots\cdots\cdots\cdots$$
$$\pi_n = a_{1n}u_1 + \cdots + a_{mn}u_m - p_n \leq 0$$

where the equality sign holds at least once.

After much experimentation, we might be lucky enough to hit upon the combination $(u_1, u_2) = (1, 10)$, which results in

$$\begin{aligned} \pi_1 &= 1(1) + 0(10) - 2 = -1 < 0 \\ \pi_2 &= 0(1) + 1(10) - 20 = -10 < 0 \\ \pi_3 &= 1(1) + 0(10) - 3 = -2 < 0 \\ \pi_4 &= 1(1) + 1(10) - 11 = 0 \\ \pi_5 &= 2(1) + 1(10) - 12 = 0 \end{aligned}$$

We might conclude that our rule had led us to the true equilibrium prices. But this is too hasty a conclusion. Perhaps there are other lucky guesses for the u's that would also give us a pattern of profits compatible with the above rule. Even worse, our result seems to be independent of the C's.

Our darkest suspicions are confirmed when we happen to try any of the following price combinations:

$$
\begin{aligned}
&(A) \qquad & (u_1, u_2) &= (0, 11)\\
&(B) & &= (1, 10)\\
&(C) & &= (2, 8)\\
&(D) & &= (2, 0)
\end{aligned}
$$

We have already seen that B satisfies our profit rule. The reader may verify that any of these other points also gives profits which are nowhere positive. Thus, he will find that for D,

$$(\pi_1, \pi_2, \pi_3, \pi_4, \pi_5) = (0, -20, -1, -9, -8)$$

Our rule leads not to one solution but to four solutions. Actually, the multiplicity of price patterns is much greater; it is infinite. Any weighted average of the prices in A and B, or in B and C, or in C and D also satisfies our rule. For example, the point $(u_1, u_2) = (1.3, 9.4)$, which is three-tenths of the way between B and C, gives a profit pattern, the algebraic signs of which are $(-, -, -, -, 0)$, etc.

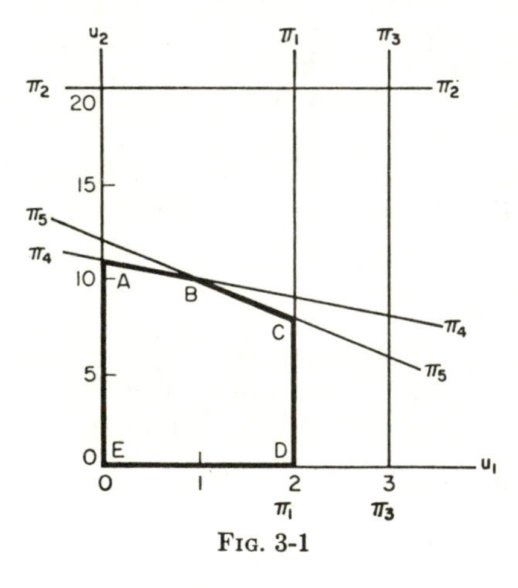

Fig. 3-1

All this is summarized in Fig. 3-1. The lines $\pi_1\pi_1, \ldots, \pi_5\pi_5$ represent the combinations of the shadow prices that will make each food show zero profitability. The "pure foods" x_1, x_2, and x_3 all yield east-west or north-south lines. The mixed foods have profit boundaries that slope downward depending upon the mixture of calories and vitamins.

Our rule of no positive profits means that we must always be below or to the left of every line. Also shadow prices cannot be negative, so that, in all, our profitability rule constrains us to the five-sided area $ABCDE$.

Our rule of no positive profits certainly needs amplification. So long as we had only one constraint—as in the simple theory of comparative advantage—the rule worked out satisfactorily. But now that we have two (or more) constraints, the rule gives us too many possible sets of equilibrium prices. What we seem to need in Fig. 3-1 is a "best" direction toward which to aim. And intuitively one feels that this best direction can only be supplied by a knowledge of the C's, or minimum nutritional requirements.

The incompleteness of our rule can be seen also from another point of view. Consider any set of shadow prices that satisfies the rule, i.e., that is such that the price of every food is at least as great as the imputed value of its nutritive contents. Then whatever diet the housewife purchases will cost her at least as much as the imputed value of the nutrients it contains. Suppose, now, that the housewife buys a nutritionally adequate diet. That diet will contain at least as much of every nutrient as the nutritional standard calls for and therefore—and this is the result we are aiming for—will cost at least as much as the shadow-price value of the nutrients required by the adequate diet.

Now suppose that strict inequality happens to hold, i.e., that at given shadow prices and market prices the housewife pays more for an adequate diet than it is worth in shadow-price terms. Then either the housewife is purchasing wastefully (not an unheard-of possibility) or there is an inconsistency in the system of prices, since what a thing is worth in a market economy is what a wise buyer has to pay or forego in order to obtain it. Let us rule out the possibility that the housewife is wasteful. Then she spends her money exclusively on foods with prices exactly equal to the shadow worth of their nutritive contents, and any excess of the cost of her purchases over the shadow worth of an adequate diet must arise from purchasing more of certain nutrients than the nutritive standard calls for.

For example, suppose $u_1 = 2$, $u_2 = 8$. Then the housewife, following the table d'hôte rule,[1] is restricted to Foods 1 and 5. The cheapest way that she can meet the nutritional requirement is to buy 400 units of Food 5 at a cost of $12 \times 400 = 4{,}800$. On the other hand, the shadow worth of the minimum adequate diet is only $2 \times 700 + 8 \times 400 = 4{,}600$. Somehow she has been euchred out of 200 money units. How this happened is clear: she has bought 100 more vitamins (worth 2 each) than the requirement called for. Furthermore, the inconsistency in this system of prices is clear: since she has an excess of vitamins over the nutritional minimum, their marginal cost (as defined previously) is 0 and not 2.

We can now see an additional requirement for the set of shadow prices: they must be such that it is possible to satisfy the nutritional requirement for a cost that does not exceed the shadow worth of the nutritional requirement. This implies that the shadow prices must be such that either it is possible to satisfy the requirements exactly, using only foods permitted under the table d'hôte rule, or, if there has to be an excess of any nutrient, the shadow price of that nutrient is zero.

[1] Never pay more for a dinner than the total price of the individual dishes.

An equivalent—and more convenient—way to state this requirement is: *Consistent with every food's costing at least as much as the shadow worth of its nutritive contents, the shadow prices must maximize the imputed value of the schedule of nutritive requirements.*

Why is this statement equivalent? Because when no food is assigned a shadow worth greater than its market price, the shadow worth of a specified schedule of nutrients cannot be greater than the market price of the cheapest diet that meets that schedule. When that shadow worth of nutrients equals the market price, then that shadow worth is as great as possible, subject to the condition.

Why is this mode of statement convenient? Because it puts the price-imputation problem in linear-programming form. Specifically, it formulates the problem as follows: The total imputed worth of the schedule of requirements is the sum of its calorie worth, c_1u_1, plus its vitamin worth, c_2u_2, ... Hence the shadow prices are to be chosen so that

$$Z' = c_1u_1 + c_2u_2 + \cdots + c_mu_m$$

is a maximum subject to

$$\begin{aligned}
a_{11}u_1 + a_{21}u_2 + \cdots + a_{m1}u_m &\leq p_1 \\
a_{12}u_1 + a_{22}u_2 + \cdots + a_{m2}u_m &\leq p_2 \\
\cdots\cdots\cdots\cdots\cdots\cdots&\cdots \\
a_{1n}u_1 + a_{2n}u_2 + \cdots + a_{mn}u_m &\leq p_n \\
u_1 \geq 0, u_2 \geq 0, \ldots, u_m &\geq 0
\end{aligned}$$

A minimum-cost problem has been turned into a maximum-rent problem! A quantities problem has been turned into a prices problem! This is the implication of the remarkable mathematical duality feature, discovered by von Neumann and other mathematicians, but quite consistent with economic reasoning. Thus we should expect the national income ("rents") paid to factors to equal in a simple economic system the value of the national output sold; or maximum Z' = minimum Z.

We note, further, that under the expanded rule the housewife will not buy an excess of any nutrient unless that nutrient has a shadow price of zero Thus this rule leads both to efficient purchasing on her part and to a consistent system of market and shadow prices.

One last question: Will the system of shadow prices that satisfies this criterion completely close the gap between the cost of the cheapest adequate diet and the shadow worth of the nutritive requirements? The remark in the last paragraph suggests that it will; a mathematical proof (which amounts simply to spelling out this remark carefully) is given in Chap. 4.

Let us apply our generalized rule to see whether it does select out the correct one of the four consistent price patterns: (0, 11); (1, 10); (2, 8);

(2, 0). Our four different totals of rents are

$$Z' = C_1u_1 + C_2u_2 = 700u_1 + 400u_2$$

$$(A) \qquad = 700(0) + 400(11) = 4{,}400$$

$$(B) \qquad = 700(1) + 400(10) = 4{,}700$$

$$(C) \qquad = 700(2) + 400(8) = 4{,}600$$

$$(D) \qquad = 700(2) + 400(0) = 1{,}400$$

Clearly the second case, B, represents the true optimum or maximum Z', which does equal the minimum $Z = 4{,}700$ that we have seen earlier.[1]

Figure 3-2 indicates this same solution for optimal prices. We are free to move in $ABCDE$ so as to maximize Z'. Contours of equal Z' are given by parallel lines with absolute slopes $C_1/C_2 = {}^{700}\!/_{400}$. The arrows (perpendicular to the contour) indicate the direction in which our optimum lies. Clearly, the best place to end up is at B, where $Z' = 4{,}700$; anywhere else will give a lower Z'; only at B will the broken line ABC be touching (but not crossing) the highest Z' contour.

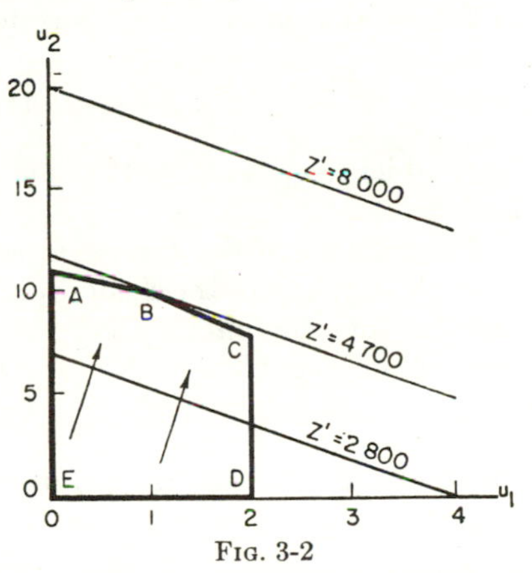

Fig. 3-2

If we imagine an Office of Price Administration (OPA) that tries to find the best prices by the deliberate use of intelligence, its task is now finished. It has posed for itself a prices-maximum problem in linear programming, rather than the quantities-minimum problem that a War Production Board (WPB) might formulate.

It is easy to verify that the problem of finding a minimum-cost adequate diet and the problem of imputing values to nutritive elements form a pair of duals with all the properties listed in Sec. 3-2. The first property held that the cost of the minimum-cost adequate diet would be equal to the maximum value imputable to the nutritional standard under the restraint that every food cost at least as much as its imputed nutritional worth. We found that the minimum-cost diet was to consume 100 units of Food 4 and 300 units of Food 5 at a cost of

$$100 \times 11 + 300 \times 12 = 4{,}700 = \text{minimized } Z$$

The solution to the valuation problem was to value a unit of calories at

[1] *ABCDE* is a convex polygon, i. e., one that contains any straight-line segment that connects two of its points. It follows, as common sense will confirm, that raising calorie requirements will, if anything, raise the shadow price of calories.

1 and a unit of vitamins at 10, giving as the total valuation of the required standard of nutrition: $700 \times 1 + 400 \times 10 = 4{,}700 =$ maximized Z'. The two are equal, as they should be.

The second property concerned the relationship between the restraints satisfied with equalities and inequalities in the direct problem, on the one hand, and the variables that took on zero and positive values in the solution of the dual problem, on the other. In this example, both restraints are satisfied exactly in the solution to the diet problem, so neither dual variable is zero. Further, since x_4 and x_5 are positive in the solution to the diet problem, the fourth and fifth constraints will be satisfied exactly in the solution to the dual. The dual thus reduces to

$$u_1 + u_2 = 11$$
$$2u_1 + u_2 = 12$$

from which $u_1 = 1$, $u_2 = 10$ could be calculated readily if not already known.

The solution in the reverse direction illustrates the underlying logic more clearly. Once the shadow prices of the nutrients are given, we can use this information to select the best diet without solving a linear-programming problem at all. For, given the best prices, that is,

$$(u_1\ u_2) = (1, 10)$$

we know that at least all but $m = 2$ of the foods can be regarded as unprofitable and their quantities set equal to zero. Thus, for

$$(u_1, u_2) = (1, 10)$$

we have $(\pi_1, \pi_2, \pi_3, \pi_4, \pi_5) = (-, -, -, 0, 0)$, so that our best diet will be $(x_1, x_2, x_3, x_4, x_5) = (0, 0, 0, x_4?, x_5?)$. We can therefore concentrate on only two foods, in this case x_4 and x_5. We must select the precise amounts of these foods so as to achieve our nutritional constraints. These constraints are also 2 (or m) in number, so that we have enough equations to determine our two unknowns.

Our general procedure is as follows:

Suppose we know our best prices. They are m in number at most; but some prices may be zero, so that we may have only $r \leq m$ effective constraints. This means there will be some r economical goods which will exactly satisfy the effective constraints. We solve these r equations for exact values of the r economical X's.

The exact solution to our arithmetic problem is to determine the amounts of X_4 and X_5 from our calorie and our vitamin equations; or from

$$1x_4 + 2x_5 = 700$$
$$1x_4 + 1x_5 = 400$$

or

$$x_4 = 100, \quad x_5 = 300$$

as substitution or elementary algebra will verify.

A sophisticated economist might notice in this problem that it would be possible to define certain composite commodities, or market baskets of X_4 and X_5, which, if properly weighted, would be found to consist entirely of calories and entirely of vitamins. He would have to be sophisticated because the relative weights could not be positive numbers. I cannot buy X_4 and X_5 and put them into a basket and hope that there will be only vitamins in the basket. But if I *buy* 2 *units of* X_4 *and sell* 1 *unit of* X_5 I will end up with 1 unit of vitamins and nothing else! My market basket, or composite commodity, has weights $(+2, -1)$. Similarly, to get pure calories, I must buy 1 unit of X_5 and sell 1 unit of X_4, leaving me with 1 calorie unit and nothing else.

What is the market cost or price of each of these composite commodities or constructed "pure" foods? For the vitamin basket it is twice the price of X_4 minus once the price of X_5, or $+2(1) - 1(12) = 10$. For the calories, it is the net algebraic cost of the foods in the second basket, or $+1(12) - 1(11) = 1$. It is not surprising that we have ended up with 1, 10, already seen to be the shadow prices u_1, u_2 and MC_1, MC_2.[1]

The reader may verify that if the NRC set the vitamin requirement $C_2 = 0$, then a best set of prices would be $(u_1, u_2) = (2, 0)$, and we would have only x_1 a nonzero quantity.[2] We would solve for the exact x_1, by the $r = 1$ equation system:

$$1x_1 = 700$$

or

$$x_1 = 700$$

3-6. THE DUAL AND DECENTRALIZATION

The dual and its connection with valuation invite us to apply linear programming to the study of markets and prices as well as to the direct study of production and allocation. We do this in two stages, using the international-trade example for concreteness. In this section we discuss the way in which the solution to the dual or valuation problem can be

[1] Any algebraist will have recognized that this argument depends on what he calls a "change of basis." We have expressed a pure calorie and a pure vitamin, vectors (1,0) and (0,1), respectively, on the basis of $X_4 = (1,1)$ and $X_5 = (2,1)$. The reader may be referred to P. A. Samuelson, *Foundations of Economic Analysis*, pp. 135–146, Harvard University Press, Cambridge, Mass., 1947, for a discussion of such composite commodities and the laws of their price formation.

[2] At $C_2 = 0$, MC_2 is ambiguous and ill defined. This is because $(u_1, u_2) = (2, h)$ are solutions to our dual problem for all $0 \leq h \leq 8$. This means that $0 \leq MC_2 \leq 8$. Our best direction in Fig. 3-2 is eastward, and the line segment CD represents the set of optimal prices for our dual problem. The point C itself is not truly a correct point for our original problem since it falsely tells us that X_4 shows a zero profit and can be bought. In the most general case of m constraints, even when we have best prices, care is necessary in selecting the proper sets of variables (x's) and equations (constraints).

used to guide production to an efficient allocation. This discussion will reinforce the identity of the solution of the linear-programming dual with the determination of equilibrium prices in familiar economic theory.

We have already seen, in Sec. 2-10, that the food-clothing price ratio $\lambda = p_1/p_2$ can lead the two countries to their most advantageous specialties without any more explicit form of coordination. The problem can be decentralized even further. Consider the maximum problem within any one of the countries, such as England. A central planning board could issue shutdown orders so as to maximize

$$z = \tfrac{1}{2} p_1 y_1 + p_2 y_2$$

subject to $$y_1 + y_2 \leq Q$$

using the notation of Sec. 3-2. But alternatively, we might abolish all planning agents and think of food and clothing industries each made up of a myriad of small independent firms. They incur costs and earn revenues from the sale of their products. This they do by converting resources into products. So far we have not spoken much about the character of the resources involved (labor, as in Ricardo's theory, etc.). But it will be obvious on reflection that Q is a measure of total resources that can be parceled out to the various firms in the two industries, and thus that all the little y's provided to the firms must add up to not more than Q. Also, the production function of, say, the 999th food producer is of the form

$$x_{1,999} \leq \tfrac{1}{2} y_{1,999}$$

and the corresponding production function of the 77th clothing firm is

$$x_{2,77} \leq y_{2,77}$$

where total output of England's food is the sum of all food firms' output,

$$x_1 = x_{1,1} + \cdots + x_{1,999} + \cdots$$

and total output of England's clothing is

$$x_2 = x_{2,1} + \cdots + x_{2,77} + \cdots$$

and where the total of all resources used by all firms cannot exceed the total of all resources available in the country; or where

$$Q \geq y_{1,1} + \cdots + y_{1,999} + \cdots + y_{2,1} + \cdots + y_{2,77} + \cdots$$

Note too that England's grand production possibility curve, as given earlier in Fig. 2-5, is simply an aggregation of these individual-firm production relations.

One minor point should be noted. Why is there an inequality in the production function? This is because a firm may be inefficient and not be getting as much product as known technology permits. It must be shown that such inefficient behavior—which is clearly inconsistent with a final optimum—will in fact be heavily penalized by a competitive market and thus be eliminated. The similar inequality with respect to the sum of the y's, which would imply wasteful unemployment of available resources, will also turn out to be prohibited by a perfect market mechanism.

To the observer with world vision, p_1, p_2 are shadow prices constructed for a purpose. To the domestic English planners interested in selecting the various x's and y's in order to maximize the value of English NP, they are real enough external prices, giving a barter ratio at which goods can be converted into each other by international trade. But the English planners could let the decentralized (but not uncoordinated) individual firms seek out the optimal solution. The planners need, in order to do this, to introduce a shadow price, or internal-accounting point price, for resources. Let this price be u. Then for any food producer, say the 999th, the profit statement reads

$$\text{Total revenue} - \text{total cost} = p_1 x_{1,999} - u y_{1,999}$$

and the "profitability per unit of output" will be

$$\pi_{1,999} = p_1 - u \frac{y_{1,999}}{x_{1,999}} \leq p_1 - 2u$$

from the production function for food. Similarly for the 77th clothing producer, unit profits are

$$\pi_{2,77} = p_2 - u \frac{y_{2,77}}{x_{2,77}} \leq p_2 - u$$

In the above profit expressions, we may omit the inequality signs whenever we are speaking of the most efficient producers.

How shall the resource price u (what we earlier called shadow price) be determined? At first let us suppose that there is an all-powerful Office of Price Stabilization (OPS) that will use high intelligence to solve this problem, but that once the best price has been established we shall try to rely on the quasi-automatic response of competing firms to determine appropriate output quantities. The planning authorities will probably set up some such rule of behavior for firms as follows:

1. If you make losses, you must contract your scale of operations until finally you go out of business.

2. If you make positive profits, you may (and hence will) expand your scale of operations at some positive rate.

3. If you just break even with zero profits, let us for the moment say you stand pat at any existing level of activity.

In terms of this rule, there are certain obvious things that the OPS must do in setting the best price or wage for resources. At the least, u must be set so as *not to lead to positive profits anywhere in the system*—in other words, set so high that even the most efficient producers in clothing and food are unable to realize surplus profits. This means we must have

$$\pi_{1,999} = p_1 - u \leq 0 \qquad \text{or} \qquad u \geq p_1$$
$$\pi_{2,77} = p_2 - \frac{u}{2} \leq 0 \qquad \text{or} \qquad u \geq 2p_2$$

It will be noted that there is no longer an inequality sign accompanying the left-hand equality signs in the above profit expression. This is because we are considering the profitability of the most efficient producers of food and clothing; not even they are permitted to have (excess) profits. These are the same inequalities for u as we found when we solved the dual in Sec. 3-5.

Our above conditions determine a minimum below which u must not go, but they do not rule out still higher shadow prices. However, it is reasonable to add the further condition that *profits are not to be everywhere negative*—that they are to be *somewhere* zero. Otherwise, no firms could stay permanently in business and the total of resources used would be zero, and all production and income would also be zero.

If the equality sign must hold for at least one of our profit expressions, it follows logically that

$$\begin{aligned} u &= p_1 \qquad \text{or} \qquad 2p_2 \qquad \text{whichever is greater} \\ &= \max\,(p_1, 2p_2) \end{aligned}$$

This means that if $p_1 = 30$ and $p_2 = 20$, we must set our shadow price equal to max (30, 40), or to 40. This yields negative profits for all food firms and zero profits for all most efficient clothing firms; profits for inefficient food firms are always negative.

If the international prices should be $p_1 = 60$, $p_2 = 20$, then

$$u = \max\,(60, 40) = 60$$

and it is clothing that has negative profits. Only in the critical case where $p_1 = 2p_2$ will both industries be capable of simultaneous operation in England.

Our OPS has solved its price problem satisfactorily. But there is one major difficulty about our setup when it comes to determining the exact quantities of resources and output. This difficulty is a consequence of the extreme constant-cost assumption involved in all simple versions of linear programming, assumptions which put the word "linear" into the name of the subject. At our final best price, firms are not forced all to contract or all to expand indefinitely. Efficient firms in the proper industry are permitted to have a large or small output. But there is nothing driving them in total toward 100 per cent use of society's resources, neither more nor less. The atomistic firms are suspended in a kind of *neutral* equilibrium: they have no incentive to do other than what they are doing. Profit considerations neither encourage nor discourage them from doing what society desires.

At the very last stage OPS must call upon WPB for a few direct quantity fiats. Once the proper prices have been promulgated by the pricing authorities, the production authority does not have to use much intelligence; but it does have to use a little. It must lead the neutral (efficient) producers—by their moustaches so to speak—to use up exactly 100 per cent of the available resources.

If we call the actual amount of total resources currently in use at time t, $y(t)$, then

$$y(t) = (y_{1,1} + \cdots + y_{1,999} + \cdots) + (y_{2,1} + \cdots + y_{2,77} + \cdots)$$

The WPB must make sure that

$$y(t) = Q$$

if a true optimum is to be reached, and it must have the intelligence to recognize this condition.[1]

[1] There is yet another difficulty that arises. If we write down the simplest dynamic adjustment process corresponding to decentralized competitive decision making, the process turns out not to be dynamically stable. Instead, it oscillates endlessly about the equilibrium position. But this leads far afield. See K. Arrow and G. Debreu, "Existence of an Equilibrium for a Competitive Economy," *Econometrica*, **22:** 265–90 (July, 1954).

4

The Algebra of Linear Programming

4-1. INTRODUCTION

Thus far linear programming has been regarded as an economic and business problem. We have seen how it arises naturally in the course of economic optimization. In this chapter we shall consider it as a problem in algebra, because the current importance of linear programming stems from the fact that when economic problems are formulated in this manner they can be solved by relatively simple, though sometimes tedious, algebraic methods.

In the diet problem of Chap. 2 we considered n different foods $X_1, \ldots, X_n$ and m different nutritive elements. We assumed that one unit of the jth food contained a_{ij} units of the ith nutritive element. Then, if a housewife purchased x_1 units of the first food, x_2 units of the second food, . . . , and x_n units of the nth food, she would be purchasing

$$y_1 \equiv a_{11}x_1 + \cdots + a_{1n}x_n$$

units of the first nutritive element,

$$y_2 \equiv a_{21}x_1 + \cdots + a_{2n}x_n$$

units of the second nutritive element, . . . ,

$$y_m \equiv a_{m1}x_1 + \cdots + a_{mn}x_n$$

units of the mth nutritive element.

We then assumed that the housewife's family required $c_1, c_2, \ldots, c_m$ units of the m nutritive elements, respectively. Thus the housewife could satisfy her family's requirements by buying any quantities $x_1, \ldots, x_n$ of the n foods, provided that the nutritive requirements were satisfied, i.e., provided that

$$\begin{array}{c} a_{11}x_1 + \cdots + a_{1n}x_n \geq c_1 \\ a_{21}x_1 + \cdots + a_{2n}x_n \geq c_2 \\ \cdots\cdots\cdots\cdots\cdots \\ a_{m1}x_1 + \cdots + a_{mn}x_n \geq c_m \end{array} \tag{4-1}$$

and, since it is meaningless in this context to buy a negative amount of a food, provided that

$$x_1 \geq 0, x_2 \geq 0, \ldots, x_n \geq 0 \tag{4-2}$$

There may be an infinite number of diets[1] which satisfy the inequalities (4-1) and (4-2). Any such diet is called a *feasible* diet. We assumed that the housewife would want to purchase the most economical feasible diet. In order to express this we introduced the market prices of the n foods, $P_1, P_2, \ldots, P_n$, respectively, and the aggregate cost of a diet,

$$Z \equiv P_1x_1 + P_2x_2 + \cdots + P_nx_n \tag{4-3}$$

Then we asserted that the housewife's problem was to find a set of food quantities $x_1, \ldots, x_n$ which satisfied inequalities (4-1) and (4-2) and made Z as small as possible.

That was an economic, or planning, problem. It can be converted to an algebraic problem simply by omitting most of the explanatory text. Thus, let $c_1, \ldots, c_m, P_1, \ldots, P_n, a_{11}, a_{12}, \ldots, a_{1n}, \ldots, a_{m1}, \ldots, a_{mn}$ be any numbers whatsoever. We may pose the algebraic problem of finding the numbers $x_1, \ldots, x_n$ which satisfy the inequalities (4-1) and (4-2) and make the expression (4-3) as small as possible. If we can solve this problem, then we can solve the diet problem or any other economic (or noneconomic) problem of the same form. This is the linear-programming minimum problem.

Similarly, if $c_1, \ldots, c_m, P_1, \ldots, P_n, a_{11}, \ldots, a_{mn}$ are any numbers whatsoever, we can set the problem of finding a set of numbers $x_1, \ldots, x_n$ such that

$$\begin{aligned} a_{11}x_1 + \cdots + a_{1n}x_n &\leq c_1 \\ a_{21}x_1 + \cdots + a_{2n}x_n &\leq c_2 \end{aligned} \tag{4-4}$$

$$\cdots\cdots\cdots\cdots\cdots\cdots$$

$$\begin{aligned} a_{m1}x_1 + \cdots + a_{mn}x_n &\leq c_m \\ x_1 \geq 0, \ldots, x_n &\geq 0 \end{aligned} \tag{4-5}$$

and

$$Z = P_1x_1 + \cdots + P_nx_n \tag{4-6}$$

is as great as possible. This is the linear-programming maximum problem.

This chapter will discuss the solution of the linear-programming maximum problem. The solution of the linear-programming minimum problem is similar. In fact, since minimizing Z is the same as maximizing $-Z$, a minimum problem can easily be converted into an equivalent maximum problem.[2]

[1] In this context a diet is just a set of numbers $x_1, \ldots, x_n$ representing food quantities.

[2] This is *not* the dual problem.

4-2. THE EXISTENCE OF SOLUTIONS

It is quite possible for a linear-programming problem to have no solution. Suppose, for example, that $n = 2$, $m = 2$, and the problem is to find x_1, x_2 such that

$$\begin{aligned} x_1 - x_2 &\leq 1 \\ -x_1 + x_2 &\leq -2 \end{aligned} \tag{4-7}$$

$$x_1 \geq 0,\ x_2 \geq 0 \tag{4-8}$$

and

$$Z = 3x_1 + 4x_2$$

is as great as possible. Equations (4-7) can be written in the form

$$\begin{aligned} x_1 &\leq x_2 + 1 \\ x_1 &\geq x_2 + 2 \end{aligned}$$

or

$$x_2 + 2 \leq x_1 \leq x_2 + 1$$

which is impossible. In this case the inequalities were inconsistent and the consequence is similar to that encountered in ordinary algebra when a system of simultaneous equations is inconsistent.

There is also a second kind of nonsolvability. Consider the problem of finding x_1, x_2 such that

$$\begin{aligned} x_1 - x_2 &\leq 1 \\ 2x_1 - x_2 &\leq 0 \\ x_1 \geq 0,\ x_2 &\geq 0 \end{aligned}$$

and

$$Z = 3x_1 + 4x_2$$

is as great as possible. In this case the inequalities will be satisfied for any positive value of x_1 and $x_2 = 10x_1$. Then $Z = 3x_1 + 40x_1 = 43x_1$ and can be made as large as we please. The inequalities are consistent, but there is no upper limit to Z. In the following we shall always assume that the inequalities are consistent, thus avoiding the first kind of nonsolvability. The second kind of nonsolvability will be dealt with in the course of the discussion.

4-3. THE STRATEGY OF LINEAR PROGRAMMING

No direct method for solving linear-programming problems is known; i.e., there is no formula which can be used to calculate the solution of a linear-programming problem directly by substituting in the values of the coefficients and other given data. This contrasts with the closely allied problem of solving simultaneous linear equations, where direct formulas do exist. For example, the simplest case of simultaneous equations is the case with two equations in two unknowns, as follows:

$$a_{11}x_1 + a_{12}x_2 = b_1$$
$$a_{21}x_1 + a_{22}x_2 = b_2$$

It can be verified easily that the solution to these equations is

$$x_1 = \frac{a_{22}b_1 - a_{12}b_2}{a_{11}a_{22} - a_{12}a_{21}}$$
$$x_2 = \frac{a_{11}b_2 - a_{21}b_1}{a_{11}a_{22} - a_{12}a_{21}}$$

if the denominator is not zero. Similar explicit formulas can be given for solving systems of any size desired.

In the absence of direct solutions to linear-programming problems we resort to iterative solutions, i.e., solutions carried out in a number of steps each of which brings us closer to the desired result. A number of methods of solution have been proposed, all of this general type. In this chapter we shall discuss two of them: the "simplex method" due to Dantzig and the "complete-description method" due to Motzkin, Raiffa, Thompson, and Thrall.

4-4. THE SIMPLEX METHOD, GENERAL ARGUMENT

The description of a linear-programming problem given in Eqs. (4-4) to (4-6) is rather inconvenient for use in the simplex method. A more manageable formulation can be obtained by the simple device of introducing "slack," or "disposal," variables which permit us to replace the inequalities of Eqs. (4-4) by equalities. The ith equation of set (4-4), for example, is

$$a_{i1}x_1 + \cdots + a_{in}x_n \leq c_i \tag{4-9}$$

Suppose we replace that inequality by the equation

$$a_{i1}x_1 + \cdots + a_{in}x_n + x_{n+i} = c_i \tag{4-10}$$

where $x_{n+i} \geq 0$. Then any set $x_1, \ldots, x_n$ which satisfies (4-9) will satisfy (4-10), and conversely. The new variable x_{n+i} is known as a slack variable because it merely accounts for the excess of the right-hand side of (4-9) over the left-hand side. We may therefore write the linear-programming maximum problem in the following form:

Find $x_1, x_2, \ldots, x_n, x_{n+1}, \ldots, x_{n+m}$ such that

$$\begin{aligned} a_{11}x_1 + \cdots + a_{1n}x_n + x_{n+1} &= c_1 \\ a_{21}x_1 + \cdots + a_{2n}x_n + x_{n+2} &= c_2 \\ \cdots\cdots\cdots\cdots\cdots\cdots\cdots\cdots \\ a_{m1}x_1 + \cdots + a_{mn}x_n + x_{n+m} &= c_m \end{aligned} \tag{4-11}$$

$$\begin{aligned} x_1 &\geq 0 \\ &\cdots\cdots \\ x_n &\geq 0 \\ x_{n+1} &\geq 0 \\ &\cdots\cdots \\ x_{n+m} &\geq 0 \end{aligned} \tag{4-12}$$

and

$$Z = P_1x_1 + \cdots + P_nx_n \tag{4-13}$$

is as great as possible.

By this device, the inequalities of (4-4) are replaced by the equalities of (4-11) at the cost of introducing m additional *nonnegative* unknowns.

We shall usually write equations like (4-11) in a slightly disguised but more symmetric form. Suppose we let

$$\begin{aligned} a_{i,n+i} &= 1 \qquad i = 1, \ldots, m \\ a_{i,n+j} &= 0 \qquad i, j = 1, \ldots, m \qquad i \neq j \end{aligned}$$

Then (4-11) can be written

$$\begin{aligned} a_{11}x_1 + \cdots + a_{1n}x_n + a_{1,n+1}x_{n+1} + \cdots + a_{1,n+m}x_{n+m} &= c_1 \\ a_{21}x_1 + \cdots + a_{2n}x_n + a_{2,n+1}x_{n+1} + \cdots + a_{2,n+m}x_{n+m} &= c_2 \\ \cdots\cdots\cdots\cdots\cdots\cdots\cdots\cdots\cdots\cdots\cdots\cdots \\ a_{m1}x_1 + \cdots + a_{mn}x_n + a_{m,n+1}x_{n+1} + \cdots + a_{m,n+m}x_{n+m} &= c_m \end{aligned} \tag{4-14}$$

Equations (4-14), (4-12), and (4-13) are the form of the linear-programming problem which we shall solve.

Any set of values $(x_1, x_2, \ldots, x_{n+m})$ which satisfies (4-11) and (4-12) and which makes (4-13) as large as possible will be called an optimal solution. There may be more than one. Any set of values which satisfies (4-11) and (4-12), whether or not it makes (4-13) as large as possible, will be called a feasible solution. The simplex method, conceptually, solves a linear-programming problem in two stages. First, it gives a procedure whereby we start with any set of values at all (a zeroth approximation) and, by iteration, find a feasible solution. Second, it gives a procedure for starting with any feasible solution and finding, by iteration, an optimal solution. We shall see that in practice these two stages can usually be combined, but it is useful to separate them in explaining and justifying the method.

The two stages can be discussed in either order, but the second stage, the one which begins with a feasible solution and ends with an optimal solution, is the core of the method, and we shall begin with it.

To see the idea behind the method, consider Eqs. (4-11). They consist of m equations in $m + n$ unknowns. Thus, if we set any n of the unknowns arbitrarily equal to zero, there will remain an ordinary set of m simultaneous equations in m unknowns. Elementary algebra teaches that under certain conditions (which we shall consider later in this chapter and in Appendix B), such a set of equations has a unique solution. Furthermore, we can select an arbitrary set of n unknowns from a set of $m + n$ in only a finite (though maybe large) number of ways. Thus there is only a finite number of these unique solutions, each involving a different set of m of the variables. We shall call a solution which involves no more than m of the variables a basic solution. A particular value of Z, given by Eq. (4-13), corresponds to each basic solution. When the values of the variables associated with a basic solution are nonnegative it is called a basic feasible solution.

Each nonbasic feasible solution, i.e., each solution in which less than n variables are zero, can be thought of as a positive weighted average of the values of the corresponding variables of two or more basic feasible solutions. The value of Z corresponding to a nonbasic solution is, because of the linear form, the same weighted average of the values of the Z's of the basic solutions which combine to form it. Thus the value of Z corresponding to a nonbasic feasible solution cannot be greater than the highest value of Z corresponding to its component basic solutions. It follows that if there is any optimal solution to a problem, at least one of the basic solutions will be optimal. Thus, in searching for an optimal solution we need to consider only basic feasible solutions of which there are only a finite number. The heart of the simplex method is a procedure for starting with any feasible basic solution and from it computing another feasible basic solution which corresponds to a higher value of Z. Since there are only a finite number of basic solutions this procedure must lead in a finite number of steps to a basic solution which corresponds to the highest possible value of Z.

In this outline of the argument we have made a number of plausible statements. We must now prove those statements and, in the course of the proof, develop the simplex method of solution.

4-5. A DIGRESSION ON SIMULTANEOUS EQUATIONS

The foregoing discussion has perhaps made it clear that linear programming is closely related to ordinary simultaneous equations. Before going into any details we shall recall some basic facts about simultaneous equations.

We can use a system of three equations in three unknowns to bring out all the facts that we require. The general form of such a system is

$$\begin{aligned} a_{11}x_1 + a_{12}x_2 + a_{13}x_3 &= b_1 \\ a_{21}x_1 + a_{22}x_2 + a_{23}x_3 &= b_2 \\ a_{31}x_1 + a_{32}x_2 + a_{33}x_3 &= b_3 \end{aligned} \tag{4-15}$$

The discussion and solution of such systems is usually conducted with the aid of the algebraic concept of a "determinant," and this is very convenient. The discussion of determinants would take us too far afield, however, and we shall deny ourselves this device with one slight exception: it will simplify our work considerably to make use of second- and third-order determinants. A second-order determinant is a symbol involving four numbers, thus:

$$\begin{vmatrix} c & d \\ e & f \end{vmatrix}$$

Where no confusion will result, we abbreviate it by writing only the first and last elements, thus: $|c, f|$. The value of a second-order determinant is defined by a simple formula depending on the four numbers which comprise it. The formula is

$$|c, f| = \begin{vmatrix} c & d \\ e & f \end{vmatrix} = cf - ed$$

For example,

$$\begin{vmatrix} 2 & 4 \\ 3 & 5 \end{vmatrix} = 10 - 12 = -2$$

$$\begin{vmatrix} 2 & 4 \\ -3 & 5 \end{vmatrix} = 10 + 12 = 22$$

$$\begin{vmatrix} 4 & 28 \\ 17 & 119 \end{vmatrix} = 476 - 476 = 0$$

Now turn to the simultaneous equations (4-15). Multiply the first equation by a_{21} and the second by a_{11} and subtract the first result from the second. We obtain

$$(a_{11}a_{21} - a_{11}a_{21})x_1 + (a_{11}a_{22} - a_{12}a_{21})x_2 + (a_{11}a_{23} - a_{13}a_{21})x_3 = a_{11}b_2 - a_{21}b_1$$

The coefficients of x_1 are zero, and the coefficients of the remaining terms are second-order determinants. We therefore have

$$|a_{11}, a_{22}|x_2 + |a_{11}, a_{23}|x_3 = |a_{11}, b_2| \tag{4-16}$$

Similarly, multiplying the first equation by a_{31} and the third by a_{11}, and subtracting the first from the third, we obtain

$$|a_{11}, a_{32}|x_2 + |a_{11}, a_{33}|x_3 = |a_{11}, b_3| \tag{4-17}$$

In these formulas we used the abbreviated form of the determinants since the first and fourth elements of each determinant form two corners of a square, the other corners of which can be found by referring to Eqs. (4-15).

Equations (4-16) and (4-17) are a pair of simultaneous equations in two unknowns, and one of the unknowns can be eliminated by the same procedure. Eliminating x_2 we obtain

$$\big|\,|a_{11}, a_{22}|, |a_{11}, a_{33}|\,\big|\,x_3 = \big|\,|a_{11}, a_{22}|, |a_{11}, b_3|\,\big| \qquad (4\text{-}18)$$

Equation (4-18) involves determinants each of whose elements is a determinant. Since determinants are merely numbers expressed in complicated form, this causes no difficulty. To understand what this compact notation means, let us write out the coefficient of x_3, in full:

$$\big|\,|a_{11}, a_{22}|, |a_{11}, a_{33}|\,\big| = \begin{vmatrix} \begin{vmatrix} a_{11} & a_{12} \\ a_{21} & a_{22} \end{vmatrix} & \begin{vmatrix} a_{11} & a_{13} \\ a_{21} & a_{23} \end{vmatrix} \\ \begin{vmatrix} a_{11} & a_{12} \\ a_{31} & a_{32} \end{vmatrix} & \begin{vmatrix} a_{11} & a_{13} \\ a_{31} & a_{33} \end{vmatrix} \end{vmatrix}$$

$$= \begin{vmatrix} a_{11} & a_{12} \\ a_{21} & a_{22} \end{vmatrix} \cdot \begin{vmatrix} a_{11} & a_{13} \\ a_{31} & a_{33} \end{vmatrix} - \begin{vmatrix} a_{11} & a_{13} \\ a_{21} & a_{23} \end{vmatrix} \cdot \begin{vmatrix} a_{11} & a_{12} \\ a_{31} & a_{32} \end{vmatrix}$$

$$= (a_{11}a_{22} - a_{12}a_{21})(a_{11}a_{33} - a_{13}a_{31}) - (a_{11}a_{23} - a_{13}a_{21})(a_{11}a_{32} - a_{12}a_{31})$$

$$= a_{11}^2a_{22}a_{33} - a_{11}a_{13}a_{22}a_{31} - a_{11}a_{12}a_{21}a_{33} + a_{12}a_{13}a_{21}a_{31} - a_{11}^2a_{23}a_{32} + a_{11}a_{12}a_{23}a_{31} + a_{11}a_{13}a_{21}a_{32} - a_{12}a_{13}a_{21}a_{31}$$

$$= a_{11}(a_{11}a_{23}a_{33} - a_{11}a_{23}a_{32} + a_{12}a_{23}a_{31} - a_{12}a_{21}a_{33} + a_{13}a_{21}a_{32} - a_{13}a_{22}a_{31})$$

The expression inside the parentheses consists of six terms, each being one of the six possible ways of selecting three of the a's of Eqs. (4-15) without using two from the same row or column. The assignment of plus and minus signs follows a complicated rule that can best be explained in terms of Fig. 4-1.

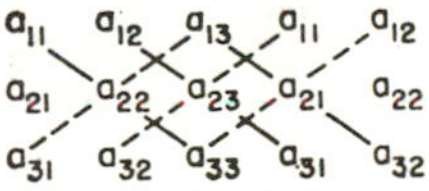

Figure 4-1 consists of the pattern of a's from Eq. (4-15) with the first two columns repeated. All possible *complete* diagonals are drawn in. Each of them corresponds to one of the terms in the parentheses. The solid diagonals, the ones that run northwest-southeast, correspond to terms with plus signs; the dashed diagonals to terms with minus signs.

$$\begin{matrix} a_{11} & a_{12} & a_{13} & a_{11} & a_{12} \\ a_{21} & a_{22} & a_{23} & a_{21} & a_{22} \\ a_{31} & a_{32} & a_{33} & a_{31} & a_{32} \end{matrix}$$

FIG. 4-1. Explanation of terms of a determinant.

This expression in parentheses is known as a third-order determinant. In summary, we define a third-order determinant by the equation

$$\begin{vmatrix} a_{11} & a_{12} & a_{13} \\ a_{21} & a_{22} & a_{23} \\ a_{31} & a_{32} & a_{33} \end{vmatrix} = a_{11}a_{22}a_{33} + a_{12}a_{23}a_{31} + a_{13}a_{21}a_{32} - a_{11}a_{23}a_{32} - a_{12}a_{21}a_{33} - a_{13}a_{22}a_{31}$$

Now, finally, if the determinant

$$\begin{vmatrix} a_{11} & a_{12} & a_{13} \\ a_{21} & a_{22} & a_{23} \\ a_{31} & a_{32} & a_{33} \end{vmatrix}$$

is not equal to zero we can divide through by it in Eq. (4.18)[1] and obtain

$$x_3 = \frac{\begin{vmatrix} a_{11} & a_{12} & b_1 \\ a_{21} & a_{22} & b_2 \\ a_{31} & a_{32} & b_3 \end{vmatrix}}{\begin{vmatrix} a_{11} & a_{12} & a_{13} \\ a_{21} & a_{22} & a_{23} \\ a_{31} & a_{32} & a_{33} \end{vmatrix}} \tag{4-19}$$

Warning: This explanation applies to third-order determinants only. It would take us too far afield to introduce the general definition of a determinant—of which the second and third orders are special cases—and no higher-order determinants will be encountered in this chapter.

The concept of the determinant is important enough to be explained twice, and it is explained again from a different point of view in Appendix B. The reader to whom the concept is new may get a firmer grip on it by consulting Appendix B.[2]

Similar formulas can be found for x_1 and x_2. If, however, the denominator determinant is zero, the solution breaks down. The coefficients $a_{11}, a_{12}, \ldots, a_{33}$ are then said to be "linearly dependent" for the following reason. Suppose $b_1 = b_2 = b_3 = 0$. This will not affect the value of the denominator determinant, but it will make the numerator determinant in Eq. (4-19) and in the similar equations for x_1 and x_2 equal to zero. Then if the denominator determinant is not equal to zero, the solution will be $x_1 = x_2 = x_3 = 0$, and this solution will be unique. But if the denominator determinant is equal to zero, Eq. (4-19) will not be valid (division by zero being illegitimate), but Eq. (4-18) will hold whatever may be the value of x_3. We may give x_3 any value we choose—1 or 10 or 100—substitute the value in Eq. (4-16), and thus solve for x_2 and x_1. There will thus be an infinite number of solutions. We can now interpret the term "linear dependence." Each one of Eqs. (4-15) can be visualized as a plane in three-dimensional space. Each solution to Eqs. (4-15) is a

[1] The factor a_{11} appears on both sides of Eq. (4-18) when written in determinantal form. We assume that it is different from zero and cancel it out.

[2] *Exercise.* Check the equation

$$\begin{vmatrix} a_{11} & a_{12} & b_1 \\ a_{21} & a_{22} & b_2 \\ a_{31} & a_{32} & b_3 \end{vmatrix} = a_{11}a_{22}b_3 + a_{12}a_{31}b_2 + a_{21}a_{32}b_1 - a_{11}a_{32}b_2 - a_{12}a_{21}b_3 - a_{22}a_{31}b_1$$

Note that the coefficient of each of the b's is a second-order determinant.

point in three-dimensional space, which lies in all three planes. If the solution to the equations is unique, the planes have just one point in common. But if there is an infinite number of solutions the three planes must have at least an entire line in common. Since the three planes have a line in common, they are said to be linearly dependent.

Now let us free b_1, b_2, b_3 from the assumption that all of them are zero while still considering the case in which the determinant on the left-hand side of Eq. (4-18) is zero. There are two possibilities. First, the determinant on the right-hand side of Eq. (4-18) might be zero. In that case, the argument we have just gone through applies, and there are an infinite number of solutions to Eqs. (4-15). Equations (4-15) are then said to be linearly dependent, or, alternatively, the numbers b_1, b_2, b_3 are said to be linearly dependent on the coefficients of x_1 and x_2 because those are the only coefficients which enter into the right-hand determinant.

Second, the determinant on the right-hand side of Eq. (4-18) might be different from zero. Then it would be impossible to solve Eq. (4-18) or, consequently, Eqs. (4-15). In that case we would say that there is a linear dependence among the coefficients a, but that the b's are independent of any pair of columns of the a's. Such a situation, as we have seen, admits no solution. It is analogous to searching for the intersection of two parallel lines.

Let us summarize and generalize the results of this discussion. Suppose we have a system of m linear equations in m unknowns. We can derive from it, by successive elimination, a single equation in one of the unknowns. If the coefficient of the remaining unknown is zero, the coefficients of the unknowns in the original set of equations are said to be linearly dependent, otherwise they are linearly independent. If the constant term in the final equation is zero, then the constant column of the original equations is said to be linearly dependent on the coefficients of $m - 1$ of the unknowns; i.e., the m original equations are said to be linearly dependent. Then three situations are possible:

1. If the coefficients of the unknowns are linearly independent, the system has a unique solution.

2. If the coefficients of the unknowns are linearly dependent and the constant column is not linearly dependent on the coefficients of $m - 1$ unknowns, the system has no solution. It is inconsistent.

3. If the coefficients of the unknowns are linearly dependent and the constant column is linearly dependent on the coefficients of $m - 1$ unknowns, then the system may have an infinite number of solutions.[1]

One final word. This whole digression has been concerned with the

[1] We say "may" rather than "will" at this point, because it is possible for such a system to be inconsistent. The exact conditions would carry us beyond our depth, and we stop here.

possibility that certain determinants might turn out to be zero. It might appear that this would be an unlikely coincidence, not deserving of all this attention. Unfortunately, however, vanishing determinants occur quite frequently, and all sound theory and application must be prepared for them.

4-6. THE SIMPLEX METHOD: FUNDAMENTAL THEOREMS

For the sake of clarity we shall consider the linear-programming problem [Eqs. (4-14), (4-12), (4-13)] in the special case where $m = 3$ and $n = 2$. Our results, however, will be valid for the general case. Thus we shall deal with this problem:

Find $x_1, x_2, \ldots, x_5$ such that

$$\begin{aligned} a_{11}x_1 + a_{12}x_2 + a_{13}x_3 + a_{14}x_4 + a_{15}x_5 &= c_1 \\ a_{21}x_1 + a_{22}x_2 + a_{23}x_3 + a_{24}x_4 + a_{25}x_5 &= c_2 \\ a_{31}x_1 + a_{32}x_2 + a_{33}x_3 + a_{34}x_4 + a_{35}x_5 &= c_3 \end{aligned} \tag{4-20}$$

$$x_1 \geq 0, \quad \ldots, \quad x_5 \geq 0 \tag{4-21}$$

$$Z = P_1x_1 + P_2x_2 = \text{as great as possible} \tag{4-22}$$

Comparing Eqs. (4-20) with Eqs. (4-15) we note that Eqs. (4-20) have two additional unknowns, x_4 and x_5. This is the characteristic which provides scope for the maximization called for in Eq. (4-22). We may eliminate two of the unknowns in Eq. (4-20) by the same process that we used in boiling down Eqs. (4-15). Let us eliminate x_1 and x_2. Eliminating x_1 we obtain

$$\begin{vmatrix} a_{11} & a_{12} \\ a_{21} & a_{22} \end{vmatrix} x_2 + \begin{vmatrix} a_{11} & a_{13} \\ a_{21} & a_{23} \end{vmatrix} x_3 + \begin{vmatrix} a_{11} & a_{14} \\ a_{21} & a_{24} \end{vmatrix} x_4 + \begin{vmatrix} a_{11} & a_{15} \\ a_{21} & a_{25} \end{vmatrix} x_5 = \begin{vmatrix} a_{11} & c_1 \\ a_{21} & c_2 \end{vmatrix}$$

$$\begin{vmatrix} a_{11} & a_{12} \\ a_{31} & a_{32} \end{vmatrix} x_2 + \begin{vmatrix} a_{11} & a_{13} \\ a_{31} & a_{33} \end{vmatrix} x_3 + \begin{vmatrix} a_{11} & a_{14} \\ a_{31} & a_{34} \end{vmatrix} x_4 + \begin{vmatrix} a_{11} & a_{15} \\ a_{31} & a_{35} \end{vmatrix} x_5 = \begin{vmatrix} a_{11} & c_1 \\ a_{31} & c_3 \end{vmatrix}$$

Now eliminate x_2:

$$\begin{vmatrix} a_{11} & a_{12} & a_{13} \\ a_{21} & a_{22} & a_{23} \\ a_{31} & a_{32} & a_{33} \end{vmatrix} x_3 + \begin{vmatrix} a_{11} & a_{12} & a_{14} \\ a_{21} & a_{22} & a_{24} \\ a_{31} & a_{32} & a_{34} \end{vmatrix} x_4 + \begin{vmatrix} a_{11} & a_{12} & a_{15} \\ a_{21} & a_{22} & a_{25} \\ a_{31} & a_{32} & a_{35} \end{vmatrix} x_5 = \begin{vmatrix} a_{11} & a_{12} & c_1 \\ a_{21} & a_{22} & c_2 \\ a_{31} & a_{32} & c_3 \end{vmatrix} \tag{4-23}$$

We now introduce a very important assumption. We assume that the constant column of Eqs. (4-20) is linearly independent of every pair (or, in general, of every set of $m - 1$) of coefficient columns on the left-hand side of those equations. We shall refer to this as the nondegeneracy assumption. It states that the constant column cannot be expressed as a linear combination of any pair of columns of coefficients on the left-hand side. Hence any solution of (4-20) must include at least *three* nonzero x's.

Moreover, there must be at least one set of three linearly independent columns on the left. For if this were not so, all three determinants on the left-hand side of Eq. (4-23) would be zero and the equation would be unsolvable. Another way to see this is to note that if there were no set of three linearly independent columns on the left, it would be possible to select two of the columns and to express each of the remaining columns as a linear combination of those two. Then in any solution of (4-20) these expressions could be used to substitute for all but two of the x's, contrary to the hypothesis that the solution of (4-20) required three nonzero x's. For further details see Appendix B. Thus we have proved:

THEOREM 1. *If a nondegenerate linear-programming problem with m restrictions and $k(>m)$ unknowns has a solution (feasible or not),*[1] *then there must be at least one set of m of the unknowns whose coefficients are linearly independent.*

In addition we can state:

THEOREM 2. *If a nondegenerate linear-programming problem with m restrictions and $k(>m)$ unknowns has a feasible solution, then it has a basic feasible solution, i.e., one which involves at most m of the unknowns at nonzero values.*

Proof: It is only the nonnegativity of the x's that causes any difficulty. We know that there are three linearly independent columns of a's. Since we are dealing with systems of three unknowns, we know from Appendix B that any column of constants (including the column of c's) can be solved for in terms of three unknowns. Hence there is no question but that (4-20) has a solution with *exactly* three nonzero x's. But we cannot yet guarantee that those three x's will be positive. We now prove, in a constructive way, that they will be.

We shall develop the proof in terms of the case where $m = 3$, $k = 5$. We suppose that Eqs. (4-20) express a nonbasic feasible solution; i.e., we suppose that $x_1, x_2, \ldots, x_5$ in Eqs. (4-20) are all nonnegative and that at least four of them are positive. Suppose, for a minute, that $x_5 = 0$. Then the last term could be omitted from the left-hand side of each of the Eqs. (4-20) and Theorem 1 would tell us that at least one of the determinants

$$\begin{vmatrix} a_{11} & a_{12} & a_{13} \\ a_{21} & a_{22} & a_{23} \\ a_{31} & a_{32} & a_{33} \end{vmatrix} \quad \text{and} \quad \begin{vmatrix} a_{11} & a_{12} & a_{14} \\ a_{21} & a_{22} & a_{24} \\ a_{31} & a_{32} & a_{34} \end{vmatrix}$$

is different from zero. Thus there is at least one nonzero determinant

[1] By a nonfeasible solution we mean a solution of (4-20), some of whose x's are negative.

on the left-hand side of Eq. (4-23) which involves the coefficients of three unknowns which are not zero in the solution. We are thus justified in assuming that the unknowns have been numbered in such a way that x_1, x_2, x_3, x_4 are all positive and

$$\begin{vmatrix} a_{11} & a_{12} & a_{13} \\ a_{21} & a_{22} & a_{23} \\ a_{31} & a_{32} & a_{33} \end{vmatrix} \neq 0$$

We now drop the assumption that $x_5 = 0$; it has served its purpose. Consider the following equations:

$$\begin{aligned} a_{11}y_1 + a_{12}y_2 + a_{13}y_3 &= a_{14} \\ a_{21}y_1 + a_{22}y_2 + a_{23}y_3 &= a_{24} \\ a_{31}y_1 + a_{32}y_3 + a_{33}y_3 &= a_{34} \end{aligned} \tag{4-24}$$

The constants in these equations have been taken from Eqs. (4-20). Therefore if we solve for the y's by boiling down we shall obtain

$$\begin{vmatrix} a_{11} & a_{12} & a_{13} \\ a_{21} & a_{22} & a_{23} \\ a_{31} & a_{32} & a_{33} \end{vmatrix} y_3 = \begin{vmatrix} a_{11} & a_{12} & a_{14} \\ a_{21} & a_{22} & a_{24} \\ a_{31} & a_{32} & a_{34} \end{vmatrix}$$

Since the left-hand-side determinant is not zero, there is a solution. Of course, some of y_1, y_2, y_3 may be negative, but inspection of Eqs. (4-24) shows that they cannot all be zero. We now perform the critical step. We multiply each of Eqs. (4-24) by some constant θ (to be specified later) and subtract each from the corresponding equation of set (4-20). The result is

$$\begin{aligned} a_{11}(x_1 - \theta y_1) + a_{12}(x_2 - \theta y_2) + a_{13}(x_3 - \theta y_3) \qquad & \\ + a_{14}(x_4 + \theta) + a_{15}x_5 &= c_1 \\ a_{21}(x_1 - \theta y_1) + a_{22}(x_2 - \theta y_2) + a_{23}(x_3 - \theta y_3) \qquad & \\ + a_{24}(x_4 + \theta) + a_{25}x_5 &= c_2 \\ a_{31}(x_1 - \theta y_1) + a_{32}(x_2 - \theta y_2) + a_{33}(x_3 - \theta y_3) \qquad & \\ + a_{34}(x_4 + \theta) + a_{35}x_5 &= c_3 \end{aligned} \tag{4-25}$$

If $\theta = 0$, Eqs. (4-25) are the same as Eqs. (4-20). If θ is slightly different from zero, then the numbers $x_1 - \theta y_1$, $x_2 - \theta y_2$, $x_3 - \theta y_3$, $x_4 + \theta$, x_5 will be slightly different from the numbers given by Eqs. (4-20). But, since the numbers given by Eqs. (4-20) were all positive (with the possible exception of x_5), so will these be (with the same possible exception), and these new numbers will also constitute a solution to the problem. Now think of θ as starting at zero and slowly decreasing. The first three parentheses may increase or decrease, depending on the signs of y_1, y_2, y_3, but the fourth parenthesis will surely decrease. Thus eventually we shall come to a feasible solution which has one fewer nonzero component than the one we started with. This process can be repeated until there are just three nonzero components.

We have now shown, as we set out to do, that if there is any feasible solution to a nondegenerate linear-programming problem, then there is a basic feasible solution. Furthermore, our proof has been constructive; we have described a procedure for starting with any feasible solution and deriving from it a basic feasible solution.

We now take into account the maximizing part of the problem, i.e., Eq. (4-22). The value of Z corresponding to the solution to Eqs. (4-20) is given by Eq. (4-22). Call it Z^0. The value of Z corresponding to the solution of Eqs. (4-25) is, by direct substitution,

$$Z(\theta) = P_1(x_1 - \theta y_1) + P_2(x_2 - \theta y_2) + P_3(x_3 - \theta y_3) + P_4(x_4 + \theta) + P_5 x_5$$

The change in Z is the difference:

$$Z(\theta) - Z^0 = \theta(-P_1 y_1 - P_2 y_2 - P_3 y_3 + P_4) \qquad (4\text{-}26)$$

The parenthesis occurring in Eq. (4-26) is of great importance because it indicates which of a number of feasible solutions corresponds to the largest value of Z. Suppose, first, that the parenthesis is zero. Then we may vary θ at will without changing Z, and, starting with a nonbasic feasible solution, we can derive a feasible solution with one fewer nonzero variable without decreasing Z. Next, suppose that the parenthesis is negative. Then, if $\theta < 0$ we shall have $Z(\theta) > Z^0$. In this case, by taking a large enough negative value of θ we can eliminate one unknown and at the same time increase Z. Finally, suppose that the parenthesis is positive. Then in order to increase Z we must take $\theta > 0$. Now two subcases arise. If none of y_1, y_2, y_3 is positive, then inspection of Eqs. (4-25) shows that there is nothing to prevent us from taking θ as large as we please, for none of the parentheses will become negative no matter how large θ is. Then we may make Z as large as we choose simply by taking θ large enough. This is one of the no-solution cases which we mentioned at the outset. The second subcase is the one in which at least one of y_1, y_2, y_3 is positive. Then as θ increases from zero, at least one of the parentheses in Eqs. (4-25) will approach and finally reach zero. When the first of those parentheses reaches zero we shall have a feasible solution with a reduced number of nonzero variables and an increased value of Z. We have now dealt with all possible cases and have demonstrated the following theorem.

THEOREM 3. *Suppose that a linear-programming problem is nondegenerate and that a nonbasic feasible solution exists. Then, either*

1. *The function to be maximized, Z, can be made as large as desired, or*
2. *A feasible solution with fewer nonzero unknowns can be derived such*

that this new solution will correspond to a value of Z at least as large as that corresponding to the nonbasic solution whose existence was posited.

An immediate consequence is the following key corollary.

COROLLARY TO THEOREM 3. *If a nondegenerate linear-programming problem has an optimal solution, then it has an optimal solution which is also basic.*

To prove the corollary we have only to notice that if there is a nonbasic optimal solution we can apply Theorem 3 to reduce the number of nonzero variables step by step until a basic feasible solution is reached. Since this procedure does not decrease the value of Z, the resultant basic feasible solution will be optimal.

This array of results permits us, in solving a linear-programming problem, to restrict our attention to basic solutions, for if there is any solution, it will be one of these.

4-7. GEOMETRIC INTERPRETATION

A geometric interpretation of the concepts we have been working with may provide welcome relief at this stage.[1] The basic ideas can be seen most easily in terms of the geometry of two dimensions. The most familiar way to represent two-dimensional quantities is by means of points referred to a pair of Cartesian coordinates. For present purposes it is more convenient to use an alternative but exactly equivalent representation. Instead of using a single point to depict a two-dimensional quantity we use a line segment that begins at the origin and ends at the ordinary Cartesian point. Thus the two-dimensional quantity (5,3) would be shown as a line segment that begins at the origin and ends at the point whose x coordinate is 5 and whose y coordinate is 3. Such a line segment is a "vector," in particular, a two-dimensional vector. We denote a vector by the coordinates of its end point. Thus, (5,3) denotes not only the point (5,3) but the vector ending there. We now define some mathematical operations that will enable us to relate vectors to one another. First, a multiple of a vector: k times any vector is the vector whose direction is the same as that of the original vector and which is k times as long. A moment with a scratch pad will convince us that 2 times (5,3) is the vector (10,6); 0.5 times (5,3) is the vector (2.5,1.5). By a slight extension, negative multiplication of a vector means going backward instead of forward along it. Thus, $-(5,3)$ is $(-5,-3)$; $-2(5,3)$ is $(-10,-6)$.

Two vectors can be added by a process which is familiar to everyone

[1] Our main discussion of the geometry underlying linear programming is contained in Appendix B, Secs. B-2 and B-3. This section is included merely to indicate the geometric significance of some of the basic concepts.

who has encountered the "parallelogram of forces" in his elementary physics. Draw the two vectors to be added, say vector **A** and vector **B**, and at the end of one of them, say **A**, draw a line parallel to the other vector and of the same length. The end point of this line is the desired sum, $\mathbf{A} + \mathbf{B}$. The reader had better make some sketches here. If he does he will find, for example, that

$$(5,3) + (2,1) = (5 + 2, 3 + 1) = (7,4);$$

or, to take another example, $(5,3) + (2,-1) = (7,2)$. Finally, these two processes of multiplication and addition can be combined. As an example, $3(5,3) + 2(2,1) = (15,9) + (4,2) = (19,11)$.

These two operations permit us to state the following fundamental theorem: If we are given any pair of two-dimensional vectors that do not lie on the same straight line through the origin, then we can write any third two-dimensional vector as a weighted sum of the two of them. For instance, if we are given the two vectors $\mathbf{A} = (5,3)$ and $\mathbf{B} = (2,1)$, then, as the reader can verify easily, any third vector (x,y) can be written as

$$(x,y) = (2y - x)\mathbf{A} + (3x - 5y)\mathbf{B}$$

for example, $(1,0) = -\mathbf{A} + 3\mathbf{B}$. Such a pair of vectors is known as a "basis," and we can restate our theorem by saying that any vector in the plane can be expressed in terms of a basis consisting of two noncolinear vectors. This is the two-dimensional case of the general theorem that underlies Theorem 2 of Sec. 4-6, above.

The geometric principle at work is that we can get from the origin to any point in the plane by going an appropriate distance in one direction assigned in advance and then another appropriate distance (perhaps zero) in another assigned direction. Each of the two assigned vectors simply sets a direction; how far we go in that direction depends on the multiplier (which may be zero or negative, of course) applied to that vector.

How is it in three dimensions? Now a scratch pad is not much help, but the mind's eye will suffice. Any three vectors which do not all lie in the same plane will provide us with three directions, and if we march appropriate distances in these three directions in succession we can arrive at the end point of any other vector in three-dimensional space. Thus, three vectors form a basis for three-dimensional space, provided that they do not all lie in the same plane. Why is this proviso necessary? Because if all three vectors lie in the same plane we can never get out of that plane by following the directions they determine, and thus can never reach the end point of any vector not in that plane. In general, any m vectors which do not all lie in any $(m - 1)$ dimensional hyperplane can be used as a basis for expressing all the vectors in an m-dimensional space;

i.e., any vector in m-dimensional space is equal to some weighted sum of m preassigned vectors in that space, provided that the preassigned vectors do not all lie in some smaller space.

Now we apply these ideas to the argument of Sec. 4-6. Look at Eqs. (4-20). The coefficients of each of the five x_i's can be thought of as determining a vector in three dimensions. Thus the five x_i's give us five vectors. The three numbers in the constant column also determine a vector in that space. The discussion of vectors that we have just gone through shows that the constant column can be expressed in terms of any three noncoplanar activity vectors. If this is done (and it can be done by solving ordinary simultaneous equations), the multipliers of the three selected vectors are nothing but the activity levels, and if these multipliers are all nonnegative, the three selected activities constitute a feasible basis. Thus the basic theorem of linear programming is nothing but an application of a simple geometrical principle.

In just the same way, if we select any three vectors as a basis, we can express any fourth activity in terms of them. This is the graphic interpretation of Eqs. (4-24), (4-25), etc. The y's in those equations are, in graphic terms, simply the multipliers of the three vectors used as a basis for expressing other vectors.

This graphic approach helps us to visualize the meaning of feasible and infeasible solutions. Think, for a moment, of the two-dimensional case. Under what conditions can a vector in two dimensions be expressed as a sum of multiples of two other vectors without using negative multipliers? A glance at our scratch pad tells us: A vector can be so expressed if it does not lie outside the angle formed by the two vectors in terms of which it is to be expressed. For example:

$$(4,2) = \tfrac{1}{2}(5,3) + \tfrac{1}{2}(3,1) \qquad \text{(feasible)}$$
$$(7,5) = 2(5,3) - (3,1) \qquad \text{(infeasible)}$$

In general, a column of constants can be expressed as a feasible combination of a set of activities if its vector (in the appropriate number of dimensions) does not lie outside the cone formed by the vectors of those activities.

All the theorems that we have deduced algebraically can be derived geometrically by using the concepts of vectors and combinations of vectors. We shall not pursue this approach further, however.

4-8. THE SIMPLEX METHOD: FINDING AN OPTIMUM

We assume that we have a basic feasible solution to start with. Later we shall discuss how to find such a starting point. To be specific, we assume that $x_1 > 0$, $x_2 > 0$, $x_3 > 0$, $x_4 = 0$, $x_5 = 0$ is a set of values which satisfies Eqs. (4-20). We shall call the variables with nonzero values the dependent, or basic, variables, and those with zero values can be called independent, or nonbasic, or excluded, variables. We shall call them excluded variables.

Then, since $x_4 = x_5 = 0$, the values of the basic variables satisfy

$$\begin{aligned} a_{11}x_1 + a_{12}x_2 + a_{13}x_3 &= c_1 \\ a_{21}x_1 + a_{22}x_2 + a_{23}x_3 &= c_2 \\ a_{31}x_1 + a_{32}x_2 + a_{33}x_3 &= c_3 \end{aligned} \tag{4-27}$$

a set of ordinary simultaneous equations. Equations (4-24) still apply. Just as we did in the last section, we may multiply each equation of set (4-24) by a constant θ_4 and subtract from the corresponding equation of set (4-27). The result is

$$\begin{aligned} a_{11}(x_1 - \theta_4 y_1) + a_{12}(x_2 - \theta_4 y_2) + a_{13}(x_3 - \theta_4 y_3) + a_{14}\theta_4 &= c_1 \\ a_{21}(x_1 - \theta_4 y_1) + a_{22}(x_2 - \theta_4 y_2) + a_{23}(x_3 - \theta_4 y_3) + a_{24}\theta_4 &= c_2 \\ a_{31}(x_1 - \theta_4 y_1) + a_{32}(x_2 - \theta_4 y_2) + a_{33}(x_3 - \theta_4 y_3) + a_{34}\theta_4 &= c_3 \end{aligned} \tag{4-28}$$

If θ_4 is slightly greater than zero this will be a nonbasic feasible solution involving the first four variables at values $x_1 - \theta_4 y_1$, $x_2 - \theta_4 y_2$, $x_3 - \theta_4 y_3$, θ_4, respectively. The value of Z corresponding to $\theta_4 = 0$ is

$$Z^0 = P_1x_1 + P_2x_2 + P_3x_3$$

The value of Z corresponding to any value of θ_4 is

$$Z(\theta_4) = P_1(x_1 - \theta_4 y_1) + P_2(x_2 - \theta_4 y_2) + P_3(x_3 - \theta_4 y_3) + P_4\theta_4$$

The effect on Z of taking θ_4 different from zero is thus given by the difference

$$Z(\theta_4) - Z^0 = \theta_4(P_4 - P_1y_1 - P_2y_2 - P_3y_3) \tag{4-29}$$

This is the same parenthesis as the one which occurred in Eq. (4-26), and again it is critical. The new feature of this situation is that θ_4 cannot now be negative, for this would signify a solution in which the fourth variable had a negative value in violation of Eq. (4-21). If the parenthesis is zero or negative, therefore, Z cannot be increased by varying θ_4 in the permissible range, i.e., by introducing the fourth variable. But if the parenthesis is positive, Z can be increased by taking $\theta_4 > 0$. Again we must consider two subcases. If the parenthesis is positive and if none of y_1, y_2, y_3 is positive, then Eqs. (4-28) show that there is no restriction on the size of θ_4 and Eq. (4-29) shows that there is no upper limit to the size of Z. This is the unsolvable case.

The second subcase is the one in which the parenthesis is positive and at least one of y_1, y_2, y_3 is also positive. In that case let

$$\theta_4 = \min\left(\frac{x_1}{y_1}, \frac{x_2}{y_2}, \frac{x_3}{y_3}\right) \tag{4-30}$$

where, however, only fractions which have positive denominators are taken into account. The result will be to replace the basic feasible solution defined by Eqs. (4-27) by a new basic feasible solution in which the fourth variable appears with the value θ_4 given by Eq. (4-30) and that one of the first three variables which corresponds to the smallest positive fraction in Eq. (4-30) has dropped out. This new basic solution will correspond to a larger value of Z. The effect is to obtain a new and superior basic feasible solution in which the fourth variable has been introduced and one of the formerly basic variables has been excluded.

The indicated procedure for solution is as follows. Suppose that we have a basic feasible solution to start with and that, to be specific, the first three variables are included. The first step is to calculate y_1, y_2, y_3 by solving Eqs. (4-24). Then calculate

$$\bar{P}_4 = P_1y_1 + P_2y_2 + P_3y_3 \tag{4-31}$$

This quantity, $\bar{P}_4$, is the sum of the negative terms in the parenthesis of Eq. (4-29). Then compute the "simplex criterion," $P_4 - \bar{P}_4$. There are two possibilities:

(1) $P_4 - \bar{P}_4 > 0$. In this case compute θ_4 by formula (4-30) and

$$\begin{aligned} x_1' &= x_1 - \theta_4 y_1 \\ x_2' &= x_2 - \theta_4 y_2 \\ x_3' &= x_3 - \theta_4 y_3 \end{aligned}$$

One of x_1', x_2', x_3' will turn out to be zero. The result is an improved basic solution, and we are ready to repeat the whole procedure, with appropriate changes in notation, to see if we can obtain still further improvement.

(2) $P_4 - \bar{P}_4 \leq 0$. In this case no improvement can be made by introducing the fourth variable. We therefore pass on to the fifth variable and, inserting its coefficients in place of the coefficients of the fourth variable in Eqs. (4-24), go through the same procedure.

If there are sixth, seventh, or later variables, the same procedure applies with respect to each of them. In this way, if there is any variable, say the ith, for which $P_i - \bar{P}_i > 0$, we shall find it and use it to obtain an improved basic solution. But suppose there is no such variable; i.e., suppose that $P_i - \bar{P}_i \leq 0$ for all variables. Then the basic solution we are considering is optimal. It is easy to show this by generalizing our previous demonstration. Suppose that we have gone through the procedure for the fourth variable, solving Eqs. (4-24) in the course of it and, for the fifth variable, including the solution of

$$\begin{aligned} a_{11}y_1' + a_{12}y_2' + a_{13}y_3' &= a_{15} \\ a_{21}y_1' + a_{22}y_2' + a_{23}y_3' &= a_{25} \\ a_{31}y_1' + a_{32}y_2' + a_{33}y_3' &= a_{35} \end{aligned} \tag{4-32}$$

4-9. THE SIMPLEX METHOD: COMPUTATION

The computations required for solving a linear-programming problem by the simplex method (or any other method) can undoubtedly be formidable. The method would indeed be impracticable for many important applications that involve large numbers of variables and restraints, except that it can be performed mechanically by punch card and by electronic calculating machines. Whether performed by hand or by machinery, however, the method can be made much simpler than the foregoing algebraic treatment may suggest. In this section we shall describe and illustrate an efficient computing procedure.

The bulk of the computing task described in the last section consisted in solving systems of simultaneous equations such as Eqs. (4-24) and (4-32). According to the procedure, given any basic solution, one such set of equations had to be solved for each variable excluded from the basis. But all these equations have an important saving grace: they have the same coefficients on the left-hand side and differ only on the right.

When all the sets of simultaneous equations for any basic solution have been solved, the result will be either that that solution is optimal (in which case the work is complete) or that some new basic solution is preferable. In the latter case, some new sets of simultaneous equations must be solved, but they have the saving grace that they have the same coefficients on the left-hand side as the sets just solved, with the exception of one column. The idea behind all systematic methods of solution is to take advantage of these two features.

The simplification possible when two sets of simultaneous equations have the same left-hand side can be seen by combining Eqs. (4-24), (4-27), and (4-32). We write

$$\begin{aligned} a_{11}u_1 + a_{12}u_2 + a_{13}u_3 &= c_1,\ a_{14},\ a_{15} \\ a_{21}u_1 + a_{22}u_2 + a_{23}u_3 &= c_2,\ a_{24},\ a_{25} \\ a_{31}u_1 + a_{32}u_2 + a_{33}u_3 &= c_3,\ a_{34},\ a_{35} \end{aligned} \tag{4-36}$$

In this setup we have indicated by commas three alternative right-hand sides. Now we eliminate the variables u_1 and u_2:

$$\begin{vmatrix} a_{11} & a_{12} \\ a_{21} & a_{22} \end{vmatrix} u_2 + \begin{vmatrix} a_{11} & a_{13} \\ a_{21} & a_{23} \end{vmatrix} u_3 = \begin{vmatrix} a_{11} & c_1 \\ a_{21} & c_2 \end{vmatrix}, \begin{vmatrix} a_{11} & a_{14} \\ a_{21} & a_{24} \end{vmatrix}, \begin{vmatrix} a_{11} & a_{15} \\ a_{21} & a_{25} \end{vmatrix}$$

$$\begin{vmatrix} a_{11} & a_{12} \\ a_{31} & a_{32} \end{vmatrix} u_2 + \begin{vmatrix} a_{11} & a_{13} \\ a_{31} & a_{33} \end{vmatrix} u_3 = \begin{vmatrix} a_{11} & c_1 \\ a_{31} & c_3 \end{vmatrix}, \begin{vmatrix} a_{11} & a_{14} \\ a_{31} & a_{34} \end{vmatrix}, \begin{vmatrix} a_{11} & a_{15} \\ a_{31} & a_{35} \end{vmatrix}$$

$$\begin{vmatrix} a_{11} & a_{12} & a_{13} \\ a_{21} & a_{22} & a_{23} \\ a_{31} & a_{32} & a_{33} \end{vmatrix} u_3 = \begin{vmatrix} a_{11} & a_{12} & c_1 \\ a_{21} & a_{22} & c_2 \\ a_{31} & a_{32} & c_3 \end{vmatrix}, \begin{vmatrix} a_{11} & a_{12} & a_{14} \\ a_{21} & a_{22} & a_{24} \\ a_{31} & a_{32} & a_{34} \end{vmatrix}, \begin{vmatrix} a_{11} & a_{12} & a_{15} \\ a_{21} & a_{22} & a_{25} \\ a_{31} & a_{32} & a_{35} \end{vmatrix}$$

Multiply Eqs. (4-24) by θ_4 and Eqs. (4-32) by θ_5 and subtract both fr Eqs. (4-27). The five variables then become $x_1 - \theta_4 y_1 - \theta_5 y_1'$, $x_2 - \theta$ $- \theta_5 y_2'$, $x_3 - \theta_4 y_3 - \theta_5 y_3'$, θ_4, θ_5. Substituting these in Eq. (4-22) obtain

$$\begin{aligned} Z(\theta_4,\theta_5) &= P_1(x_1 - \theta_4 y_1 - \theta_5 y_1') + P_2(x_2 - \theta_4 y_2 - \theta_5 y_2') + P_3(x_3 - \theta_4 y_3 \\ &\qquad - \theta_5 y_3') + P_4\theta_4 + P_5\theta \\ &= Z^0 + \theta_4(P_4 - \bar{P}_4) + \theta_5(P_5 - \bar{P}_5) \end{aligned}$$

If neither parenthesis is positive we cannot make $Z(\theta_4,\theta_5)$ greater than Z^0 by giving θ_4 and θ_5 any positive values; i.e., if we cannot obtain an improved solution by introducing any of the nonbasic variables singly, then we cannot obtain an improvement by introducing any combination of them and we must already be at the optimum.

The simplex criterion, which determines whether an excluded variable should be introduced, has a significant economic interpretation in economic problems. This will be treated in Chap. 6.

Thus far we have consistently assumed that we had a feasible solution to use as a starting point. In many cases it is easy to find a feasible solution by inspection. Consider, for example, Eqs. (4-11). If $c_1 > 0$, $c_2 > 0, \cdots, c_m > 0$, then we can write down at sight the feasible solution $x_1 = x_2 = \cdots = x_n = 0$, $x_{n+1} = c_1$, $x_{n+2} = c_2, \ldots, x_{n+m} = c_m$. Sometimes, however, things are not so easy. In Eqs. (4-20), for example, no feasible solution is evident on sight. In such a case a feasible solution may be found by the following neat device, due to Dantzig. Using the constants of Eqs. (4-20), set up the following auxiliary problem:

Find $x_1, x_2, \ldots, x_5, x_6, x_7, x_8$ such that

$$\begin{aligned} a_{11}x_1 + a_{12}x_2 + a_{13}x_3 + a_{14}x_4 + a_{15}x_5 + x_6 &= c_1 \\ a_{21}x_1 + a_{22}x_2 + a_{23}x_3 + a_{24}x_4 + a_{25}x_5 + x_7 &= c_2 \\ a_{31}x_1 + a_{32}x_2 + a_{33}x_3 + a_{34}x_4 + a_{35}x_5 + x_8 &= c_3 \end{aligned} \tag{4-33}$$

$$x_1 \geq 0, x_2 \geq 0, \ldots, x_8 \geq 0 \tag{4-34}$$

$$z = x_6 + x_7 + x_8 = \text{as small as possible} \tag{4-35}$$

Two features of the auxiliary problem should be noted. First, since its restraints have the same form as Eqs. (4-11), a feasible solution can be written down at sight. Second, the minimum possible value of z is zero, and this value occurs when $x_6 = x_7 = x_8 = 0$. But when these last three variables vanish, Eqs. (4-33) are the same as Eqs. (4-20). Thus an optimal solution to the auxiliary problem is a feasible (but not necessarily optimal) solution to the original problem.[1]

[1] An alternative, but more laborious, method for finding a feasible solution has also been given by Dantzig. See his "Maximization of a Linear Function of Variables Subject to Linear Inequalities," in T. C. Koopmans (ed.), *Activity Analysis of Production and Allocation*, pp. 345–347, John Wiley & Sons, Inc., New York, 1951.

Thus we obtained three values simultaneously for u_3, one corresponding to the solution of each of the three sets of equations. The first simplification, then, consists simply of carrying along a number of right-hand sides simultaneously.

The second simplification is known as "basis-shifting," i.e., the process of shifting from one basic solution to a second. Suppose that we have solved Eqs. (4-24) and (4-27), both of which involve the first three variables as the basic solution, and wish to solve

$$\begin{aligned} a_{11}x_1' + a_{12}x_2' + a_{14}x_4' &= c_1 \\ a_{21}x_1' + a_{22}x_2' + a_{24}x_4' &= c_2 \\ a_{31}x_1' + a_{32}x_2' + a_{34}x_4' &= c_3 \end{aligned} \tag{4-37}$$

In Eqs. (4-37) the basic solution includes the first, second, and fourth variables. We can form at once Eqs. (4-28) with $\theta_4 = x_3/y_3$. The result is

$$\begin{aligned} a_{11}\left(x_1 - \frac{x_3}{y_3}y_1\right) + a_{12}\left(x_2 - \frac{x_3}{y_3}y_2\right) + a_{14}\frac{x_3}{y_3} &= c_1 \\ a_{21}\left(x_1 - \frac{x_3}{y_3}y_1\right) + a_{22}\left(x_2 - \frac{x_3}{y_3}y_2\right) + a_{24}\frac{x_3}{y_3} &= c_2 \\ a_{31}\left(x_1 - \frac{x_3}{y_3}y_1\right) + a_{32}\left(x_2 - \frac{x_3}{y_3}y_2\right) + a_{34}\frac{x_3}{y_3} &= c_3 \end{aligned} \tag{4-38}$$

Comparing these equations with Eqs. (4-37) we see that the values of the variables given in Eqs. (4-38) constitute a solution to Eqs. (4-37). Thus,

$$\begin{aligned} x_1' &= x_1 - \frac{x_3}{y_3}y_1 \\ x_2' &= x_2 - \frac{x_3}{y_3}y_2 \\ x_4' &= \frac{x_3}{y_3} \end{aligned} \tag{4-39}$$

It is therefore not necessary to solve the simultaneous equations afresh every time we consider a new basic solution.

4-10. THE SIMPLEX METHOD: AN EXAMPLE

All the concepts and theorems we have been discussing can be illustrated by a straightforward example. Table 4-1 sets forth the situation of a firm that processes a certain raw material by the use of two major types of equipment, called stills and retorts. We assume that four different production processes are available to the firm and that they are characterized by the input coefficients and value data given in the first four columns of the table. Thus, one unit of Process 1 will treat 100 tons of raw material and will absorb 7 per cent of the weekly capacity of the firm's stills and 3 per cent of the weekly capacity of its retorts. It will yield a final product worth $1,110 but will consume $1,000 worth of raw material and will require $50 worth of other direct costs, leaving a net value of $60. The next three columns are interpreted similarly.

The firm's operations are limited by the total capacity of its stills and retorts (100 per cent in both instances, of course) and by the availability of its raw materials, which is assumed to be 1,500 tons per week. Thus, suppose that Process 2 is used. If it is operated at a level of 15 units, it

TABLE 4-1. INPUTS OF RAW MATERIALS AND OUTPUTS PER WEEK FOR MANUFACTURING BY FOUR DIFFERENT PRODUCTION PROCESSES

	Alternative production processes				Quantity of inputs available
	1	2	3	4	
Inputs:					
Raw materials, tons	100	100	100	100	1,500
Still capacity, %	7	5	3	2	100
Retort capacity, %	3	5	10	15	100
Values, dollars:					
Value of output	1,110	1,120	1,130	1,150	
Value of raw material	1,000	1,000	1,000	1,000	
Other direct costs	50	60	40	60	
Net value	60	60	90	90	

will absorb 15 × 100 = 1,500 tons of raw material, 15 × 5 = 75 per cent of still capacity, and 15 × 5 = 75 per cent of retort capacity. Furthermore, it will produce a net value of 15 × 60 = $900 per week. This is as much as can be earned by the use of Process 2 alone since it uses all the raw material available.

We now undertake to find the best combination of Processes 1, 2, 3, and 4, that is, the combination that yields the greatest net value without violating any of the restrictions assumed. To express the situation algebraically, let x_1, x_2, x_3, x_4 denote the levels of the four processes, respectively. Total net value as a function of these four variables is, then,

$$z = 60x_1 + 60x_2 + 90x_3 + 90x_4$$

This sum is to be maximized subject to the restrictions

$$\begin{aligned} 100x_1 + 100x_2 + 100x_3 + 100x_4 &\leq 1{,}500 \\ 7x_1 + 5x_2 + 3x_3 + 2x_4 &\leq 100 \\ 3x_1 + 5x_2 + 10x_3 + 15x_4 &\leq 100 \\ x_1 \geq 0,\ x_2 \geq 0,\ x_3 \geq 0,\ x_4 &\geq 0 \end{aligned} \tag{4-40}$$

To eliminate the first three inequalities, introduce slack variables x_5, x_6, x_7 so that Eqs. (4-40) become

$$\begin{aligned} 100x_1 + 100x_2 + 100x_3 + 100x_4 + x_5 &= 1{,}500 \\ 7x_1 + 5x_2 + 3x_3 + 2x_4 + x_6 &= 100 \\ 3x_1 + 5x_2 + 10x_3 + 15x_4 + x_7 &= 100 \end{aligned} \tag{4-41}$$

The economic significance of x_5, x_6, x_7 is clear: x_5 is the number of tons of raw material which could be purchased but which are not, x_6 and x_7 denote the percentage of equipment capacity left unused.

We are now ready to solve the problem by the simplex method. Since there are three restrictions in the problem (aside from the nonnegativity restraints), a basic solution will include three of the seven variables. We have already seen that if all 1,500 tons of raw material is treated by Process 2, some of both types of capacity will be left over. Hence, $x_2 = 15$, $x_6 = 25$, $x_7 = 25$ is a feasible basic solution, as the reader can verify easily. We adopt it as our starting point.

The calculations are carried out in Table 4-2. This table looks a bit cryptic, as any good worksheet should, because nothing is written down in it that is not absolutely necessary. It is easy to follow, however. Our first task is to determine formally (even though we know the answer already) the values of x_2, x_6, x_7, which constitute the trial solution we have selected, that satisfy the restraints. Thus we must solve the equations

$$\begin{aligned} 100x_2 \quad &= 1{,}500 \\ 5x_2 + x_6 &= 100 \\ 5x_2 + x_7 &= 100 \end{aligned}$$

Note that by selecting a number of included variables equal to the number of restraints we have derived an ordinary set of simultaneous equations in which the number of equations equals the number of variables.

Now look at the first three lines of Table 4-2. The coefficients of x_2 in these equations appear in the column headed x_2, the coefficients of x_6 are in its column, similarly for the coefficients of x_7, and the numbers on the right-hand side of the equations appear in the constant column. Thus these entries in the table are nothing but an abbreviated transcription of the equations which determine the feasible levels of x_2, x_6, x_7. This kind of table is known as a table of detached coefficients. In such a table each line corresponds to an equation. The variables appear as column headings rather than being repeated in each equation, and the vertical bars that separate the columns perform the roles of plus and minus signs in a readily understandable way.

Now we turn to the remaining columns of the first three lines. Our second task will be to solve some equations of the form of Eqs. (4-24) above, in which the coefficients of the included variables appear on the

TABLE 4-2. WORKSHEET FOR SIMPLEX SOLUTION OF TABLE 4-1

Line no.	Formula	x_2	x_6	x_7	Constant	x_1	x_3	x_4	x_5
1	Eqs. (4-41)	100			1500	100	100	100	1
2		5	1		100	7	3	2	
3		5		1	100	3	10	15	
4	(1)/20	5			75	5	5	5	.05
5	(2)-(4)		1		25	2	−2	−3	−.05
6	(3)-(4)			1	25	−2	5	10	−.05
7	(1)/100	1			15	1	1	1	.01
8	(5)		1		25	2	−2	−3	−.05
9	(6)			1	25	−2	5	10	−.05
10	P	60	0	0	....	60	90	90	0
11	$\bar{P}$				900	60	60	60	.6
12	Simplex C					0	30	30	−.6

Introduce x_3. Eliminate x_7.

		x_1	x_2	x_3	x_4	x_5	x_6	x_7	Constant
	x_2	1.4	1		−1	.02		−.2	10
	x_3	−.4		1	2	−.01		.2	5
	x_6	1.2			1	−.07	1	.4	35
	P	60	60	90	90	0	0	0	
	$\bar{P}$	48	60	90	120	.3	0	6	1050
	Simplex C	12	0	0	−30	−.3	0	−6	

Introduce x_1. Eliminate x_2.

		x_1	x_2	x_3	x_4	x_5	x_6	x_7	Constant
	x_1	1	.714		−.714	.014		−.143	7.143
	x_3		.285	1	1.714	−.004		.143	7.857
	x_6		−.857		1.857	−.087	1	.571	26.428
	P	60	60	90	90	0	0	0	
	$\bar{P}$	60	68.571	90	111.429	.471	0	4.286	1135.714
	Simplex C	0	−8.571	0	−21.429	−.471	0	−4.286	

left-hand side, and those of one of the excluded variables appear on the right.[1] The effect of solving these equations will be to determine the

[1] The reader may be confused by the fact that the variables in Eqs. (4-24) are called y instead of x. But, of course, the names that we give the variables, x, y, u, or anything else, do not make any difference; the guts of a system of equations are in the coefficients. In general, in this chapter the variables are called x if the right-hand

combination of levels of the included variables which has the same effect on the restraint equations as each of the excluded variables at unit level. This operation is known as expressing the excluded variables on the basis of the included ones. The last four columns are provided for expressing the excluded variables on the basis of the included ones. Each column contains in its first three lines the coefficients of one of the excluded variables as given in Eqs. (4-41). The first three lines of the columns headed x_2, x_6, x_7, x_1 summarize the equations

$$\begin{aligned} 100x_2 \qquad\; &= 100 \\ 5x_2 + x_6 &= 7 \\ 5x_2 + x_7 &= 3 \end{aligned}$$

which are recognized as a special case of Eqs. (4-24). The remaining columns are interpreted similarly.

In short, the first three lines of Table 4-2 merely record all the fundamental equations of the problem in a manner which singles out x_2, x_6, x_7 as the included variables in the first trial solution.

Our next task is to solve these sets of simultaneous equations, and we handle all of them at the same time, as suggested in the previous section. How to solve simultaneous linear equations is no part of the technique of linear programming because such equations are handled in the same way in linear-programming computations as in any other kind of problem in which they occur.[1] Sometimes the solution of simultaneous equations is so tedious that an electronic calculator has to be used; sometimes, as now, they are so simple that the solution can be written down practically at sight. In the present instance, elementary algebra suffices. Remember that each of the first three lines of Table 4-2 is a shorthand expression for a family of ordinary equations of the general type given in Eqs. (4-1). Therefore, new valid equations can be derived by multiplying any line through by a constant, by adding one line to another, or by combining these two operations. We therefore multiply through line 1 by $\frac{1}{20}$ and record the result on line 4. We subtract line 4 from line 2, placing the result on line 5, and subtract line 4 from line 3, placing the result on line 6. Finally, we multiply line 1 by $\frac{1}{100}$, entering the result on line 7, and transcribe lines 5 and 6 onto lines 8 and 9, respectively. The result—

---

side of the equations is a column of restraining constants, and y if the right-hand side is a column of coefficients of some other variable. This notation avoids some ambiguity. The physical or economic significance of these x's and y's is discussed fully in the interpretive chapters.

[1] For an excellent discussion of methods for solving simultaneous linear equations, see Paul S. Dwyer, *Linear Computations*, chaps. 4, 5, and 6, John Wiley & Sons, Inc., New York, 1951.

lines 7, 8, and 9—is the solution to the original equations. For example, written out in full, line 7 says that

$$x_2 = 15, 1, 1, 1, 0.01$$

these numbers being the value of x_2 in the basic feasible solution and in the basic expressions of x_1, x_3, x_4, x_5, respectively. Lines 8 and 9, similarly, give the values of x_6 and x_7. Note that the values of x_2, x_6, x_7 in the basic feasible solution are 15, 25, 25—as our preliminary inspection indicated.

We are now ready to apply the simplex criterion described in Sec. 4-7. On line 10 we record in each column the coefficient of the variable corresponding to that column found in the objective function. Since the slack variables do not appear in the objective function their coefficients are all zero. On line 11 we first compute the value of the objective function by multiplying the level of each included variable, given in the constant column, by its coefficient, given in line 10. The result is 900, entered in the constant column. Next we compute $\bar{P}$, defined by Eq. (4-31), for each excluded variable, by multiplying the level of each included variable in the expression for the excluded variable (given in lines 7 to 9) by its coefficient on line 10. For example, in the column for x_3 we find

$$1 \times 60 - 2 \times 0 + 5 \times 0 = 60$$

which is duly recorded on line 11. Finally, on line 12 we compute the simplex criterion by subtracting line 11 from line 10.

In Sec. 4-7 we saw that if the simplex criterion is positive in any column, then a basic feasible solution can be found with a higher value of the objective function if we add the variable in whose column the positive value occurs to the list of included variables and drop out one of the included variables in the trial set. In this example, positive values occur in the columns for x_3 and x_4. Thus either of these can be introduced with advantage, and since the positive values happen to be equal, there is no evident reason for preferring one to the other. We select x_3 to introduce, arbitrarily. Which of the previously included variables should be dropped? Equation (4-30) tells us: it is the one for which the ratio of the entry in the constant column to the entry in the x_3 column is smallest, considering only lines for which the entry in the x_3 column is positive. This ratio is $15/1 = 15$ for x_2 (looking at line 7) and $25/5 = 5$ (looking at line 9) for x_7. We skipped line 8 because the entry in the x_3 column there was negative. Since 5 is less than 15, x_7 is to be dropped. Thus we know that x_2, x_3, x_6 constitutes an improved set of included variables.

Our next task is to find the solutions to the same old sets of equations, using our revised list of included variables. Rather than start this whole business *ab initio* we use the basis-shifting technique, i.e., Eqs. (4-38), described in Sec. 4-9. From here on it is convenient to list the columns corresponding to the various variables in their natural order rather than in the scrambled order used to get the initial solution. Thus we start the second panel of Table 4-2 with a new set of column headings. In this panel there is one line for each of the included variables in the new set.

The first computation now is to find the levels in the various columns of the newly introduced variable x_3. As seen from Eqs. (4-39) these are obtained simply by dividing the entries on the x_7 line of the old solution, the line for the variable being dropped, by the entry on that line in the column for x_3, the variable being introduced. Thus we must divide each number on line 9, which corresponds to x_7, by 5. This has been done, and the results, rearranged in accordance with the rearrangement of the columns, have been recorded in the line for x_3 in the second panel.

Next we fill in the entries for the remaining included variables x_2 and x_6. This is done column by column using the following restatement of Eqs. (4-39):

New constant column = old constant column − θ
× old solution for the variable being introduced

where θ = the value just found for the level of the variable being introduced in the constant column

New column for any variable = old column for that variable − θ_i
× old solution for the variable being introduced

where θ_i = level of the newly introduced variable in that column

These are not easy formulas to read on first acquaintance, but they are easy to apply and will become clear as soon as we have illustrated them. The formula for the new constant column works as follows:

$$\text{New constant column} = \begin{bmatrix} 15 \\ 25 \\ 25 \end{bmatrix} - 5 \begin{bmatrix} 1 \\ -2 \\ 5 \end{bmatrix} = \begin{bmatrix} 15 \\ 25 \\ 25 \end{bmatrix} - \begin{bmatrix} 5 \\ -10 \\ 25 \end{bmatrix} = \begin{bmatrix} 10 \\ 35 \\ 0 \end{bmatrix}$$

Note that the old constant column has been taken from lines 7 to 9, and the old solution for x_3, the variable being introduced, has been taken from those same lines and multiplied by 5, the level of x_3 in the new constant column. The result tells us that the new value of x_2 is 10, the new value of x_6 is 35, and the new value of x_7 is 0, so that it drops out as we intended. Since the values of x_2 and x_6 are recorded in the constant column of the second panel and we knew already the value of x_3 in that column, we have no further interest in x_7; therefore work is completed.

The calculation of the other columns is similar. For example, to compute the column for x_4 we note that we have already found $x_3 = 2$ in that column and write

$$\text{New column for } x_4 = \begin{bmatrix} 1 \\ -3 \\ 10 \end{bmatrix} - 2\begin{bmatrix} 1 \\ -2 \\ 5 \end{bmatrix} = \begin{bmatrix} 1 \\ -3 \\ 10 \end{bmatrix} - \begin{bmatrix} 2 \\ -4 \\ 10 \end{bmatrix} = \begin{bmatrix} -1 \\ 1 \\ 0 \end{bmatrix}$$

The first column in this calculation is the column for x_4 on lines 7 to 9, the second column is the same as the second column in the previous computation, and the coefficient 2 has already been accounted for. Again x_7 takes on a value of zero and drops out; the values for x_2 and x_6 are recorded on the appropriate lines of the x_4 column in the second panel. The other columns are to be filled in similarly. Of course, in practice it is not necessary or advisable to do all the transcribing that we have done for expository reasons.

The result of all this is the set of levels of the new set of included variables which would have been obtained if we had set up and solved a new set of simultaneous equations. But since, in general, basis shifting is much less work than solving simultaneous equations, this less transparent method for deriving the solutions is preferable in practice.

We are now ready to apply the simplex criterion to the new set of included variables and do so on the last three lines of the panel just as we did on lines 10 to 12. The result is that the simplex criterion is positive only for x_1, which must, accordingly, be introduced. Equation (4-30) tells us that x_2 is going to drop out.

The rest is repetition. In the third panel we shift the basis from x_2, x_3, x_6 to x_1, x_3, x_6, using the procedure already described. Then we apply the simplex criterion and this time find that no excluded variable gives a positive value to the criterion. Thus this third basis is optimal, and the maximum attainable weekly net value is \$1,135.714, found on the $\bar{P}$ line of the constant column of the third panel. The optimal levels of the variables are those given in the constant column. The variables which do not occur in this column, i.e., the excluded variables, are, of course, to be set at zero. The problem is solved.

4-11. THE SIMPLEX METHOD: DEGENERATE CASE

We have now set forth the principles of the simplex method on the assumption that the problem is not degenerate. That is, we have assumed that the constant column is linearly independent of every set of $m - 1$ columns of coefficients on the left-hand side of Eqs. (4-20). It happens that degeneracy, in this sense, is a fairly frequent occurrence in economic problems. Fortunately, it does not require a major modification in the procedure of solution. Suppose, then, that degeneracy does occur so that when the eliminations are carried out as in Eqs. (4-18) and

(4-23) the right-hand determinant is zero. The first set of equations of the simplex method [Eqs. (4-36)] can still be solved. The trouble is that one or more of x_1, x_2, x_3 in the solution will be zero, and when that solution is substituted into Eq. (4-30), the shifting constant θ may also be zero. If this occurs, the significance is either that the basic solution being tested is optimal or that it is impossible to go from it to a superior solution in one step.

The signal that a problem is degenerate flashes in the course of the regular basis-shifting procedure. What occurs is that at some stage in the iteration when some excluded variable is to be added to the basic set in accordance with the simplex criterion, it displaces not one (as would be proper) but several of the variables of the basic set. This happens when several of the ratios x_i/y_i in Eq. (4-30) are tied for being the smallest. If we dropped all these tied variables from the basic set, we should be left with an insufficient number of basic variables to continue the work. We must therefore drop only one of them, and we need some rule for breaking the tie. One simple rule is to drop that one of the tied variables having the smallest subscript. Another acceptable rule is to drop the tied variable whose subscript turns up first in a table of random numbers. Both of these rules are, of course, arbitrary, but they have the desired effect of permitting the work to continue in the standard way.

When any such rule is applied it may be found that at the next stage of iteration the θ of Eq. (4-30) equals zero, so that no improvement in the objective function is possible. Indeed, zero improvement may occur for several iterations before the system breaks out of deadlock and starts to move ahead once more. But the breakout will eventually occur unless the process of iteration leads us back to some basic set which had turned up before, thus leading us around in a circle.

Recycling, as the return to a previously rejected basis is called, is a theoretical possibility which has never been observed in a practical problem. Even this theoretical danger can be avoided, but a more elaborate rule for breaking the tie would be needed to do so, and we shall not go into this refinement.[1]

4-12. THE COMPLETE DESCRIPTION METHOD: INTRODUCTION

The simplex method is only one of four or five methods which have been proposed for solving linear-programming problems. Although it is the

[1] Complete discussions of the degeneracy problem can be found in G. B. Dantzig, A. Orden, and P. Wolfe, "The Generalized Simplex Method for Minimizing a Linear Form under Linear Inequality Restraints," *Pacific Journal of Mathematics*, 5:183–195 (June, 1955), and in A. Charnes, W. W. Cooper, and A. Henderson, *Introduction to Linear Programming*, pp. 62–67, John Wiley & Sons, Inc., New York, 1953.

most frequently used method and appears to be the most efficient method in many instances, the comparative merits of the procedures have not yet been definitely established. In order not to tie the problem of linear programming to any single approach, we now describe an alternative procedure, the "complete-description method."

Refer now to the original definition of a linear-programming problem [Eqs. (4-4) to (4-6)]. All the difficulties of the problem arise from the fact that Eqs. (4-4) incorporate so many restrictions, all of which have to be satisfied. In the case $m = 1$, where there is only one restriction, the problem is very easy. We could then try out, for example, the solution $x_1 = c_1/a_{11}$, $x_2 = x_3 = \cdots = x_n = 0$, in which case, $Z = P_1c_1/a_{11}$. Then we could try out $x_2 = c_1/a_{12}$, $x_1 = x_3 = \cdots = x_n = 0$, so that $Z = P_2c_1/a_{12}$. We could go through the whole list of n variables in this way and finally try the solution $x_1 = x_2 = \cdots = x_n = 0$, for which $Z = 0$. Then we could throw out all the trials in which the x value turned out to be negative. In this way we would obtain a complete listing of all the basic feasible solutions, i.e., a complete description of the problem. We know already that if there were an optimal solution, it would be included in this list. We would only have to look down the list and pick out the solution or solutions corresponding to the largest value of Z to solve the problem.

If $m > 1$, things are not so easy. We can use the insight gained by considering the case $m = 1$ to conceive of the following strategy: Consider the first restriction alone and find all the basic feasible solutions which satisfy it. Then consider the second restriction and find all the basic feasible solutions which satisfy it as well as the first restriction. Continue in this manner, taking up all m restrictions one at a time. The result will be a complete description of the permissible ranges of the variables, and if there are any optimal solutions, the list will include some of them. The final step is to compute the value of Z corresponding to each solution in the list and select those solutions which have the maximum value of Z.

4-13. THE COMPLETE DESCRIPTION METHOD: BASIC ALGEBRA

The essential issue in the complete description method is as follows: Suppose that we have considered $k < m$ restrictions and have compiled a set of solutions which satisfy all of them. We now have to take up the $(k + 1)$st restriction. The problem is to determine which solutions should be deleted from the list and what new solutions should be added to it.

Our standard is that, after considering each restriction, we shall want a tentative list of solutions which includes all the basic feasible solutions to all the restrictions thus far considered. In order to be able to recognize basic feasible solutions in this context we make a distinction. Consider any possible solution, say $x'_1, x'_2, \ldots, x'_n$, in relation to any one restriction, say,

$$a_{i1}x_1 + a_{i2}x_2 + \cdots + a_{in}x_n \leq c_i$$

If this solution satisfies the restriction, then either

$$a_{i1}x'_1 + a_{i2}x'_2 + \cdots + a_{in}x'_n = c_i$$

or

$$a_{i1}x'_1 + a_{i2}x'_2 + \cdots + a_{in}x'_n < c_i$$

If the first relationship holds, we say that $x'_1, \ldots, x'_n$ is an "extreme" solution to the ith restriction; otherwise it is an "internal" restriction. The extreme solutions are the important ones, for they indicate the limits on the permissible values of the variables.

Let us now think of taking up the restrictions one at a time. We have already seen how to get extreme solutions to the first restriction considered in isolation. They have the form

$$x_i = \frac{c_1}{a_{1i}}, x_j = 0 \qquad j \neq i, i = 1, \ldots, n$$

Note that each of these solutions has only one nonzero element. Such solutions will be referred to below as corner solutions.

Next consider the first two restrictions simultaneously. Any solution which is an extreme solution to both of them must satisfy the following two equations:

$$a_{11}x_1 + \cdots + a_{1n}x_n = c_1$$
$$a_{21}x_1 + \cdots + a_{2n}x_n = c_2$$

We know from elementary algebra that if a pair of equations like this can be satisfied at all, it can be satisfied by two unknowns, but not, usually, by fewer. Hence we shall call a solution $x'_1, \ldots, x'_n$ a basic solution to the first two restraints if it is an extreme solution to both restraints and has no more than two nonzero components.

In general we shall call any solution a basic solution to k restraints if it is an extreme solution to all k restraints and involves no more than k nonzero components. Since basic solutions, in this sense, mark out the boundary of the permissible ranges of the variables, we are interested in this type of solution and only in this type.

We are now ready to consider the procedure for revising the list of basic solutions when k restrictions have been considered and the $(k + 1)$st restriction is to be taken up.

To be more concrete, consider the problem of maximizing

$$Z = P_1x_1 + P_2x_2 + \cdots + P_5x_5$$

subject to the restrictions

$$\begin{gathered}
a_{11}x_1 + a_{12}x_2 + a_{13}x_3 + a_{14}x_4 + a_{15}x_5 \le c_1 \\
a_{21}x_1 + a_{22}x_2 + a_{23}x_3 + a_{24}x_4 + a_{25}x_5 \le c_2 \\
a_{31}x_1 + a_{32}x_2 + a_{33}x_3 + a_{34}x_4 + a_{35}x_5 \le c_3 \\
a_{41}x_1 + a_{42}x_2 + a_{43}x_3 + a_{44}x_4 + a_{45}x_5 \le c_4 \\
x_1 \ge 0, x_2 \ge 0, \ldots, x_5 \ge 0
\end{gathered}$$

Suppose also that we have a set of solutions that satisfy the first two restrictions and want to take into account the third restriction, namely,

$$a_{31}x_1 + a_{32}x_2 + a_{33}x_3 + a_{34}x_4 + a_{35}x_5 \le c_3$$

It is convenient to write this restriction in the form

$$d^{(3)} = a_{31}x_1 + a_{32}x_2 + a_{33}x_3 + a_{34}x_4 + a_{35}x_5 - c_3 \le 0 \tag{4-42}$$

Now consider any solution in the list, say $x_1', x_2', x_3', x_4', x_5'$. Substitute these values in Eq. (4-42). If the result is positive, that solution does not satisfy the third restraint and should be dropped from the list. If the result is zero or negative, the point should be retained. Thus we determine the deletions.

The additions are of two types. First, suppose

$$x_2 = x_3 = \cdots = x_5 = 0$$

Then from Eq. (4-42), $x_1 = c_3/a_{31}$, $x_2 = \cdots = x_5 = 0$ satisfies the third restraint. If, moreover, the fraction is nonnegative and if it is not greater than c_2/a_{21} and c_1/a_{11}, the corresponding fractions derived from the first two restraints, then this solution is feasible and satisfies the first three restraints and should be added to the list. In a similar manner we consider $x_1 = 0$, $x_2 = c_3/a_{32}$, $x_3 = x_4 = x_5 = 0$, etc. These solutions, in which only one variable is different from zero, are "corner solutions," as we mentioned above.

The other type of addition we shall call "vertex solutions." Suppose that there were a pair of solutions to the first two restrictions, one of which gave rise to a positive and the other to a negative value of $d^{(3)}$; i.e., denoting the two solutions by $x_1', \ldots, x_5'$ and $x_1'', \ldots, x_5''$, suppose that

$$\begin{aligned}
d' &= a_{31}x_1' + a_{32}x_2' + a_{33}x_3' + a_{34}x_4' + a_{35}x_5' - c_3 > 0 \\
d'' &= a_{31}x_1'' + a_{32}x_2'' + a_{33}x_3'' + a_{34}x_4'' + a_{35}x_5'' - c_3 < 0
\end{aligned}$$

Multiply the first of these by $-d''$ (which will be a positive number) and the second by d' and add:

$$\begin{aligned}
0 = a_{31}(d'x_1'' - d''x_1') + a_{32}(d'x_2'' - d''x_2') + a_{33}(d'x_3'' - d''x_2') \\
+ a_{34}(d'x_4'' - d''x_4') + a_{35}(d'x_5'' - d''x_5') - c_3(d' - d'')
\end{aligned}$$

The quantities in parentheses will all be nonnegative since none of the quantities which enter into them are negative; for example,

$$d'x_1'' - d''x_1' = d'x_1'' + (-d'')x_1' \geq 0$$

Finally, divide both sides by $d' - d''$ and transpose:

$$a_{31}\frac{d'x_1'' - d''x_1'}{d' - d''} + \cdots + a_{35}\frac{d'x_5'' - d''x_5'}{d' - d''} = c_3 \tag{4-43}$$

Let

$$x_1^* = \frac{d'x_1'' - d''x_1'}{d' - d''}, \ldots, x_5^* = \frac{d'x_5'' - d''x_5'}{d' - d''}$$

We now consider whether the solution $x_1^*, \ldots, x_5^*$ should be added to the list of basic solutions. We note that it is a solution. In the first place, none of $x_1^*, \ldots, x_5^*$ is negative. In the second place, it satisfies the first two restrictions. To see this, consider the first restriction. By assumption,

$$\begin{aligned} a_{11}x_1' + \cdots + a_{15}x_5' &\leq c_1 \\ a_{11}x_1'' + \cdots + a_{15}x_5'' &\leq c_1 \end{aligned} \tag{4-44}$$

Multiply the first of these by $-d''$, multiply the second by d', add, and divide by $d' - d''$. The result is

$$a_{11}x_1^* + \cdots + a_{15}x_5^* \leq c_1$$

A similar argument applies to the second restriction. But although $x_1^*, \ldots, x_5^*$ will be a solution, it may not be basic. To tell whether $x_1^*, \ldots, x_5^*$ will be basic, we notice three things. First, $x_1^*, \ldots, x_5^*$ will be an extreme solution to the third restraint. Second, a component of $x_1^*, \ldots, x_5^*$ will be zero only if the corresponding component is zero in both the solutions $x_1', \ldots, x_5'$ and $x_1'', \ldots, x_5''$ which combine to form it. Third, $x_1^*, \ldots, x_5^*$ will be an extreme solution to any restriction previous to the third if and only if both solutions which combine to form it are extreme solutions to that restriction. [This can be proved simply by dropping the inequality signs from Eqs. (4-44).] Thus the number of nonzero components in $x_1^*, \ldots, x_5^*$ will be equal to the number of components which have a nonzero value in either $x_1', \ldots, x_5'$ or $x_1'', \ldots, x_5''$. Further, $x_1^*, \ldots, x_5^*$ will be an extreme solution to all restraints to which $x_1', \ldots, x_5'$ and $x_1'', \ldots, x_5''$ are extreme solutions, plus 1. If the number of restraints to which $x_1^*, \ldots, x_5^*$ is an extreme solution is as great as the number of nonzero components in $x_1^*, \ldots, x_5^*$, or greater, then $x_1^*, \ldots, x_5^*$ is a basic solution and belongs in the list.

In this manner the list of basic solutions is modified by considering the third restriction, and we are ready to go on to the fourth. When all the restrictions have been taken into account we compute the value of Z for each solution in the final list and determine the optimum by inspection.

To complete the description of the method we need only state that it is started by calculating the corner points for the first restraint.

4-14. COMPLETE DESCRIPTION METHOD: EXAMPLE

Let us illustrate the complete-description method by applying it to the problem we have just solved by the simplex method. The solution is carried out in Table 4-3. The three restrictions, written in the form of

TABLE 4-3. WORKSHEET FOR SOLUTION OF TABLE 4-1 BY COMPLETE DESCRIPTION

Line	Coordinates				Constant	Extremities							$d^{(2)}$	$d^{(3)}$	Notes
	x_1	x_2	x_3	x_4	Z	R_1	R_2	R_3	O_1	O_2	O_3	O_4			
R-1	100	100	100	100	1500										Restraints
R-2	7	5	3	2	100										
R-3	3	5	10	15	100										
Z	60	60	90	90											Objective
(S-1)	15				......	x				x	x	x	5		Corner points of R-1
S-2		15			900	x			x		x	x	−25	−25	
(S-3)			15		......	x			x	x		x	−55	50	
(S-4)				15	......	x			x	x	x		−70	125	
S-5	12 1/2	2 1/2			900	x	x				x	x		−50	S-1 and S-2
S-6	13 3/4		1 1/4		937 1/2	x	x			x		x		−46 1/4	S-1 and S-3
S-7	14			1	930	x	x			x	x			−43	S-1 and S-4
S-8	14 3/7				857 1/7		x			x	x	x		−57 1/7	Corner of R-2
S-9		10	5		1050	x		x	x	x		x			S-3 and S-2
S-10	7 1/7		7 6/7		1135 5/7	x		x		x		x			S-3 and S-6
S-11		12 1/2		2 1/2	975	x		x	x		x				S-4 and S-2
S-12	10 5/12			4 7/12	1037 1/2	x		x		x	x				S-4 and S-7
S-13			10		900			x	x	x		x			Corner of R_3
S-14				6 2/3	600			x	x	x	x				

SOLUTION: S-10.

NOTE: Blanks in the coordinate columns indicate zeros.

detached coefficients, are recorded in the first three lines. The coefficients of the function to be maximized are written in the fourth line. The remaining 14 lines, which we shall discuss immediately, constitute a list of solutions.

The columns headed x_1, x_2, x_3, and x_4, obviously, give the four coordinates of each solution listed. It should be noted that slack variables are not required for the complete-description method. The columns headed $d^{(2)}$ and $d^{(3)}$ give the results of applying formula (4-42) and the analogous formula for the second restraint to each solution. Note that it is not necessary to fill in these columns completely.

The remaining columns are used for bookkeeping. An x in the column headed R_i signifies that the solution on whose line the x occurs is an extreme solution to the ith restraint. Similarly an x in the column headed 0_j indicates that $x_j = 0$ in the solution on whose line it appears. These columns are useful, as we shall see, in determining whether combined solutions are basic. The "notes" column is used to record the derivation of each solution.

We can now trace through the table. Lines S-1 to S-4 (i.e., solutions 1 to 4) are the corner points of the first restraint. Each of them is an extreme solution of the first restraint—hence the x's in the R_1 column, and for each (since they are corner solutions), three of the variables are zero.

We are now ready to consider the second restraint, R_2. For each of the four points we calculate $d^{(2)}$ by the formula

$$d^{(2)} = 7x_1 + 5x_2 + 3x_3 + 2x_4 - 100$$

We note that S-1 and S-2, S-1 and S-3, and S-1 and S-4 form pairs whose $d^{(2)}$ values have opposite signs. Now look, for example, at the pair S-1 and S-2. The columns $0_1, \ldots, 0_4$ tell us that x_3 and x_4 are zero for both members of the pair. They will therefore be zero if we average the pair so as to satisfy R_2 by applying the analog to formula (4-43). This average will therefore have only two nonzero components, x_1 and x_2. It will satisfy two restraints, R_1 and R_2. It will therefore be a basic solution and should be added to the list. It is found and entered as solution S-5, and x's are recorded in the columns for which it is an extreme solution. Solutions S-6 and S-7 arise from the other two pairs.

Solution S-8 is the only corner solution of R_2 which also satisfies R_1, as is shown by the fact that the coordinate in the x_1 column is lower than the coordinate of the corresponding corner solution of R_1. Finally, we delete S-1, indicated by placing a parenthesis around its designation, because it had a positive value in the $d^{(2)}$ column.

We now go on to restraint R_3. We calculate $d^{(3)}$ for all solutions in the revised list and examine all pairs with opposite signs in the $d^{(3)}$ column. The pair S-2 and S-3, for example, if averaged, will have a nonzero x_2 coordinate and a nonzero x_3 coordinate, since only one member of the pair has an x in each of the columns 0_2 and 0_3. It will be an extreme solution to R_1 (since both members of the pair have x's in that column) and to R_3

by virtue of formula (4.43). Thus the pair S-2 and S-3 gives rise to a solution which is an extreme solution to two restraints and involves only two nonzero coordinates. This solution is therefore basic and appears as S-9. The analysis of the pair S-3 and S-5, however, has a different result. This result is an extreme solution to two restraints, R_1 and R_3. But it has only one zero component, namely, x_4. The number of nonzero components is therefore greater than the number of restraints to which it is an extreme solution. It is therefore not basic and is not computed.

In this manner, the complete list of basic solutions to all the restraints is built up. The final list includes eleven solutions, S-2 to S-14, with the exception of S-3 and S-4.[1] Finally Z is computed for each solution in the final list. This is done for any solution by multiplying each of its coordinates by the number in the same column of the Z line and adding the results. The last step is to pick out the solution with the highest Z value. It is S-10.

It is instructive to compare this solution with the simplex solution to the same problem, Table 4-2. In Table 4-2 we calculated only three of the basic solutions, while in Table 4-3 we found all eleven basic solutions plus three other solutions which were discarded in the course of the work. The first trial solution of Table 4-2 appears as S-2 in Table 4-3. Note that the values of the two slack variables in Table 4-2 are the negatives of $d^{(2)}$ and $d^{(3)}$ on the S-2 line of Table 4-3. Similarly, the second trial solution of Table 4-3 appears as S-9 of Table 4-3. If $d^{(2)}$ were computed for S-9 (which did not turn out to be necessary), it would be found to be -35, the negative of the disposal value in Table 4-2. S-10 is, of course, the final trial solution of Table 4-2, and its $d^{(2)}$ value, if computed, would check.

The complete-description method thus requires the calculation of many more solutions than the simplex method, which moves steadily toward the optimum. The simplex method, however, requires much more calculation to obtain each point. In the present example, the complete-description method appeared to be the more convenient of the two. The weight of opinion, however, is currently that the simplex method will usually be superior because, if there are very many restrictions, the number of solutions required by the complete-description method can build up alarmingly.

4-15. DUALISM

We conclude this algebraic discussion by considering a rather surprising aspect of linear programming which is important both mathemati-

[1] Solutions S-3 and S-4 are deleted because they give rise to positive values of $d^{(3)}$.

cally and economically. Consider the linear-programming problem of finding $x_1, \ldots, x_4$ such that

$$Z = P_1x_1 + \cdots + P_4x_4 \tag{4-45}$$

is as great as possible subject to the conditions

$$\begin{aligned} a_{11}x_1 + \cdots + a_{14}x_4 &\leq c_1 \\ a_{21}x_1 + \cdots + a_{24}x_4 &\leq c_2 \\ a_{31}x_1 + \cdots + a_{34}x_4 &\leq c_3 \end{aligned} \tag{4-46}$$

$$x_1 \geq 0, \ldots, x_4 \geq 0 \tag{4-47}$$

The same data can be used to construct a different problem, as follows:
Find u_1, u_2, u_3 such that

$$w = c_1u_1 + c_2u_2 + c_3u_3 \tag{4-48}$$

is as small as possible subject to the conditions

$$\begin{aligned} a_{11}u_1 + a_{21}u_2 + a_{31}u_3 &\geq P_1 \\ a_{12}u_1 + a_{22}u_2 + a_{32}u_3 &\geq P_2 \\ a_{13}u_1 + a_{23}u_2 + a_{33}u_3 &\geq P_3 \\ a_{14}u_1 + a_{24}u_2 + a_{34}u_3 &\geq P_4 \end{aligned} \tag{4-49}$$

$$u_1 \geq 0, u_2 \geq 0, u_3 \geq 0 \tag{4-50}$$

In deriving the second problem from the first we have interchanged the roles of $P_1, \ldots, P_4$ with those of c_1, c_2, c_3, used each row of coefficients of the first problem as a column of coefficients in the second, changed the sense of the inequalities in the first set of restrictions, and substituted a minimum for a maximum problem. Any two linear-programming problems which are related in this manner are known as duals of each other. It is clear that every linear-programming problem has a dual.

Although this chapter is not devoted to applications, it may help to mention two interpretations of duality in order to indicate why this phenomenon is of interest. The first occurs in the theory of games. A two-person zero-sum game may be solved by the methods of linear programming. If this is done, the problem of finding the best strategy for one player is the dual of the problem of finding the best strategy for the other. The second application occurs when linear programming is used to find an optimal allocation of scarce resources. In this case, the problem of imputing values to the resources is the dual of the allocation problem. In short, in many instances duals are significant rather than artificial constructs.

The principal features of the relations of a linear-programming problem to its dual are these: First, if a linear-programming problem has a solution, so does its dual. Second, the maximum value of Z corresponding to the direct problem equals the minimum value of w corresponding to the dual. Third, if the direct problem has been solved, then the solution of the dual is greatly facilitated, and vice versa.

In order to establish these facts, we shall assume that the direct problem [Eqs. (4-45) to (4-47)] has been solved. Then we shall give a formula for computing some numbers, u_1, u_2, u_3, from the solution to the direct problem. Finally we shall show that these numbers are the solution to the dual.

In order to formulate the solution to the direct problem, introduce slack variables x_5, x_6, x_7 and write Eqs. (4-46) in the form

$$\begin{aligned} a_{11}x_1 + \cdots + a_{14}x_4 + a_{15}x_5 + a_{16}x_6 + a_{17}x_7 &= c_1 \\ a_{21}x_1 + \cdots + a_{24}x_4 + a_{25}x_5 + a_{26}x_6 + a_{27}x_7 &= c_2 \\ a_{31}x_1 + \cdots + a_{34}x_4 + a_{35}x_5 + a_{36}x_6 + a_{37}x_7 &= c_3 \end{aligned}$$

where $a_{15} = a_{26} = a_{37} = 1$ and all the other new coefficients are zero. Equation (4-45) then becomes

$$Z = P_1x_1 + \cdots + P_4x_4 + 0x_5 + 0x_6 + 0x_7$$

and three new lines are added to Eqs. (4-49), namely,

$$\begin{aligned} a_{15}u_1 + a_{25}u_2 + a_{35}u_3 &\geq P_5 = 0 \\ a_{16}u_1 + a_{26}u_2 + a_{36}u_3 &\geq P_6 = 0 \\ a_{17}u_1 + a_{27}u_2 + a_{37}u_3 &\geq P_7 = 0 \end{aligned} \tag{4-51}$$

Note that these three lines really amount to $u_1 \geq 0$, $u_2 \geq 0$, $u_3 \geq 0$ and therefore add no new restrictions to the problem. Then, assuming that no degeneracy occurs, precisely three of the x's in the optimal basic solution will be positive and all the rest will be zero. For convenience we assume that the first three x's are the positive ones. If one or more slack variables is included in the optimal solution, this assumption would require a renumbering of the variables but would cause no other difficulty. We are thus assuming that the optimal solution, after suitable renumbering of variables, satisfies

$$\begin{aligned} a_{11}x_1 + a_{12}x_2 + a_{13}x_3 &= c_1 \\ a_{21}x_1 + a_{22}x_2 + a_{23}x_3 &= c_2 \\ a_{31}x_1 + a_{32}x_2 + a_{33}x_3 &= c_3 \end{aligned} \tag{4-52}$$

We now select as u_1, u_2, u_3, our proposed solution to the dual problem, the numbers resulting from the solution of

$$\begin{aligned} a_{11}u_1 + a_{21}u_2 + a_{31}u_3 &= P_1 \\ a_{12}u_1 + a_{22}u_2 + a_{32}u_3 &= P_2 \\ a_{13}u_1 + a_{23}u_2 + a_{33}u_3 &= P_3 \end{aligned} \tag{4-53}$$

The idea behind this selection is that we look back at Eqs. (4-49) and make equations out of those inequalities that correspond to nonzero x's in the basic solution to the direct problem. Note also that if, after renumbering, one of the nonzero x's is, say, the disposal variable for the second constraint, then the second equation of (4-53) will read simply $u_2 = 0$. Since the dual variables frequently represent shadow prices or valuations on limiting resources, this remark has the following interpretation: Any resource that has excess capacity in an optimal solution must have a zero shadow price. We now show that the solutions to Eqs. (4-53) satisfy the restrictions (4-49) and (4-51). Since Eqs. (4-53) are the same as the first three equalities of (4-49), those are taken care of. Turn now to the fourth inequality of (4-49). Equations (4-24) can be applied to its left-hand members, yielding

$$(a_{11}y_1 + a_{12}y_2 + a_{13}y_3)u_1 + (a_{21}y_1 + a_{22}y_2 + a_{23}y_3)u_2 + (a_{31}y_1 + a_{32}y_2 + a_{33}y_3)u_3 \geq P_4$$

Rearranging terms we obtain

$$(a_{11}u_1 + a_{21}u_2 + a_{31}u_3)y_1 + (a_{12}u_1 + a_{22}u_2 + a_{32}u_3)y_2 + (a_{13}u_1 + a_{23}u_2 + a_{33}u_3)y_3 \geq P_4$$

These three parentheses are simply the left-hand sides of Eqs. (4-53). So the fourth restriction of set (4-49) can be written

$$P_1y_1 + P_2y_2 + P_3y_3 \geq P_4$$

or, using Eq. (4-31),

$$\bar{P}_4 \geq P_4$$

This, of course, is simply the simplex criterion, and since our hypothesis was that x_1, x_2, x_3 formed an optimal solution, this inequality holds. In the same manner the remaining conditions of set (4-49), if there are any, and of set (4-51) can be established. Thus the values u_1, u_2, u_3, defined by Eqs. (4-53), satisfy Eqs. (4-49) and (4-51) of the dual problem. It remains to show that they satisfy Eq. (4-48). To do this, let $\bar{u}_1$, $\bar{u}_2$, $\bar{u}_3$ denote any set of numbers that satisfies Eqs. (4-49) and (4-51). If we substitute these numbers into the first three equations of Eqs. (4-49) we obtain

$$\begin{aligned} a_{11}\bar{u}_1 + a_{21}\bar{u}_2 + a_{31}\bar{u}_3 &\geq P_1 \\ a_{12}\bar{u}_1 + a_{22}\bar{u}_2 + a_{32}\bar{u}_3 &\geq P_2 \\ a_{13}\bar{u}_1 + a_{23}\bar{u}_2 + a_{33}\bar{u}_3 &\geq P_3 \end{aligned} \qquad (4\text{-}54)$$

Multiply the first of these by x_1, the second by x_2, the third by x_3, and add. Applying Eqs. (4-52) we obtain

$$c_1\bar{u}_1 + c_2\bar{u}_2 + c_3\bar{u}_3 \geq x_1P_1 + x_2P_2 + x_3P_3 \qquad (4\text{-}55)$$

Comparing this with Eq. (4-45) we see that the right-hand side is the maximum value of Z, which corresponds to the solution of the direct problem. The left-hand side is the value of w corresponding to $\bar{u}_1$, $\bar{u}_2$, $\bar{u}_3$. Hence Eq. (4-55) states that the value of w corresponding to any $\bar{u}_1$, $\bar{u}_2$, $\bar{u}_3$ which satisfies Eqs. (4-49) and (4-50) will be at least as great as the Z corresponding to the solution of the direct problem. From Eqs. (4-53) we note that if u_1, u_2, u_3 are substituted for $\bar{u}_1$, $\bar{u}_2$, $\bar{u}_3$, the inequality sign will be replaced by an equality. Thus the values defined by Eqs. (4-53) correspond to the minimum possible value of w as well as satisfying Eqs. (4-49) and (4-51). They are therefore an optimal solution to the dual problem, as we set out to prove. Simultaneously, we have proved that the minimum value of w equals the maximum value of Z.

The dualism feature has an interesting consequence that we shall find useful in our discussion of general equilibrium in Chap. 13. Consider any linear-programming problem. We have already seen that it is perfectly possible to have a feasible solution even though no optimal feasible solution may exist. This may happen because it is possible that no matter which feasible solution we obtain, there may exist some other feasible solution that gives rise to a larger value of the objective function. In such a case the objective function can be made indefinitely large, and there is no optimum. This situation was discussed in Sec. 4-6, especially on page 77.

Of course, the same situation exists with respect to the dual. The existence of a feasible solution does not guarantee the existence of an optimal feasible solution, i.e., one which makes w so small that no other feasible solution makes it smaller. But if a linear-programming problem and its dual both have feasible solutions, the situation is otherwise. In this case, as we shall now show, both the problem and its dual have optimal solutions. The proof follows at once from the fact that, as we have just seen in Eq. (4-55), any feasible solution to the dual gives an upper limit to the objective function of the direct problem and any feasible solution to the direct problem provides a lower limit to the objective function of the dual. Hence the objective function of the direct problem cannot be made infinitely large without violating the restraints, and, since we have assumed at least one feasible solution, there is a maximizing feasible solution. The proof that there is a minimizing feasible solution to the dual is similar.

EXERCISES

4-1. Solve the diet problem of Chap. 2 by the simplex method.

4-2. Solve the diet problem by the complete-description method.

4-3. Suppose that a market research firm wishes to meet the quotas given in the first column of the following table:

DATA FOR MARKET RESEARCH PROBLEM: QUOTAS AND PROBABILITIES OF RESPONSE

Type of household	Quota	Probability of response to calls		
		Morning	Afternoon	Evening
1 person.	50	0.1	0.1	0.5
Married, no children.	100	0.5	0.4	0.7
Married, children.	150	0.75	0.6	0.9

How should calls be distributed among the three types of family by the three times of day so as to be expected to fulfill the quotas with a minimum of calls when the contracts with enumerators impose the following requirements: (1) the number of morning calls must not exceed the number of afternoon calls; (2) the number of evening calls must not exceed half the number of afternoon calls? Partial answer: The minimum number of calls that satisfies these conditions is 540.

5

The Transportation Problem

5-1. A SIMPLE CASE

One special case of linear programming is of particular interest both because of its economic applications and because of its computational simplicity. This is the "transportation problem," originated by F. L. Hitchcock[1] and solved by him several years before the general concept of linear programming was formulated. This problem has numerous economic and business applications which have nothing to do with transportation, but its designation, which stems from its original formulation, is still used. The most remarkable mathematical characteristic of this problem is that whereas linear-programming problems typically require such elaborate calculations that high-speed computing machinery is needed, the transportation problem is frequently best solved by nothing more complicated than pencil and paper.

The essence of the problem is conveyed by a simple example. Suppose that a manufacturer has three factories, which we shall call A, B, and C, and supplies five localities, which we shall call 1, 2, 3, 4, 5. Suppose also that the costs of shipping a ton of product from each factory to each locality are given, that the capacities of the factories are known, and that the number of tons to be supplied to each locality is fixed. The data that we shall use in this example are given in Table 5-1. The problem is to find a pattern of shipments that involves the least possible total transportation cost consistent with these conditions. This calls for a decision regarding the number of tons (if any) to be shipped by each plant to each locality. Clearly the shipments planned from each plant must not exceed its capacity, while the shipments planned to each locality must equal its requirements. There are many possible routings that meet these conditions; the transportation problem is the problem of finding one that does so with the least possible total transportation cost.

According to Table 5-1 the capacities of the three factories are 50, 100,

[1] "The Distribution of a Product from Several Sources to Numerous Localities," *Journal of Mathematics and Physics*, **20**:224–230 (1941).

and 150 tons per month. The quantities to be shipped to each locality are given in the last column. Transportation rates are given in the body of the table. Thus, the cost of shipping a ton of product from Factory C to locality 1 is \$30. An important characteristic of this example and of many transportation problems (in the technical sense of the phrase) is that the total capacity of all the plants just equals the total requirements of all the consumers.

TABLE 5-1. TRANSPORTATION COSTS PER TON CAPACITIES AND REQUIREMENTS

	Factory			Requirement
	A	*B*	*C*	
Locality: 1	\$10	\$20	\$30	25
2	15	40	35	115
3	20	15	40	60
4	20	30	55	30
5	40	30	25	70
Capacity.....	50	100	150	300

A good start on solving this problem can be made by setting it up in the usual linear-programming form. Let x_{ij} denote the nonnegative number of tons shipped from Factory i to Locality j, and let c_{ij} denote the transportation cost per ton for shipments between these termini. Then the total transportation cost, the sum to be minimized, is

$$T = \Sigma_i \Sigma_j c_{ij} x_{ij} \qquad i = A, B, C; j = 1, 2, 3, 4, 5 \tag{5-1}$$

To achieve this minimum we can select any values of x_{ij} which satisfy three sorts of restriction. First, the shipments planned for each factory must not exceed the capacity of that factory. Now the total requirements which must be met by the three factories are just equal to the total capacity of the three factories. Thus, if the shipments planned for any factory were less than its capacity, the shipments planned for some other factory would have to exceed capacity. In consequence, a stronger restriction must be met than the one which we started with, namely, the shipments planned for each factory must equal the capacity of that factory. This set of restrictions can be written

$$\Sigma_j x_{ij} = k_i \qquad i = A, B, C; k_i = \text{capacity of factory } i \tag{5-2}$$

The second restriction is that the total shipments to each locality must equal the requirements of that locality. Symbolically,

$$\Sigma_i x_{ij} = r_j \qquad j = 1, 2, 3, 4, 5; r_j = \text{requirements of locality } j \tag{5-3}$$

The third restriction is that the shipment variables cannot be negative, i.e.,

$$x_{ij} \geq 0 \qquad i = A, B, C; j = 1, 2, 3, 4, 5 \tag{5-4}$$

These equations suggest why the computations required by the transportation problem are exceptionally simple: all the choice variables enter the restraining equations with the same coefficient, namely, unity.

Considering this as an ordinary linear-programming problem, we have to minimize a weighted sum T of 15 choice variables x_{ij}. There are 15 variables because i takes on three values, j takes on five values, and all combinations are permissible. These 15 variables are apparently subject to 8 restraining equations, one for each of three factories plus one for each of five localities. Actually, however, one of the restraining equations is redundant. For suppose that any set of x_{ij} satisfies all the capacity restrictions and meets the requirements of the first four localities. Then, since the total capacity is equal to the total requirements, the volume of shipments corresponding to that set of x_{ij} will be equal to the total volume of requirements of all five localities. Furthermore, since the shipments going to the first four localities are equal to their requirements, the volume of shipments left over must be just sufficient to meet the requirements of the last locality. In setting up the problem, then, we could leave out one of the restrictions, any one we please, except for considerations of symmetry.

Effectively, then, there are seven rather than eight restraining equations. We now apply the general principle of linear programming which tells us that an optimum program exists in which the number of activities at positive levels is no greater than the number of restraining equations, in this case seven.

To state this conclusion in general form, assume a transportation problem in which there are $m > 1$ points of origin and $n > 1$ destinations. We define an activity to be the making of a shipment from a specific point of origin to a specific destination. There would then be mn activities to be considered. The levels of these activities would have to satisfy m restrictions relating to origins and n restrictions relating to destinations, a total of $m + n$. But if any $m + n - 1$ restrictions are satisfied, the remaining restriction would have to be satisfied also, by an argument similar to the one we gave for the special case. Thus only $m + n - 1$ restrictions will be effective, and a minimum-cost set of routes will exist in which only $m + n - 1$ of the activities are used at positive levels.

We now return to the numerical example. The strategy of solution is the same as that used in ordinary linear programming. A "basic" solution is a set of routes in which the number of routes used at positive levels is equal to the number of restraining equations, i.e., the number of origins plus destinations less 1. We start with any basic solution and then, by an iterative routine, derive from it successively better basic solutions until an optimal one is obtained. The criterion for improvement is analogous to the ordinary simplex criterion. In order to see how any proposed solution can be improved, we examine all the routes not used in that solution and see whether any of these is more economical than its indirect equivalent made up of routes used in the solution. We shall see that this process of test and improvement is very much easier than the procedure given for solving a general linear-programming problem.

The first step is to obtain a starting basic solution. This can be done at sight. Looking at the Factory *A* column of Table 5-1 we see that unit transportation costs are least if Factory *A* ships to Locality 1. Therefore we provisionally choose to have Factory *A* supply the full requirement of Locality 1, or 25 tons. Since 25 tons of Factory *A*'s capacity remains, we ship it to the next cheapest locality, Locality 2. After this allocation, Locality 2 requires 90 more tons, and comparing costs in row 2 we see that it can be supplied most economically by Factory *C*. Continuing in this manner, the provisional allocation shown in Table 5-2 results. This allocation satisfies all the restrictions, but does not necessarily correspond to minimal transportation costs.

TABLE 5-2. INITIAL SHIPMENT PLAN, TONS

Locality	Factory		
	A	*B*	*C*
1	25		
2	25	...	90
3	...	60	
4	...	...	30
5	...	40	30

Now we must test each of the routes not used in the initial shipping plan, i.e., each of the excluded activities, and for each determine whether the cost of shipping a ton along the route is cheaper than achieving the same result indirectly by means of the routes included in the initial shipping plan. If the direct cost of an excluded route is lower than the indirect cost of achieving its result, then total transportation cost can be reduced by introducing that route into the shipping plan and dropping one of the included routes. These observations are just a restatement of the familiar simplex criterion.

Conceptually the process of testing excluded routes consists of three stages: (1) determining the combination of included routes which accomplishes the same result as each of the excluded routes, (2) computing the cost of each of these combinations, (3) comparing the cost of using each combination with the cost of using the corresponding direct route. In practice the first two stages are performed simultaneously by means of the simple arithmetic that we now derive.

We denote a route or activity by writing the points of origin and destination in parentheses. Thus (B-1) denotes the activity of making a shipment from Factory B to Locality 1. We note from Table 5-2 that activity (B-1) does not occur in the initial solution. How can we express it in terms of the activities which do occur? A helpful, though somewhat artificial, way to phrase this question is: How can we make a shipment from Factory B to Locality 1 by following only routes which appear in the basic solution? Since the basic solution does not include the direct route from Factory B to Locality 1, our shipment would have to follow a zigzag path. In fact, tracing through Table 5-2, we see that the shipment would have to move from Factory B to Locality 5, from Locality 5 to Factory C, from Factory C to Locality 2, from Locality 2 to Factory A, and finally from Factory A to Locality 1. We have already introduced a symbol to denote a shipment from a factory to a locality; let us use the same symbol with a minus sign to indicate a shipment over the same route in the reverse direction. Then we can write the equivalence

$$(B\text{-}1) = (B\text{-}5) - (C\text{-}5) + (C\text{-}2) - (A\text{-}2) + (A\text{-}1)$$

This "equation" can be interpreted as follows: If we try to ship 1 ton from Factory B to Locality 1, some compensating adjustments must be made in the other routes. For example, Factory B's shipments to some other destination must be reduced by 1 ton (capacity restriction), and if (B-5) is reduced, shipments to Locality 5 from some other source must be increased (requirements restriction). If shipments from Factory C to Locality 5 are increased by 1 ton, Factory C's shipments elsewhere must be reduced by 1 ton, etc., until the chain of adjustments can be completed with a reduction in shipments to Locality 1. These compensating adjustments must take place over routes present in the initial plan for two reasons: no shipment not taking place can be reduced; and if we increase a shipment not taking place, we introduce *two* new routes into the plan instead of one. Thus the right-hand side above gives the feasible compensations that must be made to accommodate a 1-ton shipment from Factory B to Locality 1.[1] Clearly we could work out all the routes

[1] At this point the careful reader may want to know if such a scheme of compensating adjustments always exists for every excluded route, and if it is unique. For proof that the answers to both of these questions are affirmative, we refer the reader to G. B. Dantzig's basic paper, "Application of the Simplex Method to a Transportation Problem," in T. C. Koopmans (ed.), *Activity Analysis of Production and Allocation*, John Wiley & Sons, Inc., New York, 1951. Alternatively, we may note that this formula is simply a special case of the basis-shifting procedure of the simplex method and is covered by the discussion of that method in Chap. 4 above.

in this same way. This would be equivalent to solving the linear equations for expressing excluded activities in terms of included ones in ordinary linear programming, but computationally it is very much easier. Even this, however, is not necessary. It will be recalled that one of the main purposes of computing all the equivalent combinations in ordinary programming is to help compare the cost of performing an activity directly with the cost of performing it indirectly by means of the equivalent combination of included activities. In the transportation problem we can make this comparison of costs without actually going through the work of expressing the excluded activities in terms of the included ones.

The cost of a direct shipment over the route (B-1) is, from Table 5-1, \$20 per ton. We now need to find the cost of a shipment over the zigzag route. This is done by considering the five legs of the journey separately. A unit shipment from Factory B to Locality 5 costs \$30. The next leg of the journey is to ship from Locality 5 to Factory C. Now, the initial shipment plan included a shipment of 30 tons from Factory C to Locality 5. Rather than indulging in a cross-hauling between Factory C and Locality 5 it is more economical to accomplish the results of a shipment from Locality 5 to Factory C by reducing the volume of shipments from the factory to the locality. This, then, saves money. From Table 5-1 we see that it saves \$25 per ton. Incidentally, the fact that reverse shipments save money justifies the use of the minus sign in symbolizing them. Continuing in this manner for the other three legs of the journey, we find

$$\text{Cost of indirect shipment from Factory } B \text{ to Locality 1} \\ = \$30 - \$25 + \$35 - \$15 + \$10 = \$35$$

Comparing this result with Table 5-1 we see that if we restrict ourselves to routes contained in the initial shipment plan, a shipment from Factory A to Locality 1 (which is direct) costs \$10 per ton and a shipment from Factory B to Locality 1 (which is indirect) costs \$35, or \$25 more than a shipment from Factory A. Thus if we introduce a direct shipment of 1 ton over route (B-1) there is a transport cost of \$20. But the compensating adjustments yield a saving of \$35. Obviously, then, it is advantageous to introduce route (B-1). Each ton that can be shipped over this route will yield a net saving of \$15.

We now want to show that a shipment from Factory B to any locality costs \$25 more per ton than a shipment from Factory A to the same locality if the only routes used are those contained in the initial shipment plan. To see this we note that a shipment from Factory B to any locality, say j, can be achieved indirectly by shipping from Factory B to Locality 1 (this will be an indirect shipment, but that does not affect the argument), from Locality 1 to Factory A, and from Factory A to Locality j. In symbols,

$$(B\text{-}j) = (B\text{-}1) - (A\text{-}1) + (A\text{-}j)$$

To evaluate the cost of this, recall that we have already defined c_{ij} to denote the cost of a direct shipment from Factory i to Locality j. Let $\bar{c}_{ij}$ denote the cost of indirect shipment between these two points. Of course, if the route $(i\text{-}j)$ is included in the initial shipment plan, $\bar{c}_{ij} = c_{ij}$. Using these symbols, the formula for $(B\text{-}j)$ gives as the cost of that shipping route:

$$\bar{c}_{Bj} = \bar{c}_{B1} - \bar{c}_{A1} + \bar{c}_{Aj}$$

or

$$\bar{c}_{Bj} - \bar{c}_{Aj} = \bar{c}_{B1} - \bar{c}_{A1}$$

i.e., the difference between the cost of indirect shipment from Factory B to any locality and from Factory A to the same locality is the same constant, namely, $\bar{c}_{B1} - \bar{c}_{A1}$, for all localities.

By the same argument we could show that the same constancy of differences holds between any pair of factories. This fact makes it very easy to calculate the initial indirect costs of shipment between every factory and every locality. The calculation is made in Tables 5-3 and 5-4.

TABLE 5-3. INITIAL DIRECT AND INDIRECT COSTS, INCOMPLETE, PER TON

Locality	Factory		
	A	B	C
1	$10		
2	15	$40	$35
3	....	15	10
4	....	60	55
5	....	30	25

TABLE 5-4. INITIAL DIRECT AND INDIRECT COSTS, PER TON

Locality	Factory		
	A	B	C
1	$10	$35*	$30
2	15	40	35
3	−10	15	10
4	35*	60**	55
5	5	30	25

* Indirect costs exceed direct costs.
** Maximum excess of indirect over direct costs.

In Table 5-3, which is the first step in calculating a table of indirect costs, we have entered the transportation costs for every route which appears in the initial shipment plan, given in Table 5-2. We note that the cost of shipment from Factory *B* to Locality 5 is $5 greater than the cost of shipment from Factory *C* to the same locality. We have just seen that this same difference must hold for every locality. Thus the indirect cost of route (*B*-4) is $60, that of route (*C*-3) is $10, and that of route (*B*-2) is $40. These indirect costs have been entered in Table 5-4. At this stage we can see, without repeating the previous argument and simply by comparing columns *A* and *B* in the partially completed Table 5-4, that a shipment from Factory *B* to any locality costs $25 more than a shipment from Factory *A* to the same locality. We can now fill in the indirect shipping costs for routes (*A*-3), (*A*-4), and (*A*-5); in every case they are $25 less than the entries in the same line of column *B*. We can also fill in the indirect cost for route (*B*-1) and, finally, for route (*C*-1), which must be $20 greater than the cost for (*A*-1). Note that the indirect cost for route (*A*-3) is negative. This is not objectionable. Clearly, indirect costs are much more easily obtained in the transportation problem than in ordinary linear programming.

The procedure just described depended on the fact that the difference in indirect shipment cost between shipments starting in any two factories is the same in these two factories for all localities. It is also sometimes helpful to use the fact that the difference between cost of shipment to any two localities is the same for all factories. The proof of this fact is similar to the one which we have given.

Table 5-4 permits us to compare the direct cost of any shipment (all values in Table 5-1), c_{ij}, with the indirect cost, $\bar{c}_{ij}$. We have placed an asterisk next to every shipment from a particular factory to a particular locality for which the indirect cost is greater than the direct cost and a double asterisk next to the route for which this difference is greatest. We have now completed our examination of the initial routing.

The next stage is to derive a revised routing which includes the route marked with a double asterisk, i.e., route (*B*-4), since this is the revision that promises the greatest saving per ton.[1] To do this we must express route (*B*-4) in terms of the routes included in the initial shipment plan.[2]

[1] There is no guarantee that the route with the double asterisk is *really* the best single adjustment to make in the initial plan, in the sense of its being the one that leads to the optimum with the smallest number of revisions. It is certainly a plausible adjustment, however. By looking further ahead we *might* do slightly better. We could calculate which direct route would save us the largest total of transportation costs—this depends on both the per-ton saving and on the number of tons that can be shipped over the route within the restrictions of the problem. We forego this possibility of gain in order to avoid the additional calculations that would be needed.

[2] Our earlier work in expressing route (*B*-1) in terms of included routes, though useful for exposition, does not contribute at all to the solution of the problem.

The procedure is the same as before. We determine how a shipment may be made from Factory B to Locality 4 using only routes which were included in the initial shipment plan. Clearly the path is from Factory B to Locality 5, from Locality 5 to Factory C, and from Factory C to Locality 4. In symbols,

$$(B\text{-}4) = (B\text{-}5) - (C\text{-}5) + (C\text{-}4)$$

In order to introduce direct shipments from Factory B to Locality 4 and still satisfy the restrictions, we must reduce shipments over the equivalent indirect route. Table 5-2 shows that shipments over the indirect route can be reduced to the extent of 30 tons per month because, at that level, shipments from Factory C to Locality 4 will be entirely eliminated and cannot be decreased further. The critical number is the smallest positive shipment on the right-hand side of the above equation. Thus, an improved routing is to ship 30 tons per month directly from Factory B to Locality 4, to reduce by 30 tons per month the shipments from Factory B to Locality 5 and from Factory C to Locality 4, and to increase by 30 tons per month the shipments from Factory C to Locality 5. In general, in substituting a direct route for an indirect one, we reduce shipments over all routes which appear with positive signs in the formula for the indirect route and increase shipments over all routes which appear with negative signs. The resulting shipment plan is shown in Table 5-5.

TABLE 5-5. SECOND SHIPMENT PLAN, TONS

Locality	Factory		
	A	*B*	*C*
1	25		
2	25	. . .	90
3	. . .	60	
4	. . .	30	
5	. . .	10	60

It is instructive, though not an essential part of the computation, to see how this rerouting has affected total transportation costs. Total transportation costs corresponding to the initial shipment plan can be found by multiplying the volume of shipments over each route (found in Table 5-2) by the transportation costs over those routes (found in Table 5-1). The result is $8,275. By using Table 5-5 in place of Table 5-2 we find that the transportation cost of the second shipment plan is $7,375. The saving is $900. We can account for this magnitude by noting that the cost of direct shipment from Factory B to Locality 4 was $30 a ton as compared with \$60 for indirect shipment. The rerouting has thus saved \$30 on each of 30 tons, or a total of \$900, as we found.

The second shipment plan is thus an improvement on the first. To see whether further improvement is possible we simply repeat the process. We construct a second indirect-cost table on the basis of the second shipment plan, following the same procedure as was used in constructing the first indirect-cost table. The result is shown in Table 5-6. Only one entry in Table 5-6 is greater than the corresponding entry in Table 5-1;

TABLE 5-6. SECOND SET OF INDIRECT COSTS, PER TON

Locality	Factory		
	A	*B*	*C*
1	\$10	\$35*	\$30
2	15	40	35
3	−10	15	10
4	5	30	25
5	5	30	25

* Indirect costs exceed direct costs.

that is, only one indirect routing is more costly than the corresponding direct routing. This is a shipment from Factory *B* to Locality 1 and is indicated by the asterisk in the table. Total costs can be reduced by making this shipment directly. In preparation for this we express the route (*B*-1) in terms of the routes which are used in the second shipment plan. From Table 5-5 we find the formula

$$(B\text{-}1) = (B\text{-}5) - (C\text{-}5) + (C\text{-}2) - (A\text{-}2) + (A\text{-}1)$$

The minimum shipment of the positive-signed terms (*B*-5), (*C*-2), (*A*-1) is seen from Table 5-5 to be 10 tons. In order to ship 10 tons directly from Factory *B* to Locality 1 and still satisfy all the restrictions, we should have to ship 10 fewer tons over the indirect route. This would exactly eliminate direct shipments from Factory *B* to Locality 5 and is, therefore, the maximum extent to which direct shipments from Factory *B* to Locality 1 can be introduced. The rerouting, then, is to ship 10 tons directly from Factory *B* to Locality 1 and to compensate by sending 10 fewer tons over routes (*B*-5), (*C*-2), and (*A*-1) and sending 10 more tons over routes (*C*-5) and (*A*-2). The resultant routing is shown in Table 5-7.

Is this third shipping plan optimal? We repeat the previous analysis. The revised indirect costs, calculated in the familiar manner, are shown in Table 5-8. A comparison of Table 5-8 with Table 5-1 shows that in no case are indirect costs greater than direct costs. This indicates that no further improvement is possible and that the routing shown in Table 5-7 corresponds to the smallest possible transportation cost. This cost is $7,225.

TABLE 5-7. THIRD SHIPMENT PLAN, TONS

Locality	Factory		
	A	B	C
1	15	10	
2	35	...	80
3	...	60	
4	...	30	
5	...	...	70

TABLE 5-8. THIRD INDIRECT-COST TABLE, PER TON

Locality	Factory		
	A	B	C
1	$10	$20	$30
2	15	25	35
3	5	15	25
4	20	30	40
5	5	15	25

It is interesting to note that the solution to this problem is not unique. Table 5-9 gives an alternative routing with a total cost of $7,225. There are still other optimal plans as well, but that is a matter of indifference. In general, if the final indirect-cost table includes no indirect costs larger

TABLE 5-9. ALTERNATE OPTIMAL SHIPMENT PLAN, TONS

Locality	Factory		
	A	B	C
1	...	25	
2	35	...	80
3	...	60	
4	15	15	
5	...	...	70

than the corresponding direct costs, but does contain some that are equal, then there is a multiplicity of optimal routing plans.

The allocation of output that we have arrived at is optimal, from the point of view of minimizing transportation costs, whether the three factories are under unified management (as we have assumed) or not. The question then arises: Will this allocation be achieved in a free market or not? That, of course, depends on the structure of the market, but the reader undoubtedly suspects already that the allocation corresponding to a competitive equilibrium is optimal. The reason is that any other allocation will permit cost-saving arbitrages precisely analogous to the reroutings employed in computing our solution.

5-2. APPLICATION TO COMPARATIVE ADVANTAGE

What really underlies this problem is the familiar principle of comparative advantage. To make this clear we shall work through an example of the classic application of comparative advantage to international trade. In constructing this example, which is purely fictitious, we have pointedly ignored tariffs and transportation costs and have chosen numbers which illustrate a technical aspect of the transportation problem, the possibility of degeneracy.

TABLE 5-10. DATA FOR INTERNATIONAL-TRADE EXAMPLE

	Labor per acre for same yield in:			Demand (thousands of acres)
	United Kingdom	France	Spain	
Crop:				
Wheat	20	14	17	13,700
Barley	15	12	12	5,800
Oats	12	10	11	7,000
Acres available	7,000	12,400	7,100	26,500

The data for this example are contained in Table 5-10. In this table we assume that 20 units of labor per acre is required to produce a ton of wheat per acre in the United Kingdom and that the same yields per acre can be obtained by using 14 units of labor per acre in France and 17 units in Spain. The data on barley and on oats are to be interpreted similarly. The bottom line of the table gives the total acreages available for these three crops in the three countries, and the last column gives the total number of acres cultivated with the assumed intensity required to satisfy the demand for the three crops. Note that the total number of acres available and the total number of acres required to satisfy demand are the same, namely, 26,500,000 acres. The problem is to allocate the three crops among the three countries in such a way that the over-all requirement for labor is as small as possible.

The procedure is just the same as before. We start with an arbitrary allocation which satisfies the row and column restrictions. Such an allocation is shown in Table 5-11. But there is a difficulty in Table 5-11. Recall that a basic solution to a transportation problem involves a number of entries at positive levels equal to the number of rows plus the number of columns minus 1. In this case, with three rows and three columns, a basic solution should have five positive entries. But in this case, because of the coincidence that the number of acres available in the United Kingdom is just equal to the number of acres required to supply the demand for oats, there are only four positive entries in Table 5-11. This makes it impossible to construct an indirect-cost table.

TABLE 5-11. TRIAL ALLOCATION OF PRODUCTION, THOUSANDS OF ACRES

Crop	United Kingdom	France	Spain
Wheat.............	.....	6,600	7,100
Barley.............	.....	5,800	
Oats..............	7,000		

This is an instance of the algebraic degeneracy referred to in Sec. 4-10, because we have found it possible to satisfy five algebraic conditions by using fewer than five of the variables at our disposal. Degeneracies in transportation problems are not always as transparent or as easy to detect as this one, but the consequence is always the same, namely, that at some stage it will be impossible to construct an indirect-cost table. Fortunately, although degeneracies complicate the theory of linear programming, they do not prevent solution or even make the calculations more difficult.

We now deal with the degeneracy by means of a device which is necessary for purposes of theoretical justification, but not, as we shall see, for purposes of practical computation. The device is to change the data so as to obtain a new, related problem free of degeneracy. This can be done in several ways. Our choice is to change the demand for oats to $7{,}000 + \delta$ thousands of acres worth and to assume that the acreage available in Spain is $7{,}100 + \delta$ thousands of acres. We shall conceive of δ as a small nonnegative number and shall discuss its magnitude more precisely later. The data, altered in this fashion, are shown in Table 5-12. The initial allocation of production and the corresponding indirect costs are shown in Table 5-13. Comparison of the indirect costs with Table 5-12 indicates that Spain should produce barley. The economic interpretation of this comparison is that in the allocation being tested, Spain obtains barley by growing wheat (at a cost of 17 labor units per acre), shipping wheat to France (thereby saving France 14 labor units per acre), and importing French barley (which costs 12 labor units per acre).

The net cost is then 17 − 14 + 12 = 15 labor units, as shown in Table 5-13. But Spain can grow barley for 12 labor units per acre, according to Table 5-12. Hence a revision is necessary.

TABLE 5-12. ALTERED DATA, INTERNATIONAL-TRADE EXAMPLE

	Labor per acre for same yield in:			Demand (thousands of acres)
	United Kingdom	France	Spain	
Crop:				
Wheat	20	14	17	13,700
Barley	15	12	12	5,800
Oats	12	10	11	7,000 + δ
Acres available	7,000	12,400	7,100 + δ	26,500 + δ

TABLE 5-13. INTERNATIONAL-TRADE EXAMPLE, FIRST ALLOCATION

Crop	Acreage			Indirect-cost table, labor units		
	United Kingdom	France	Spain	United Kingdom	France	Spain
Wheat	.....	6,600 − δ	7,100 + δ	16	14	17
Barley	.....	5,800	....	14	12	15*
Oats	7,000	δ	....	12	10	13**

* Maximum excess of indirect over direct costs.

** Indirect costs exceed direct costs.

To revise the allocation we note that the indirect equivalent of Spanish production of barley is given by the formula

$$(S,B) = (S,W) - (F,W) + (F,B).$$

in which initial letters are used to indicate countries and products and the symbolism is the same as was used before. Spanish production of barley must therefore be offset by decreases in Spanish production of wheat and in French production of barley. French production of barley can be reduced by 5,800 acres; therefore Spanish barley can be introduced to this extent.

Table 5-14 shows the reallocation of production and the corresponding indirect costs. The comparison test shows that Spain should produce oats. The indirect equivalent of Spanish oat production is given by the formula

$$(S,O) = (S,W) - (F,W) + (F,O)$$

Thus if Spain is to produce oats, Spanish wheat production and French oat production must be decreased. French oat production can be reduced by δ thousand acres. Table 5-15 shows the reallocation and the corresponding indirect costs. Comparison with Table 5-12 shows that the allocation of production given in Table 5-15 is optimal. The problem is solved, except for the unfortunate appearance of δ in the final allocation.

TABLE 5-14. INTERNATIONAL-TRADE EXAMPLE, SECOND ALLOCATION

Crop	Acreage			Indirect-cost table, labor units		
	United Kingdom	France	Spain	United Kingdom	France	Spain
Wheat...	.....	12,400 + δ	1,300 + δ	16	14	17
Barley...	.....	....	5,800	11	9	12
Oats.....	7,000	δ	....	12	10	13*

* Indirect costs exceed direct costs.

We did not specify δ except to state that it was small and nonnegative. In working through the calculation, we carried the δ along, treating it at every stage as if it were an ordinary number. Thus the allocation of production in Table 5-15 would be optimal for any value of δ for which it was technically feasible. In particular, it would be optimal for $\delta = 0$, in which case the altered problem is the same as the original problem that we had to solve. We then assign this value to δ.

TABLE 5-15. INTERNATIONAL-TRADE EXAMPLE, THIRD ALLOCATION

Crop	Acreage			Indirect-cost table, labor units		
	United Kingdom	France	Spain	United Kingdom	France	Spain
Wheat...........	.....	12,400	1,300	18	14	17
Barley...........	.....	......	5,800	13	9	12
Oats.............	7,000	......	δ	12	8	11

Now we can see why the δ, while necessary in tracing through our reasoning, is not necessary in the actual work of solution. We did not have to wait until the final step before setting $\delta = 0$; we could have done this right at the beginning. In that case we would not have needed Table 5-12. Instead, in calculating Table 5-11 from Table 5-10 we would have written a zero in the (F,O) space instead of a blank,[1] and thereafter we

[1] The choice of the blank in which to insert the zero is arbitrary. The important thing is to have a zero somewhere; its location has no essential effect on the result.

would have treated the zero in every respect in the same way as an activity at positive level. This would have provided the necessary five entries, and the work would have proceeded smoothly. Degeneracies, then, can be handled by taking either explicit or virtual account of a δ variation and do not affect a transportation problem or its solution in any essential manner.

The allocation of production arrived at in Table 5-15 conforms with expectations and is easily interpreted. The United Kingdom, under our assumptions, has a comparative advantage in oat production since the gap between labor requirements for oat production and for the production of the other crops is greater there than in the other countries. Similarly, inspection of Table 5-10 shows that Spain has a comparative advantage in barley production. The optimal solution which we found conforms to these two comparative advantages.

It should be noted that the international-trade example, while an illustration of the transportation problem, has nothing to do with transportation costs. Neither, in any fundamental sense, does the transportation problem, in spite of its misleading name. The essential characteristic of the transportation problem is that restraints are imposed on both inputs and outputs and that, subject to those restraints, some number which depends on the allocation of the inputs is to be either minimized (as in both our examples) or maximized. Any economic or business problem of this general structure is a transportation problem.

5-3. OTHER INTERPRETATIONS OF THE PROBLEM

Another straightforward application of the same procedure is the so-called personnel-assignment problem. It goes like this: Suppose there are 10 men to be assigned to 10 positions in an organization and that a numerical estimate can be made of the contribution which each man could make to the organization if assigned to each job. These values, of course, would reflect both the importance of the jobs and the capabilities of the men. Then the problem is to assign the men to the jobs in such a way that the total value of the group to the organization is as great as possible. This problem may be varied by considering categories of men and categories of jobs rather than individual men and jobs, but the principle is the same.

Still another group of applications concerns the allocation of raw materials. Suppose that a soap company has a number of different kinds of vegetable oil that may be substituted for one another in certain proportions to form a number of different soap products. Suppose that the total amount of each kind of oil to be used is predetermined and also the total amount of each kind of soap and, further, that some measure of the value of each oil in each use is known. Then a problem of the same general kind results. In short, the transportation problem is of a quite general nature.

5-4. IMPLIED VALUES: THE DUAL

The solution of a transportation problem implicitly places values on the various inputs and outputs involved, just as does any other linear-programming problem. This aspect is somewhat obscured by the usual formulation of the transportation problem, which we presented in Eqs. (5-1) to (5-4). Let us now set up the problem in a slightly altered form which will enable us to derive the dual form, and thereby perceive the value implications.

The three-factory, five-locality example can be stated in the following form. We require a schedule of shipments which will minimize the total transportation cost,

$$T = \Sigma_i \Sigma_j c_{ij} x_{ij} \qquad (5\text{-}1)$$

subject to the restrictions that no factory be required to ship a greater amount than its monthly capacity, i.e., that

$$\Sigma_j x_{ij} \leq k_i \qquad (5\text{-}5)$$

and that each locality receives at least its monthly requirement, i.e., that

$$\Sigma_i x_{ij} \geq r_j \qquad (5\text{-}6)$$

where the symbols have the same definitions as before. We retain the assumption that total capacity equals total requirements, or that

$$\Sigma_i k_i = \Sigma_j r_j \qquad (5\text{-}7)$$

Formulas (5-5) and (5-6) have inequality signs where their analogs in the previous statement of the problem, formulas (5-2) and (5-3), had equality signs. Apart from that, the problem is exactly as it was before. The change in the problem is, however, formal rather than substantive; i.e., any solution to the first form of the problem is a solution to the second, and conversely. The reason for this is contained in Eq. (5-7), which guarantees that the inequality signs in Eqs. (5-5) and (5-6) will never be effective. Consider Eq. (5-5) and suppose that for Factory A we had $\Sigma_j x_{Aj} < k_A$; then when we added up the outputs of all the factories we should have to have $\Sigma_i \Sigma_j x_{ij} < \Sigma_i k_i$, since, by virtue of Eq. (5-5), neither of the other factories could make up the deficit created by the fact that Factory A was producing at less than capacity. Because of Eq. (5-7) this would mean that total output is less than total requirements, which is forbidden by Eq. (5-6). It follows that it is impossible for inequality signs actually to occur in any admissible solution, even though they appear in the restatement of the problem.

In Eqs. (5-5) and (5-6) the inequality signs run in opposite directions. In order to put this problem in standard linear-programming form we multiply Eq. (5-5) through by -1, which has the effect of changing the sense of the inequality. Thus, and finally, the restrictions take the form

$$\Sigma_i x_{ij} \geq r_j \tag{5-6}$$
$$-\Sigma_j x_{ij} \geq -k_i \tag{5-8}$$

Equations (5-1), (5-6), and (5-8) now constitute a standard linear-programming problem. This problem involves 15 variables, x_{ij}, and 8 restraints, 5 in Eq. (5-6) and 3 in Eq. (5-8). The restraining equations are written out in full in Table 5-16. The sum to be minimized is given on the last line of that table.

TABLE 5-16. RESTRAINING EQUATIONS FOR GENERALIZED FORM OF TRANSPORTATION PROBLEM

Variable	x_{A1}	x_{A2}	x_{A3}	x_{A4}	x_{A5}	x_{B1}	x_{B2}	x_{B3}	x_{B4}	x_{B5}	x_{C1}	x_{C2}	x_{C3}	x_{C4}	x_{C5}		Constant
Eq. (5-6)	1					1					1					$\geq$	$r_1 = 25$
		1					1					1				$\geq$	$r_2 = 115$
			1					1					1			$\geq$	$r_3 = 60$
				1					1					1		$\geq$	$r_4 = 30$
					1					1					1	$\geq$	$r_5 = 70$
Eq. (5-8)	-1	-1	-1	-1	-1											$\geq$	$-k_1 = -50$
						-1	-1	-1	-1	-1						$\geq$	$-k_2 = -100$
											-1	-1	-1	-1	-1	$\geq$	$-k_3 = -150$
c_{ij}	10	15	20	20	40	20	40	15	30	30	30	35	40	55	25		= minimum

Now we form the formal dual of the direct problem that has just been stated. It will be recalled that the relationships between any problem and its dual are as follows:

1. The dual of a minimum problem is a maximum problem.
2. The dual problem has one restraining inequality for each variable in the direct problem, and one variable for each restraining inequality in the direct problem.
3. The inequalities in the dual problem have the opposite sense to the inequalities in the direct problem.
4. The coefficients of the objective function of the direct problem appear as restraining constants in the dual, and the restraining constants of the direct problem are the coefficients of the objective function of the dual.
5. The restraining equations of the dual problem are constructed from those of the direct problem in the following manner. Each restraining equation in the dual problem corresponds to one of the variables in the direct problem. Each variable in the direct problem appears with some

coefficient, possibly zero, in each equation in the direct problem. **Each** variable in the dual problem corresponds to one of the equations in the direct problem. Suppose that the ith dual variable corresponds to the ith direct equation and that the jth dual equation corresponds to the jth direct variable. Then the coefficient of the ith dual variable in the jth dual equation is the same as the coefficient of the jth direct variable in the ith direct equation.

These rules relating the dual problem to the corresponding direct problem are not, of course, susceptible of proof. They merely define the dual. The essential algebraic properties of the dual were discussed in Chap. 4.

To construct the dual of the transportation problem let v_j be the dual variable corresponding to the requirement that the jth locality receive its full quota, so that, for example, v_1 corresponds to the top line in Table 5-16. Similarly, let u_i be the dual variable corresponding to the restriction that Factory i cannot be required to ship more than its capacity. Then the dual problem involves 8 variables and 15 restraints. This problem is set forth in Table 5-17 and in the following equations.

To maximize

$$S = \Sigma_j r_j v_j - \Sigma_i k_i u_i \tag{5-9}$$

subject to the restrictions

$$\begin{aligned} v_j - u_i &\leq c_{ij} \qquad i = A, B, C; j = 1, 2, 3, 4, 5 \\ u_i, v_j &\geq 0 \end{aligned} \tag{5-10}$$

Thus far we have performed only a formal algebraic manipulation. But the economic interpretation is evident. Let us interpret u_i as the value of the product f.o.b. Factory i and v_j as its value delivered at Locality j. Then Eqs. (5-10), which may be written

$$v_j \leq u_i + c_{ij}$$

state that for any factory-locality pair the value at the locality must be no greater than the value at the factory plus the transportation cost; i.e., transportation of the commodity must never rise to an unimputed surplus of value over and above f.o.b. value and transportation cost. Equation (5-9), the expression to be maximized, is the total excess of delivered value over value at the factory. In fact, over routes actually used, delivered value equals factory value plus transport cost. Over routes not used, delivered value falls short of factory value plus transport cost (or equals it, in which case the route could be included in an alternative optimal shipping plan).

Since we have already solved the direct problem, the solution of the dual is trivial. We use the principle that in Eqs. (5-10) the equality holds for those pairs of values of i and j which enter into the optimal shipment plan, while either equality or inequality holds for the rest. Table 5-7 gives the pairs which enter into the shipment plan as (A-1),

TABLE 5-17. DUAL OF THE THREE-FACTORY TRANSPORTATION PROBLEM

Corresponding direct variable	Dual variable								Constant
	v_1	v_2	v_3	v_4	v_5	u_A	u_B	u_C	
x_{A1}	1					−1			≤10
x_{A2}		1				−1			≤15
x_{A3}			1			−1			≤20
x_{A4}				1		−1			≤20
x_{A5}					1	−1			≤40
x_{B1}	1						−1		≤20
x_{B2}		1					−1		≤40
x_{B3}			1				−1		≤15
x_{B4}				1			−1		≤30
x_{B5}					1		−1		≤30
x_{C1}	1							−1	≤30
x_{C2}		1						−1	≤35
x_{C3}			1					−1	≤40
x_{C4}				1				−1	≤55
x_{C5}					1			−1	≤25
Objective function. .	25	115	60	30	70	−50	−100	−150	= maximum

(A-2), (B-1), (B-3), (B-4), (C-2), and (C-5). Hence we have these seven simple equations:

$$
\begin{aligned}
v_1 - u_A &= 10 \\
v_2 - u_A &= 15 \\
v_1 - u_B &= 20 \\
v_3 - u_B &= 15 \\
v_4 - u_B &= 30 \\
v_2 - u_C &= 35 \\
v_5 - u_C &= 25
\end{aligned}
$$

This is a set of seven equations in eight unknowns. We can therefore give an arbitrary value to one of the unknowns, and, just for definiteness, we set $u_C = 0$. Then, substituting in these equations, the values shown in Table 5-18 result.

Table 5-18 contains two sorts of useful economic information. (1) The values of u measure the comparative locational advantages of the three factory sites. The product of Factory A is worth \$20 per ton more than the product of Factory C, simply because of proximity to points of consumption. This differential is not entirely obvious from Table 5-1, where it appeared that Factory C was actually closer than Factory A to the

TABLE 5-18. IMPLICIT VALUES, THREE-FACTORY EXAMPLE

	Factory			Locality				
	A	B	C	1	2	3	4	5
Variable	u_A	u_B	u_C	v_1	v_2	v_3	v_4	v_5
Value per ton, dollars	20	10	0	30	35	25	40	25

second largest market. (2) The values of v are the delivered prices that correspond to the most economical allocation of output from the viewpoint of minimum aggregate transportation cost. The results are not the same as those of either a single- or a multiple-basing-point pricing system, though they have some aspects relating them to both. Our results are similar to those of basing-point prices in that the realized prices at each factory site (the u's) are influenced by the transportation costs pertaining to the other factory sites. Thus prices f.o.b. Factory A are greater than prices at the site of Factory C because C suffers from locational disadvantages and cannot dispose of its output without loss otherwise. This disparity in realized f.o.b. prices among factories is also observed in the single-basing-point system. Since this disparity is present in the optimal allocation, one may conclude validly that its presence in single-basing-point systems is not an adequate reason for condemning those systems. But neither is it an argument in their favor. On the other hand, our values contrast with a pure basing-point system because the price structure is such that no factory can supply all localities without accepting a realized price lower than the value of the product at the factory. The delivered prices do not correspond to any simple multiple-basing-point system either. Of course, our problem diverges in one important respect from the kind of situation in which basing-point systems normally arise. We have assumed that there is no excess capacity, whereas basing-point systems are typically a response to the existence of excess capacity at one or more of the producing points.[1] The implications of our method of analysis for the allocative efficiency of basing-point systems are not affected by this reservation, however.

[1] Our formulation of the transportation problem can be adapted easily to cases in which excess capacity exists. Simply introduce a fictitious destination for which requirements equal the amount of excess capacity and assume that the cost of transportation from each point of origin to that destination is zero. The shipping plan that minimizes transportation cost for this modified problem will also minimize transportation cost for the original problem. The shipments to the fictitious destination are, of course, to be interpreted as capacity disposal activities.

It should be remarked that the alternative optimal shipping plan given in Table 5-9 corresponds to precisely the same structure of implicit values, as the reader can easily verify.

These results recall the remarks that we made at the end of Sec. 5-1 about the relationship of our solution to competitive equilibrium. The prices that we have obtained are the competitive equilibrium prices that would result from the uncoordinated efforts of the three factories to sell their entire outputs at the maximum possible prices. In this sense the economic system can be viewed as a gigantic analog computer and director in which the computations of optimal values and the directing of resources are simultaneous and interdependent.

5-5. THE DUAL OF THE INTERNATIONAL-TRADE EXAMPLE

The analysis of the pricing implications of the international-trade example follows the same lines. We let u_i denote the rental value of acreage in country i, and v_j the value per ton of crop j. Values are here measured in terms of the cost data given in Table 5-10, that is, in terms of labor units. The only novel aspect of the trade example is that we have to remember to count oat production in Spain (at zero level, of course) as one of the activities appearing in the final solution, in order to have sufficient equations. Table 5-19 shows the resulting values with rentals in the United Kingdom being set, arbitrarily, at zero. The interpretation of these figures is that land should rent for 1 labor unit per acre in Spain and for 4 labor units per acre in France, while wheat should sell for 18 labor units per ton, barley for 13, and oats for 12. The word "should" indicates only that these are the prices which correspond to the most efficient allocation of production in the light of the data of this problem.

TABLE 5-19. IMPLICIT VALUES, INTERNATIONAL-TRADE EXAMPLE

	Rental value of land in:			Value per ton of:		
	United Kingdom	France	Spain	Wheat	Barley	Oats
Variable........	u_{UK}	u_F	u_S	v_w	v_b	v_o
Value, dollars....	0	4	1	18	13	12

5-6. TECHNICAL NOTE

The transportation problem is, as we have seen, peculiarly simple to solve. The reason for this is revealed by the equations that we solved implicitly, without writing them down, in arriving at the initial shipment plans. In the three-factory illustration the variables that occurred at positive levels in the initial plan were x_{A1}, x_{A2}, x_{B3}, x_{B5}, x_{C2}, x_{C4}, x_{C5}. The equations connecting these variables can be written

$$\begin{aligned}
x_{A1} &= 25\\
x_{A1} + x_{A2} &= 50\\
x_{A2} + x_{C2} &= 115\\
x_{C4} &= 30\\
x_{C2} + x_{C4} + x_{C5} &= 150\\
x_{C5} + x_{B5} &= 70\\
x_{B5} + x_{B3} &= 100
\end{aligned}$$

Now notice that the first equation involves only one unknown and serves to determine that unknown; the second equation introduces one additional unknown; and each subsequent equation adds just one more unknown. Thus the unknowns can be determined seriatim without elaborate computation. Such a system of equations is known as "triangular"—a name suggested by the appearance of the setup—and is a godsend whenever it occurs.

There is still an additional simplifying feature in the transportation problem: whenever an unknown appears, its coefficient is unity. As a result, if the requirements and capacities (the right-hand sides of the equations) are integral numbers, so will be the levels of the activities in the solution, because each level will be determined in its turn by simple addition and subtraction of integers.

It is easy to see that any transportation problem must work out in this simple way. Consider a transportation problem with n row restraints and m column restraints and focus attention on any activity level in the array. Now set this level equal to its row or column total, whichever is smaller. If the row total is smaller, this operation will determine all the other activity levels in that row: they will be zero. Similarly, if the column total is smaller, all the other activities in that column will be zero. In either case the result is to leave a transportation problem with one less row or one less column. This process can be repeated until only one row or one column is left, at which time all the activity levels will have been determined (not necessarily optimally, of course). And, since these levels will have been determined seriatim, the equations connecting them and implicit in this process must have been triangular in form.

In this chapter we have not dealt rigorously with many of the theoretical issues that arise in the transportation problem. For a more thorough treatment the reader is referred to the definitive developments by G. B. Dantzig and T. C. Koopmans.[1]

[1] *Activity Analysis of Production and Allocation*, Chaps. 14 and 23.

EXERCISES

5-1. A race-horse owner has four horses, named Smith, Malthus, Ricardo, and Mill, and plans to enter them in four races. If he wishes to enter one horse in each race and cannot enter any horse in more than one race, how should he enter them so as to make his expected total purse winnings as great as possible? The data are given below.

HORSERACE EXERCISE: PROBABILITIES AND PURSES

	Probability of winning			
	Race 1	Race 2	Race 3	Race 4
Horse:				
Smith	0.40	0.30	0.20	0.60
Malthus	0.20	0.20	0.15	0.30
Ricardo	0.15	0.10	0.00	0.20
Mill	0.10	0.00	0.00	0.20
Purse	$1,000	$1,500	$2,000	$1,000

Partial answer: Maximum expected purse winnings are $1,150.

5-2. Set up, solve, and *interpret* the dual to Exercise 5-1.

6

Linear-programming Analysis of the Firm

6-1. THE LINEAR-PROGRAMMING CONCEPT OF THE FIRM

We have already seen that linear programming is a method for calculating the best plan for achieving stated objectives in a situation in which resources are limited. As such it is a method for solving the classic problem of economizing, whether in the context of an entire economy, in that of a government program, or in a single firm. In this chapter we consider the application of linear programming to the problems of the individual firm.

The problem of the optimal utilization of limited resources by the individual firm has long been studied in economics, particularly in the analysis of the theory of production in the short run. The received doctrine provides us with a convenient point of departure. Consider, then, a competitive firm which cannot influence the prices of the factors that it uses or the products that it produces. The model usually employed for discussing the behavior of such a firm conceives of it as a unit of control whose objective is a maximum time rate of profit and those variables of choice are the time rates of consumption of various inputs and of production of various outputs. The choices of production and consumption rates are interdependent; otherwise output would be chosen very large (perhaps infinite) and input very small. The necessary interrelationships are expressed by means of a production function, and if we regard a production function as specifying all technologically feasible combinations of inputs and outputs, the problem of production amounts to selecting the most profitable point of the production function.

For most purposes of economic analysis we assume that this problem is solved. We do not care how it is solved in practice, although presumably it is usually not done by formulating the production function mathematically and finding its optimum by the methods of the differential calculus. More plausibly, the solution process is one of trial and error or survival of the fittest. But all this is neither here nor there.

The important thing is that the problem is solved, by one means or another, and we can use the solution (which, however arrived at, satisfies the conditions deduced from the differential calculus) to characterize the behavior of markets and their responses to changes in prices, techniques, tastes, or other conditions.

The conventional theory can be justified in terms like these, and often has been, because the analysis of the firm is but a step in the analysis of markets. But what if we are interested in the firm per se? What if we are interested in giving a prescription to the firm as to how to solve its optimizing problem? Then the method of finding the firm's optimum ceases to be irrelevant and we must look at the assumptions about the production function with a more critical eye.

The production function is a description of the technological conditions of production, and the economist takes no direct responsibility for ascertaining it. Instead he regards it as falling within the purview of the technologist or engineer. But there seems to have been a misunderstanding somewhere because the technologists do not take responsibility for production functions either. They regard the production function as an economist's concept, and, as a matter of history, nearly all the production functions that have actually been derived are the work of economists rather than of engineers.

The engineers do not, on these grounds, stand convicted of neglect of duty. The fact is that engineers look at things somewhat differently from economists and the production function does not usually enter explicitly into an engineer's analysis. This is true for two reasons. First, the activities of firms, even small firms, are generally too complex to be considered as a whole. The engineer can analyze an assembly line without studying the shipping room or billing department, and therefore he has no occasion to formulate a production function for the firm as a whole. Second, the variables of choice as they appear to the engineer are not the ones on which economic analysis is based. His problem, typically, is not whether to use slightly more capital and slightly less labor but how many units to install of a new machine that costs slightly more but requires less tending and has other distinguishing characteristics as well. Putting it roughly, the choice is not among various time rates of input and output but, more directly, among different ways of doing things, each of which implies its own characteristic pattern of input and output rates.

The distinction we are drawing between the engineer's approach and the production-function approach does not bear directly on the facts of life or on empirical realism. We are stating merely that the production function short-circuits certain aspects of the problem that the engineer cannot afford to neglect. The economist cannot afford to neglect them either when he wants to look inside the firm. Moreover, there is some advantage in talking the engineer's language. For, among other things, this is the language in which engineering and accounting data are expressed. This is the point of view taken in linear programming.

Our point of view, then, will be that the essential choices made by a firm do not deal directly with levels of input and output, but rather concern the extent to which "different ways of doing things" are used. This concept of a "way of doing things" is by no means easy to define. The underlying idea is that a change in a way of doing things implies a change in the composition of ingredients or results or both. Thus, to use the agricultural illustrations so prevalent in the theory of production, if a farmer sows his acre with a new variety of seed (thus decreasing the input of the old seed and increasing that of the new), or if he switches to a new crop entirely, or if he sows the old seed more thickly than last year—in any of these cases we should say that he is doing things differently. But if he should plant twice as many acres with the old seed using the old density of sowing, etc., then he is "doing the same things in the same way," except, as we shall say, at a higher level or intensity.

This attitude toward choices implies an assumption closely related to the assumption of constant returns to scale. We assume that if acreage is doubled while the treatment given each acre is unchanged, then output will be doubled and so will be all associated inputs. As an empirical matter, this might or might not hold true. In this agricultural example one feels that constant rates of return to an unchanged way of doing things will hold approximately but cannot hold exactly. Exact proportionality is impossible because, to suggest a few reasons, (1) yields toward the edge of a field are different from yields in the center and the proportion of edge to total changes with scale; (2) this edge problem also enters into the difficulties of cultivation; (3) the length of fencing is not proportional to acreage; (4) travel time to, from, and within the plot is not proportional to acreage. These examples should illustrate the kind of nonproportional effects that are inevitable and also the possibility that the deviations from proportionality may be unimportant. In our formulation we shall assume that deviations from proportionality are unimportant and shall neglect them.

We can now define a concept which incorporates the idea of a "way of doing things." This concept is that of a "process," or "activity," which we define as a set of ratios obtaining among rates of consumption of various inputs and rates of production of various outputs. We conceive of a firm as making choices among a number of processes (each presumably corresponding to different physical operations, but that is not part of our definition). Each process, we assume, can be operated at any positive level so long as the necessary inputs are available. By changing the level of a process we mean, of course, changing the time rates of consumption of all its inputs and the rates of production of all its outputs in the same proportion. We repeat: The assumption that the level of a process can be varied is an assumption of fact which is not likely to be exactly true. We assume also that a firm may use several processes simultaneously so long as necessary supplies are available. This assumption is subject to the same empirical qualifications as the preceding one.

Still another assumption is that a firm has only a finite number of processes available to it. In cases in which this assumption does not apply, the traditional marginal analysis of smooth curves is likely to be more appropriate than the methods of linear programming with which we are concerned.[1] Agriculture, obviously, provides many examples in which the number of processes available is, to all intents and purposes, infinite. On the other hand, there are many practical cases, even in agriculture, in which the number of processes available is finite, sometimes very small.

If a firm employs several processes, then it follows from our definitions that the firm's total consumption of factors and total production of products will be the sum of the quantities of factors consumed by the various processes and the sum of the products produced by the individual processes, respectively. Thus a change in the proportions among the quantities of factors consumed and products produced by the firm can result only from a change in the levels at which the various activities are utilized.

In this view of production the quantities of inputs and outputs of the firm cannot be altered directly, but only indirectly by means of changes in the levels of various processes. Thus linear programming does not seek to determine directly the optimal quantity of each factor and product but, instead, the optimal level of each activity. From these levels, the factor and product quantities follow in due course.

Two simple examples will illustrate this approach.

6-2. An Automobile Example. Let us consider a hypothetical automobile company equipped for the production of both automobiles and trucks. We shall assume that this company's plant is organized into four departments, namely, sheet-metal stamping, engine assembly, automobile final assembly, and truck final assembly, and that raw materials and all other components are purchased from supply companies.

The capacity of each department of the plant is limited, of course. We assume that the metal-stamping department can turn out sufficient stampings for 25,000 automobiles or 35,000 trucks per month or for some

[1] It would be misleading to contrast the linear-programming model with marginal analysis in general. Linear programming *is* marginal analysis, appropriately tailored to the case of a finite number of activities. "Traditional" marginal analysis is tailored to the case of a differentiable production function.

appropriate combination of automobiles and trucks. To illustrate what we mean by an "appropriate combination," note that, for example, 15,000 automobiles represents 60 per cent of capacity of this department (15,000 = 60 per cent of 25,000) and 14,000 trucks represents 40 per cent of capacity. Thus, 15,000 automobiles plus 14,000 trucks could be produced with the department operating at full capacity, and this would be an appropriate combination.

In a like manner we assume that the engine-assembly department has monthly capacity for 33,333 automobile engines or 16,667 truck engines or, again, some combination of fewer automobile and truck engines.

TABLE 6-1. MONTHLY CAPACITIES OF DEPARTMENTS OF AUTOMOBILE PLANT, UNITS PER MONTH

Department	Product	
	Automobiles	Trucks
Metal stamping	25,000	35,000
Engine assembly	33,333	16,667
Automobile assembly	22,500	
Truck assembly	......	15,000

The automobile assembly line is assumed to accommodate 22,500 automobiles per month, and the truck assembly line 15,000 trucks. These assumptions are summarized in Table 6-1.

We can now define two processes or activities: the production of automobiles and the production of trucks. The process of producing an automobile would have as an output one automobile and as inputs 1/25,000 = 0.004 per cent of metal-stamping capacity, 1/33,333 = 0.003 per cent of engine assembly capacity, and 1/22,500 = 0.00444 per cent of automobile assembly capacity. In a similar way we deduce from Table 6-1 that the process of producing one truck requires as inputs 0.00286 per cent of metal-stamping capacity, 0.006 per cent of engine assembly capacity, and 0.00667 per cent of truck assembly capacity. These data are summarized in Table 6-2.

The scope for choice in this firm consists in deciding how many automobiles and how many trucks to produce each month subject to the restriction that no more than 100 per cent of the capacity of any department be used. Clearly, if automobiles alone are produced, at most 22,500 units per month can be made, automobile assembly being the bottleneck. Similarly, if only trucks are produced, the maximum output is 15,000 units because of the limitation on truck assembly. Which of these alternatives should be adopted or whether some combination of trucks and automobiles should be produced depends on the relative profitability of manufacturing automobiles and trucks. Let us assume that the sales value of an automobile is $300 greater than the total cost of purchased materials, labor, and other direct costs attributable to its manufacture. And similarly, let us assume that the sales value of a

TABLE 6-2. PER CENT OF CAPACITY REQUIRED PER UNIT OF AUTOMOBILE AND TRUCK PRODUCTION

Department	Process	
	Automobile production	Truck production
Metal stamping	0.004	0.00286
Engine assembly	0.003	0.006
Automobile assembly	0.00444	0
Truck assembly	0	0.00667

truck is $250 more than the direct cost of manufacturing it. Then the manufacturing profit of the plant for any month is 300 times the number of automobiles produced plus 250 times the number of trucks. The statement of the problem is now complete: Find the combination of automobile and truck outputs that yields the greatest manufacturing

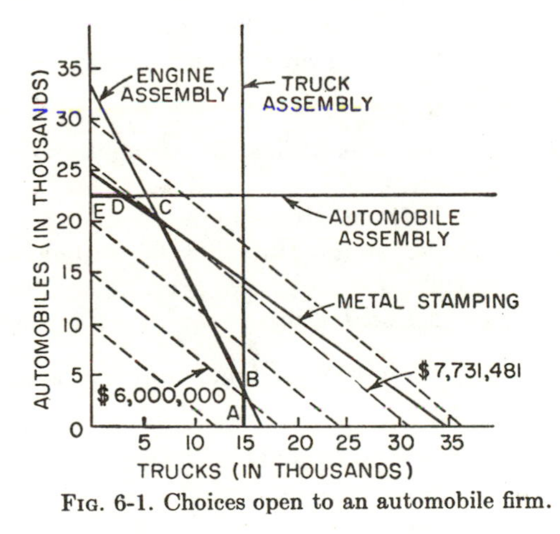

FIG. 6-1. Choices open to an automobile firm.

profit without requiring more than 100 per cent of the capacity of any department.

Because this example involves only two activities, the situation can be portrayed graphically, and we have done this in Fig. 6-1. In this figure the number of trucks produced is measured horizontally and the number of automobiles is measured vertically. The line labeled "metal stamping," which runs from 25,000 automobiles and no trucks to 35,000 trucks and no automobiles, represents all combinations of automobile and truck outputs that require 100 per cent of metal-stamping capacity. In a like manner, the "engine assembly," "truck assembly," and "automobile assembly" lines show the other three capacity limitations. Then the area between the origin and the broken line segment *ABCDE* shows all combinations of automobiles and trucks that can be produced, i.e., that do not violate any of the capacity restrictions. The dashed lines on the figure are "isoprofit" lines; i.e., each of them is the graph of one of the equations

$$300 \times \text{automobiles} + 250 \times \text{trucks} = \text{constant}$$

for a particular constant (for example, 4.5 million dollars), so that all combinations of automobiles and trucks that lie on the same dashed line yield the same manufacturing profit. Graphically speaking, then, the problem is to find the point in the polygon *OABCDE* that lies on the highest possible dashed line. Clearly this is point *C*. This simple graph solves the problem handily in this simple case, but obviously such a graph will not suffice if more than two processes have to be considered. We therefore formulate the problem algebraically.

Let x_1 denote the number of automobiles produced per month, and x_2 the number of trucks. Table 6-2 gives the proportion of the capacity of each department required for each automobile and truck. Using those data we can write

$0.004x_1 + 0.00286x_2 =$ per cent utilization of metal-stamping department

$0.003x_1 + 0.006x_2 =$ per cent utilization of engine assembly line

$0.00444x_1 =$ per cent utilization of automobile assembly line

$0.00667x_2 =$ per cent utilization of truck assembly line

Since the capacity limitation is 100 per cent, in every case the algebraic restrictions on x_1 and x_2 are

$$\begin{aligned} 0.004x_1 + 0.00286x_2 &\leq 100 \\ 0.003x_1 + 0.006x_2 &\leq 100 \\ 0.00444x_1 &\leq 100 \\ 0.00667x_2 &\leq 100 \end{aligned}$$

In addition, since we do not permit the production of negative numbers of automobiles and trucks, we require $x_1, x_2 \geq 0$. Subject to these conditions we wish to find x_1, x_2 so that

$$\text{Manufacturing profit} = 300x_1 + 250x_2 = \text{as great as possible}$$

Methods for finding such a pair of values are discussed in Chap. 4 and later in this chapter. By applying these methods we can show that the maximum profit for the automobile firm is obtained when 20,370 automobiles and 6,481 trucks are produced per month.[1] The reason for this is easy to see. Suppose the company produced as many automobiles as possible, namely, 22,500 (point E in the figure). There would be unused capacity in all departments except automobile assembly. The firm could manufacture as many as 3,500 trucks without reducing its automobile output (point D), but at this stage metal-stamping capacity would be fully utilized. Further increases in truck output could be achieved only by reducing automobile output. Would further expansion of truck production be worthwhile? Each truck requires 0.00286 per cent of metal-stamping capacity, or the same as five-sevenths of an automobile. Thus every 7 trucks manufactured require a reduction of 5 in the output of automobiles in order to stay within the metal-stamping capacity limitation. But the profit on 7 trucks is $1,750 (i.e., 7 × $250), while the profit on 5 automobiles is only $1,500 (that is, 5 × $300). So it is worthwhile to substitute trucks for automobiles.

The substitution of truck output for automobile production increases the utilization of engine assembly capacity. When this capacity is fully absorbed (point C) the picture changes. From there on, as can be seen from Table 6-2, a cutback of two automobiles per month is required to free sufficient capacity for one truck. This is not worthwhile, and the optimal schedule is given by the values of x_1 and x_2 which fully utilize both metal-stamping and engine-assembly capacity. Those are the values that we have given.

This example conforms to the simplest possible linear-programming model of the firm, and it is important to recount explicitly the simplifying assumptions that have been made. The most obvious assumption is that the total capacity requirement for the production of automobiles is directly proportional to the number of automobiles made, and the same holds true for trucks. Further, although automobiles and trucks compete with each other for the use of productive facilities, they do not interfere with each other, so that if both automobiles and trucks are made, the total capacity requirement for the production program is simply the sum of the requirement for automobiles and that for trucks. Actually, changeover time is a factor in costs.

[1] This is point C on Fig. 6-1. It is the point where the first two inequalities listed above are exactly satisfied, and that, diagrammatically, is where the engine assembly line crosses the metal-stamping line on Fig. 6-1. It is easy to see that there will be unused capacity on both the truck and automobile assembly lines. This is visible in the figure, and substitution in the formulas will confirm it. The manufacturing profit corresponding to this production program is $7,731,481.

There is another important set of assumptions which was carried, so to speak, in the margin. Consider the production of automobiles. In addition to the assumed amounts of the four capacities listed, it requires raw materials, components purchased from outside suppliers (which are also raw materials from the point of view of the firm), financing, and labor. In short, it requires certain direct expenses or variable costs. The production of an automobile results in the creation of a salable commodity and, eventually, in the receipt of a certain gross revenue. We shall refer to the net revenue from the production of an automobile as the excess of the gross revenue per unit over direct costs per unit. Now we note that our model has assumed that this unit net revenue is a constant, independent of the number of automobiles and trucks produced. Mathematically speaking, this is the only assumption which we have made about direct expenses, gross revenue, and net revenue; i.e., we have assumed only that

$$\begin{aligned}\text{Unit net revenue} &= \text{unit gross revenue} - \text{unit direct costs} \\ &= \text{constant}\end{aligned}$$

But, economically speaking, this results from assuming that

1. Unit gross revenue = constant.
2. Unit direct costs = constant.

The first of these assumptions is an assumption of pure competition, i.e., an assumption that all units which can be produced can be sold at the going price and none can be sold at a higher price. Linear-programming models can be constructed for somewhat more complicated marketing situations, and we shall discuss one such model later.

The second assumption postulates that marginal direct costs are constant up to the point at which capacity is fully utilized and thereafter are infinite. In effect, we have divided total costs of production into two categories. First, there are the expenses of providing the stated amounts of productive capacity. Second, there are the operating costs that are regarded as variable and simply proportional to the level of operation. In later examples we shall show how the revenue and cost assumptions can be incorporated explicitly in the mathematical formulation of the model. In models as simple as the present one, the explicit introduction of the assumptions would add complications without any real advantage.

Our second example will bear a closer resemblance to the type of problem considered in the familiar marginal analysis.

6-3. AN INCREASING-COST EXAMPLE

Consider a manufacturing firm which produces a single product. We shall imagine that this firm owns a main plant with a capacity of 100 units of output per hour and some stand-by obsolete equipment capable of producing 25 units per hour. We shall assume that direct material cost is \$0.50 per unit for production in the main plant and \$0.55 per unit for the stand-by plant. Direct labor cost is \$0.60 per unit in the main plant and \$0.70 per unit in the stand-by plant for schedules of operation up to 40 hours a week. After 40 hours, time and a half must be paid until 60 hours a week are reached. Conditions do not permit operations at more than 60 hours a week. Further alternatives, such as multishift operation, can easily be imagined, but the foregoing are sufficient for our purposes. We have now defined four productive activities, as follows:

1. Production in main plant on regular time
   Unit marginal cost: \$1.10
   Capacity: 4,000 units per week
2. Production in stand-by plant on regular time
   Unit marginal cost: \$1.25
   Capacity: 1,000 units per week
3. Production in main plant on overtime
   Unit marginal cost: \$1.40
   Capacity: 2,000 units per week
4. Production in stand-by plant on overtime
   Unit marginal cost: \$1.60
   Capacity: 500 units per week

Clearly, these four activities will enter operation successively as demand conditions call them forth.

This situation gives rise to an increasing-cost curve of the familiar sort. The average cost of any output x, denoted by $AC(x)$, can be derived from the marginal costs as follows. For $x \leq 4{,}000$,

$$AC(x) = \text{marginal cost} = \$1.10$$

For $4{,}000 < x \leq 5{,}000$ we have

$$AC(x) = \frac{1.10(4{,}000) + 1.25(x - 4{,}000)}{x} = 1.25 - \frac{600}{x}$$

For $5{,}000 < x \leq 7{,}000$ we have

$$\begin{aligned} AC(x) &= \frac{1.10(4{,}000) + 1.25(1{,}000) + 1.40(x - 5{,}000)}{x} \\ &= \frac{1.40x - 1{,}350}{x} \\ &= 1.40 - \frac{1{,}350}{x} \end{aligned}$$

For $7{,}000 < x \leqslant 7{,}500$ we have

$$AC(x) = \frac{1.10(4{,}000) + 1.25(1{,}000) + 1.40(2{,}000) + 1.60(x - 7{,}000)}{x}$$

$$= 1.60 - \frac{2{,}750}{x}$$

Figure 6-2 shows these marginal- and average-cost curves. Note that although the average-cost curve in the figure looks like a sequence of broken-line segments, it is really composed of very flat hyperbolas for all outputs above 4,000 units per week.

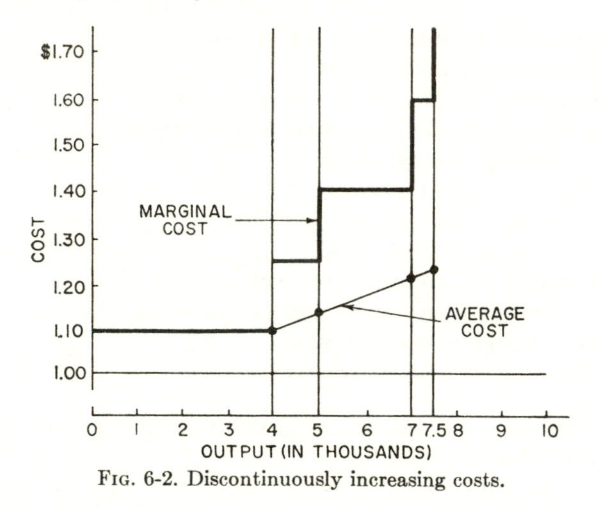

FIG. 6-2. Discontinuously increasing costs.

This simple example was introduced to show how conditions of varying marginal cost can be comprehended within the framework of linear programming. Suppose we postulate that a change in marginal cost must reflect some physical or economic change in the conditions of production. Then, if each change in productive conditions is thought to give rise to a new productive activity, a linear-programming model will describe the situation. If the number of variations to be considered is large, the linear-programming model will become unwieldy.[1] Apart from this there is no difficulty about it, provided marginal costs are increasing. Note that we do not need to specify that the four activities be introduced in the order listed. It will come out as part of the solution to any maximum-profit problem that no high-cost process should be used until all lower-cost processes have been exhausted.

[1] We mean here, conceptually unwieldy. As far as practical computations go, rising marginal costs present no difficulties if they are describable by step functions as in Fig. 6-2.

These two examples illustrate the linear-programming point of view. This point of view, as compared with that of the traditional smooth marginal analysis, involves a shift in the focus of attention. Instead of seeking the optimal combination of inputs and outputs, we seek the optimal combinations of levels of activities. Thus the linear-programming analysis provides more information than the marginal approach; it not only defines a goal in terms of optimal quantities of inputs and outputs, but it also gives specific directions for achieving this goal in terms of the various activities available to the firm. It should be noted, though, that linear programming requires more detailed information about the physical conditions of production than does the marginal approach.

On the other hand, a question of applicability arises. The marginal analysis assumes that the quantities of each factor and product can be varied infinitesimally and individually. Marginal analysis grows out of the concept of "dosing," best exemplified in agricultural production. In linear-programming terminology, the usual marginal analysis assumes that an infinite number of different processes are available, each derived from a similar one by a slight alteration in the proportions of inputs and outputs. Linear programming, by contrast, assumes that the production function can be decomposed into a finite number of activities, each characterized by physically determined ratios among inputs and outputs. Linear programming grows out of the limitations of production with the use of machinery. Which method is to be used in any specific analysis depends in part upon which assumptions conform best to the facts of the case at hand. By considering increasing numbers of processes, linear programming can be made to approach the continuous case as closely as may be desired.

6-4. A CHEMICAL EXAMPLE

Still working within the framework of the simplest set of economic assumptions, let us illustrate a slightly more complex production model. In the automobile example we discussed a multiproduct firm whose products competed with each other for the use of the limited means of production. Now we shall discuss an instance in which the products not only compete with each other in production, but stand in a complementary productive relationship as well. Such situations are likely to arise in vertically integrated firms.

Let us then imagine a chemical firm which produces three chemicals called simply C_1, C_2, and C_3. The firm uses four types of chemical equipment which we shall designate as E_1, E_2, E_3, and E_4. We assume that the firm has available three productive activities, as follows:

1. The production of \$100 worth of C_1 per month. This requires 10

per cent of the available capacity of E_1, 5 per cent of the available capacity of E_2, and $70 in direct costs.

2. The production of $100 worth of C_2 per month. This requires 4 per cent of the available capacity of E_2, 5 per cent of the available capacity of E_3, $30 worth of C_1, and $50 in direct costs.

3. The production of $100 worth of C_3 per month. This requires 2 per cent of the available capacity of E_2, 10 per cent of the available capacity of E_4, $10 worth of C_1, $50 worth of C_2, and $15 in direct costs.

These production opportunities are summarized in Table 6-3, in which a positive entry indicates that a commodity is produced and a negative entry indicates that a commodity or factor of production is consumed by the activity in question.

TABLE 6-3. MONTHLY PRODUCTION AND CONSUMPTION PER UNIT OF ACTIVITY IN A CHEMICAL PLANT

	Unit of measure	Activity		
		1	2	3
Item:				
C_1	Dollars	100	−30	−10
C_2	Dollars	0	100	−50
C_3	Dollars	0	0	100
E_1	Per cent of capacity	−10	0	0
E_2	Per cent of capacity	−5	−4	−2
E_3	Per cent of capacity	0	−5	0
E_4	Per cent of capacity	0	0	−10
Net revenue...	Dollars	−70	−50	−15

In Table 6-3 we exhibit the production opportunities open to the firm. But selling, not production, is the firm's objective. Accordingly we must introduce three selling activities, one for each of the products, as follows:

4. The sale of $100 worth of C_1 per month. This consumes $100 worth of C_1 and produces a gross revenue of $100.

5. The sale of $100 worth of C_2 per month. This consumes $100 worth of C_2 and produces a gross revenue of $100.

6. The sale of $100 worth of C_3 per month. This consumes $100 worth of C_3 and produces a gross revenue of $100.

The expanded schedule of activities is shown in Table 6-4.

We now express the situation of this firm algebraically and, to do so conveniently, shall use the concepts of "vectors" and "matrices."[1]

[1] The reader who is not familiar with these concepts may refer to Appendix B, in which their basic properties are outlined and illustrated. We have tried to make the present discussion self-contained, however, and do not assume familiarity with matrix algebra.

A vector is simply a list of numbers. Thus each of the activities defined above and listed in Table 6-4 is a column vector. We shall exclude the "net revenue" line of Table 6-4, so that each of the column vectors will be considered to have seven elements. We denote by $\mathbf{A}_i (i = 1, \ldots, 6)$

TABLE 6-4. PRODUCTION AND SELLING ACTIVITIES IN A CHEMICAL PLANT

	Unit of measure	Activity					
		1	2	3	4	5	6
Item:							
C_1	Dollars	100	−30	−10	−100	0	0
C_2	Dollars	0	100	−50	0	−100	0
C_3	Dollars	0	0	100	0	0	−100
E_1	Per cent of capacity	−10	0	0	0	0	0
E_2	Per cent of capacity	−5	−4	−2	0	0	0
E_3	Per cent of capacity	0	−5	0	0	0	0
E_4	Per cent of capacity	0	0	−10	0	0	0
Net revenue. .	Dollars	−70	−50	−15	100	100	100

the column vector of the ith activity. For example, for Activity 1 we have

$$\mathbf{A}_1 = \begin{bmatrix} 100 \\ 0 \\ 0 \\ -10 \\ -5 \\ 0 \\ 0 \end{bmatrix}$$

A matrix can be defined as a row of column vectors.[1] Let $\mathbf{A}$ denote the matrix consisting of the six activity vectors:

$$\mathbf{A} = (\mathbf{A}_1 \quad \mathbf{A}_2 \quad \mathbf{A}_3 \quad \mathbf{A}_4 \quad \mathbf{A}_5 \quad \mathbf{A}_6)$$

Then $\mathbf{A}$ will be simply the first seven rows of Table 6-4.

Let $x_i (i = 1, \ldots, 6)$ denote the level of the ith activity. Thus if \$1,000 worth of C_2 is produced, $x_2 = 10$, and if \$500 worth of C_3 is sold, $x_6 = 5$. Let y_1, y_2, y_3 denote the total production of C_1, C_2, C_3, respectively. Negative production will indicate consumption. And let y_4, y_5, y_6, y_7 denote the production of E_1, E_2, E_3, E_4, respectively. We must now express the relationship between the x's and the y's, i.e., between the activity levels and the quantities of input and output.

[1] A matrix can also be defined as a column of row vectors or as a rectangular array of numbers. These three definitions, clearly, all amount to the same thing. We chose the one that suited our context.

The program of the firm, i.e., its complete schedule of activities, is completely specified by the list of activity levels $x_1, x_2, \ldots, x_6$. We shall call such a list an activity vector and, sometimes, denote it by $\mathbf{x}$ (without a subscript) for short. Such an activity vector determines all the inputs and outputs; for consider its first component, x_1. This signifies that Activity 1 is being operated at level x_1; and as a result, looking at Table 6-4, $100x_1$ units of C_1 are produced, $-10x_1$ units of E_1 are produced (i.e., $10x_1$ units of E_1 are consumed), and $-5x_1$ units of E_2 are produced. Thus if Activity 1 is operated at level x_1 we can describe the resultant schedule of inputs and outputs by writing

$$\mathbf{A}_1x_1 = \begin{bmatrix} 100 \\ 0 \\ 0 \\ -10 \\ -5 \\ 0 \\ 0 \end{bmatrix} x_1 = \begin{bmatrix} 100x_1 \\ 0x_1 \\ 0x_1 \\ -10x_1 \\ -5x_1 \\ 0x_1 \\ 0x_1 \end{bmatrix}$$

Similarly, if Activity 2 is operated at level x_2 we can obtain the vector of inputs and outputs resulting from this activity by multiplying the vector $\mathbf{A}_2$, element by element, by the level x_2. We write this as $\mathbf{A}_2x_2$. Finally, the over-all schedule of inputs and outputs is the sum of the inputs and outputs of the six individual activities, or,

$$\begin{bmatrix} y_1 \\ y_2 \\ y_3 \\ y_4 \\ y_5 \\ y_6 \\ y_7 \end{bmatrix} = \mathbf{A}_1x_1 + \mathbf{A}_2x_2 + \mathbf{A}_3x_3 + \mathbf{A}_4x_4 + \mathbf{A}_5x_5 + \mathbf{A}_6x_6 \tag{6-1}$$

For short we shall let $\mathbf{y}$ stand for the column of the seven y's. We also introduce a more convenient notation for the right-hand side. The matrix $\mathbf{A}$, it will be recalled, is simply the matrix made up of the six activity vectors. The vector $\mathbf{x}$ is the vector composed of the six activity levels. We then define the matrix product $\mathbf{Ax}$ to mean the result of multiplying each of the six vectors $\mathbf{A}_i$ by the corresponding level x_i and adding. Then $\mathbf{Ax}$ is just the right-hand side of Eq. (6-1), and we can write the relationship between the list of inputs and outputs, $\mathbf{y}$, and the list of activity levels, $\mathbf{x}$, as

$$\mathbf{y} = \mathbf{Ax}$$

We must next consider limitations on the operations of the firm because, apart from them, the firm could obtain infinite profits per unit of time. One type of restriction results from the fact that in a static equilibrium solution, net change in inventories must be zero; i.e., $y_1 = y_2 = y_3 = 0$. This means that the whole gross output of each chemical must be absorbed either as input to another production process or as input to its selling activity. Even if we were to permit net accumulation of inventories, to do so would require direct costs and would add nothing to net revenue. Hence no optimal or maximum net-revenue solution would ever involve adding to inventory. Of course, a model that extended over several time periods might well involve varying inventories. Another type of restriction results from the presence of four kinds of fixed equipment. The total utilization of each of the four kinds of equipment must be no greater than 100 per cent. Symbolically, $y_i \geq -100$ for $i = 4, 5, 6, 7$. These considerations lead to a vector of restrictions, which we call **s**. Writing it out in full,

$$\mathbf{s} = \begin{bmatrix} 0 \\ 0 \\ 0 \\ -100 \\ -100 \\ -100 \\ -100 \end{bmatrix}$$

The restrictions within which the firm must operate can now be summarized as

$$\mathbf{y} = \mathbf{A}\mathbf{x} \geq \mathbf{s} \tag{6-2}$$

A combination of activities **x** is *feasible* for the firm if and only if it satisfies Eq. (6-2) and the requirement that no activity can be operated at a negative level.

Before passing on, let us illustrate this restriction. It consists of seven lines. The fifth line, to choose an example, is

$$y_5 = -5x_1 - 4x_2 - 2x_3 + 0x_4 + 0x_5 + 0x_6 \geq -100$$

The left-hand member is the symbol for the production of equipment services of type E_2. The middle member is made up of the coefficients found on the fifth row of Table 6-4, each multiplied by the level of the corresponding activity. The whole sum is the output of services of type E_2 and is essentially negative, indicating that such services are consumed rather than produced. The right-hand member sets a limit to the rate at which such services can be consumed.

Equation (6-2), it should be noted, actually permits inventories to be accumulated, though not drawn down; i.e., it permits the net output resulting from Activities 1, 2, and 3 to exceed sales. It seems unlikely that such inventory accumulation would turn out to be advisable in a static problem, but there is no harm in permitting it and it simplifies the exposition a little.

Now we have specified what the firm *can* do and how that range of choice is related to the technical characteristics of the available activities and to the resource limitations. We next consider the specification of the best program within that range of choice. The net revenue resulting from any program is the sum of the inflows and outflows of money that result from the activities included in that program. Thus, referring to the last line of Table 6-4, if Activity 1 is operated at level x_1 (meaning the production of $\$100x_1$ worth of C_1), it causes an outflow of $\$70x_1$; if Activity 5 is operated at level x_5 (meaning the sale of $\$100x_5$ worth of C_2), it causes an inflow of $\$100x_5$. Thus the total net revenue, which we shall denote by r, is

$$r = -70x_1 - 50x_2 - 15x_3 + 100x_4 + 100x_5 + 100x_6$$

Note that the net revenue r is the sum of the cross products of the activity levels $x_1, \ldots, x_6$, i.e., of the activity vector $\mathbf{x}$, with a list of coefficients -70, -50, -15, 100, 100, 100. We have already learned to call such lists vectors. Let us call this vector $\mathbf{v}$ and its components $v_1, \ldots, v_6$, respectively. Then r is the sum of the cross products of the components of vectors $\mathbf{v}$ and $\mathbf{x}$. Such a sum of cross products of two vectors is called an "inner product" and is written $[\mathbf{v},\mathbf{x}]$. Using this notation, we can write[1]

$$r = [\mathbf{v},\mathbf{x}]$$

We now assume that the objective of the firm is to achieve a production program $\mathbf{x}$ that will make net revenue r as great as possible, subject to the resource and other limitations imposed. Thus we can state the production problem succinctly. It is to find a program vector $\mathbf{x}$ which satisfies

1. $\mathbf{x} \geq 0$. (No activity can be carried on at a negative level.)
2. $\mathbf{Ax} \geq \mathbf{s}$. (No more than the total supply of any commodity or equipment can be consumed.)
3. $r = [\mathbf{v},\mathbf{x}] =$ as great as possible. (Net revenue is to be maximized.)

The rest is a matter of calculation in accordance with the methods presented elsewhere in this volume. The calculations, however, have an

[1] A statistician might use the notation $r = \Sigma \mathbf{vx}$ to express the same relationship, but we prefer the inner-product notation as being more consistent with the vectorial point of view.

economic as well as an algebraic significance, and we shall examine now the economic ideas underlying the numerical solution.

As a preliminary to this examination we must restate the problem in a convenient form for solution. The object of the restatement is to avoid some of the inequalities which resulted from the economic analysis and which are awkward mathematically. In our analysis, for example, we found the inequality

$$5x_1 + 4x_2 + 2x_3 \leq 100 \tag{6-3}$$

which expressed the requirement that the production program $x_1, x_2, \ldots, x_6$ could not consume more than 100 per cent of the capacity of E_2. We now define a new activity $\mathbf{A}_8$ which consists of allowing 1 per cent of the capacity of E_2 to go unused. Let x_8 denote the level of $\mathbf{A}_8$. Then x_8 is the amount of unused capacity of type E_2, or

$$x_8 = 100 - (5x_1 + 4x_2 + 2x_3)$$

The requirement expressed by inequality (6-3) can be expressed equally well by

$$\begin{gathered} 5x_1 + 4x_2 + 2x_3 + x_8 = 100 \\ 0 \leq x_1, x_2, x_3, x_8 \end{gathered}$$

This is only a formal and not a conceptual change, but it has definite mathematical advantages. To gain these advantages we define

$$\begin{aligned} \mathbf{A}_7 &= \text{activity of allowing } 1\% \text{ of } E_1 \text{ to be unused} \\ \mathbf{A}_8 &= \text{activity of allowing } 1\% \text{ of } E_2 \text{ to be unused} \\ \mathbf{A}_9 &= \text{activity of allowing } 1\% \text{ of } E_3 \text{ to be unused} \\ \mathbf{A}_{10} &= \text{activity of allowing } 1\% \text{ of } E_4 \text{ to be unused} \\ x_7 &= \text{level of } \mathbf{A}_7 \\ x_8 &= \text{level of } \mathbf{A}_8 \\ x_9 &= \text{level of } \mathbf{A}_9 \\ x_{10} &= \text{level of } \mathbf{A}_{10} \end{aligned}$$

The problem now becomes that of maximizing

$$r = [\mathbf{v},\mathbf{x}] = -70x_1 - 50x_2 - 15x_3 + 100x_4 + 100x_5 + 100x_6 + 0x_7 + 0x_8 + 0x_9 + 0x_{10}$$

subject to

(a) $$0 \leq x_1, x_2, \ldots, x_{10}$$

and (b)

$$\begin{bmatrix} 100 & -30 & -10 & -100 & 0 & 0 & 0 & 0 & 0 & 0 \\ 0 & 100 & -50 & 0 & -100 & 0 & 0 & 0 & 0 & 0 \\ 0 & 0 & 100 & 0 & 0 & -100 & 0 & 0 & 0 & 0 \\ -10 & 0 & 0 & 0 & 0 & 0 & -1 & 0 & 0 & 0 \\ -5 & -4 & -2 & 0 & 0 & 0 & 0 & -1 & 0 & 0 \\ 0 & -5 & 0 & 0 & 0 & 0 & 0 & 0 & -1 & 0 \\ 0 & 0 & -10 & 0 & 0 & 0 & 0 & 0 & 0 & -1 \end{bmatrix} \times \begin{bmatrix} x_1 \\ x_2 \\ x_3 \\ x_4 \\ x_5 \\ x_6 \\ x_7 \\ x_8 \\ x_9 \\ x_{10} \end{bmatrix} = \begin{bmatrix} 0 \\ 0 \\ 0 \\ -100 \\ -100 \\ -100 \\ -100 \end{bmatrix}$$

In interpreting condition b we use the definition of the product of a matrix and a column vector. It may help to restate the definition in the following way: The product of a matrix and a column vector is itself a column vector. The first element of the product vector is the inner product of the first row of the matrix with the given column vector; the second element of the product is the inner product of the second row of the matrix with the given column vector; etc. Thus condition b is a set of seven simultaneous linear equations in 10 unknowns. If arbitrary values are assigned to any three of the unknowns, the result will be a set of seven equations in the remaining seven unknowns. To be sure, if three arbitrary values are assigned, some of the unknowns may turn out to be negative, thus violating condition a, but condition b can be satisfied easily enough.

We shall now go through the simplex method for finding an optimal program of this example. The algebra of the simplex method was discussed in Chap. 4. Our present objective is to reveal the economic interpretation of this method. The procedure for finding an initial set of activity levels (i.e., an activity vector $\mathbf{x}$) that satisfies both restrictions a and b begins with setting three of the x's equal to zero. Economically this consists in deciding not to use three of the activities. For instance, if we set $x_8 = x_9 = x_{10} = 0$, the reader should verify that the effect is to decide to manufacture and sell all three of the chemicals, to permit a surplus of equipment E_1, and to utilize fully E_2, E_3, and E_4. Is it possible to carry out this decision? This question can be answered by setting $x_8 = x_9 = x_{10} = 0$ and solving the equations of condition b for the remaining $x_1, x_2, \ldots, x_7$. If all the x's in the solution should turn out to be nonnegative, the program is possible; otherwise it is not. A possible program is known technically as a "feasible" program.

Making this calculation for $x_8 = x_9 = x_{10} = 0$, straightforward algebra yields

$$\begin{array}{lll} x_1 = 0 & x_4 = -7 & x_6 = 10 \\ x_2 = 20 & x_5 = 15 & x_7 = 100 \\ x_3 = 10 & & \end{array}$$

Since x_4 turned out to be negative, i.e., since the program required selling *minus* \$700 worth of C_1, the program is not possible. We shall not go into the mechanics of searching for a feasible program. A number of mathematical procedures are available, but in a small problem like this one the best method is simply trial and error. Thus, if we set $x_7 = x_8 = x_{10} = 0$, restriction b reduces to

$$\begin{bmatrix} 100 & -30 & -10 & -100 & 0 & 0 & 0 \\ 0 & 100 & -50 & 0 & -100 & 0 & 0 \\ 0 & 0 & 100 & 0 & 0 & -100 & 0 \\ -10 & 0 & 0 & 0 & 0 & 0 & 0 \\ -5 & -4 & -2 & 0 & 0 & 0 & 0 \\ 0 & -5 & 0 & 0 & 0 & 0 & -1 \\ 0 & 0 & -10 & 0 & 0 & 0 & 0 \end{bmatrix} \begin{bmatrix} x_1 \\ x_2 \\ x_3 \\ x_4 \\ x_5 \\ x_6 \\ x_9 \end{bmatrix} = \begin{bmatrix} 0 \\ 0 \\ 0 \\ -100 \\ -100 \\ -100 \\ -100 \end{bmatrix}$$

and solution of these equations yields

$$\begin{array}{ll} x_1 = 10 & \text{i.e., produce \$1,000 worth of } C_1 \\ x_2 = 7.5 & \text{i.e., produce \$ 750 worth of } C_2 \\ x_3 = 10 & \text{i.e., produce \$1,000 worth of } C_3 \\ x_4 = 6.75 & \text{i.e., sell \$ 675 worth of } C_1 \\ x_5 = 2.5 & \text{i.e., sell \$ 250 worth of } C_2 \\ x_6 = 10 & \text{i.e., sell \$1,000 worth of } C_3 \\ x_9 = 62.5 & \text{i.e., leave unused 62.5\% of capacity of type } E_3 \end{array}$$

Since none of the x's are negative, this program is feasible.

In general there will be many feasible programs, i.e., programs which satisfy both restrictions; and the essence of the problem is to select from all the feasible programs the one which yields the greatest possible profit. The theory of linear programming assures us that an optimal program can be found by searching through all feasible programs of a certain type, namely, those feasible programs which involve using no more processes than there are equations in restriction b. Programs of this type are known as basic programs. In this case there are seven equations in restriction b, and the feasible program we have just found involves no more than seven processes. So this is a basic program.

The idea underlying the search is this: It begins with any basic and feasible program. A rule is provided for testing whether that choice is optimal. (Indeed, the economic significance of this rule is our primary interest in going through the detailed calculation.) If the program under test does not turn out to be optimal, there is a procedure for finding a better program which is also basic. The improved program is then tested, and the procedure is repeated as many times as is necessary until an optimal program is found. An optimal program always will be found (if one exists) since each revision results in improvement and there are only a finite number of programs among which to search.

We now need the rule for testing whether a given program, in particular the one just found, is optimal. Recall that the program is completely specified by selecting three processes and setting their levels equal to zero. In the present case we set the levels of $\mathbf{A}_7$, $\mathbf{A}_8$, and $\mathbf{A}_{10}$ equal to zero. Now consider $\mathbf{A}_7$. If we add a small amount of it to the program and readjust the levels of the previously used processes so that the resulting eight-process program is feasible, we might either increase or decrease net revenue. If this increases net revenue, the program under test could not be optimal, for we have found a better one. Turning to the contrary case, assume that the introduction of $\mathbf{A}_7$ does not increase profits, and neither does the introduction of $\mathbf{A}_8$ or $\mathbf{A}_{10}$. Then it can be shown that if profits cannot be increased by introducing $\mathbf{A}_7$, $\mathbf{A}_8$, $\mathbf{A}_{10}$ individually, they cannot be increased by introducing any combination of them and, therefore, cannot be increased at all.

All that is needed then is a procedure for determining the effect on profits of introducing each of the excluded processes separately. Consider the effect of introducing one unit of $\mathbf{A}_7$, i.e., of allowing 1 per cent of E_1 capacity to go unused. The levels of the previously used seven processes must now be changed so that in the aggregate they consume 1 per cent less of E_1 capacity while still satisfying all the other equations of condition b. In more general terms, if any excluded process is introduced, then the previously used processes must adjust by amounts which just offset the consumption of materials by the process being introduced. Now suppose we introduce $\mathbf{A}_7$ in amount Δx_7. Let Δx_1 denote the necessary change in the level of $\mathbf{A}_1$, Δx_2 the change in the level of $\mathbf{A}_2$, etc. Then the set of ratios

$$\left[\frac{\Delta x_1}{\Delta x_7}, \frac{\Delta x_2}{\Delta x_7}, \ldots, \frac{\Delta x_9}{\Delta x_7}\right]$$

is known as the "equivalent combination" to $\mathbf{A}_7$ because it is the combination of previously used processes that consumes the same amount of all resources as a unit of $\mathbf{A}_7$. The equivalent combination is perfectly determinate,[1] as we shall see.

We have already introduced the numbers $v_1, \ldots, v_6$ to denote the net revenue per unit of Activities 1 to 6, respectively. Let us define v_7, v_8, v_9, and v_{10} analogously for Activities 7 to 10. These four values are all zeros, but that does not affect the argument. In these terms the effect of introducing Δx_7 units of $\mathbf{A}_7$ is to increase total net revenue by $v_7\,\Delta x_7$ and to decrease total net revenue, because of the readjustment of the previously used processes, by $v_1\,\Delta x_1 + v_2\,\Delta x_2 + \cdots + v_9\,\Delta x_9$. The net effect on over-all net revenue is favorable if

$$v_1\,\Delta x_1 + v_2\,\Delta x_2 + \cdots + v_9\,\Delta x_9 < v_7\,\Delta x_7$$

or, dividing by Δx_7, if

$$\frac{v_1\,\Delta x_1}{\Delta x_7} + \frac{v_2\,\Delta x_2}{\Delta x_7} + \cdots + \frac{v_9\,\Delta x_9}{\Delta x_7} < v_7$$

otherwise it is not favorable.

This is the test we have been looking for. The $v_7, v_1, v_2, \ldots, v_9$ are elements of the $\mathbf{v}$ vector previously defined, so all that remains is to determine the equivalent combination. In the matrix notation previously introduced, $\mathbf{A}_i$ denoted a vector whose elements showed how much of each resource 1 unit of the ith process produced or consumed. The $\mathbf{A}_i$'s are just the columns of restriction b. Thus,

$$\begin{aligned} \mathbf{A}_1 &= [100 \quad 0 \quad 0 \quad -10 \quad -5 \quad 0 \quad 0] \\ \mathbf{A}_7 &= [\;\;0 \quad 0 \quad 0 \quad -1 \quad 0 \quad 0 \quad 0] \end{aligned}$$

and so forth.[2]

Now consider a set of changes such as $\Delta x_1, \Delta x_2, \ldots, \Delta x_9$. These changes may be positive or negative or zero. The aggregate effect of these changes is to change the consumption of resources by

$$\mathbf{A}_1\,\Delta x_1 + \mathbf{A}_2\,\Delta x_2 + \cdots + \mathbf{A}_9\,\Delta x_9$$

The equivalent combination to $\mathbf{A}_7$ is the set of changes $\Delta x_1, \Delta x_2, \ldots, \Delta x_9$ which consumes the same amount of all resources as a unit of $\mathbf{A}_7$, namely, the combination which satisfies

$$\mathbf{A}_1\,\Delta x_1 + \mathbf{A}_2\,\Delta x_2 + \cdots + \mathbf{A}_9\,\Delta x_9 = \mathbf{A}_7$$

This is a set of seven equations in seven unknowns and can be solved straightforwardly.

[1] Except in certain special cases not relevant to the present discussion.

[2] We have written these vectors horizontally to save space. We adopt the convention that a list of numbers included in brackets is to be interpreted as a column vector, even if written horizontally.

To recapitulate: In order to test the optimality of any program, consider separately each process excluded from that program. Compute the equivalent combination for each excluded process. Find the net profitability of each equivalent combination and compare it with the profitability[1] of the excluded process. If any excluded process is more profitable than its equivalent combination, then the program under test is not optimal and that excluded process should be added to the program. If every equivalent combination is at least as profitable as the corresponding excluded process, then the program under test is optimal.

We now apply this test to the program we have found. Naturally the test is purely formal and hides the economic considerations which justify it. The first step in the test is to compute the equivalent combination for $\mathbf{A}_7$. The equations are abstracted directly from the data in condition b and are

$$\begin{bmatrix} 100 & -30 & -10 & -100 & 0 & 0 & 0 \\ 0 & 100 & -50 & 0 & -100 & 0 & 0 \\ 0 & 0 & 100 & 0 & 0 & -100 & 0 \\ -10 & 0 & 0 & 0 & 0 & 0 & 0 \\ -5 & -4 & -2 & 0 & 0 & 0 & 0 \\ 0 & -5 & 0 & 0 & 0 & 0 & -1 \\ 0 & 0 & -10 & 0 & 0 & 0 & 0 \end{bmatrix} \begin{bmatrix} \Delta x_1 \\ \Delta x_2 \\ \Delta x_3 \\ \Delta x_4 \\ \Delta x_5 \\ \Delta x_6 \\ \Delta x_9 \end{bmatrix} = \begin{bmatrix} 0 \\ 0 \\ 0 \\ -1 \\ 0 \\ 0 \\ 0 \end{bmatrix}$$

This is a simple set of linear equations with the solution

$$\begin{array}{lll} \Delta x_1 = 0.1 & \Delta x_3 = 0 & \Delta x_5 = -0.125 \\ \Delta x_2 = -0.125 & \Delta x_4 = 0.1375 & \Delta x_6 = 0 \\ & & \Delta x_9 = 0.625 \end{array}$$

The profitabilities of the seven processes in the program are

$$\begin{array}{lll} v_1 = -70 & v_4 = 100 & v_6 = 100 \\ v_2 = -50 & v_5 = 100 & v_9 = 0 \\ v_3 = -15 & & \end{array}$$

Cross-multiplying these two sets of numbers and adding, the profitability of the equivalent combination for $\mathbf{A}_7$ is found to be 0.5. The profitability of $\mathbf{A}_7$ is $v_7 = 0$, so that in this case the equivalent combination is more profitable than the excluded process.

Process $\mathbf{A}_8$ is considered next. Its equivalent combination is

$$\begin{array}{lll} \Delta x_1 = 0 & \Delta x_4 = -0.075 & \Delta x_6 = 0 \\ \Delta x_2 = 0.25 & \Delta x_5 = 0.25 & \Delta x_9 = -1.25 \\ \Delta x_3 = 0 & & \end{array}$$

[1] In subsequent discussion we shall use the word "profit" as a synonym for "net revenue."

and the profitability of that combination is 5.0. This again exceeds the profitability of the direct process, namely, $v_8 = 0$. And finally with respect to $\mathbf{A}_{10}$, the profitability of the equivalent process is 1.5, which exceeds $v_{10} = 0$, the profitability of the direct process.

Thus the program we have found is not only feasible but optimal. It will yield a net revenue of $700 per month, and that is the greatest profit which can be obtained in the circumstances of this firm.

Inspection of this result shows that the firm should use its full capacity of E_1, E_2, and E_4, but only 37.5 per cent of its capacity of E_3. This conclusion is not entirely surprising if we note (from Table 6-3 or 6-4) that E_2 is used for manufacturing all three chemicals, while each of the other types of equipment is used in only a single activity. Now, making $100 worth of C_1 requires 5 per cent of the capacity of E_2 and yields a net revenue (value of output minus direct costs) of $30. Hence when E_2 is used to manufacture C_1 it produces a net revenue of $6 per per cent of capacity. Similarly when E_2 is used to produce C_2 it yields $5 per per cent of capacity, and when it is used to manufacture C_3 it yields $12.50 per per cent of capacity. Clearly, then, C_3 should be manufactured to the greatest extent possible (the limitation lying in the capacity of E_4) and the preferred use of the remaining capacity of E_2 is the manufacture of C_1. The limitation on the output of C_1 is the capacity of E_1, and when this is reached there is still more capacity of E_2 left over. C_2 should then be manufactured to the extent made possible by this remaining capacity, and when this limit is reached there is still some excess capacity of E_3 which must be left unused.

In the last paragraph we used a common-sense analysis to arrive at the conclusions which we had already reached by means of linear programming. It will be noted that this common-sense reasoning was rather tortuous and that it involved at several stages a careful weeding out of possibilities, depending on which of the possible limitations on output was effective. This method of reasoning was possible at all only because we had a small number of possibilities to take into account and at each stage things fell out in the simplest possible way. Even in this simple case, the common-sense reasoning was delicate, and a more complicated problem could hardly be analyzed without recourse to formal linear programming.

It is worth drawing attention to the three selling activities introduced in Table 6-4. These are examples of the "slough-off," or "disposal," activities which play an important role in the simplex solution of linear-programming problems. Such activities consist in getting rid of or not using the various items which enter into the problem. In the numerical solution of this chemical example, seven disposal activities appear: the three selling activities plus one disposal activity for each of the four types of equipment. Thus in this example, although there are only three productive processes, a total of 10 different activities must be considered. In the optimum program we found that four disposal activities must be used: the three selling activities plus the disposal of 62.5 per cent of the capacity of E_3. Optimal solutions to linear-programming problems quite typically include disposal activities along with actively productive ones.

6-5. A SUIT-MANUFACTURING EXAMPLE[1]

We have now presented the basic framework for analyzing the production problem of the firm by means of linear programming. We have seen that this mode of analysis becomes applicable when a firm engages in a number of interrelated processes whose relationships can be approximated to a satisfactory degree by a set of linear equations and inequalities. All the examples discussed so far have depended on the fact that the various processes available to the firm competed for the use of resources whose availability to the firm was subject to an absolute limitation. It is true that the problem of production frequently takes this form, especially for planning in the relatively short run. But other situations, too, can give rise to problems of the linear-programming type. One class of such problems arises when a number of products are produced jointly and, for some reason, must be sold in certain definite proportions. Then if the productive conditions are such that the products need not be produced in precisely the proportions required for sale, a linear-programming problem exists even if there are no limitations on the supply of the factors of production at constant cost.

Consider, for example, a manufacturer of men's suits.[2] His raw material consists of large bolts of cloth on which he must arrange his patterns so as to obtain as many suits as possible from each bolt. A suit may consist of a coat and a pair of trousers or a coat and two pairs of trousers. In any event, because of the difficulty of arranging irregular clothing patterns on a rectangular bolt of cloth, there is no guarantee that any arrangement can be found which will produce coats and trousers in exactly the desired proportions without excessive wastage, or, what is the same thing, that it is efficient to obtain the desired proportions from each bolt individually.

Suppose, to be concrete, that with respect to a certain model of suit a

[1] This section presents still another elementary example and may be omitted if the concepts seem sufficiently familiar.

[2] The problem of manufacturing a suitable variety of sizes will be neglected for simplicity. We shall also imagine that a suit consists of only two or three pieces instead of the many different cuts which have to be sewn together in actuality. These assumptions have no effect on the reasoning.

cutter reports to the manufacturer that he can arrange the patterns in four different ways, as shown in Table 6-5.

TABLE 6-5. OUTPUT OF COATS AND TROUSERS PER BOLT IN A CLOTHING FIRM

Pattern	Trousers	Coats
1	90	35
2	80	55
3	70	70
4	60	90

Suppose also that this manufacturer deals in two-pants suits. The manufacturer must decide how many bolts to cut according to each pattern.[1]

The problem is thus to find a combination of patterns which will yield as many complete suits as possible from a definite number of bolts of cloth, say 100. This is equivalent to requiring that the manufacturer cut as many trousers as possible subject to the requirements that (1) he obtain at least one coat for every two pairs of trousers, and (2) he use no more than 100 bolts of cloth.

Now set up the problem in terms of six alternative activities as follows:

Activity	*Definition*
1	Cutting a bolt according to Pattern 1
2	Cutting a bolt according to Pattern 2
3	Cutting a bolt according to Pattern 3
4	Cutting a bolt according to Pattern 4
5	Discarding or sloughing off a bolt
6	Discarding or sloughing off a coat

Denote by x_1, x_2, x_3, x_4, x_5 the number of units used of the first five activities, respectively. For arithmetic convenience, let $5x_6$ denote the number of units of the sixth activity.

Using this notation, the number of pairs of trousers produced is

$$90x_1 + 80x_2 + 70x_3 + 60x_4$$

This is to be maximized. The restraints under which this maximization

[1] In the actual clothing industry this problem does not ordinarily arise. All sections of better-quality suits are taken from the same part of the same bolt in order to ensure uniform weave and dye. For less expensive suits all sections are also taken from the same bolt, though it is not clear whether the precaution is justified in this case. Nevertheless, this highly simplified example brings out the issues met in joint production of commodities desired in rigid ratios.

is to be achieved are

$$x_1, x_2, \ldots, x_6 \geq 0$$
$$x_1 + x_2 + x_3 + x_4 + x_5 = 100$$
$$35x_1 + 55x_2 + 70x_3 + 90x_4 - 5x_6 = \tfrac{1}{2}(90x_1 + 80x_2 + 70x_3 + 60x_4)$$

Gathering terms, we can write this last equation:

$$10x_1 - 15x_2 - 35x_3 - 60x_4 + 5x_6 = 0$$

or

$$2x_1 - 3x_2 - 7x_3 - 12x_4 + x_6 = 0$$

We may now restate the problem in linear-programming form, as follows:

To maximize

$$90x_1 + 80x_2 + 70x_3 + 60x_4$$

subject to

$$\begin{bmatrix} 1 & 1 & 1 & 1 & 1 & 0 \\ 2 & -3 & -7 & -12 & 0 & 1 \end{bmatrix} \begin{bmatrix} x_1 \\ x_2 \\ x_3 \\ x_4 \\ x_5 \\ x_6 \end{bmatrix} = \begin{bmatrix} 100 \\ 0 \end{bmatrix}$$

We know that since the number of restraining equations is two, no more than two of the six processes have to be used. Our problem is to determine which two. Now each process has a value which is equal to the number of pairs of trousers obtained by using a unit of the process. These values are given in Table 6-6.

TABLE 6-6. VALUES OF SIX PROCESSES IN A CLOTHING FIRM

Process	*Value*
1	90
2	80
3	70
4	60
5	0
6	0

Each of these six processes may be performed either directly or by an equivalent combination. This may be seen as follows. Consider Process 2. It consists in converting a bolt of cloth into 55 coats and a certain number of pairs of trousers. We can obtain the same result by cutting $\frac{3}{7}$ of a bolt by Pattern 1 and $\frac{4}{7}$ of a bolt by Pattern 3 for

$$\tfrac{3}{7}(35) + \tfrac{4}{7}(70) = 55 \text{ coats}$$

and thus a combination of Processes 1 and 3 converts a bolt into the same number of coats as Process 2. But the direct Process 2 produced

80 pairs of trousers, while the equivalent combination of Processes 1 and 3 produced only $78\frac{4}{7}$ pairs. Thus a direct process and its equivalent combination do not necessarily, or generally, yield equivalent results in terms of a measure of value.

The same reasoning applies to a disposal process such as Process 5, since Process 5 merely consumes a bolt of cloth without producing any coats. But if any production plan is altered by cutting 2 additional bolts by Pattern 1 and 1 less bolt by Pattern 3, the consumption of cloth will be increased by 1 bolt without changing the output of coats.

Thus 2 units of Process 1 minus 1 unit of Process 3 is an equivalent combination to 1 unit of Process 5. But Process 5 does not produce any trousers, while 2 units of Process 1 minus 1 unit of Process 2 yields a net increase of 110 pairs of trousers. Here again, when the item we are trying to maximize (output of trousers) is considered, the direct process and its equivalent combination are not equal.

The existence of equivalent combinations is fundamental to the solution. Consider any two processes. By using them we can construct an equivalent combination for each of the remaining four processes. We can also compare the "value" (i.e., the number of pairs of trousers produced) of each direct process with the value of its equivalent combination. Now suppose that we make these comparisons and find that in each case the value of the equivalent combination is at least as great as the value of the direct process. Then, according to the simplex criterion, the two processes that were used in constructing the equivalent combination are the two which should be used by the manufacturer. On the other hand, if any pair does not pass this test it is not the best pair for the manufacturer to use.

By means of the simplex method the pair of processes which meets this test is found to be Processes 1 and 2. The best use that the manufacturer can make of his 100 bolts is to cut 60 according to Pattern 1 and 40 according to Pattern 2, obtaining 4,300 complete suits.

In order to see how solutions to such problems respond to changes in the conditions it is interesting to look for the best plan for manufacturing suits with one pair of trousers. The analysis is the same as before except that now the second restraint on the maximizing process becomes

$$35x_1 + 55x_2 + 70x_3 + 90x_4 - 5x6 = 90x_1 + 80x_2 + 70x_3 + 60x_4$$

which simplifies to

$$11x_1 + 5x_2 - 6x_4 + x_6 = 0$$

In linear-programming form, then, we must maximize

$$90x_1 + 80x_2 + 70x_3 + 60x_4$$

subject to

$$\begin{bmatrix} 1 & 1 & 1 & 1 & 1 & 0 \\ 11 & 5 & 0 & -6 & 0 & 1 \end{bmatrix} \begin{bmatrix} x_1 \\ x_2 \\ x_3 \\ x_4 \\ x_5 \\ x_6 \end{bmatrix} = \begin{bmatrix} 100 \\ 0 \end{bmatrix}$$

Exactly the same considerations apply as before, and this time we find that the optimal plan includes Activities 2 and 4, but not Activity 3, which, all by itself, would produce coats and trousers in the desired proportions.

The reason for this can be seen by making a graph of the situation.

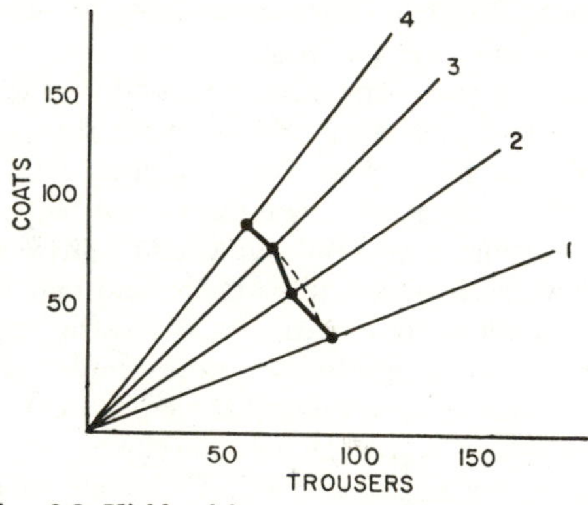

FIG. 6-3. Yields of four patterns for cutting suits.

Figure 6-3 shows the alternatives open to the manufacturer. The four points on the heavy solid broken line show the number of coats (ordinate) and trousers (abscissa) produced by cutting 1 bolt according to each of the patterns. Cutting ½ bolt or 2 bolts or any other number of bolts according to Pattern 1 would yield points corresponding to radius vector 1, and the other radius vectors show similarly the yields obtained by cutting any desired number of bolts according to one or another of the patterns. Let us call these points P_1, P_2, P_3, P_4. The lines connecting them have an important significance. If the manufacturer takes a bolt and cuts it partly according to Pattern 1 and partly according to Pattern 2, the yield will be somewhere along the line connecting P_1 and P_2, and similar interpretations apply to the other lines shown.[1]

[1] *Proof:* Suppose Pattern 1 yields c_1 coats and t_1 trousers per bolt, Pattern 2 yields c_2 coats and t_2 trousers, and z is a number in the interval 0 to 1. Then z bolts cut by Pattern 1 and $1 - z$ bolts cut by Pattern 2 yield $c_1z + c_2(1 - z) = c_2 + (c_1 - c_2)z$ coats and $t_1z + t_2(1 - z)$ trousers. By elementary algebra, the relationship between

Now notice that P_3 lies below the line connecting P_2 and P_4. This means that if the manufacturer cuts part of a bolt according to P_2 and part according to P_4, and if these parts are chosen correctly, he can get coats and trousers in the same proportion (1:1) as if he cut the whole bolt according to P_3. And furthermore, since P_3 lies below this line, it means that he can get more coats and trousers per bolt by using the appropriate combination of P_2 and P_4 than by using P_3. Thus the equivalent combination of P_3 is preferable to the direct use of P_3. If, however, P_3 yielded 72 coats and 72 trousers instead of 70 coats and 70 trousers, the reverse would be the case. P_3 would then be above the line P_2P_4 and solution of the linear-programming problem in this altered form would show that the most economical plan for the manufacturer would be to cut all his cloth according to P_3.

In analyzing this problem we came across some rather peculiar equations like

$$2x_1 - 3x_2 - 7x_3 - 12x_4 + x_6 = 0$$

which did not have any apparent economic significance. These equations crept into the discussion because we did not treat the outputs of coats and trousers symmetrically. They can be avoided, if desired, by formulating the problem in an alternative, but equivalent, manner. Let us consider the two-pants-suit case from a different point of view. We can suppose the manufacturer to ask himself: How can I manufacture a definite number of coats (say 1,000) and a definite number of pairs of pants (say 2,000) from the smallest possible amount of cloth?

As before, let x_1, x_2, x_3, x_4 be the numbers of bolts cut according to Patterns 1, 2, 3, and 4, respectively; let x_6 be one-fifth the number of coats discarded; and let x_7 be one-tenth the number of pairs of trousers discarded. Then the manufacturer wants to find positive numbers, x_1, x_2, x_3, x_4, x_6, x_7, such that

$$\begin{aligned} 90x_1 + 80x_2 + 70x_3 + 60x_4 - 10x_6 &= 2{,}000 \\ 35x_1 + 55x_2 + 70x_3 + 90x_4 - 5x_7 &= 1{,}000 \end{aligned}$$

and $x_1 + x_2 + x_3 + x_4$ is as small as possible.

In linear-programming form, he has to minimize

$$z = x_1 + x_2 + x_3 + x_4$$

---

the number of coats and the number of trousers corresponding to z can be seen to be

$$\frac{\text{Coats} - c_1}{\text{Trousers} - t_1} = \frac{c_2 - c_1}{t_2 - t_1}$$

This is the equation of the straight line connecting P_1 and P_2 in Fig. 6-3.

subject to

$$\begin{bmatrix} 90 & 80 & 70 & 60 & 0 & -10 \\ 35 & 55 & 70 & 90 & -5 & 0 \end{bmatrix} \begin{bmatrix} x_1 \\ x_2 \\ x_3 \\ x_4 \\ x_6 \\ x_7 \end{bmatrix} = \begin{bmatrix} 2,000 \\ 1,000 \end{bmatrix}$$

We now look for a pair of patterns (or activities) which satisfies the following requirement: if we compute the equivalent combination for each remaining process in terms of this pair, then in every case the direct process requires at least as much cloth as its equivalent combination. Just as before, we find that Patterns 1 and 2 satisfy this requirement. Thus the linear-programming problem may be expressed in either a maximization or a minimization form, whichever is more convenient.

The analysis of this coats-and-pants problem was very easy, and, indeed, the most efficient method of solution was simply to construct an elementary graph, as shown in Fig. 6-3. We achieved this simplicity by imagining that a suit consisted of only two components, whereas in fact it comprises left and right front coat panels, left and right rear coat panels, left and right sleeves, collar fillets, four trouser panels, etc., and all these components must be cut in rigid proportions. Even such a straightforward problem as the one here in mind would quickly elude graphic analysis. On the other hand, linear-programming analysis applies no matter now many components are considered, although the arithmetic may become very laborious.

6-6. SOME GENERAL CONCLUSIONS

We have now seen three applications of the linear-programming method—the automobile example, the chemical example, and the suit example—designed to bring out different features of the method. All these applications are instances of a single type of problem or, more precisely, of two equivalent problems. Before proceeding to other aspects of the method and to other types of problem it will be well to recapitulate the main consequences of our analysis.

Let us restate the general formulation of the linear-programming problem as applied to the competitive firm in terms of the matrix notation explained in Sec. 6-4. Suppose that a firm has n activities available. These include methods of production, selling, disposal of unneeded factor services, and anything else the firm can do. Each of these activities, for example the ith, can be described completely by a column vector of k elements;[1] e.g.,

$$\mathbf{A}_i = [a_{i1}a_{i2} \cdots a_{ik}]$$

where k is the number of restrictions to which the firm must conform. These restrictions can take on a practically infinite variety of meanings. In the examples we have encountered restrictions that expressed such requirements as that (1) the capacities of the various components of fixed plant be limited; (2) the inputs for some activities may be produced within the firm, in which case activities that produce those inputs must be in balance with activities that consume them; (3) certain outputs may be usable only in rigidly fixed proportions. In addition to restrictions of these types, there may be others, not exemplified thus far, such as (4) quality specifications for some or all products; (5) minimum quantities of inputs or outputs, set by contractual obligations or other considerations; (6) financial constraints. It obviously does not pay to try to list all the sorts of restrictions that a firm might encounter. But whatever may be the economic significance of the restrictions, we can think of them as requiring some functions of the activity levels to satisfy certain equalities or inequalities. Furthermore, if the restrictions include inequalities, we can convert them to equalities by adding disposal activities to the list.

In linear programming we assume that the functions that express the restraints are all at least approximately linear. There is ample empirical evidence to indicate that this is the case in many practical instances. In this case a typical restriction can be written

$$a_{1j}x_1 + a_{2j}x_2 + \cdots + a_{nj}x_n = s_j$$

and all the restrictions together can be written

$$\mathbf{Ax} = \mathbf{A}_1x_1 + \mathbf{A}_2x_2 + \cdots + \mathbf{A}_nx_n = \mathbf{s}$$

using the notation developed in Sec. 6-4.

The economic significances of the coefficients a_{ij} are, naturally, as varied as those of the restrictions themselves, and it does not pay to try to enumerate them. In every case, however, a_{ij} is the effect of a unit increase in the ith activity level on the function subject to the jth restraint. We have illustrated such coefficients many times in this chapter.

We have not heretofore defined the concept "activity level," which came up in the preceding paragraph and in earlier passages, but its meaning is obvious. The intent is to define activities in terms of the propor-

[1] As before, we use brackets to signify that this is a column vector.

tions in which they produce and consume the various commodities or, more generally and more precisely, in terms of their effects on the functions involved in the restraints. Now two physical events may produce and consume commodities in exactly the same proportions but in very different amounts. We wish to form our definition so that we can say that these two events are instances of the same activity and, further, that the event that involves the larger amounts represents a higher level of the activity than the other. In short, we need a metric for the level of an activity. The linearity assumption provides us with such a metric easily, for, since all the inputs and outputs of an activity are linked together in strict proportionality, we may choose the absolute level of any of the inputs or outputs as a complete specification of the extent to which an activity is being used. Thus, in the automobile example, we defined the number of units of the truck-producing activity to be equal to the number of trucks produced. We might just as validly have defined it to be equal to the number of units of truck-assembly capacity that were consumed.

The activity vector, or program vector, $\mathbf{x}$ specified the levels of all the activities available to the firm. Some of the components of $\mathbf{x}$ may be zero; none can be negative, for negative levels of operation are meaningless. If a program vector satisfies all the restrictions, i.e., if $\mathbf{x}$ satisfies

$$\mathbf{x} \geq 0$$
$$\mathbf{A}\mathbf{x} = \mathbf{s}$$

then it is said to be feasible. A firm may choose among all feasible program vectors. Let us assume that the profits earned by the firm are a linear function of the activity levels. Using the notation of Sec. 6-4, we write this as

$$r = v_1x_1 + v_2x_2 + \cdots + v_nx_n = [\mathbf{v},\mathbf{x}]$$

where v_i denotes the contribution of a unit level of the ith activity to the profitability of the firm and $\mathbf{v}$ is the vector comprised of the v_i. The profitabilities v_i can be positive, negative, or zero, as we have seen.

The profit-maximizing firm, then, seeks that feasible program vector $\mathbf{x}$ which makes the profitability r as great as possible. We can now set forth the two main theorems of linear programming, both of which have already been exemplified.[1]

THEOREM 1. *If in any problem there is any optimal feasible program, then there is an optimal feasible program which involves no more than k activities at nonzero levels, where k is the number of restrictions to which the solution must conform.*

[1] Proofs are given in Chap. 4.

This theorem is fundamental to our method of searching for an optimal feasible solution, and we have invoked it in each of our examples.

The second theorem gives the criterion for telling whether a given program is optimal. It depends on the concept of an "equivalent combination," which we now recall.

Still letting k denote the number of restrictions in a problem, consider any $k + 1$ activities, e.g., $\mathbf{A}_1, \mathbf{A}_2, \ldots, \mathbf{A}_k, \mathbf{A}_q, q \geq k + 1$. Then, except in certain algebraically peculiar cases which we ignore, it is possible to find a set of levels for $\mathbf{A}_1, \mathbf{A}_2, \ldots, \mathbf{A}_k$ such that these k activities operated at the specified levels have the same effect on the restraints as does $\mathbf{A}_q$ operated at unit level. For instance, if all the restraints impose limits on the quantities of factor services to be consumed, then $\mathbf{A}_1, \mathbf{A}_2, \ldots, \mathbf{A}_k$ operated at this set of levels will consume precisely the same amount of each factor service in limited supply as $\mathbf{A}_q$ operated at unit level. If we denote these activity levels by $y_1, y_2, \ldots, y_k$, we may express this fact by saying that we can find a set of y's such that

$$\mathbf{A}_1 y_1 + \mathbf{A}_2 y_2 + \cdots + \mathbf{A}_k y_k = \mathbf{A}_q \tag{6-4}$$

Formally, Eq. (6-4) is nothing but a system of k simultaneous linear equations in k variables, the y's, so that determining these y's involves nothing more than some rather extensive arithmetic. We refer to the vector $\mathbf{y}$ whose components are the solution to Eq. (6-4) as the "equivalent combination" to $\mathbf{A}_q$ in terms of $\mathbf{A}_1, \ldots, \mathbf{A}_k$. A number of equivalent combinations were deduced in the course of the chemical example. It may be remarked that a basic solution to a linear-programming problem is simply an equivalent combination to the constant column, $\mathbf{s}$. In contrast with a feasible solution, however, an equivalent combination may contain some negative elements.

The contribution of an equivalent combination to the profits earned by a firm is the algebraic sum of the values of the k activities that it contains, each multiplied by its level (positive, negative, or zero) in the combination. Thus if $\bar{v}_q$ denotes the value (i.e., the contribution to profits) of the equivalent combination to $\mathbf{A}_q$ in terms of $\mathbf{A}_1, \mathbf{A}_2, \ldots, \mathbf{A}_k$, we define

$$\bar{v}_q = v_1 y_1 + v_2 y_2 + \cdots + v_k y_k$$

We may then, by comparing v_q with $\bar{v}_q$, see whether the value of $\mathbf{A}_q$ is greater or less than the value of its equivalent combination on the basis of any k other activities.

Now consider any feasible program

$$\mathbf{x} = [x_1, x_2, \ldots, x_n]$$

We shall suppose that, at most, k of the x_i's are greater than zero, since

Theorem 1 assures us that if there is any optimal feasible program, there will be one of this type. Assume $k < n$; otherwise there will be no scope for choice. This assumption is generally appropriate because, if for no other reason, the introduction of a disposal activity for each inequality restriction will assure that there will be more activities available than restraints. Thus we are assuming that some of the x_i's in $\mathbf{x}$ are zero.

Now we must think of two cases. In the first case, precisely k of the x_i's are greater than zero. Then we shall call every activity which corresponds to a nonzero x_i an "included" activity, and every activity which corresponds to a zero x_i an "excluded" activity. This terminology is reasonable, since the included activities are just the ones which would actually be used in the program denoted by $\mathbf{x}$.

In the second case, fewer than k of the x_i's are greater than zero. We shall again call all the activities which correspond to nonzero x_i's "included" activities. In addition we shall add to the list of included activities sufficient activities which correspond to x_i's with zero values so as to have a list of k included activities. How these additional activities are to be selected is immaterial for our present definitional purposes. Thus, in either case, we have k included activities and $n - k$ excluded activities. We can now state Theorem 2.

THEOREM 2. *A feasible program is an optimal feasible program if and only if it contains a list of included activities such that no excluded activity is more profitable than its equivalent combination in terms of those included activities.*

This is the theorem that we have used in solving the chemical- and suit-manufacturing examples. The economic significance of comparing the value of an activity with the value of its equivalent combination has been set forth at length in the treatment of those examples.

At the outset of this section we mentioned that there were two equivalent problems illustrated by our examples. One of them is the problem of maximizing profit, which we have just described in detail. The other is the problem of minimizing cost, best exemplified by the suit-manufacturing example, in which the objective was to consume as little cloth as possible in the manufacture of a given number of suits.

There is no need to repeat the detailed analysis of the maximum-profit problem for the minimum-cost case. The only change in formulation required is to revise the criterion of Theorem 2 to read as follows:

THEOREM 2*a*. *A feasible program is an optimal feasible program in the sense of minimum cost if and only if it contains a list of included activities such that no excluded activity costs less per unit of operation than its equivalent combination in terms of those included activities.*

We can state the upshot of our examples and discussion in two sentences. (1) If a firm's situation can be described adequately by a linear-programming model, then for that firm to maximize its time rate of profits, it should use a number of activities which does not exceed the number of the restrictions that limit its operations. (2) No activity excluded from an optimal operating program for such a firm should be more profitable than its equivalent combination in terms of activities included in that program. These two theorems together are the linear-programming analog of the "equate your marginal productivities" dictum in the orthodox marginal analysis.

7

Application to the Firm; Valuation and Duality

7-1. MARGINAL PRODUCTIVITY IN LINEAR PROGRAMMING

We have now discussed a method for finding the optimal production program for a firm in perfect competition. It turned out that the method, though initially strictly mathematical in form, really depended on an economic criterion, namely, the criterion that no activity should be used if a more profitable activity or combination of activities was available to the firm. In addition to yielding this intuitively reasonable result, our analysis led to a systematic procedure for applying the criterion.

It is economically obvious that the method can do more than find an optimal production program. The method implies values to be placed on the various scarce resources which define the opportunities open to the firm. This can be seen by recalling that, in addition to finding the optimal program, the linear-programming solution yields the net revenue that will result from applying that program. This net revenue, the maximum that can be obtained under the circumstances faced by the firm, is a measure of the economic, or business, worth of the firm considered as a whole. Furthermore, if we have the maximum net revenue obtainable with any list of resources and then set up and solve the corresponding problem with the same list of resources except that the quantity of one of the resources has been decreased by one unit, we shall find that the second problem may yield a smaller net revenue than the first. The decrease in net revenue is, of course, the marginal revenue product of the resource whose quantity was decreased.[1] Thus a linear-programming solution implies a value to be imputed to each unit of each fixed resource.

The method just suggested for calculating the marginal revenue prod-

[1] An alternative definition would be in terms of a unit increase in the availability of a resource. The only case in which these two definitions will yield different results is that in which the resource is just on the border line of being redundant. We would then have to distinguish between downward and upward marginal revenue products.

ucts of the scarce resources, to which we shall refer from now on as their values,[1] is cumbersome. One of the remarkable properties of the linear-programming solution is that the values of the fixed resources emerge in the course of determining the optimal program and do not require additional computation. To see how this occurs we shall first present a simple method for finding the values of the scarce resources and then shall see why even this simple computation is unnecessary.

Consider a firm that has k types of fixed resources available to it and let $s_1, s_2, \ldots, s_k$ denote the quantities available of the various resources. We define, as before, the value of the unit of any fixed resource to be the decrease in maximum obtainable net revenue which would result from subtracting a unit of that resource from the list of resources available to the firm. We shall denote these unit values by $u_1, u_2, \ldots, u_k$.

As usual we think of the firm as having a certain number of activities, $A_1, A_2, \ldots$, available to it. Each activity absorbs resources and produces a net revenue which we have previously defined to be the excess of the total value of output per unit of the activity over the direct expenses for the purchase of factors in the open market. We know that the optimal program for a firm with k types of fixed resource will generally require the use of k activities at positive levels. Let us assume that in the firm under consideration the activities $A_1, A_2, \ldots, A_k$ appear at the positive levels $x_1, x_2, \ldots, x_k$, respectively, in the optimum program, and that the net revenues per unit of these activities are $v_1, v_2, \ldots, v_k$. Since revenue can arise only from the operation of the various activities, the total net revenue must be the sum of the net revenues yielded by the activities used, or

$$r = v_1x_1 + v_2x_2 + \cdots + v_kx_k = \text{maximum total net revenue} \quad (7\text{-}1)$$

Our first task is to show that the total value imputed to the fixed resources, $u_1s_1 + u_2s_2 + \cdots + u_ks_k$, equals the maximum total net revenue r. To do this we inquire how much the total revenue would increase if one more unit of one of the resources was available. We may break this question into two parts. First, we ask how much each of the k activity levels $x_1, \ldots, x_k$ would change if one more unit of one of the resources was available. This is clearly the same as finding an equivalent combination, i.e., the combination of activity levels that absorbs one more unit of the resource in question and nothing else.[2] We shall denote this set of changes from an initial optimum by $\Delta x_1, \ldots, \Delta x_k$. Second, we ask what the increase in revenue is that results from these

[1] It would be more accurate, but unduly awkward, to call this the value of the services of the scarce resources.

[2] For the concept of "equivalent combination" see Sec. 6-6.

changes in activity levels. The answer to the second question is clearly

$$\Delta r = v_1 \Delta x_1 + v_2 \Delta x_2 + \cdots + v_k \Delta x_k \tag{7-2}$$

or the sum obtained by multiplying each change by the unit revenue of the activity to which it applies and adding. It should be noted that some of the changes may be negative.

Now we turn to the first question, the determination of $\Delta x_1, \ldots, \Delta x_k$. Since we shall have to do some algebra, it will be worthwhile to simplify the notation a little by imagining that the firm is subject to just three restraints, that is, $k = 3$, and that the activities in the optimal program are A_1, A_2, A_3. As usual we assume that a typical activity A_i absorbs a_{i1} units of the first fixed factor per unit of operation, a_{i2} units of the second, and a_{i3} units of the third.

Suppose that the factor whose quantity has been increased by one unit is the first. Then the set of changes in activity levels must absorb one unit of the first fixed factor and no units of any other. Thus Δx_1, Δx_2, Δx_3 must satisfy

$$\begin{aligned} a_{11} \Delta x_1 + a_{21} \Delta x_2 + a_{31} \Delta x_3 &= 1 \\ a_{12} \Delta x_1 + a_{22} \Delta x_2 + a_{32} \Delta x_3 &= 0 \\ a_{13} \Delta x_1 + a_{23} \Delta x_2 + a_{33} \Delta x_3 &= 0 \end{aligned} \tag{7-3}$$

The resultant change in net revenue is $v_1 \Delta x_1 + v_2 \Delta x_2 + v_3 \Delta x_3$, where Δx_1, Δx_2, Δx_3 satisfy Eqs. (7-3). Since u_1 denotes the marginal revenue product of the first factor, we have

$$u_1 = \Delta r = v_1 \Delta x_1 + v_2 \Delta x_2 + v_3 \Delta x_3 \tag{7-4}$$

The response to the availability of an additional unit of the second factor would be a similar set of changes. Let us call them $\Delta x_1''$, $\Delta x_2''$, $\Delta x_3''$. These changes would satisfy the equations

$$\begin{aligned} a_{11} \Delta x_1'' + a_{21} \Delta x_2'' + a_{31} \Delta x_3'' &= 0 \\ a_{12} \Delta x_1'' + a_{22} \Delta x_2'' + a_{32} \Delta x_3'' &= 1 \\ a_{13} \Delta x_1'' + a_{23} \Delta x_2'' + a_{33} \Delta x_3'' &= 0 \end{aligned} \tag{7-5}$$

and define the marginal value of the second factor as

$$u_2 = v_1 \Delta x_1'' + v_2 \Delta x_2'' + v_3 \Delta x_3'' \tag{7-6}$$

Finally, letting the activity changes that absorb a unit increase in the third factor be $\Delta x_1'''$, $\Delta x_2'''$, $\Delta x_3'''$, we have the equations

$$\begin{aligned} a_{11} \Delta x_1''' + a_{21} \Delta x_2''' + a_{31} \Delta x_3''' &= 0 \\ a_{12} \Delta x_1''' + a_{22} \Delta x_2''' + a_{32} \Delta x_3''' &= 0 \\ a_{13} \Delta x_1''' + a_{23} \Delta x_2''' + a_{33} \Delta x_3''' &= 1 \end{aligned} \tag{7-7}$$

$$u_3 = v_1 \Delta x_1''' + v_2 \Delta x_2''' + v_3 \Delta x_3''' \tag{7-8}$$

The total value of s_1 units of the first factor, s_2 units of the second factor, and s_3 units of the third factor will be denoted by

$$z = u_1s_1 + u_2s_2 + u_3s_3$$

Multiply Eq. (7-4) by s_1, Eq. (7-6) by s_2, Eq. (7-8) by s_3, and add. The result is

$$\begin{aligned} z = {} & v_1(s_1\,\Delta x_1 + s_2\,\Delta x_1'' + s_3\,\Delta x_1''') \\ & + v_2(s_1\,\Delta x_2 + s_2\,\Delta x_2'' + s_3\,\Delta x_2''') \\ & + v_3(s_1\,\Delta x_3 + s_2\,\Delta x_3'' + s_3\,\Delta x_3''') \end{aligned} \tag{7-9}$$

All that remains is to determine these three parentheses. Let us call them x_1^*, x_2^*, x_3^*, respectively. Multiply Eqs. (7-3) by s_1, term by term, Eqs. (7-5) by s_2, and Eqs. (7-7) by s_3. Then add the corresponding equations in the three sets. The result is a new set of three equations, which is a weighted combination of the old sets. Note that the a coefficients are the same in the three sets of equations. Thus, in the first equation of the combined set we find, for example, that a_{11} is multiplied by $s_1\,\Delta x_1 + s_2\,\Delta x_1'' + s_3\,\Delta x_1''' = x_1^*$. Carrying out this work completely, the combined set of equations is found to be

$$\begin{aligned} a_{11}x_1^* + a_{21}x_2^* + a_{31}x_3^* &= s_1 \\ a_{12}x_1^* + a_{22}x_2^* + a_{32}x_3^* &= s_2 \\ a_{13}x_1^* + a_{23}x_2^* + a_{33}x_1^* &= s_3 \end{aligned} \tag{7-10}$$

Equations (7-10) tell us all we need to know about x_1^*, x_2^*, x_3^*, since they are precisely the equations for a set of levels of A_1, A_2, A_3 that absorb the available supplies of the limiting resources; i.e., the three starred quantities satisfy the same equations as the optimal activity levels x_1, x_2, x_3. In consequence, $x_i^* = x_i$, $i = 1, 2, 3$. Substituting this result in Eq. (7-9), we get

$$z = v_1x_1 + v_2x_2 + v_3x_3 = r$$

by virtue of Eq. (7-1).

We have now proved

$$r = u_1s_1 + u_2s_2 + \cdots + u_ks_k \tag{7-11}$$

i.e., the total value imputed to the fixed resources by our definition of resource value just exhausts the maximum total net revenue.

7-2. DETERMINATION OF RESOURCE VALUES

Consider any activity in the final solution, say activity A_1. Per unit of operation it absorbs a_{11} units of the first fixed factor, a_{12} units of the second fixed factor, . . . , a_{1k} units of the kth fixed factor. The value of resources absorbed by this activity is, then,

$$a_{11}u_1 + a_{12}u_2 + \cdots + a_{1k}u_k = \text{(say) } w_1 \tag{7-12}$$

per unit of use, and the value produced is v_1. We shall refer to w_1 as the imputed cost of activity A_1 and, similarly, to

$$w_2 = a_{21}u_1 + a_{22}u_2 + \cdots + a_{2k}u_k$$

as the imputed cost of activity A_2, etc. If the level of activity A_1 were reduced by one unit, a revenue equal to v_1 would be sacrificed, but at the same time resources capable of producing a revenue equal to w_1 would be released for use by the other activities. This would be worth doing if w_1 were greater than v_1. But activity A_1 was assumed to appear in the optimal solution. Therefore it is not worthwhile to cut back the level of activity A_1, and v_1 must at least be as great as w_1. The same reasoning applies, of course, to all the activities in the final solution. Now the activities in the final solution absorb the total available supplies of all fixed resources, i.e.,

$$\begin{aligned}
a_{11}x_1 + a_{21}x_2 + \cdots + a_{k1}x_k &= s_1 \\
a_{12}x_1 + a_{22}x_2 + \cdots + a_{k2}x_k &= s_2 \\
\cdots\cdots\cdots\cdots\cdots\cdots\cdots \\
a_{1k}x_1 + a_{2k}x_2 + \cdots + a_{kk}x_k &= s_k
\end{aligned}$$

As above we use u_j to denote the marginal revenue product of the jth resource, v_i to denote the given net revenue of the ith activity, and w_i to denote the imputed cost of the ith activity. If we multiply the first equation of this set by u_1, the second by u_2, etc., and add, the right-hand side becomes r by Eq. (7-11). Regrouping terms on the left-hand side of the sum, the coefficient of x_1 is seen to be w_1, the coefficient of x_2 is w_2, etc. Thus,

$$r = w_1x_1 + w_2x_2 + \cdots + w_kx_k \tag{7-13}$$

Subtracting Eq. (7-13) from Eq. (7-1), we get

$$(v_1 - w_1)x_1 + (v_2 - w_2)x_2 + \cdots + (v_k - w_k)x_k = 0 \tag{7-14}$$

Our discussion of the relation between v_1 and w_1 showed that if the ith activity appeared at a positive level in the optimal solution, i.e., if x_i were positive, then $v_i - w_i$ could not be negative. Therefore none of the terms on the left-hand side of Eq. (7-14) is negative, and the equation can be satisfied only if every term is zero. We have established that if x_i, the level of the ith activity, is positive, then the resources used up in the activity, valued at marginal productivity, will exactly absorb the net revenue; i.e.,

$$v_i = w_i = a_{i1}u_1 + a_{i2}u_2 + \cdots + a_{ik}u_k \tag{7-15}$$

If k activities are actually used in the optimal program (the usual case), then Eq. (7-15) provides k linear equations for determining the k unknowns, $u_1, \ldots, u_k$, from the known values $v_1, \ldots, v_k$. Although these equations can be solved easily, we shall show in a moment that it is unnecessary to solve them. But first let us see how these general formulas apply to the automobile example.

In the automobile firm analyzed on pages 133–138 of Chap. 6, the optimal program was found to consist of manufacturing 20,370 automobiles and 6,481 trucks per month. The program also involved underutilizing automobile assembly capacity by 9 per cent and underutilizing truck assembly capacity by 57 per cent. The list of fixed resources available to the firm was as follows:

Fixed resources	Quantity available	Value per unit
Metal-stamping capacity	$100\% = s_1$	u_1
Engine-assembly capacity	$100\% = s_2$	u_2
Automobile-assembly capacity	$100\% = s_3$	u_3
Truck-assembly capacity	$100\% = s_4$	u_4

The four activities which appeared in the optimal program and their input coefficients are given in Table 7-1.

Our problem is to determine the unit values u of the four types of resource. To do this we require, in addition to the data just listed, the net revenues per unit of the four activities in the optimal program. In setting up this example we assumed that the net value above direct costs from producing an automobile is \$300 and that the net value above direct costs from producing a truck is \$250. Activities A_3 and A_4 are both disposal activities, so that their net revenues are both zero.

TABLE 7-1. OPTIMAL ACTIVITIES AND THEIR COEFFICIENTS IN AN AUTOMOBILE PLANT

Resource (capacity type)	Activity			
	A_1 Automobile production	A_2 Truck production	A_3 Automobile-assembly disposal	A_4 Truck-assembly disposal
Metal stamping	0.00400	0.00286	0	0
Engine assembly	0.00300	0.00600	0	0
Automobile assembly	0.00444	0	1	0
Truck assembly	0	0.00667	0	1

We can apply Eq. (7-15) to these data, obtaining

$$\begin{aligned}
0.0040u_1 + 0.0030u_2 + 0.00444u_3 \qquad\qquad &= 300 \\
0.00286u_1 + 0.0060u_2 \qquad\qquad + 0.00667u_4 &= 250 \\
u_3 \qquad\qquad &= 0 \\
u_4 &= 0
\end{aligned}$$

From this set of four equations we find

$$u_1 = \$68{,}056 \qquad u_3 = 0$$
$$u_2 = \$\ 9{,}259 \qquad u_4 = 0$$

Note that the values of the two underutilized types of capacity are zero; economically speaking, they are free goods. This follows from the fact that decreasing the availability to the firm of one of these already underutilized capacities will not decrease net revenue at all. Stated differently, we have arrived at an opportunity-cost type of valuation. Since either the automobile or the truck division of the firm could use more automobile-assembly capacity if it desired, without sacrificing other opportunities, no charge is to be made against the divisions for the use of automobile (or truck) assembly capacity.

The chemical example is only a little more complicated. In that example the optimal program turned out to be to produce \$1,000 worth of C_1, \$750 worth of C_2, and \$1,000 worth of C_3. This program utilized the full capacities of equipment E_1, E_2, and E_4 and underutilized equipment E_3 by 62.5 per cent. The valuation problem is to determine the value per per cent of capacity of each type of equipment. Let

$$u_1 = \text{value of } 1\% \text{ of } E_1$$
$$u_2 = \text{value of } 1\% \text{ of } E_2$$
$$u_3 = \text{value of } 1\% \text{ of } E_3$$
$$u_4 = \text{value of } 1\% \text{ of } E_4$$

Since E_3 is underutilized, $u_3 = 0$. The data for determining the other three values are found in Table 6-3 (p. 142).

This problem differs from the automobile example and from the setup that led to Eq. (7-15) in that two of the activities consume as inputs not only fixed resources, but also outputs of other activities. We must therefore reconsider both Eq. (7-15) and the concept of the net revenue of an activity. Recall that Eq. (7-15) amounted to

Net revenue per unit of an activity = *total value of fixed factors absorbed by a unit of that activity*

If we add back direct costs to both sides of this equation we have

Gross revenue per unit of an activity = *total value of fixed factors absorbed per unit of that activity* + *unit direct costs*

Furthermore, we must count as unit direct costs not only cash disbursements for factors acquired on the open market, but also the value of final products absorbed by the activity and which, therefore, cannot be sold on the open market. Considering activity A_2, for example, we see that the unit gross revenue, or, what is the same thing, the unit gross value of output is \$100. Unit direct costs are \$50 in cash expenditures plus \$30 worth of C_1, giving a total of \$80. Applying these concepts, we equate total gross unit revenue to total unit costs for the three activities as follows:

$$\begin{aligned} 10u_1 + 5u_2 \qquad\qquad &+ 70 = 100 \\ 4u_2 \qquad\qquad &+ 80 = 100 \\ 2u_2 + 10u_4 &+ 75 = 100 \end{aligned}$$

Terms involving u_3 have been omitted from these equations since u_3 is already known to be zero. The solution is

$$\begin{aligned} u_1 &= 0.5 \qquad & u_3 &= 0 \\ u_2 &= 5.0 \qquad & u_4 &= 1.5 \end{aligned}$$

The method of analysis applied in these two instances is obviously general. Once an optimal program has been found, values can be found for all the owned factors of production by the straightforward solution of a system of simultaneous linear equations. Actually, however, the determination of the resource values implicit in a linear-programming solution is even easier than that; the values emerge as a by-product of finding the optimum program. This can be seen by looking at the details of the solution to the chemical example.

The optimal program for the chemical firm was found by selecting three activities, A_7, A_8, and A_{10}, that were not to be used. We then tested the optimality of that program by computing the profitability of the equivalent combination to each excluded activity and comparing it with the profitability of the excluded activity. Let us define the "gain" from excluding an activity to be the excess of the profitability of the equivalent combination over the profitability of the activity itself. Applying this term to the calculations in the chemical example, we found that

$$\begin{aligned} \text{Gain from excluding } A_7 &= 0.5 \\ \text{Gain from excluding } A_8 &= 5.0 \\ \text{Gain from excluding } A_{10} &= 1.5 \end{aligned}$$

Comparing these with the valuations we see that

$$\begin{aligned} \text{Gain from excluding } A_7 &= u_1 \\ \text{Gain from excluding } A_8 &= u_2 \\ \text{Gain from excluding } A_{10} &= u_4 \end{aligned}$$

This is no coincidence. A_7, recall, was the activity of discarding 1 per cent of type E_1 capacity. The gain from excluding A_7, therefore, is the negative of the economic loss resulting from a reduction of 1 per cent in type E_1 capacity; it equals the value of a unit of type E_1 capacity to the firm. The other gains can be interpreted similarly as measures of the values of the resources to which they relate. Note that we have a disposal activity for each resource. In the optimal program any disposal activity whose level is set at zero usually yields, as above, a positive value for its resource. If a disposal activity appears at positive level the corresponding resource has zero value; i.e., it is a free good to the firm.

In short, the problems of optimum allocation and of valuation are inseparable. In the work of finding an optimal program, certain "gains" have to be computed, and these "gains" are nothing but the values to be imputed to the various types of scarce resources. Although we calculated values separately for expository reasons, this is not necessary in actual practice.

7-3. DUALISM OF PRICE AND PROGRAMMING

The foregoing considerations show the intimate connection between the problem of valuation, i.e., that of determining the resource values u, on the one hand, and the problem of allocation, i.e., that of finding an optimal production program, on the other. This, of course, is just what one would expect on the basis of usual economic reasoning. But mathematically the relationship of the valuation and allocation problems is even closer than may be evident from the discussion thus far. In fact we shall see that the two problems are mathematically identical.

As a first step in making this clear, let us formulate the familiar allocation problem. Suppose that a firm has k types of fixed resource in quantities $s_1, s_2, \ldots, s_k$ and n possible activities $A_1, A_2, \ldots, A_n$ whose respective net revenues per unit are $v_1, v_2, \ldots, v_n$. A production program is specified by a set of activity levels, $x_1, x_2, \ldots, x_n$, that fulfills the following requirements:

1. No activity level is negative; i.e.,

$$0 \leq x_1, x_2, \ldots, x_n$$

2. No more than the available supply of any resource is required. Expressed algebraically, this is

$$\begin{aligned} a_{11}x_1 + a_{21}x_2 + \cdots + a_{n1}x_n &\leq s_1 \\ a_{12}x_1 + a_{22}x_2 + \cdots + a_{n2}x_n &\leq s_2 \\ \cdots\cdots\cdots\cdots\cdots\cdots & \\ a_{1k}x_1 + a_{2k}x_2 + \cdots + a_{nk}x_n &\leq s_k \end{aligned}$$

where a_{ij} is the number of units of the jth resource used by one unit of the ith activity.

The net revenue resulting from any program $x = [x_1, x_2, \ldots, x_n]$ is, as usual,

$$r = v_1x_1 + v_2x_2 + \cdots + v_nx_n$$

The allocation problem is simply to find a program vector x that satisfies requirements 1 and 2 and makes r as great as possible.

We now formulate the valuation problem in analogous terms. Instead of a set of activity levels we seek a set of unit values for fixed resources $u_1, u_2, \ldots, u_k$ that fulfills the following requirements:

1*a*. No resource value is negative; i.e.,

$$0 \leq u_1, u_2, \ldots, u_k$$

This is the same as requirement 1, formulated for the allocation problem.

This requirement is justified by applying Theorem 2 of Chap. 6 to the disposal activities. Consider, for example, the first scarce factor. The profitability of the disposal activity for that factor (i.e., the net revenue resulting directly from disposing of a unit of that factor) is zero. The profitability of the equivalent combination to that disposal activity is the net revenue resulting from the combination of levels of $A_1, A_2, \ldots, A_k$ that consumes 1 unit of the first factor and nothing else. We have seen that this profitability is u_1. By Theorem 2 of Chap. 6 (page 164), if activities $A_1, \ldots, A_k$ constitute an optimal program, the profitability of the equivalent combination must be at least as great as the profitability of the disposal activity. Hence, $0 \leq u_1$. The same reasoning applies to each of the other resources; in every case the imputed value of the resource must be either positive or zero. Economically, this reflects the fact that as long as we assume the possibility of costless disposal, no resource can have negative value.

To obtain a requirement which we shall call 2*a*, parallel to 2, we note that $u_1, \ldots, u_k$ constitutes a set of k prices. If we select any set of k activities we can choose the k prices so that in each activity of the set the imputed cost of the resources required by a unit of the activity, calculated from these prices, is equal to the net revenue of a unit of the activity.[1] If the set of k activities selected is an optimal set, then, as we have just seen, the set of k prices will satisfy requirement 1*a*, and the simplex criterion will show that for each activity not in the optimal set the imputed cost of the resources absorbed will be at least as great as the unit net revenue. This is true because each activity not in the optimal set can be compared with an equivalent combination of activities in the

[1] For some sets of activities some of the resultant prices may be negative, but such a set cannot be optimal.

optimal set. The imputed value of resources absorbed by the equivalent combination will be equal to the resulting net revenue, since this holds for each activity in the combination. At the same time, the imputed cost of the activity will be the same as the imputed cost of its equivalent combination, since they absorb the same quantities of all resources, while the net revenue of the activity will be no greater than the net revenue of the equivalent combination and will generally be smaller. In other words, the gain from excluding a nonoptimal activity will never be negative. In summary,

Imputed cost of an activity not in the optimal program
= imputed cost of its equivalent combination
= net revenue of the equivalent combination
$\geq$ net revenue of the activity

We have now justified the second requirement:

2*a*. The imputed cost of a unit of each activity is at least as great as its unit net revenue. This may be written

$$\begin{aligned} v_1 &\leq a_{11}u_1 + a_{12}u_2 + \cdots + a_{1k}u_k \\ v_2 &\leq a_{21}u_1 + a_{22}u_2 + \cdots + a_{2k}u_k \\ &\cdots\cdots\cdots\cdots\cdots\cdots \\ v_n &\leq a_{n1}u_1 + a_{n2}u_2 + \cdots + a_{nk}u_k \end{aligned} \tag{7-16}$$

Thus the two requirements on the resource values, u, have the same form as the two requirements on the activity levels, x. It remains only to find a feature in the valuation problem that is analogous to the profit-maximizing objective in the allocation problem.

Let us suppose, now, that $u_1, \ldots, u_k$ is any set of values which satisfies requirements 1*a* and 2*a* and that $x_1, \ldots, x_n$ is an optimal set of activity levels. Multiply the first equation in set (7-16) by x_1, the second by x_2, etc. Since the x's are nonnegative, the inequality signs will be undisturbed. Adding the inequalities, the left-hand side will become

$$v_1x_1 + v_2x_2 + \cdots + v_nx_n = r$$

On the right-hand side the coefficient of u_1, for example, becomes

$$a_{11}x_1 + a_{21}x_2 + \cdots + a_{n1}x_n = y_1 \text{ (say)}$$

where y_1 denotes the total quantity of the first fixed resource required by the program $x_1, \ldots, x_n$. Clearly $y_1 \leq s_1$. In a similar manner the coefficient of u_2 is y_2 and $y_2 \leq s_2$, etc. Thus,

$$\begin{aligned} r &= v_1x_1 + v_2x_2 + \cdots + v_nx_n \\ &\leq y_1u_1 + y_2u_2 + \cdots + y_ku_k \\ &\leq s_1u_1 + s_2u_2 + \cdots + s_ku_k \end{aligned} \tag{7-17}$$

The first member of Eq. (7-17) is the maximum total net revenue which the firm can obtain. The last member is the total value of the fixed resources computed on the basis of $u_1, \ldots, u_k$. We shall denote this value by r', so that Eq. (7-17) may be summarized as $r \leq r'$ for all values $u_1, \ldots, u_k$ that satisfy the requirements. In words, if the resources are valued in such a way that the value of the resources required by every available activity is at least as great as its net revenue, then the total value of the resources owned by the firm will be at least as great as the maximum total net revenue.

But we have already seen that if the resources are assigned values equal to their marginal productivities, then the total value of the resources will be just equal to the maximum total net revenue. It follows that resource values equal to the marginal productivities make the aggregate value of the resources as small as it possibly can be, subject to the two requirements. Thus, instead of searching directly for the marginal productivities, we may obtain them by searching for a set of values which minimizes r', the aggregate value of the resources, that is, we may minimize

$$r' = s_1u_1 + s_2u_2 + \cdots + s_ku_k$$

subject to the requirements 1*a* and 2*a*.

Matters now stand as follows: On the one hand, the allocation problem is the problem of finding a set of activity levels $x_1, \ldots, x_n$ that satisfy requirements 1 and 2 and that make total net revenue r as great as possible. It is a linear-programming maximizing problem. If we solve this problem we obtain as by-products the maximum net revenue obtainable in the circumstances of the firm and the marginal productivity values of the fixed resources.

But on the other hand, we may solve the valuation problem without first solving the allocation problem. This requires finding a set of values $u_1, \ldots, u_k$ that satisfy requirements 1*a* and 2*a* and that make the total value of the fixed resources, r', as small as possible. It is a linear-programming minimizing problem. The strong parallelism between the two problems shows that if the valuation problem is solved, the optimal set of activity levels will emerge as a by-product and the minimum value r' that is obtained will be equal to the maximum value r associated with the corresponding allocation problem.

Thus we may address ourselves to either the valuation or the allocation problem, as we please. Whichever one we choose, we shall obtain the solution to the other simultaneously as a by-product. The two problems are identical in form, but it frequently turns out that one of them is appreciably simpler computationally than the other, so there is a real advantage in having this choice available This phenomenon, the parallelism and inseparability of the valuation and allocation problems, has been called the "dualism of pricing and allocation."

7-4. DUALISM IN THE CHEMICAL EXAMPLE

Let us exemplify the dualism by solving the chemical example in the other order and seeking *ab initio* values to be imputed to the four types of equipment. The 10 possible activities have already been defined and

TABLE 7-2. ACTIVITIES AND COEFFICIENTS IN A CHEMICAL FIRM

Inputs		Activity									
		Manufacturing, dollars			Selling, dollars			Disposal, per cent			
Item	Unit	A_1	A_2	A_3	A_4	A_5	A_6	A_7	A_8	A_9	A_{10}
C_1	\$1	−100	30	10	100						
C_2	\$1		−100	50		100					
C_3	\$1			−100			100				
E_1	1%	10						1			
E_2	1%	5	4	2					1		
E_3	1%		5							1	
E_4	1%			10							1
Cash receipts.....	\$1	−70	−50	−15	100	100	100	0	0	0	0
Net revenue......	\$1	30	20	25	...	...	...	0	0	0	0

discussed. They are listed, together with their coefficients, in Table 7-2. Substituting the coefficients in this table, restriction 2*a* becomes

$$
\begin{array}{llllll}
10u_1 + 5u_2 & & & \geq 30 \\
& 4u_2 + 5u_3 & & \geq 20 \\
& 2u_2 & + 10u_4 & \geq 25 \\
u_1 & & & \geq 0 \\
& u_2 & & \geq 0 \\
& & u_3 & \geq 0 \\
& & u_4 & \geq 0
\end{array}
\qquad (7\text{-}18)
$$

We require the set of values u_1, u_2, u_3, u_4 that minimizes

$$r' = 100(u_1 + u_2 + u_3 + u_4)$$

since the total supply of each type of capacity is 100 per cent, subject to Eqs. (7-18). The last four equations in set (7-18) are unnecessary since they are merely a restatement of requirement 1*a*. Just as in the solution of the allocation problem we introduce some new nonnegative variables to take up the slack in Eqs. (7-18) and convert them to equalities. These new variables will be

$$
\begin{aligned}
u_5 &= \text{"loss" on activity } A_1 \\
u_6 &= \text{"loss" on activity } A_2 \\
u_7 &= \text{"loss" on activity } A_3
\end{aligned}
$$

We now replace Eqs. (7-18) by

$$
\begin{aligned}
10u_1 + 5u_2 \qquad\qquad - u_5 \qquad\qquad &= 30 \\
4u_2 + 5u_3 \qquad\qquad - u_6 \qquad &= 20 \\
2u_2 \qquad + 10u_4 \qquad\qquad - u_7 &= 25
\end{aligned}
\qquad (7\text{-}19)
$$

The left-hand sides of Eqs. (7-19) are the imputed costs of the three manufacturing activities. They depend on the seven values $u_1, \ldots, u_7$. Our problem is to find a set of values, all nonnegative, which satisfies Eqs. (7-19) and which makes r' as small as possible.

We proceed, as in any other linear-programming problem, by making a guess and then seeing whether it can be improved. In making our guess we are guided by the basic theorem of linear programming, namely, that the number of nonzero values in the final solution can be made equal to the number of equations which have to be satisfied. In this case, there are three equations. Accordingly we guess that u_1, u_2, and u_4 will not be zero.[1] With the other u's set equal to zero, Eqs. (7-19) become

$$
\begin{aligned}
10u_1 + 5u_2 \qquad &= 30 \\
4u_2 \qquad &= 20 \\
2u_2 + 10u_4 &= 25
\end{aligned}
$$

whence $\qquad u_1 = 0.5 \qquad u_2 = 5.0 \qquad u_4 = 1.5$

Since these values are all positive, requirement 1*a* is satisfied.

We must now test whether there exists any set of values satisfying requirements 1*a* and 2*a* that makes r' smaller than does the set just given. In Chap. 6, we derived Theorem 2*a*, which provided a test of optimality for a program that satisfied a minimum-cost objective. Now we require a test for a set of imputed resource values that satisfies a minimum-aggregate-value objective. The two problems are so analogous that we have only to adapt Theorem 2*a* to the present set of concepts. To do this we note that the imputed values with which we are working are of two kinds: positive values, of which there are three, and zero values, of which there are four. For each zero value we may find an equivalent combination, as follows: Suppose we raised the imputed value of any zero-valued resource to $1. This would change the imputed costs

[1] Our previous experience with this example assures us that this "guess" is correct. By adopting it we can see the principles at work without excessive arithmetic.

of the various activities. We may calculate this change in imputed costs and then calculate the change in the imputed values of the positive-valued resources that would have the same effect on the imputed costs of the activities. This is the equivalent combination to the zero value that was supposed to be raised.

This will be clear if we restate it algebraically. Suppose, as before, that a_{ij} denotes the number of units of the jth resource used by a unit of the ith activity. Then, given any set of values, $u_1, \ldots, u_7$, the imputed costs of the three activities are

$$\begin{aligned} a_{11}u_1 + a_{12}u_2 + a_{13}u_3 + a_{14}u_4 + \cdots + a_{17}u_7 &= w_1 \\ a_{21}u_1 + a_{22}u_2 + a_{23}u_3 + a_{24}u_4 + \cdots + a_{27}u_7 &= w_2 \\ a_{31}u_1 + a_{32}u_2 + a_{33}u_3 + a_{34}u_4 + \cdots + a_{37}u_7 &= w_3 \end{aligned} \tag{7-20}$$

If only u_1, u_2, and u_4 are different from zero, these equations become

$$\begin{aligned} a_{11}u_1 + a_{12}u_2 + a_{14}u_4 &= w_1 \\ a_{21}u_1 + a_{22}u_2 + a_{24}u_4 &= w_2 \\ a_{31}u_1 + a_{32}u_2 + a_{34}u_4 &= w_3 \end{aligned} \tag{7-21}$$

and, as we saw, since $w_1 = v_1$, $w_2 = v_2$, $w_3 = v_3$, these equations determine u_1, u_2, u_4. Now suppose $u_3 = 1$, all other values remaining unaltered. Each of the imputed costs w will change. Let the change in w_1 be δw_1, so that the new imputed cost of activity A_1 is $w_1 + \delta w_1$, etc. Then, letting $u_3 = 1$ in the first of Eqs. (7-20), we have

$$a_{11}u_1 + a_{12}u_2 + a_{13}u_3 + a_{14}u_4 = w_1 + \delta w_1$$

By subtracting the first of Eqs. (7-21) from this, we get

$$a_{13} = \delta w_1$$

and by performing the same operations on the other two equations of (7-20), we get

$$\begin{aligned} a_{23} &= \delta w_2 \\ a_{33} &= \delta w_3 \end{aligned}$$

Thus, in general, the effect on imputed costs of raising the value of any resource by \$1 is given by the input coefficients of that resource. In the present example,

$$\begin{aligned} \delta w_1 &= a_{13} = 0 \\ \delta w_2 &= a_{23} = 5 \\ \delta w_3 &= a_{33} = 0 \end{aligned}$$

Now we determine the equivalent combination to u_3, i.e., the set of changes in u_1, u_2, u_4 which will have the same effect as a \$1 increase in

u_3. Let δu_1 be the change in u_1, δu_2 the change in u_2, etc. These changes must satisfy

$$\begin{aligned} a_{11}(u_1 + \delta u_1) + a_{12}(u_2 + \delta u_2) + a_{14}(u_4 + \delta u_4) &= w_1 + \delta w_1 = w_1 + a_{13} \\ a_{21}(u_1 + \delta u_1) + a_{22}(u_2 + \delta u_2) + a_{24}(u_4 + \delta u_4) &= w_2 + \delta w_2 = w_2 + a_{23} \\ a_{31}(u_1 + \delta u_1) + a_{32}(u_2 + \delta u_2) + a_{34}(u_4 + \delta u_4) &= w_3 + \delta w_3 = w_3 + a_{33} \end{aligned}$$

Subtracting Eqs. (7-21) from these equations, we get

$$\begin{aligned} a_{11}\,\delta u_1 + a_{12}\,\delta u_2 + a_{14}\,\delta u_4 &= a_{13} \\ a_{21}\,\delta u_1 + a_{22}\,\delta u_2 + a_{24}\,\delta u_4 &= a_{23} \\ a_{31}\,\delta u_1 + a_{32}\,\delta u_2 + a_{34}\,\delta u_4 &= a_{33} \end{aligned} \tag{7-22}$$

These three equations determine δu_1, δu_2, δu_4, and the changes so determined are the equivalent combination we seek. As applied to the chemical example, Eqs. (7-22) are

$$\begin{aligned} 10\,\delta u_1 + 5\,\delta u_2 \qquad\qquad &= 0 \\ 4\,\delta u_2 \qquad\qquad &= 5 \\ 2\,\delta u_2 + 10\,\delta u_4 &= 0 \end{aligned}$$

giving, as the equivalent combination to u_3,

$$\delta u_1 = -0.625 \qquad \delta u_2 = 1.25 \qquad \delta u_4 = -0.25$$

Now, the aggregate of resource values is

$$r' = s_1u_1 + s_2u_2 + s_3u_3 + \cdots + s_7u_7 \tag{7-23}$$

In the present case,

$$r' = 100(u_1 + u_2 + u_3 + u_4)$$

If the u's have the values that we guessed, then $r' = 700$, as an easy substitution will show. If, now, we set $u_3 = 1$ and offset this change by subtracting the equivalent combination from u_1, u_2, u_4, then r' takes on the value

$$100(u_1 - \delta u_1 + u_2 - \delta u_2 + 1 + u_4 - \delta u_4)$$

This increases r' by s_3 ($s_3 = 100$, in this case) and decreases it by $s_1\,\delta u_1 + s_2\,\delta u_2 + s_4\,\delta u_4$. If s_3 is less than $s_1\,\delta u_1 + s_2\,\delta u_2 + s_4\,\delta u_4$, the over-all effect is to decrease r', so that it would be advantageous to give a positive value to u_3. Any combination of changes such as we are discussing, of course, leaves the imputed costs of the individual activities unaltered.

In the present case, $s_3 = 100$, and

$$s_1\,\delta u_1 + s_2\,\delta u_2 + s_4\,\delta u_4 = 100(-0.625 + 1.25 - 0.25) = 37.5$$

Therefore the over-all effect on r' of giving u_3 a positive value would be adverse.

We can now state a general criterion for testing whether any set of imputed values is optimal in the sense of minimizing r'. Recall that s_j, the coefficient of u_j in Eq. (7-23), is the available quantity of the jth resource. Assume that $\delta u_1, \delta u_2, \delta u_3, \ldots$ is the equivalent combination to u_j and let us call $s_1\,\delta u_1 + s_2\,\delta u_2 + s_3\,\delta u_3 + \cdots$ the change in aggregate valuation equivalent to a change in u_j. Then we can state Theorem 2*a* as follows:

A nonnegative set of imputed values is optimal if and only if the available quantity of each resource assigned an imputed value of zero is at least as great as the change in aggregate valuation equivalent to a change in the imputed value of that resource.

To prove this assertion we need only note that it is a restatement of Theorem 2*a*. The common sense of this rule is that if we raise any zero value to $1 or any other positive number and then correct the previously positive values so as to leave the imputed costs of all activities equal to their net revenues, then the over-all effect will be to increase the aggregate value of resources, r', if the conditions of the rule are satisfied. But we have already seen that the marginal-productivity set of values that we seek makes r' as small as possible, subject to requirements 1*a* and 2*a*.

We have already applied this criterion to u_3. To finish applying the criterion we make similar calculations for u_5, u_6, u_7, obtaining the values that appear in Table 7-3.[1] The available quantity of each resource is

Table 7-3

Resource value	Available quantity of resource	Equivalent change in aggregate valuation
u_3	100	37.5
u_5	0	−10
u_6	0	− 7.5
u_7	0	−10

greater than the equivalent change in aggregate valuation. Therefore the set of values that we have selected is optimal. It is, of course, the same set that we obtained by first solving an allocation problem and then deriving a set of resource values from that solution.

We have now seen that the valuation problem can be solved without first solving the allocation problem, and, in the present example, it has turned out that the valuation problem is the easier one arithmetically. We have yet to see that the allocation problem can be solved as a by-product of the valuation problem. To complete this final step, subtract the

[1] The arithmetic of this calculation is less tedious than the description. At any rate we omit both.

last column of Table 7-3 from the middle column, to obtain the values that appear in Table 7-4. These numbers, it will be noted at once, are the optimum levels of the disposal of equipment E_3 and of the three manufacturing processes, respectively. Thus these excesses are analogous to the "gains" computed in the allocation problem, and, as we said earlier, the allocation problem stands in the same relation to the valuation problem as the valuation problem stands in relation to the allocation problem. The duality between the two problems is complete.

TABLE 7-4

Resource value	Excess of available quantity of resource over equivalent change in aggregate valuation
u_3	62.5
u_5	10
u_6	7.5
u_7	10

Economically, the duality shows that just as an optimal-production program implies definite prices for the services of the scarce resources, so a properly selected set of prices implies an optimal-production program. It reaffirms in a new context the fundamental economic theorem that a pricing system can serve as an efficient guide to production under appropriate technological conditions. The essential characteristics of the prices which correspond to an efficient allocation of resources are as follows:

1. They are nonnegative.
2. If the resources used by each activity are valued by these prices, the net revenue of each activity in an efficient program will be imputed completely to the resources absorbed.
3. If the resources used by each activity are valued by these prices, there will exist no activity whose net revenue is greater than the value of the resources it absorbs.
4. These prices measure the marginal productivities of the scarce resources.

7-5. CONCLUSIONS

The importance of this exploration of the value implications of linear programming is twofold. In the first place, we have broadened the contact between linear programming and ordinary economic analysis by showing that the principle of allocation according to linear programming stands in the same relation to a set of prices as do the principles of allocation employed in more familiar economic analysis. We have shown that the procedure of process substitution considered in linear programming leads to results consistent with the procedure of factor substitution considered in the conventional analysis and may, indeed, be used in place of the conventional analysis to impute factor values.

In the second place, these results have important implications for cost accounting. In analyzing the allocation problem we found the over-all production program that maximized profits and we assumed that the management of the firm, equipped with this program, would pass orders down the line to have it implemented. Thus we conceived of a highly centralized management. But many firms, especially large ones, operate on a rather decentralized principle. In such firms, accounting or control prices are established by a central planning authority or committee and the individual department heads are expected to do as well as they can in the light of those prices. (Of course, in such firms the major decisions of the department managers are subject to review and revision by higher-ups.) Decentralized administration of this type has many advantages, but it is important to note that its success depends upon the establishment of proper accounting prices. If factors are used predominantly in fixed proportions, the estimation of marginal productivities of individual factors cannot be used directly as a guide to the establishment of accounting prices. Nor is it generally feasible administratively to permit department heads to bid against each other for the use of fixed resources, after the manner of competitive markets. But we have now seen that linear programming can be used for this purpose and is thus a significant aid as well as an alternative to cost-accounting guidance.

These values also, of course, are useful guides for decisions concerning the acquisition or disposition of fixed resources. The stock of a fixed resource should be increased, for example, only when the value of its services as computed by linear programming is at least as great as its rental value as calculated from its cost of acquisition and expectation of useful life.

7-6. TECHNICAL POSTSCRIPT

Most of the difficult reasoning encountered in this chapter can be avoided by the use of a little matrix algebra. Suppose that an optimal program involves activities $A_1, \ldots, A_k$ at levels $x_1, \ldots, x_k$, respectively. Let $\mathbf{B}$ be the matrix $\mathbf{B} = [a_{ij}]$ of k rows and k columns, and let $\mathbf{x}$ be the column vector of activity levels. Then $\mathbf{Bx} = \mathbf{s}$, where $\mathbf{s}$ is the vector of factor limitations.

Now let $\mathbf{y}_i$ be the vector of activity levels that absorbs 1 unit of the ith factor and nothing else. This vector is the solution of $\mathbf{By}_i = \mathbf{I}_i$, where $\mathbf{I}_i$ is a k element vector with 1 in the ith position and zeros elsewhere. If $\mathbf{B}$ is nonsingular, $\mathbf{y}_i = \mathbf{B}^{-1}\mathbf{I}_i$. Let $\mathbf{v} = [v_1, \ldots, v_k]$ be the vector of net revenues of included activities, and let $\mathbf{v}'$ be its transpose. Then the net revenue afforded by the program $\mathbf{y}_i$ is the sum of the net revenues of its components, or $[\mathbf{v},\mathbf{y}_i] = \mathbf{v}'\mathbf{B}^{-1}\mathbf{I}_i$. We take this as our definition of the value, or marginal revenue product, of the ith factor. Thus,

$$\mathbf{u}_i = [\mathbf{v},\mathbf{y}_i] = \mathbf{v}'\mathbf{B}^{-1}\mathbf{I}_i$$

This is the same as Eq. (7-4).

Let $\mathbf{u}$ be the column vector of factor values, and $\mathbf{u}'$ be its transpose. Since the matrix of the vectors $\mathbf{I}_i$ is the identity matrix $\mathbf{I}$,

$$\mathbf{u}' = \mathbf{v}'\mathbf{B}^{-1}\mathbf{I} = \mathbf{v}'\mathbf{B}^{-1} \tag{7-24}$$

Postmultiplying by $\mathbf{s}$, the vector of factor limitations, we get

$$\mathbf{u}'\mathbf{s} = \mathbf{v}'\mathbf{B}^{-1}\mathbf{s} \tag{7-25}$$

But, since $\mathbf{Bx} = \mathbf{s}$, we have $\mathbf{x} = \mathbf{B}^{-1}\mathbf{s}$, and, substituting in Eq. (7-25), $\mathbf{u}'\mathbf{s} = \mathbf{v}'\mathbf{x} = r$. This is Eq. (7-11). Postmultiply Eq. (7-24) by $\mathbf{B}$. The result is $\mathbf{u}'\mathbf{B} = \mathbf{v}'$, or, taking transposes, $\mathbf{B}'\mathbf{u} = \mathbf{v}$. This is Eq. (7-15).

Up to now we have neglected the activities excluded from the optimal program. Let A_q be such an activity. The equivalent combination to A_q is $\mathbf{B}^{-1}A_q$. The net revenue flowing from this equivalent combination is $\mathbf{v}'\mathbf{B}^{-1}A_q$. Since we are assuming that activities $A_1, \ldots, A_k$ form an optimal selection, we have, by the simplex criterion, $\mathbf{v}'\mathbf{B}^{-1}A_q \geq v_q$, where v_q is the unit net revenue of A_q. Using Eq. (7-24), $\mathbf{u}'A_q \geq v_q$; i.e., the imputed cost of any excluded activity is at least as great as its net revenue. We have already found that if A_q is an included activity, then $\mathbf{u}'A_q = v_q$. This justifies Eqs. (7-16). If A_q is a disposal activity, $v_q = 0$, and $\mathbf{u}'A_q = \mathbf{u}'\mathbf{I}_q = u_q$, where q now is the designation of the corresponding factor, and u_q is its unit imputed value. Inserting these in the previous result, $u_q \geq 0$, which justifies requirement 1a.

The rest of the results in this chapter are dualism formulas, for which see Chap. 4.

8

Nonlinear Programming

8-1. THE PROBLEM OF NONLINEAR PROGRAMMING

In the preceding two chapters we have dealt with programming for firms which could regard the prices of the products they produce and of the factors and services they consume as given constants, unaffected by any decisions they might make. Thus we have excluded from consideration the whole field of monopoly, monopolistic competition, and similar market forms. We now address to ourselves the question: Can programming analysis be applied to firms not in perfectly competitive markets?

The same issue can be raised in a broader framework. The objective functions that we have been maximizing and minimizing throughout the book are, as we have noticed before, nothing but measures of the utility of the programs to which they correspond. The marginal utility of any activity is the rate of change of the objective function with respect to changes in the level of that activity. Heretofore we have been assuming that these marginal utilities were constant. But this assumption is galling to any economist worthy of the name. Diminishing marginal utility (or, if you prefer, diminishing marginal rates of substitution) has been one of the most fruitful postulates of economic theory. We ask then: Can programming methods be used in problems in which marginal utility (or rate of substitution) is not constant?

A simple example will bring out the difficulties that arise as soon as the marginal utilities of activities are considered to be variable. Consider the automobile firm of Chap. 6 and suppose that the demand curve for its automobiles slopes downward but that the demand curve for its trucks is horizontal. To be specific, suppose that the firm's market conditions are specified by the demand equations

$$v_1 = 625 - \frac{x_1}{60}$$

$$v_2 = 250$$

that is, v_1, the net revenue per automobile sold, is a linearly declining function of x_1, the volume of automobile sales, but v_2, the net revenue per truck sold, is constant. We assume the same technical conditions and restrictions as in Chap. 6. Then the only part of the problem that is changed is the shape of the curves of constant net revenue. The altered situation is depicted in Fig. 8-1, which should be compared with Fig. 6-1 (page 135). The two figures are identical except that the isoprofit lines of Chap. 6 have been turned into isoprofit parabolas.[1] The essential new

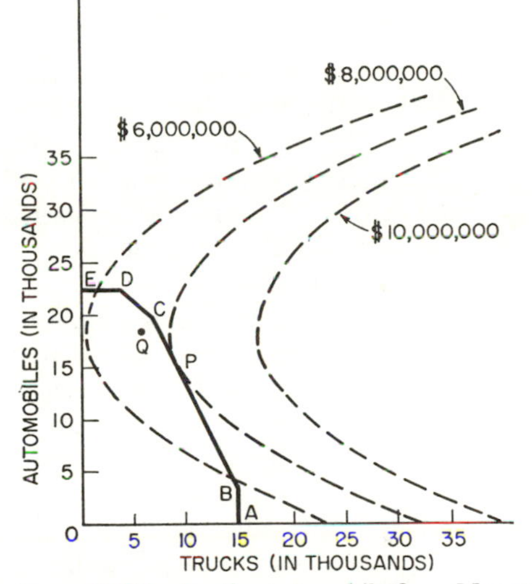

Fig. 8-1. Profit possibilities for the automobile firm, Monopolistic Case.

feature is that if the isoprofit contours are straight lines, then the highest attainable one must touch the accessible region at one of its vertexes, but if the contours are curved, as we are now assuming, then the highest attainable one may touch the accessible region anywhere on its boundary. In the example, it is evident from the figure that the optimal program corresponds to point P, which is not at any vertex.[2]

Even worse possibilities have to be taken into account. Suppose that the automobile company faced a sloping demand curve for its trucks, too.

[1] The formula for the isoprofit contours is

$$(x_1 - 18{,}750)^2 - 15{,}000x_2 = \text{constant}$$

[2] The optimal program is to produce 15,000 automobiles and slightly more than 9,000 trucks. The capacity of the engine-assembly department is fully utilized; excess capacity remains in all the other departments. Instead of using four activities at positive levels (counting disposal activities, of course), the optimal program now calls for five.

The pair of curves, for example, might be

$$v_1 = 625 - \frac{x_1}{60}$$

$$v_2 = 400 - \frac{x_2}{30}$$

Then, as the reader can verify readily, the curves of constant net revenue are the nest of ellipses given by the formula

$$(x_1 - 18{,}750)^2 + 2(x_2 - 6{,}000)^2 = \text{constant}$$

The maximum possible profit for these demand curves is attained by the following output: $x_1 = 18{,}750$ automobiles, $x_2 = 6{,}000$ trucks. This optimum is shown by point Q in Fig. 8-1, which does not even lie on the boundary. Note that it leaves excess capacity in all four departments; i.e., that six activities (including disposal activities) are used at positive levels.

We see, thus, that the introduction of diminishing marginal utilities has altered the geometry of the problem fundamentally. No longer can we limit our search for the optimum to a finite number of vertexes of the accessible region. We cannot even confine our attention to the boundary of the accessible region, for the optimum may occur anywhere.

These variants of the automobile example illustrate a type of problem that fits neither the linear-programming model nor the familiar marginal approach to profit maximization. It is instructive to consider these problems from the viewpoint of the marginal analysis. Let us take up first the case in which both demand curves slope. Then, disregarding the resource limitations, the procedure is to find the output point at which marginal net revenue is zero for both commodities. The marginal net revenue of automobiles when x_1 are sold is

$$\frac{d}{dx_1} x_1 v_1 = 625 - \frac{x_1}{30}$$

and this is zero when $x_1 = 18{,}750$. Similarly the optimal output of trucks is $x_2 = 6{,}000$. So the marginal analysis gives the right answer; i.e., it did this time, because when we disregarded the resource limitations we made use of our illegitimate foreknowledge that the resource limitations were ineffective. Without this foreknowledge we should not have known what restraints to impose in applying the marginal technique.

To reaffirm this, consider the case in which the demand curve for trucks was horizontal. Obviously, unless we impose restraints, the optimum will call for producing an infinite number of trucks. We must therefore introduce one or more of the restraints into the maximization procedure. But which ones? Until we know this we are stuck, and the conventional procedure provides no clue.

In general, then, the conventional methods of analyzing the firm succeed if we know which limitations are binding and which can be disregarded. Usually this information is denied us, and a new method is required for such problems. This is the problem of nonlinear programming.

The upshot of this introductory discussion is that when we try to solve a nonlinear problem we are deprived of the most powerful tool used in the linear case—the advance knowledge that the optimum would have to occur at one of a finite number of vertex points. The sad consequence is that no sure-fire practical method for solving nonlinear-programming problems has yet been found. This fact does not end the possibilities of fruitful discussion, however. To an economist the important question is not how to find an optimal program, but rather what will be the characteristics of such a program when found.[1] Fortunately, a good deal is known about this. The conditions for the optimum of a nonlinear-programming problem, we shall see, are a generalization of the familiar marginal-equality conditions for the optimum of a firm and constitute a bridge between them and the conditions for an optimum in linear programming.

8-2. THE KUHN-TUCKER OPTIMALITY CONDITIONS

Let us state the nonlinear programming problem in the most general form in which we shall consider it.[2] Suppose that a firm has available k activities, not counting disposal activities. As before, we denote a program by a vector (or list) of activity levels $x_1, x_2, \ldots, x_k$ and assume that the firm desires to maximize some net-revenue function

$$r = r(x_1, x_2, \ldots, x_k)$$

of the activity levels. We assume also that the choice of activity levels is limited by a set of n linear-inequality restraints; thus,

$$\begin{aligned} a_{11}x_1 + a_{12}x_2 + \cdots + a_{1k}x_k &\leq c_1 \\ a_{21}x_1 + a_{22}x_2 + \cdots + a_{2k}x_k &\leq c_2 \\ \cdots\cdots\cdots\cdots\cdots\cdots\cdots\cdots \\ a_{n1}x_1 + a_{n2}x_2 + \cdots + a_{nk}x_k &\leq c_n \end{aligned} \tag{8-1}$$

Furthermore, of course, none of the x_i $(i = 1, \ldots, k)$ can be negative.

[1] Furthermore, in practical applications, special approximative methods can be tailor-made to solve individual problems as they arise.

[2] For a more general derivation see H. W. Kuhn and A. W. Tucker, "Nonlinear Programming," in J. Neyman (ed.), *Proceedings of the Second Berkeley Symposium on Mathematical Statistics and Probability*, pp. 481–492, University of California Press, Berkeley, Calif., 1951.

Now we think of a particular program, say $x_1^0, x_2^0, \ldots, x_k^0$, which we assume to be optimal, i.e., to maximize r subject to the restraint of Eqs. (8-1). Two of the properties of this program are trivial: none of its elements is negative, and it satisfies Eqs. (8-1). Now we search for less obvious properties.

Let us exclude for the present the possibility that

$$x_1^0 = x_2^0 = \cdots = x_k^0 = 0$$

that is, the case in which the best thing to do is nothing at all. Then some of the x_i^0 will be positive. To be specific, let us assume that the activities are numbered in such a way that $x_i^0 = 0$ for $i = 1, 2, \ldots, k_1$, and $x_i^0 > 0$ for $i = k_1 + 1, \ldots, k$. We do not exclude the possibility that $k_1 = 0$, that is, that all activities are used at positive levels.

Now think of substituting $x_1^0, x_2^0, \ldots, x_k^0$ in Eqs. (8-1). Some of these restraints may turn out to be binding; i.e., the left-hand side may be exactly equal to the right-hand side. Let us suppose that the restraining conditions have been listed in such an order that the first n_1 of them are binding. If $n_1 = 0$, none of the restrictions is binding; if $n_1 = n$, they all are. We include both of these possibilities and also any intermediate case.

This setup permits us to describe an admissible variation in the program. Let δx_i denote a small change in the level of x_i from the optimal level x_i^0. Then $\delta x_1, \delta x_2, \ldots, \delta x_k$ is a specification of a set of changes in the whole program. It is admissible if the new program, $x_i^0 + \delta x_i$, satisfies all the restraining conditions. Let us consider variations which satisfy the following conditions:

$$\delta x_i \geq 0 \qquad i = 1, 2, \ldots, k_1 \tag{8-2}$$

and

$$a_{j1}\,\delta x_1 + a_{j2}\,\delta x_2 + \cdots + a_{jk}\,\delta x_k \leq 0 \qquad j = 1, 2, \ldots, n_1 \tag{8-3}$$

Condition (8-2) assures us that none of the activity levels that were zero in the x_i^0 program will become negative as a result of the variation, and condition (8-3) guarantees that none of the restraints that were binding before the variation will be violated if the variation is applied. But neither of these conditions on the δx_i protects us from violating the other restrictions of the problem. Fortunately, all the other conditions have some play in them (because they were satisfied with strict inequalities by the x_i^0 program) so we can imagine the δx_i as being small enough to satisfy those conditions, too. The reason this is satisfactory is that the hypothesis that x_i^0 is optimal implies that there is no permissible direction of change[1] which will increase r. Conditions (8-2) and (8-3) suffice

[1] Two variations are in the same "direction" if their components δx_i are proportional with a positive ratio, element by element.

to define the permissible directions of change, i.e., the directions in which at least a small variation can be made without violating the restrictions of the problem.

Now we assume that $r(x_1, \ldots, x_k)$ is continuous and has continuous partial derivatives in the neighborhood of $x_1^0, \ldots, x_k^0$. If we apply small variations $\delta x_1, \ldots, \delta x_k$, the effect on r will be given by the total differential[1]

$$\delta r = \sum_i \frac{\partial r}{\partial x_i} \delta x_i$$

where the partial derivatives are evaluated at $x_1^0, \ldots, x_k^0$.

We now make use of the fact that $x_1^0, \ldots, x_k^0$ is an optimal program. This tells us that there is no permissible direction of change in which the net-revenue function increases. Algebraically,

$$\delta r = \sum_i^k \frac{\partial r}{\partial x_i} \delta x_i \leq 0 \tag{8-4}$$

for all sets of variations satisfying

$$a_{j1}\,\delta x_1 + a_{j2}\,\delta x_2 + \cdots + a_{jk}\,\delta x_k \leq 0 \qquad j = 1, \ldots, n_1 \tag{8-3}$$

and

$$-\delta x_i \leq 0 \qquad i = 1, \ldots, k_1 \tag{8-2}$$

At this stage we cite and apply a very useful theorem due to J. Farkas.

FARKAS' THEOREM.[2] *If the inequality*

$$y_0 = b_{01}z_1 + b_{02}z_2 + \cdots + b_{0n}z_n \leq 0$$

is satisfied for every set of numbers $z_1, z_2, \ldots, z_n$ that satisfies all the inequalities

$$\begin{aligned} y_1 &= b_{11}z_1 + b_{12}z_2 + \cdots + b_{1n}z_n \leq 0 \\ y_2 &= b_{21}z_1 + b_{22}z_2 + \cdots + b_{2n}z_n \leq 0 \\ &\cdots\cdots\cdots\cdots\cdots\cdots \\ y_m &= b_{m1}z_1 + b_{m2}z_2 + \cdots + b_{mn}z_n \leq 0 \end{aligned}$$

then there exist some nonnegative multipliers $u_1, u_2, \ldots, u_m$, independent of $z_1, \ldots, z_n$, such that

$$y_0 = u_1y_1 + u_2y_2 + \cdots + u_my_m$$

The inequalities (8-4), (8-3), (8-2) satisfy the hypothesis of Farkas' theorem. Hence there exist nonnegative numbers, which we can denote

[1] Neglecting terms of second and higher order in the variations.

[2] This theorem is fundamental to linear as well as to nonlinear programming. Further discussion and a proof are given in Appendix B, Sec. 15.

by $u_1, u_2, \ldots, u_{n_1}, w_1, w_2, \ldots, w_{k_1}$, such that

$$\sum_i^k \frac{\partial r}{\partial x_i}\,\delta x_i \equiv \sum_j^{n_1} u_j \sum_i^k a_{ji}\,\delta x_i - \sum_i^{k_1} w_i\,\delta x_i$$

This expression is an identity, true for all values of $\delta x_1, \ldots, \delta x_k$ without restriction. Now change the order of summation in the double sum, transpose it to the left-hand side of the equation, and collect terms in δx_i to obtain

$$\sum_i^k \left(\frac{\partial r}{\partial x_i} - \sum_j^{n_1} a_{ji}u_j\right)\delta x_i \equiv -\sum_i^{k_1} w_i\,\delta x_i$$

for all values of $\delta x_1, \ldots, \delta x_k$. Let us define $u_j = 0$ for $j = n_1 + 1, n_1 + 2, \ldots, n$, so as to be able to extend the range of summation for j over all n restraining inequalities of (8-1), and write the last equation in the weaker but more useful form

$$\sum_i^k \left(\frac{\partial r}{\partial x_i} - \sum_j^{n} a_{ji}u_j\right)\delta x_i \leq 0 \tag{8-5}$$

for all $\delta x_1, \ldots, \delta x_k$ such that

$$\sum_i^{k_1} w_i\,\delta x_i \geq 0 \tag{8-6}$$

A little more trickery, and we shall have some results. Let us consider the effects of varying one activity at a time; i.e., let us let $\delta x_i = 1$ for some particular value of i while setting all the other variations equal to zero. There are two cases to be considered: (1) $i \leq k_1$, and (2) $i > k_1$. In case 1, from the definition of k_1, $x_i^0 = 0$, we see that Activity i is not used in the optimal program under consideration. Let $\delta x_i = 1$, and let all the other variations be zero. Then inequality (8-6) is satisfied (remember that w_i is nonnegative) and, substituting in inequality (8-5), we find

$$\frac{\partial r}{\partial x_i} - \sum_j^n a_{ji}u_j \leq 0 \qquad i = 1, 2, \ldots, k_1 \tag{8-7}$$

Now consider case 2. In this case, Activity i is used at a positive level in the optimal program, and δx_i is beyond the range of summation of inequality (8-6). Thus if we set all the variations equal to zero except the ith, inequality (8-6) will be satisfied, whatever the value of δx_i. Then from inequality (8-5) we have

$$\left(\frac{\partial r}{\partial x_i} - \sum_j^n a_{ji}u_j\right)\delta x_i \leq 0 \qquad i = k_1 + 1, \ldots, k$$

for all values of δx_i: Trying just the two convenient values $\delta x_i = 1$ and $\delta x_i = -1$, we see that this condition can be satisfied only if the expression in parentheses is precisely zero.

The algebra is done (for a while). We have only to gather together our results and interpret them. We have proved the following theorem: *If a certain program $x_1^0, x_2^0, \ldots, x_k^0$ maximizes an objective function $r(x_1, \ldots, x_k)$ subject to restrictions* (8-1), *then there must exist some nonnegative numbers $u_1, u_2, \ldots, u_n$ such that for all values of i, $i = 1, 2, \ldots, k$, either*

1. $x_i^0 = 0$ *(Activity i is not used) and*

$$\frac{\partial r}{\partial x_i} - \sum_j^n a_{ji}u_j \leq 0 \tag{8-7}$$

or

2. $x_i^0 > 0$ *(Activity i is used) and*

$$\frac{\partial r}{\partial x_i} - \sum_j^n a_{ji}u_j = 0 \tag{8-8}$$

The theorem tells us merely that $u_1, u_2, \ldots, u_n$ are "numbers," but by now we are familiar enough with this kind of problem to recognize them as imputed values. We note three obvious facts to bear out this interpretation. First, each of the u_j corresponds to one of the restrictions in Eqs. (8-1). Second, none of the u_j can be negative. Third, $u_j = 0$ if $j > n_1$; that is, if the jth restriction is satisfied with room to spare, then the jth resource is a free good. Adopting this interpretation, we see that

$$\sum_j^n a_{ji}u_j = a_{1i}u_1 + a_{2i}u_2 + \cdots + a_{ni}u_n$$

is the total imputed value of all resources used per unit of the ith activity—in other words, its imputed cost. At the same time, $\partial r/\partial x_i$ is clearly the marginal effect of the ith activity on the objective function. Let us call this the marginal-revenue product of the ith activity. Then we can restate our theorem in economic terms, as follows: If x_1^0, x_2^0,

$\ldots, x_k^0$ is a program that maximizes a nonlinear objective function subject to n linear restrictions, then there exists a set of n nonnegative imputed values such that the unit imputed cost of each activity used exactly equals its marginal-revenue product and the unit imputed cost of each activity not used is at least as great as its marginal-revenue product. Furthermore, the imputed value corresponding to each restriction that is overfulfilled by the optimal program is zero.

We have used a great display of mathematics to confirm our economic common sense. But it is confirmed.[1]

8-3. SUFFICIENCY OF THE KUHN-TUCKER CONDITIONS

In the last section we saw that any optimal solution to a nonlinear-programming situation satisfies the mathematical Kuhn-Tucker conditions, given in Eqs. (8-7) and (8-8), conditions that have a sound economic meaning. We now raise the converse question: If a program satisfies the Kuhn-Tucker conditions, is it optimal?

The answer is affirmative with some reservations. Suppose that some program, $x_1^0, x_2^0, \ldots, x_k^0$, together with some set of nonnegative imputed values, $u_1, u_2, \ldots, u_n$, satisfies the conditions of the last section, but that $x_1^0, x_2^0, \ldots, x_k^0$ is not known to be optimal. Then it is easy to see that $x_1^0, x_2^0, \ldots, x_k^0$ must be a local optimum, i.e., that there is no permissible direction in which the program can be changed a small amount so as to increase net revenue. For, suppose that $x_1, \ldots, x_k$ is some alternative permissible program close enough to $x_1^0, x_2^0, \ldots, x_k^0$ so that we can express the difference in net revenues as

$$\delta r = r(x_1, \ldots, x_k) - r(x_1^0, \ldots, x_k^0) = \sum_i^k \frac{\partial r}{\partial x_i}(x_i - x_i^0) \quad (8\text{-}9)$$

From Eqs. (8-7) and (8-8) we see that for every i, $i = 1, \ldots, k$, either $x_i^0 = 0$ or $\partial r/\partial x_i - \sum_j^n a_{ji}u_j = 0$. Multiplying the two together we find, for all i,

$$x_i^0 \frac{\partial r}{\partial x_i} = x_i^0 \sum_j^n u_j a_{ji}$$

Substituting this result in Eq. (8-9) we obtain

[1] *Exercise.* In the case where $r(x_1, \ldots, x_k)$ is a linear function of $x_1, \ldots, x_k$, deduce the simplex criterion from Eqs. (8-7) and (8-8).

$$\delta r = \sum_i^k \frac{\partial r}{\partial x_i} x_i - \sum_i^k x_i{}^0 \sum_j^n u_j a_{ji}$$

Also from Eqs. (8-7) and (8-8), making use of the fact that $x_1, \ldots, x_k$ is permissible so $x_i \geq 0$, we have

$$x_i \frac{\partial r}{\partial x_i} \leq x_i \sum_j^n u_j a_{ji}$$

Inserting this also in the expression for δr we find

$$\delta r \leq \sum_i^k x_i \sum_j^n u_j a_{ji} - \sum_i^k x_i{}^0 \sum_j^n u_j a_{ji}$$
$$\leq \sum_j^n u_j \left(\sum_i^k a_{ji} x_i - \sum_i^k a_{ji} x_i{}^0 \right)$$

Now consider this last summation for j, term by term. There are two classes of terms. In the first class $u_j = 0$. These contribute nothing to the sum. In the second class $u_j > 0$. All such terms correspond to effective restrictions, i.e., to values of j for which $\Sigma a_{ji} x_i{}^0 = c_j$. For such terms, since $x_1, \ldots, x_k$ is permissible, $\Sigma a_{ji} x_i \leq \Sigma a_{ji} x_i{}^0$ and the expression in parentheses is nonpositive. In summary, there are no positive terms on the right-hand side of this equation, $\delta r \leq 0$, and the net revenue yielded by $x_1{}^0, \ldots, x_k{}^0$ is at least as great as the net revenue yielded by any nearby program. A program that satisfies the Kuhn-Tucker conditions is, at least, locally optimal.

This is as much as can be said without making use of nonlocal properties of $r(x_1, \ldots, x_k)$. Suppose, to take the favorable case, that we know that

$$r(x_1, \ldots, x_k) \leq r(x_1{}^0, \ldots, x_k{}^0) + \sum_i^k \frac{\partial r}{\partial x_i}(x_i - x_i{}^0) \qquad \text{(8-10)}$$

for all permissible $x_1, \ldots, x_k$, where the partial derivatives are evaluated at $x_1{}^0, \ldots, x_k{}^0$. Geometrically, Eq. (8-10) asserts that, throughout the permissible region, $r(x_1, \ldots, x_k)$ lies on or below the tangent plane to it drawn at $x_1{}^0, \ldots, x_k{}^0$. In this case,

$$r(x_1, \ldots, x_k) - r(x_1{}^0, \ldots, x_k{}^0) \leq \sum_i^k \frac{\partial r}{\partial x_i}(x_i - x_i{}^0)$$

and we have already seen that the sum on the right-hand side is nonpositive for permissible $x_1, \ldots, x_k$ if the Kuhn-Tucker conditions are satisfied at $x_1^0, \ldots, x_k^0$. Thus if the objective function satisfies Eq. (8-10) and if the Kuhn-Tucker conditions are satisfied by any program, then that program corresponds to a *maximum maximorum*. Fortunately, Eq. (8-10) is likely to be satisfied in many economic circumstances because it corresponds to the results of decreasing marginal revenue (or utility) or of increasing marginal costs. It is possible for a program that satisfies the Kuhn-Tucker conditions to be the optimal program in the permissible region even though Eq. (8-10) is not satisfied.

8-4. THE AUTOMOBILE COMPANY AGAIN

We illustrate these results by applying them to the example of the automobile company. Consider first the version in which the demand curve for automobiles sloped downward but the curve for trucks did not. Let x_1 denote the sales of automobiles and x_2 denote the sales of trucks. Then the function to be maximized is

$$\begin{aligned} r(x_1,x_2) &= x_1v_1 + x_2v_2 = x_1\left(625 - \frac{x_1}{60}\right) + 250x_2 \\ &= 625x_1 + 250x_2 - \frac{x_1^2}{60} \end{aligned}$$

The restraints were discussed in Sec. 6-2 and were

$$\begin{aligned} 0.00400x_1 + 0.00286x_2 &\leq 100 \\ 0.00300x_1 + 0.00600x_2 &\leq 100 \\ 0.00444x_1 \qquad\qquad &\leq 100 \\ 0.00667x_2 &\leq 100 \end{aligned}$$

As a first try at solving this problem, let us assume (incorrectly, it will turn out) that the first restraint is effective and that the others are not. This converts the problem into a standard calculus exercise in finding a maximum subject to a single restraint. We form the Lagrangean expression

$$L(x_1,x_2) = r(x_1,x_2) + \lambda(0.004x_1 + 0.00286x_2 - 100)$$

compute the partial derivatives,

$$\begin{aligned} \frac{\partial L}{\partial x_1} &= 625 - \frac{x_1}{30} \qquad\qquad + 0.004\lambda \\ \frac{\partial L}{\partial x_2} &= 250 \qquad\qquad\qquad + 0.00286\lambda \\ \frac{\partial L}{\partial \lambda} &= -100 + 0.004x_1 + 0.00286x_2 \end{aligned}$$

and solve for x_1, x_2, and λ by equating these partials to zero. The result is $x_1 = 8{,}250$, $x_2 = 23{,}450$, $\lambda = -87{,}500$. Comparison with the four restrictions shows that this solution is infeasible; it violates the second restraint by requiring about 165 per cent of available engine-assembly capacity. We must thus try again, and this time we assume that the second restraint is the only effective one. The Lagrangean expression becomes

$$L(x_1,x_2) = r(x_1,x_2) + \lambda(0.003x_1 + 0.006x_2 - 100)$$

Following the same procedure we find, as a trial optimum, $x_1 = 15{,}000$, $x_2 = 9{,}167$, $\lambda = -41{,}667$. Checking the four restraint equations shows that this solution satisfies the second restraint exactly and the first, third, and fourth with room to spare. It is therefore feasible. To test whether it is optimal (locally) we must see whether it satisfies Eq. (8-8). Equation (8-7) is inapplicable in this case because both available activities are used at positive levels. We thus need imputed values for the four resources $u_1, \ldots, u_4$. Since there is excess capacity for metal stamping, automobile assembly, and truck assembly, their imputed values u_1, u_3, u_4, respectively, all are zero. It remains for us to determine u_2, the imputed value of engine assembly, the fully utilized resource.

To find u_2 we use the requirement that the marginal-revenue product of truck production must equal its unit imputed cost.[1] The marginal-revenue product of truck production is

$$\frac{\partial r}{\partial x_2} = 250$$

The unit imputed cost of truck production is, since $u_1 = u_4 = 0$,

$$0.00286u_1 + 0.006u_2 + 0.00667u_4 = 0.006u_2$$

Equating these two we obtain $u_2 = 41{,}667$. Note that $u_2 = -\lambda$. This is not just coincidence. The mathematician's Lagrange multipliers and the economist's imputed values are intimately related concepts.

It is now trivial to verify that the solution $x_1 = 15{,}000$, $x_2 = 9{,}167$, $u_1 = 0$, $u_2 = 41{,}667$, $u_3 = 0$, $u_4 = 0$ satisfies Eq. (8-8) and, indeed, all the conditions for an optimum. The problem is solved.

As a final illustration, consider the problem of the automobile firm, assuming declining demand curves for both automobiles and trucks. The net-revenue function, using the data assumed in Sec. 8-1, is

$$r(x_1,x_2) = x_1v_1 + x_2v_2 = 625x_1 + 400x_2 - \frac{x_1^2}{60} - \frac{x_2^2}{30}$$

To maximize this function subject to the restraints let us guess that none of the resource restrictions is binding and equate the partial derivatives

[1] It would be just as valid to form the analogous equality for automobile production. This alternative calculation is left as an exercise.

of $r(x_1,x_2)$ to zero. We find $x_1 = 18{,}750$, $x_2 = 6{,}000$. This solution is easily seen to be feasible with an excess of every one of the resources. Thus $u_1 = u_2 = u_3 = u_4 = 0$. Equation (8-8) is obviously satisfied, and this solution is optimal.

8-5. LESS AND MORE GENERAL FORMULATIONS

In this chapter we have made no assumptions about the objective function $r(x_1, \ldots, x_k)$ except that it is differentiable for positive values of the arguments. In particular it might be linear, in which case the nonlinear-programming problem becomes a linear one. This remark suffices to show that many of the results presented in the earlier chapters are special cases of the ones contained in this chapter. (Recall, especially, footnote 1, p. 194.)

Two distinctions between the linear-programming problem and the present more general situation are worth mentioning, however. First, the explicit solution methods that work so powerfully in the linear case do not apply in general for the reasons noted in Sec. 8-1. This endows the linear case with its special practical importance. Second, in our discussion of the linear-programming problem we were able to deduce the conditions satisfied by an optimal program and even to find computational procedures for determining such a program without raising the "dual" problem of implicit valuation. In the general case this independent discussion of programming and evaluation proved impossible. In order to make any progress at all, we had to introduce the dual variables (the u's), and the result of our labors was a statement of conditions involving both the activity levels and the imputed values. If, as we noted, the programming and valuation problems are intimately related in the linear case, they are inextricably connected in the nonlinear one.

Now let us face in the other direction. Kuhn and Tucker, in their fundamental paper cited above, actually solved a still more general problem than the one we have dealt with. Let us state their problem and, without proof, their results. Suppose that it is desired to maximize some objective function $r(x_1, \ldots, x_k)$, subject to n inequality restraints on the activity levels; thus,

$$\begin{aligned} f_1(x_1, \ldots, x_k) &\leq 0 \\ f_2(x_1, \ldots, x_k) &\leq 0 \\ \cdots\cdots\cdots&\cdots \\ f_n(x_1, \ldots, x_k) &\leq 0 \end{aligned} \tag{8-11}$$

We assume only that all these functions are differentiable and that the feasible region defined by conditions (8-11) is convex. By this last we mean that if two programs $x_1^0, \ldots, x_k^0$ and $x_1', \ldots, x_k'$ both satisfy conditions (8-11), so does the intermediate program $\sigma_0 x_1{}^0 + \sigma_1 x_1'$, . . . , $\sigma_0 x_k{}^0 + \sigma_1 x_k'$ for all nonnegative σ_0, σ_1 satisfying $\sigma_0 + \sigma_1 = 1$. Verbally this requires that if any two programs are both feasible, then so is any internal average of the two.

We have already treated the case in which the restraining functions $f_j(x_1, \ldots, x_k)$ were linear. In that case it was unnecessary to distinguish between the average imputed cost of any activity and its marginal imputed cost because average costs were constant. With the more general restraining functions, however, we must make the distinction. To do this we consider the partial derivatives evaluated for any particular program, $x_1{}^0, \ldots, x_k{}^0$:

$$\frac{\partial r}{\partial x_i} = \text{marginal net revenue of } i\text{th activity}$$

and $$\frac{\partial f_j}{\partial x_i} = \text{marginal effect of } i\text{th activity on } j\text{th restriction}$$

Then if we have a set of imputed values $u_1, \ldots, u_n$ associated with the n restrictions, clearly

$$\sum_j^n u_j \frac{\partial f_j}{\partial x_i} = u_1 \frac{\partial f_1}{\partial x_i} + \cdots + u_n \frac{\partial f_n}{\partial x_i}$$

is the marginal imputed cost of the ith activity.

We can now state the Kuhn-Tucker conditions for an optimal solution of this more general problem. If $x_1{}^0, \ldots, x_k{}^0$ maximizes $r(x_1, \ldots, x_k)$ subject to conditions (8-11) and to $x_1{}^0 \geq 0, \ldots, x_k{}^0 \geq 0$, then there must exist a set of nonnegative imputed values $u_1, \ldots, u_n$ such that the following conditions are satisfied:

1. $f_j(x_i{}^0, \ldots, x_k{}^0) \leq 0 \qquad j = 1, \ldots, n$

i.e., the solution is feasible.

2. If $f_j(x_1{}^0, \ldots, x_k{}^0) < 0$, then $u_j = 0$

i.e., if the jth restraint is satisfied with a surplus, then the associated imputed value is zero.

3. $\dfrac{\partial r}{\partial x_i} - \sum_j^n u_j \dfrac{\partial f_j}{\partial x_i} \leq 0 \qquad i = 1, \ldots, k$

i.e., the marginal imputed cost of every activity is at least as great as its marginal net revenue.

4. If $\frac{\partial r}{\partial x_i} - \sum_j^n u_j \frac{\partial f_j}{\partial x_i} < 0$, then $x_i = 0$

i.e., if the marginal imputed cost of any activity exceeds its marginal net revenue, that activity is not used.

The conditions deduced in Sec. 8-2 are clearly the special case of these in which $\partial f_j/\partial x_i = a_{ji}$.

The converse question arises here also, of course. If we have a feasible program $x_1^0, \ldots, x_k^0$ and can find some nonnegative imputed values $u_1, \ldots, u_n$ such that conditions 1 to 4 are satisfied, is that program optimal? Just as in the case in which the restraints were linear, such a program is locally optimal, at least; but before we can declare it to be the best attainable program, some additional assumptions must be satisfied. For example, suppose that

$$r(x_1, \ldots, x_k) \leq r(x_1^0, \ldots, x_k^0) + \sum_i^k \frac{\partial r}{\partial x_i}(x_i - x_i^0)$$

for all admissible $x_1, \ldots, x_k$ (the same assumption that we made in Sec. 8-3) and

$$f_j(x_1, \ldots, x_k) \geq f_j(x_1^0, \ldots, x_k^0) + \sum_i^k \frac{\partial f_j}{\partial x_i}(x_i - x_i^0) \qquad j = 1, \ldots, n \quad \text{(8-12)}$$

Assumption (8-12), by the way, is not a new assumption but follows from the convexity of the feasible region, which we postulated earlier. If these conditions hold in addition to conditions 1 to 4, then Kuhn and Tucker show (and the proof is not hard) that $x_1^0, \ldots, x_k^0$ yields the greatest net revenue that can be obtained.

In this most general formulation, linearity has completely vanished. Since assumptions are good things to eliminate, this is an advance. But we have already noted the cost; convenient methods of practical solution go by the board along with the linearity assumptions. This fact does not entirely rob nonlinear programming of its importance for practical guidance, however, because mathematicians have ways for getting around the difficulty. The situation is analogous to the problem of integration in the calculus. There are many functions for which no formal integral can be found (the Gaussian "normal" curve is a notorious example), but approximate integrals for such functions, accurate to any desired degree, can be found by numerical methods. In the same spirit, practical nonlinear-programming problems can frequently be solved to a high degree of precision by a variety of mathematical tricks. The problem is difficult technically, however, and the methods used throw no light on the logical or economic significance of nonlinear programming, so we shall not go into this subject.

8-6. COMPARISON WITH THE CONVENTIONAL THEORY OF PRODUCTION

Now, having banished the linearity assumptions of linear programming, have we come back to the orthodox marginal-productivity analysis of maximization? The comparison between general programming, as we may call the last formulation, and the conventional analysis of production is surprisingly intricate because, although the two methods of analysis deal with the same problem, they focus attention on different variables and therefore meet rather obliquely.

We may take Sune Carlson's presentation as typical of the conventional approach. He writes:[1]

> If we denote the quantity of output by y, and the quantities of the variable productive services, m in number, by $v_1, \ldots, v_m$, we write
>
> $$y = \phi(v_1, \ldots, v_m)$$
>
> This is our *production function.* The production function, it must be remembered, is defined in relation to a given plant; that is, certain fixed services.
>
> A given amount of output may frequently be produced from a number of different service combinations. It may also be true that the same combination of productive services gives varied amounts of output, depending upon how efficiently the productive services are organized. . . . If we want the production function to give only one value for the output from a given service combination, the function must be so defined that it expresses the *maximum product* obtainable from the combination at the existing state of technical knowledge. Therefore, the purely *technical* maximization problem may be said to be solved by the very definition of our production function.

If we compare this with the formulation given in Eqs. (8-11), we see the same ingredients, very differently expressed. Carlson focuses attention on the flows of inputs and outputs[2] because these are the variables that convey the impact of the firm in question on the markets in which it operates. In the programming formulation these flows play a subsidiary role (buried in the activity-level variables) because, just by reason of being freely variable, the variable inputs and outputs do not delimit the

[1] Sune Carlson, *A Study on the Pure Theory of Production*, pp. 14–15, The University of Chicago Libraries, Chicago, 1939. We have changed Carlson's notation slightly in order to avoid confusion with other uses of the symbols.

[2] The fact that there is one output in the equation quoted is fortuitous, not an essential characteristic of the conventional approach.

field of choice and are not "scarce" to the firm in question. On the other hand, in the programming formulation, most explicitly when the restraints are linear, the quantities of the fixed productive factors are central to the problem because they are essential data in determining what the firm can and cannot do, while in the conventional formulation these same fixed factors are regarded as being somewhat aside from the problem just because their quantities are fixed and predetermined. Another way to state the same contrast is to note that the programming approach takes the prices of the variable factors as given, and therefore not deserving explicit attention, and is concerned with imputing values to the fixed factors, while the conventional analysis regards the costs of fixed factors as "sunk" and concentrates on the influences that determine the flows and prices of the variable factors. In short, the programming approach is inward-looking, into the firm; the conventional approach is outward-looking, out into the market, even when it deals with a single firm.

It is worth reiterating that although the two approaches emphasize different questions, they both answer the full range. The conventional approach has been used to impute values to all the cooperating factors of production, fixed and variable; the programming method does provide a firm's demand schedules for variable factors. We mustn't be fooled by appearances.

The contrast between the choice of explicit variables in the two approaches is something more than the appearance of a difference. Consider a firm in which the number of available activities is greater than the number of variable inputs. If we solve the programming problem for that firm in given market circumstances, we shall know the quantities of the inputs that that firm will consume under those circumstances. Thus the programming approach solves the conventional problem. But is the converse true? On the contrary. Carlson's production function (i.e., the conventional one) cannot even be written down until a programming problem has been solved. That is what Carlson meant by the last sentence quoted. In order to derive the production function we must consider various definite combinations of inputs, and for each of them (i.e., holding the variable inputs temporarily constant) we must determine the program that maximizes output. This is the "purely technical maximization problem" to which Carlson refers. It may, perhaps, be regarded as "purely technical" so long as the firm has a single output (as in the equation given), but as soon as multiple products are admitted, an undeniably economic question of resource allocation arises. In short, the use of production functions presumes that a large proportion of the problem of allocation has been solved before the analysis begins.

It is illuminating to pause here to note that the definition of an activity used, say, in Chaps. 6 and 7, does not apply in the most general formulation of the programming problem. In the most general conditions the input and output flows generated by an activity are not assumed to be directly proportional to the level of that activity. What, then, is the significance of an activity? Each activity is merely the designation of one of the decision variables of the production plan. These are the variables which can be decided upon independently and which jointly determine all the inflows and outflows of the enterprise.[1] It is for this reason that a programming problem (of this general kind) underlies the production problem no matter which way it is formulated.

We see, thus, that the production function as conventionally defined summarizes the solutions to the underlying programming problems for various values of the variable factors. The numerous restraints and the inequality signs that clutter up a programming problem are absent from the conventional formulation, not because they are inapplicable, but because it is assumed that they have already been handled. Perhaps economists would not have gotten into the habit of making this assumption so glibly if they had realized what, and how much, they were assuming.

[1] *Exercise.* The conventional statement of the firm's maximization problem, using Carlson's notation, is as follows: maximize $r(y, v_1, \ldots v_m)$, subject to the production function $y = \phi(v_1, \ldots, v_m)$. (1) Show that this problem can be regarded as a special case of the general programming problem; and (2) thence deduce the familiar marginal equality conditions of profit maximization from the Kuhn-Tucker conditions 1 to 4.

9

The Statical Leontief System

One of the most interesting developments in the field of economics of recent years is the model of industrial interdependence known as input-output. Largely the creation of Professor Wassily W. Leontief of Harvard University, the theory of input-output has at least three important aspects: (1) It is of interest to the economic theorist because it provides the simplest form of Walrasian general equilibrium; its form is so simple that it holds out the hope of empirical statistical measurement. (2) Input-output is of interest to the national-income economist because it provides a more detailed breakdown of the macroaggregates and money flows. (3) The theory of input-output can also be regarded as a peculiarly simple form of linear programming: in the simplest Leontief system, in which no substitutions of inputs are technologically feasible, the optimizing solution is the one and only efficient solution possible; but in more general models, in which substitution is possible, the system can be made determinate only by solving an appropriately formulated linear-programming problem (or by requiring the solution to satisfy some restrictive outside conditions).

The present chapter gives a relatively brief exposition of the statical, or "flow," model of the Leontief system, while the next chapter takes a more advanced point of view. Subsequent chapters deal with dynamic models involving time and stocks of capital, and also with more general statical models of the type met with in the neoclassical theories of Leon Walras and J. B. Clark.[1]

9-1. INPUT-OUTPUT FLOW TABLES

We begin with a brief exposition of what input-output is generally about. Leontief imagines an economy in which goods like iron, coal,

[1] See Wassily W. Leontief, *The Structure of American Economy, 1919–1939*, 2d ed., Oxford University Press, New York, 1951; Wassily W. Leontief and others, *Studies in the Structure of the American Economy*, Oxford University Press, New York, 1953; J. Cornfield, W. D. Evans, and M. Hoffenberg, "Full Employment Patterns, 1950," parts I and II, *Monthly Labor Review*, **64**:163–190, 420–432 (February–March, 1947).

alcohol, etc., are produced in their respective industries by means of a primary factor such as labor and by means of other inputs such as iron, coal, alcohol, etc. He rejects as unrealistic the Austrian economists' view that you can identify certain industries as being in "earlier" stages of production and certain other industries as being in "later" stages. Thus, Leontief argues against the inevitability of finding an industry such as agriculture which sells only to an industry such as manufacturing, buying nothing from it; he denies that you can follow a loaf of bread from the early stages through a one-directional hierarchy of industries with value being added to the bread by virtue of primary factors employed in each of the industries.[1]

Instead Leontief finds that the real world requires you to recognize the "whirlpools" of industrial relationships characteristic of general models of interdependence. For the production of coal, iron is required; for the production of iron, coal is required; no man can say whether the coal industry or the iron industry is earlier or later in the hierarchy of production.

9-1-1. A Two-industry Example. Perhaps Leontief's model will be most clear if we imagine a grossly oversimplified economy in which there are two industries, agriculture and manufacturing. Each directly requires some of a primary factor called labor in its productive process, and each requires in its production process inputs of the other.[2]

Table 9-1 provides a simplified picture of our economy. Agriculture and manufacturing are the first two entries, and each of their rows will show what happens to their total output. The third row is given to the primary factor, labor, of which the community has a total of 50 units (thousands of man-years) per year. These 50 units of labor are allocated as inputs to the two industries in the respective amounts 10 and 40.

The first-row total shows that all together the agricultural output totals 250 units (millions of tons) per year. Of this total 50 units go directly to final consumption, i.e., to households and government, as shown in the third column of row 1. What happens to the remaining 200 units of agricultural output? They are required as inputs to help

[1] This "triangularity" (or something like it) *may* appear as an empirical fact. If it does, naturally all problems are much simplified.

[2] Leontief usually assumes that an industry does not use any of its own good as an input in producing itself. Thus, he would not include in his measurements the coal burned in steam engines inside of coal mines; instead, he would use as the total of coal production only the "net" amount of coal produced. This is a harmless convention in the statical model, but we find it convenient to include the possibility that the industry does require some of its own product as a necessary input in its production process. The importance of this is that in a dynamic model in which production takes time, the stocks of coal to be used in coal mining must be available *before* any new coal can be produced.

make possible the community's production of manufactured and agricultural goods. Thus 175 units of agricultural output is required as materials inputs in order to make possible manufacturing production: this is shown in the second column of the first row. The remainder of agricultural output, 25 units, is required in agriculture itself, e.g., oats used to feed horses that pull harvesters of wheat and oats, and is shown in column 1 of row 1. Similarly row 2 shows the allocation of the total output of manufacturing industries, 120 units (thousands of dozens) per year, among final consumption and intermediate inputs needed in society's two industries. In row 2, columns 1, 2, and 3 show allocations of 40, 20, and 60 units of manufactured goods per year to agriculture, manufacturing, and final consumption (households and government). The reader should remember that all the items in Table 9-1 are *flows*, i.e., physical units per year.

TABLE 9-1

Industries	Inputs to agriculture	Inputs to manufacturing	Final demand	Total outputs
Agriculture	25	175	50	250
Manufacturing	40	20	60	120
Labor services	10	40	0	50

There are some preliminary remarks to be made about this miniature *tableau économique*. Since the entries in any row are all measured in the *same* physical units, it makes good sense to add across the rows. The "total outputs" column gives the over-all input of labor and output of each commodity. On the other hand, items in the same column are not measured in the same units, so that it would be nonsense to add down the columns. But each column, thought of as a whole (i.e., as a "vector"), does have meaning. The first column describes the input or cost structure of the agricultural industry: the 250-unit agricultural output was produced with the use of 25 units of agricultural goods, 40 units of manufactured goods, and 10 units of labor. Similarly the second column describes the observed input structure of the manufacturing industry. To put it slightly differently, a column gives *one point* on the production function of the corresponding industry. The "final demand" column shows the commodity breakdown of what is available for consumption and government expenditure. We make the convention that labor is not directly consumed. Direct labor services are assumed away solely to restrict this example to two final commodities.

Suppose, however, that we had deliberately chosen the physical units in which each commodity is measured so that at some given base prices, one unit costs $1 million.[1] Then each entry in Table 9-1 becomes a (million) dollar value and we can interpret the columns literally as cost figures. In these special units it does make sense to add down the columns, and the sum gives the total cost of producing the industry's output. Since the output is also measured now in millions of dollars' worth, total output is the same as total revenue. Thus agricultural revenue (*at the base prices*) is $250 million, and cost of production is $75 million. In manufacturing, revenue is $120 million and cost $235 million. Thus in agriculture there was a profit of $175 million, and in manufacturing there was a loss of $115 million. Clearly, then, the prices mentioned in the footnote are *not* long-run competitive equilibrium prices (price is not equal to average cost, and there are profits and losses). We shall soon see how equilibrium prices can be found; but as soon as prices change the column sums again become nonsense, for the entries cease to be in millions of dollars' worth. At any other prices, costs and revenues have to be separately computed from the physical flow items.

Those items in Table 9-1 that show the sales of the two industries to themselves and to each other might be described as "non-GNP" items. The "final demand" column represents the output side of GNP, and the labor row represents the factor-cost side. The interindustrial sales have no "welfare" significance at all. Social benefits come from final consumption, and social costs come from the use of labor. The economy can be viewed as a machine that uses up labor (and has 50 units of labor per year at its disposal) and produces final consumption. We know from observation that with its 50 units of labor the economy is capable of turning out an annual flow of 50 units of agriculture and 60 units of manufacture. Part of our problem will be to calculate what *other* menus of final consumption society could produce with its 50 units of labor and its present technology.

Suppose we were to change the agriculture-to-manufacturing sales figures from 175 to 185 and, say, decrease the agriculture-to-agriculture item from 25 to 15, thus keeping final demands and row totals the same. As compared with Table 9-1, this would indicate a less productive manufacturing industry (more input for same output) and a more productive agricultural technology (less input for same output). But at the moment the two societies would be equally well off—they supply the same amount of labor and enjoy the same final consumption. However, the two societies would have different technologies, and therefore we would confidently expect them to have *different* possible menus of final consumption (which happen to cross at the observed point). In fact, we suspect that a soci-

[1] Thus, given our earlier choice of units, we deduce that the wage rate was $1,000 per year, the price of agricultural goods $1 per ton, and that of manufactured products $1,000 per dozen.

ety that might prefer a final-demand mix weighted more heavily than 60-50 on the manufacturing side would do better with the technology of Table 9-1; while for demand mixes with a heavier dose of farm products, the amended technology, containing a more efficient agricultural industry, would be superior (leaving aside any nonpecuniary advantages of manufacturing over agricultural employment, or vice versa). We must formulate all this more precisely.

So far Table 9-1 is just a handy way of arranging data about certain transactions in an economic system. It is more detailed than a simple statement of national-income totals would be; it is less detailed than a complete accounting which distinguishes firms within each industry. But it is certainly a useful bird's-eye view, primarily because of its double-entry character: each entry can be viewed as a revenue (the row aspect) or as a cost (the column aspect). An additional advantage of such a tabulation is that even for a very large and complicated economy the required numbers can actually be found or estimated, at some expense.[1]

9-1-2. Technological Assumptions. To convert Table 9-1 from a descriptive device to an analytical tool requires some strong assumptions. We have already spoken loosely of the technology of the system. Of course Table 9-1 far from describes the technological possibilities available to the society. What we need are the production functions which do exactly that. If we call agriculture Industry 1, manufacturing Industry 2, and give labor the subscript 0, then Table 9-1 becomes schematically Table 9-2, which establishes a notation for us. Items in the same column are the inputs in the same production function.

Table 9-2

	Inputs to Industry 1	Inputs to Industry 2	Final consumption	Total output of industries
Industry 1	x_{11}	x_{12}	C_1	X_1
Industry 2	x_{21}	x_{22}	C_2	X_2
Labor services	x_{01}	x_{02}	. . .	X_0

Thus we could write production functions

$$\begin{aligned} X_1 &= F^1(x_{11},x_{21},x_{01}) \\ X_2 &= F^2(x_{12},x_{22},x_{02}) \end{aligned} \tag{9-1}$$

since X_1 and X_2 are the total outputs. In addition, we can always add

[1] See W. Duane Evans and Marvin Hoffenberg, "The Interindustry Relations Study of 1947," *Review of Economics and Statistics*, **34**:97–142 (May, 1952) for a description of the compilation of a table involving 500 different industries.

across the rows, so we know that

$$\begin{aligned} x_{11} + x_{12} + C_1 &= X_1 \\ x_{21} + x_{22} + C_2 &= X_2 \\ x_{01} + x_{02} \qquad &= X_0 \end{aligned} \tag{9-2}$$

Now the observations in Table 9-1 tell us very little about the production functions—just that $250 = F^1(25,40,10)$ and $120 = F^2(175,20,40)$. If we care to assume the existence of constant returns to scale, then we can also be sure that, say, $500 = F^1(50,80,20)$ and $24 = F^2(35,4,8)$, and so forth. We can also assume that the isoquant surfaces have the usual convexity, that is to say, we can assume generalized diminishing returns. The distinctive feature of input-output analysis is that Leontief makes both these assumptions and the far stronger one of fixed coefficients of production; i.e., he supposes that it takes a certain minimal input of each commodity (possibly zero) per unit of output of each commodity. The word "minimal" is of some importance;[1] if it takes 2 tons of ore to yield 1 ton of iron, no doubt the same iron could be produced from even more ore, but as long as iron ore has value, nobody will be silly enough to use more than the absolutely required 2 tons.

This special Leontief production function can be written in the usual form (9-1). Let a_{ij} be the required minimal input of Commodity i per unit of output of Commodity j (in our example $i = 0$, 1, or 2, and $j = 1$ or 2). Then,[2]

$$\begin{aligned} X_1 &= \min\left(\frac{x_{11}}{a_{11}}, \frac{x_{21}}{a_{21}}, \frac{x_{01}}{a_{01}}\right) \\ X_2 &= \min\left(\frac{x_{12}}{a_{12}}, \frac{x_{22}}{a_{22}}, \frac{x_{02}}{a_{02}}\right) \end{aligned} \tag{9-3}$$

The reader can verify that if each of the x_{ij} is multiplied by a constant, the corresponding X_j is multiplied by the same constant, so that we do indeed have constant returns to scale. The isoquant surfaces are nested right-angled corners, so that we do indeed have convexity. If any a_{ij} is zero (a commodity not needed at all in a certain industry), we can either omit the corresponding term from the right-hand side of (9-3), or else we can think of x_{ij}/a_{ij} as $+\infty$, in which case it will certainly never be the limiting smallest number in (9-3).[3]

An alternative way of writing (9-3) is to note that since X_1 equals the smallest of x_{11}/a_{11}, x_{21}/a_{21}, x_{01}/a_{01}, it must be $\leq$ all three of the ratios.

[1] Primarily when we pass to the dynamic case (see Chap. 11).

[2] The notation min $(a,b,\ldots,z)$ in Eq. (9-3) means "the smallest of the numbers a, b, . . . , z."

[3] To be quite precise, perhaps we should replace the sign $=$ in (9-3) by the sign $\leq$. But clearly no sensible industry will ever waste some of all inputs simultaneously.

Hence we have $X_j \leq x_{ij}/a_{ij}$, or, written out in full,

$$\begin{aligned} x_{11} \geq a_{11}X_1,\ x_{21} \geq a_{21}X_1,\ x_{01} \geq a_{01}X_1 \\ x_{12} \geq a_{12}X_2,\ x_{22} \geq a_{22}X_2,\ x_{02} \geq a_{02}X_2 \end{aligned} \tag{9-4}$$

with equality holding at least once in each row (in fact with equality holding everywhere if none of the commodities concerned are free goods).

Under these narrow assumptions the flow data of Table 9-1 do completely describe the technology of our model economy. Assuming that no goods are free, we can divide each item in the first column of Table 9-1 by the first-row total and each item in the second column by the second-row total, and since from Eqs. (9-4) $x_{ij}/X_j = a_{ij}$, we get Table 9-3.

TABLE 9-3

	Inputs to Industry 1	Inputs to Industry 2	Final demand	Total output of industries
Industry 1	0.10	1.46	50	250
Industry 2	0.16	0.17	60	120
Labor services	0.04	0.33	...	50

From Table 9-3 we can read that it takes an input of 0.10 unit of Commodity 1 to fabricate 1 unit of Commodity 1, that it takes 0.33 unit of labor to make 1 unit of Commodity 2, etc. We have included also the final demands, the total outputs, and the total labor supply. We shall shortly see that from the first three columns of Table 9-3 we could *deduce* the fourth column except for the value of X_0.

If we convert Table 9-2 as we did Table 9-1, we get Table 9-4.

TABLE 9-4

	Inputs to Industry 1	Inputs to Industry 2	Final demand	Total output
Industry 1	a_{11}	a_{12}	C_1	X_1
Industry 2	a_{21}	a_{22}	C_2	X_2
Labor services	a_{01}	a_{02}	...	X_0

9-2. A LINEAR-PROGRAMMING INTERPRETATION

This model of production is really only a special case of the linear-programming or activity-analysis model studied in Chaps. 2 to 8. There we were talking about a firm and here about whole industries, but the technique is the same. The first column of Table 9-3 tells us that Industry 1 has one (and only one) process which converts 0.10 unit of Commodity 1,

0.16 unit of Commodity 2, and 0.04 unit of the primary factor, labor, into 1 unit of Commodity 1. This process can be expanded or contracted in proportion to any extent, as long as enough inputs are available to feed it. Another way of describing this process is to say that it has a "net" output of 0.90 unit of Commodity 1 (the gross output of 1 unit minus the 0.10 unit needed as input) and inputs of 0.16 of Commodity 2 and 0.04 of labor. We might choose to call the level that yields 1 unit of net output of Commodity 1 the "unit-level operation" of the process. To do this we have to expand output and inputs by the factor $^{10}/_{9}$, and we get 1 unit of net output (gross output of $^{10}/_{9}$ minus inputs of $^{10}/_{9} \times ^{1}/_{10}$, or $^{1}/_{9}$) and net inputs of 0.177 and 0.044 unit of Commodity 2 and labor, respectively. As still another alternative we could normalize on an input of 1 unit of labor, which would yield a gross output of 25 units and a net output of $25 - 2.5 = 22.5$ units of Commodity 1, an input of 4 units of Commodity 2, and, of course, 1 unit of labor. All these ways of looking at the process are essentially equivalent. Industry 2 also has a process whose vital statistics can be read from the second column of Table 9-3. The basic resource limitation on the system is the availability of no more than 50 units of labor.

Thus we are in a straightforward linear-programming situation. [We shall discuss later whether or not there is anything to be maximized or minimized here: one possible problem is to achieve given final consumption with minimum use of labor—remember that final consumption is the only social benefit, and use of labor (if labor is irksome) the only social cost in this model.] It is even more than straightforward, it is especially simple. In the case of a firm the main problem was to decide which activities to use and which not. Here there is only one activity producing each commodity. If some of each commodity is desired for final consumption, or if each commodity is needed as an input for some desired commodity,[1] then we know at once that all processes must be used, and the problem reduces to the simpler one of choosing the levels.

There is already one restriction on the a_{ij} coefficients of Table 9-4 that we can state. If a technology is to be viable at all, each of the "own" input coefficients, a_{11} and a_{22}, must be less than unity. Otherwise there would be negative *net* outputs ($1 - a_{11}$ and $1 - a_{22}$). A production process in which it took more than 1 ton of coal to make 1 ton of coal (if you can call this "making") is not a method of production at all, but just a hard way of running down preexisting stocks of coal. Note that if a table such as Table 9-3 or 9-4 is deduced from an observed flow table such as Table 9-1 or 9-2, viability conditions will *automatically* be satisfied. The diagonal elements must, by the rules of arithmetic, be less

[1] Or needed as an input for a commodity that is needed as an input for a desired commodity, etc.

than the sums of their own rows, and so dividing will always leave $a_{ii} < 1$. This reflects the fact that if we ignore the existence of stocks, an observable economy must be productive in this sense. There is another related but more subtle condition of viability that we shall reach shortly.[1]

9-2-1. Feasible Final Demands. In our linear-programming formulation of the Leontief system, how do the balance relations (9-2) look? We repeat them here, for ready reference:

$$\begin{aligned} x_{11} + x_{12} + C_1 &\leq X_1 \\ x_{21} + x_{22} + C_2 &\leq X_2 \\ x_{01} + x_{02} \qquad &\leq X_0 \end{aligned} \tag{9-2'}$$

The first of these says, for example, that the total output of X_1 was allocated either as input to Industry 1, or as input to Industry 2, or as final consumption, since in Table 9-2 X_1 was defined as the sum of the other three items. But now if we shift our point of view and think of X_1 as the total output of Commodity 1, we must change the sign $=$ to the sign $\leq$. The available output certainly can't be less than the sum of its alternative uses, but it could, physically, be greater. Again we have confidence that unless the commodity is a free good, there will be no waste or excess tolerated. But in Eqs. (9-2) we have kept the inequality because in the dynamic extension of this model it will play a key role.

Now the production process of Industry 1 turns out 1 unit of Commodity 1, gross, when operated at unit level. Therefore, to produce a gross output of X_1 units of the commodity, the process must be operated at level or intensity X_1. Similarly with the process of Industry 2 we can identify the total output X_2 with the process intensity. Whatever the process intensities X_1 and X_2, we can account for the output X_1 as follows: $a_{11}X_1$ will be used up in Industry 1 itself, and $a_{12}X_2$ will be used up in Industry 2 (all this from Table 9-4). What's left, namely, $X_1 - a_{11}X_1 - a_{12}X_2$, must, according to Eqs. (9-2), be at least equal to the final consumption C_1. An exactly similar relation must hold for X_2. For labor, the accounting relation in Eqs. (9-2) is even simpler. Labor is not produced, but is available in amounts up to X_0; the use of labor is $a_{01}X_1$ in Industry 1 and $a_{02}X_2$ in Industry 2. Thus we get

$$\begin{aligned} (1 - a_{11})X_1 - a_{12}X_2 &\geq C_1 \\ -a_{21}X_1 + (1 - a_{22})X_2 &\geq C_2 \end{aligned} \tag{9-5}$$

$$a_{01}X_1 + a_{02}X_2 \leq X_0 \tag{9-6}$$

[1] Clearly there is no similar reason why, as in Table 9-3, it shouldn't take 1.46 (or more) units of Commodity 1 to make a unit of Commodity 2. In fact, we can always make this happen by choosing to measure Commodity 1 in small units and Commodity 2 in large units. For "own" inputs this juggling with units leaves the a_{ii} unchanged!

Suppose society (by fiat or by market processes) specifies a set of final demands C_1 and C_2. The question immediately arises: Is this bill of goods achievable? Is this bill of goods within society's net-production-possibility schedule? In the first instance this means: Does society have at its disposal enough labor to produce the specified final demands? But we can also step outside the model for a moment and imagine that the two industries have capacity limits, so that we must verify that the bill of goods can be produced without gross output overstepping the available capacity. (The reason that this thought is outside the model is that

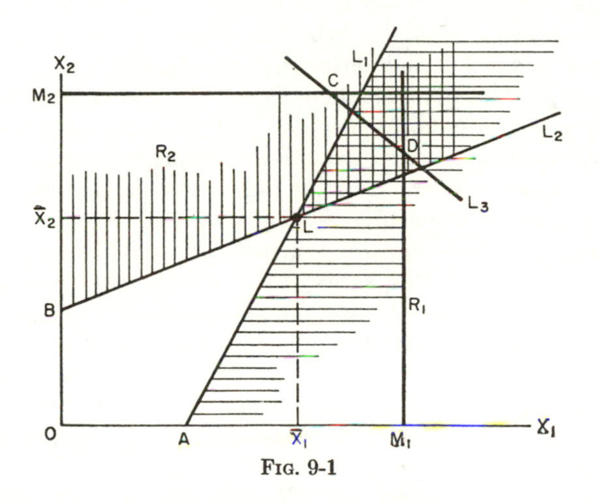

Fig. 9-1

if there are capacities, there are usually ways of increasing them. This is the province of the dynamic model.) In both cases we must first find out what gross outputs are needed to yield final demands C_1 and C_2. The answer is easy: Any gross outputs which satisfy both inequalities (9-5) will do.

Graph X_1 on the horizontal axis and X_2 on the vertical, as in Fig. 9-1. The line $(1 - a_{11})X_1 - a_{12}X_2 = C_1$ is drawn as L_1, and it is easily verified that the region where the $\geq$ holds consists of the line and all the area R_1 to the right of the line, horizontally shaded. Note that the distance OA is $C_1/(1 - a_{11})$ (positive because $1 - a_{11} > 0$, as discussed earlier). The slope of L_1, $dX_2/dX_1 = (1 - a_{11})/a_{12}$, is also positive (if $a_{12} = 0, L_1$ is vertical). We draw only the positive quadrant, since negative gross outputs have no meaning.[1]

The line L_2 is where $-a_{21}X_1 + (1 - a_{22})X_2 = C_2$, and the vertically

[1] The reader should verify that if $a_{11} > 1$, L_1 would begin to the left of the origin, have a negative slope, and the shaded region R_1 would lie completely outside the positive quadrant, indicating that no production is possible.

shaded region R_2 is where the inequality is satisfied. The distance $OB = C_2/(1 - a_{22})$, and the slope of L_2 is $a_{21}/(1 - a_{22})$.

But the gross outputs which will yield both C_1 and C_2 are those which lie in *both* shaded regions and therefore in the crosshatched region of Fig. 9-1, the cone-shaped angular region between L_1 and L_2 extending outward from the intersection point L. Any gross-output levels, or process intensities, in this region will enable society to consume C_1 and C_2 of the two commodities.

The intersection point L has special properties. At the outputs represented by L, both *equalities* hold in Eqs. (9-5) and nothing of either commodity is being wasted. What is more, any other output in the crosshatched region has *both* outputs bigger than at L. Clearly the efficient way to produce net outputs of C_1 and C_2 is to produce the smallest compatible gross outputs, namely, $\bar{X}_1$ and $\bar{X}_2$ at L. But does L represent a feasible set of process intensities, or industry outputs? Now this question is easy to answer. First, if Industry 1 has a capacity limit, say M_1, we can compare to see if $\bar{X}_1 \leq M_1$. If so, all is well; if not then L is not feasible, and surely no program with $X_1 > \bar{X}_1$ is feasible, and we would have to conclude that society can't make available final consumptions of C_1 and C_2. Second, we can do the same with the capacity limit M_2 of Industry 2. Finally, there is the labor-supply restriction. The input of labor at L is $a_{01}\bar{X}_1 + a_{02}\bar{X}_2$. We have to compare this with X_0. If Eq. (9-6) is satisfied, the program is feasible. If not, the program will require too much labor, and of course so will any program with still higher gross outputs.

Graphically the restriction $X_1 \leq M_1$ represents a vertical line and the area to the left; $X_2 \leq M_2$ is a horizontal line and the area below. The labor restriction (9-6) $a_{01}X_1 + a_{02}X_2 \leq X_0$ is a downward-sloping line (L_3 in Fig. 9-1) and the area on the origin side of it. The region of feasible gross outputs is then the polygon OM_2CDM_1 drawn in Fig. 9-1. If L lies inside this polygon, as it does in Fig. 9-1, then the final demands specified are attainable. If L lies outside, then society cannot produce such large final consumption. If, again as in Fig. 9-1, L lies *strictly inside* the feasible region, then both outputs can be expanded beyond L and society *could* increase *both* final consumptions beyond C_1 and C_2.

Must a point like L always exist? A look at Fig. 9-1 shows that if the lines L_1 and L_2 were parallel, i.e., if they had equal slopes, there would be no point L. After all L is a point of intersection, and parallel lines don't meet. In fact, if L_2 had a *bigger* slope than L_1, there would also be no point L. The two lines would intersect, to be sure, but not in the positive quadrant. If L_2 rose more steeply than L_1, the two lines would simply fan out. Not only would there be no point L, but there would be no crosshatched region. The regions R_1 and R_2 would have no points in common. It would be impossible to satisfy Eqs. (9-5) for meaningful positive outputs. We can go one step further: it would be impossible to satisfy Eqs. (9-5) for *any* positive final demands, no matter how small. That is to say, *no* final demands would be producible at all. We would be in much the same fix as if a_{11} or a_{22} were to exceed unity.

9-2-2. The Hawkins-Simon Conditions. What is the condition that L exist, or that some bills of goods should be producible?[1] It is that the slope of L_2 be less than the slope of L_1. But that says:

$$\frac{a_{21}}{1 - a_{22}} < \frac{1 - a_{11}}{a_{12}}$$

or

$$(1 - a_{11})(1 - a_{22}) - a_{12}a_{21} > 0 \tag{9-7}$$

Another way to write this is in determinant form:

$$\begin{vmatrix} 1 - a_{11} & -a_{12} \\ -a_{21} & 1 - a_{22} \end{vmatrix} > 0 \tag{9-7a}$$

This is the more subtle restriction on the input coefficients referred to at the end of the previous section. It can be given the following interpretation: Just as we earlier required that it not take a direct input of more than one ton of coal to make one ton of coal, Eq. (9-7) or (9-7a) assures us further that if we add up the direct and indirect inputs of coal that go into a ton of output (the coal to make coal, the coal to make coal to make coal, the coal to make steel to make coal, the coal to make steel to make coal to make coal, the coal to make steel to make steel to make coal, etc., ad infinitum), that all this will be less than one ton. Clearly if a ton of coal "contains," directly *and* indirectly, more than a ton of coal, self-contained production is not viable. The inequality of (9-7a) together with the earlier $1 - a_{11} > 0$, $1 - a_{22} > 0$ comprise what are called the Hawkins-Simon conditions.[2] These conditions extend to systems of more than two commodities, with strings of bigger and bigger determinants appearing in Eq. (9-7a). The interpretation is always that all subgroups of commodities should be "self-sustaining," directly and indirectly.

9-3. SOLVING AN INPUT-OUTPUT SYSTEM

In searching for the gross outputs which would yield specified final consumptions (which we had to do in order to test feasibility) we could have proceeded differently. We could have started with the given final

[1] If one bill involving all goods is producible, then any bill of goods is producible, provided only there are enough labor and enough capacity.

[2] D. Hawkins and H. A. Simon, "Note: Some Conditions of Macroeconomic Stability," *Econometrica*, 17:245–248 (July–October, 1949).

demands C_1 and C_2. To C_1 we would have added the "first-round" input requirements of Commodity 1, namely, $a_{11}C_1 + a_{12}C_2$, and to C_2 we would have added the first-round inputs of Commodity 2, namely, $a_{21}C_1 + a_{22}C_2$. Then we would have proceeded to the second round, which is more complicated. There are second-round inputs of Commodity 1 into each of the four first-round inputs, namely, $a_{11}(a_{11}C_1 + a_{12}C_2) + a_{12}(a_{21}C_1 + a_{11}C_2)$. For Commodity 2 the second-round inputs are $a_{21}(a_{11}C_1 + a_{12}C_2) + a_{22}(a_{21}C_1 + a_{22}C_2)$. Then on to the third-round, etc., ad infinitum. The rule by which the kth-round inputs can be found from the $(k-1)$st round is easy: if $X_i^{(k)}$ represents the kth-round inputs, then

$$
\begin{aligned}
X_1^{(k)} &= a_{11}X_1^{(k-1)} + a_{12}X_2^{(k-1)} \\
X_2^{(k)} &= a_{21}X_1^{(k-1)} + a_{22}X_2^{(k-1)}
\end{aligned}
$$

Our method of solution is much less laborious. By considering the two equations or inequalities simultaneously as in Fig. 9-1, we cut through the infinite chain of fictitious "rounds."

It is a theorem (if it weren't, all would be lost) that if the productive system is viable—the Hawkins-Simon conditions again—the infinite sum of all the rounds converges to a limit, and this limit is the same as the simultaneous solution we worked out earlier. The convergence of the infinite process is a relatively delicate thing to prove. But we can prove the equivalence of the two solutions with only a little algebra.

In the first place, it is evident from the form of the first- and second-round terms written out above that when all is said and done the solution is going to be of the form

$$
\begin{aligned}
\bar{X}_1 &= C_1 + a_{11}C_1 + a_{11}^2C_1 + a_{12}a_{21}C_1 + \cdots \\
&\quad + a_{12}C_2 + a_{11}a_{12}C_2 + a_{12}a_{22}C_2 + \cdots \\
&= (1 + a_{11} + a_{11}^2 + a_{12}a_{21} + \cdots)C_1 \\
&\quad + (a_{12} + a_{11}a_{12} + a_{12}a_{22} + \cdots)C_2 = A_{11}C_1 + A_{12}C_2 \\
\bar{X}_2 &= (a_{21} + a_{21}a_{11} + a_{22}a_{21} + \cdots)C_1 \\
&\quad + (1 + a_{22} + a_{21}a_{12} + a_{22}^2 + \cdots)C_2 = A_{21}C_1 + A_{22}C_2
\end{aligned}
\tag{9-8}
$$

i.e., the gross outputs are linear functions of the final demands. The A's are coefficients whose values are *defined* by Eqs. (9-8); we want to evaluate them in some simpler form.

There is another way we could look at this. We could build up $\bar{X}_1$ in two steps. First there is the final consumption itself, C_1. Then we could imagine the system to be presented with the first-round derived demands as a sort of secondary final demand. The gross output necessary to support this supplementary demand will be $A_{11}(a_{11}C_1 + a_{12}C_2) + A_{12}(a_{21}C_1 + a_{22}C_2)$. Thus we have

$$
\begin{aligned}
\bar{X}_1 &= C_1 + A_{11}(a_{11}C_1 + a_{12}C_2) + A_{12}(a_{21}C_1 + a_{22}C_2) \\
&= (1 + A_{11}a_{11} + A_{12}a_{21})C_1 + (A_{11}a_{12} + A_{12}a_{22})C_2
\end{aligned}
\tag{9-9}
$$

and similarly,

$$\begin{aligned}\bar{X}_2 &= C_2 + A_{21}(a_{11}C_1 + a_{12}C_2) + A_{22}(a_{21}C_1 + a_{22}C_2) \\ &= (A_{21}a_{11} + A_{22}a_{21})C_1 + (1 + A_{21}a_{12} + A_{22}a_{22})C_2 \end{aligned} \qquad (9\text{-}9a)$$

We now have two ways of computing $\bar{X}_1$ and $\bar{X}_2$. Since they are always to give the same result, they must be identical, coefficient by coefficient. Thus we get four equations to solve for the four unknowns A_{11}, A_{12}, A_{21}, A_{22}.

For example:

$$\begin{aligned} A_{11} &= 1 + a_{11}A_{11} + a_{21}A_{12} \\ A_{12} &= \qquad a_{12}A_{11} + a_{22}A_{12} \end{aligned}$$

or

$$\begin{aligned} (1 - a_{11})A_{11} - a_{21}A_{12} &= 1 \\ -a_{12}A_{11} + (1 - a_{22})A_{12} &= 0 \end{aligned}$$

By substitution or elimination of determinants these two equations can be solved to give

$$\begin{aligned} A_{11} &= \frac{1 - a_{22}}{(1 - a_{11})(1 - a_{22}) - a_{12}a_{21}} \\ A_{12} &= \frac{a_{12}}{(1 - a_{11})(1 - a_{22}) - a_{12}a_{21}} \end{aligned} \qquad (9\text{-}10)$$

and the other pair of equations can be written down and solved to give

$$\begin{aligned} A_{21} &= \frac{a_{21}}{(1 - a_{11})(1 - a_{22}) - a_{12}a_{21}} \\ A_{22} &= \frac{1 - a_{11}}{(1 - a_{11})(1 - a_{22}) - a_{12}a_{21}} \end{aligned} \qquad (9\text{-}10a)$$

Again we have used a simultaneous solution to circumvent an infinite multiplier chain.[1]

Now go back to our original technique in Fig. 9-1. The point $L(\bar{X}_1,\bar{X}_2)$ was the solution of a pair of simultaneous equations,[2] namely,

$$\begin{aligned} (1 - a_{11})\bar{X}_1 - a_{12}\bar{X}_2 &= C_1 \\ -a_{21}\bar{X}_1 + (1 - a_{22})\bar{X}_2 &= C_2 \end{aligned} \qquad (9\text{-}11)$$

Multiply the first equation by $1 - a_{22}$, the second by a_{12}, and add. After

[1] The capital A's will be recognized as the elements of the inverse matrix of

$$\begin{bmatrix} 1 - a_{11} & -a_{12} \\ -a_{21} & 1 - a_{22} \end{bmatrix}$$

In the input-output literature, "input matrix" usually means our a's, and "inverse matrix" usually means our A's.

[2] Compare these with Eqs. (9-5). The latter express the problem: Given C_1, C_2, find the gross outputs X_1, X_2 so that the net output of the economy is precisely C_1, C_2. The desired final product, C_1, C_2, is attainable if neither $\bar{X}_1$ nor $\bar{X}_2$ is negative.

solving for $\bar{X}_1$ there results

$$\bar{X}_1 = \frac{1 - a_{22}}{(1 - a_{11})(1 - a_{22}) - a_{12}a_{21}} C_1 + \frac{a_{12}}{(1 - a_{11})(1 - a_{22}) - a_{12}a_{21}} C_2 \tag{9-12}$$

Then multiply the first equation by a_{21}, the second by $1 - a_{11}$, and add, to get

$$X_2 = \frac{a_{21}}{(1 - a_{11})(1 - a_{22}) - a_{12}a_{21}} C_1 + \frac{1 - a_{11}}{(1 - a_{11})(1 - a_{22}) - a_{12}a_{21}} C_2 \tag{9-12a}$$

Compare this with (9-8) and with (9-10) and (9-10*a*). They are identical. We have thus proved that the "rounds" method and straightforward simultaneous solution of (9-11) yield the same result.

This proof has taught us something else. Equation (9-12) shows that the gross outputs are built up *linearly* out of the final demands C_1 and C_2. The numbers A_{ij} and their values as given in (9-10) and (9-10*a*) have a simple meaning. A_{ij} is the total direct and indirect gross output of Commodity i needed to support 1 unit of final consumption of Commodity j. $A_{11}C_1$ is the amount of X_1 needed to support C_1, $A_{12}C_2$ is the amount needed to support C_2. Therefore $\bar{X}_1 = A_{11}C_1 + A_{12}C_2$.

9-3-1. A Numerical Example. Let us use the results of the preceding section on the numbers of Table 9-3. (Of course, apart from rounding errors we already know what the result will be, namely, $\bar{X}_1 = 250$, $\bar{X}_2 = 120$, with a labor utilization of 50.) First of all, the Hawkins-Simon conditions are satisfied:

$$1 - a_{11} = 0.90 > 0,\ 1 - a_{22} = 0.83 > 0,\ (1 - a_{11})(1 - a_{22}) - a_{12}a_{21}$$
$$= (0.9)(0.83) - (1.46)(0.16) = 0.7470 - 0.2336 > 0$$

The region R_1 of Fig. 9-1 is given by

$$0.9X_1 - 1.46X_2 \geq 50$$

or

$$X_2 \leq \frac{0.9}{1.46} X_1 - \frac{50}{1.46}$$

R_2 is given by

$$-0.16X_1 + 0.83X_2 \geq 60$$

or

$$X_2 \leq \frac{0.16}{0.83} X_1 + \frac{60}{0.83}$$

To find the coordinates of the point L we have to solve

$$0.9X_1 - 1.46X_2 = 50$$
$$-0.16X_1 + 0.83X_2 = 60$$

We can solve these directly, or else compute the coefficients A_{ij} according to (9-10) and (9-10*a*):

$$A_{11} = \frac{0.83}{0.5134} = 1.61 \qquad A_{21} = \frac{0.16}{0.5134} = 0.31$$
$$A_{12} = \frac{1.46}{0.5134} = 2.84 \qquad A_{22} = \frac{0.9}{0.5134} = 1.75$$

Then from (9-12),

$$\bar{X}_1 = (1.61)(50) + (2.84)(60) = 80.5 + 170.4 = 250.9$$
$$\bar{X}_2 = (0.31)(50) + (1.75)(60) = 15.5 + 105.0 = 120.5$$

as we expected (the difference from 250 and 120 comes from rounding $^{250}/_{175}$ to 1.46 and $^{20}/_{120}$ to 0.17 in Table 9-3). To compute the total labor input we use (9-6):

$$(0.04)(250.9) + (0.33)(120.5) = 10.04 + 39.77 = 49.81$$

which is close enough to $X_0 = 50$.

The same calculations can be made for any proposed final demands, which can thus be tested for feasibility.

To conclude this section we compute in Table 9-5 the first few rounds of derived demands to see how their sum converges to $\bar{X}_1$ and $\bar{X}_2$. From

Table 9-5

	Final demand	1st round	2d round	3d round	4th round	Sum
Commodity 1...	50	92.6	35.5	29.6	15.7	223.4
Commodity 2...	60	18.0	17.8	8.7	6.2	110.7

Table 9-5 it can be seen that the contributions of successive rounds dwindle away. After four rounds, the remaining error in $\bar{X}_1$ and $\bar{X}_2$ is of the order of magnitude of 10 per cent.

9-4. THE PRODUCTION POSSIBILITY SCHEDULE

Any proposed bill of final consumption can, by the methods already described, be converted into the required gross outputs for each industry. The fixed-factor restriction in (9-6), together with capacity constraints if any, define a gross-output-possibility schedule. Thus we can test whether any particular set of final demands can be produced, given the labor endowments of the economy. Can we do better? Can we present explicitly the whole menu of producible final demands, the net-output or consumption-possibility schedule? We can.

To do so we need a different graphical representation of the input-output technology. In Fig. 9-2 we again graph Commodity 1 horizontally and Commodity 2 vertically, only now we are really going to be concerned with net outputs, so we label the axes C_1 and C_2. Now what is the *net* effect of unit operation of Industry 1? There is a *net addition* of $1 - a_{11}$ units of Commodity 1 and a *net subtraction* of a_{21} units of Commodity 2. This is shown by the point P_1 in the diagram, lying to the right of the origin to indicate a net output of Commodity 1 and below the origin to indicate a net input of Commodity 2. Because of constant returns to scale, doubling or halving the intensity or gross output of the industry will double or halve the net output and input. Thus the net result of the industry operated at any level is given by points like P_1 on the *ray* through P_1. In exactly the same way Industry 2 at unit level produces the net result shown by P_2: net outputs $1 - a_{22}$ of Commodity 2 and $-a_{12}$ of Commodity 1. Also the *ray* through P_2 contains the net results of Industry 2 operated at *any* level of gross output.

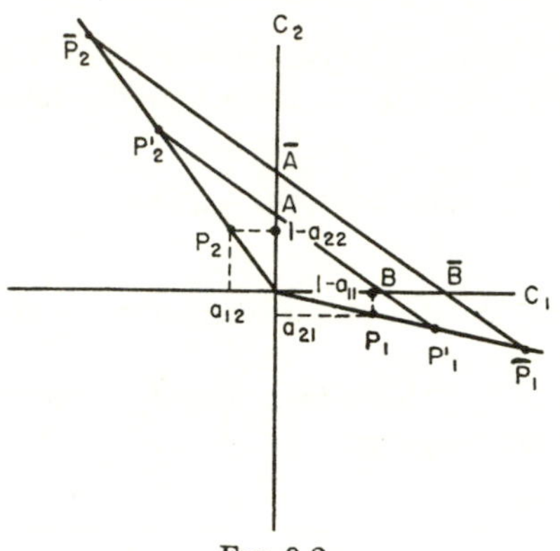

Fig. 9-2

At P_1 the input of the fixed factor, labor, into Industry 1 is a_{01} units. Where along the ray will the net input of labor be exactly 1 unit? At a gross output of $1/a_{01}$ units [net output of $(1 - a_{11})/a_{01}$ units of Commodity 1, $-a_{21}a_{01}$ units of Commodity 2], as represented by the point P_1'. Similarly P_2' uses up 1 unit of labor in Industry 2. Given any point P on the ray through P_1, the ratio OP/OP_1 is the corresponding gross output at P and the ratio OP/OP_1' is the corresponding input of labor.

Now suppose there were only 1 unit of labor to be divided between Industries 1 and 2 in some proportion to be decided. What net results could be obtained? We already know that if all the unit of labor is devoted to Industry 1, we get P_1', and if all of it is used in Industry 2, we get P_2'. If we compute the net result of dividing up the labor in all possible proportions, we get *all the points on the line connecting* P_1' and P_2'. If we divide the labor 50-50, we get the point exactly halfway between P_1' and P_2'; if we divide the labor 0.60 to Industry 1 and 0.40 to Industry 2, we get the point on the line 40 per cent of the way from P_1' to P_2'; etc.

Thus if there were only 1 unit of labor available to society ($X_0 = 1$), the *net* outputs producible would be those on the line between P_1' and P_2'. But since negative net output is meaningless (barring the presence of stocks to be run down), the only really available net outputs would be those on the line segment AB. Of course, since we could always throw away some net output, any point in the triangle OAB would be attainable, although only those on the frontier AB would be efficient.

Now whatever the available supply of labor X_0, we can find points $\bar{P}_1$ and $\bar{P}_2$ on the two rays which would just exhaust it. The coordinates of $\bar{P}_1$, for instance, would be $C_1 = [(1 - a_{11})/a_{01}]X_0$, $C_2 = -(a_{21}/a_{01})X_0$. Then to find the possible net outputs attainable by allocating X_0 between the two industries we simply draw the straight line between $\bar{P}_1$ and $\bar{P}_2$. Again we are interested only in the segment $\overline{AB}$ which represents nonnegative net outputs. OAB shows the attainable net outputs and $\overline{AB}$ itself is the consumption-possibility schedule we have been seeking.

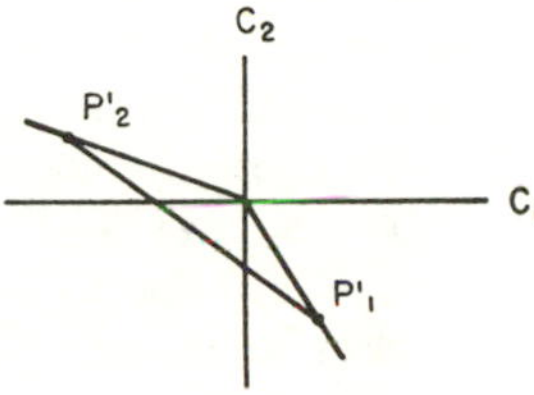

Fig. 9-3

Given the available supply of labor, society can choose to have final consumption of the two commodities in any amounts on the frontier $\overline{AB}$. All points on $\overline{AB}$ use up all the available labor; points inside the triangle leave some labor unused.[1]

Suppose that the two rays of Fig. 9-2 had come out as they are drawn in Fig. 9-3. Note that if we join P'_1 and P'_2 by a straight line, there are no points in common with the first quadrant. In fact, there is no way of joining a pair of points, one on each ray, so that the straight line crosses the positive quadrant. Economically speaking, by *no* combination of the two industries can a positive net output be produced. This is reminiscent of our earlier discussion of the Hawkins-Simon condition. In both Figs. 9-2 and 9-3 we have tacitly assumed both that $1 - a_{11} > 0$ and that $1 - a_{22} > 0$: each industry can produce positive net output of its own commodity. What additional condition must hold to ensure the configuration of Fig. 9-2 rather than that of Fig. 9-3? It will be seen that in Fig. 9-2 the slope of the P_1 ray is flatter (algebraically greater, since both slopes are negative) than that of the P_2 ray, and as a result the connecting lines go northeast of the origin. In Fig. 9-3 the P_1 ray has swung around to have steeper slope than the P_2 ray. The crucial characteristic of Fig. 9-2 is that $-a_{21}/(1 - a_{11})$, the slope of the P_1 ray, is greater than $-(1 - a_{22})/a_{12}$, the slope of the P_2 ray; i.e.,

$$-\frac{a_{21}}{1 - a_{11}} > -\frac{1 - a_{22}}{a_{12}}$$

or

$$(1 - a_{11})(1 - a_{22}) - a_{12}a_{21} > 0$$

This is exactly the Hawkins-Simon condition again.

[1] *Exercise.* How would capacity limitations on the separate industries affect the consumption-possibility frontier? How would the existence of a second fixed factor, say land, affect the consumption-possibility frontier?

This graphical study has already told us nearly all we need to know about the consumption-possibility schedule. We know, for example, that the frontier is a straight line (with three commodities it would be a plane, etc.), and we could get from Fig. 9-2 a pretty good idea of how it would change if one of the a_{ij} coefficients were to change. It remains only to develop an analytical expression for the frontier in terms of the data: the a_{ij} and X_0. This can be done in two ways: a straightforward but laborious way, and an indirect but quick way. We choose the latter here, and leave it to the reader to go through the straightforward computations and show that they lead to the same result.

The straightforward approach is to go back to the C_1, C_2 plane in Fig. 9-2. We know the coordinates of the points $\bar{P}_1$ and $\bar{P}_2$, namely, at $\bar{P}_1$, $C_1 = [(1 - a_{11})/a_{01}]X_0, C_2 = (-a_{21}/a_{01})X_0$; while at $\bar{P}_2, C_1 = -(a_{12}/a_{02})X_0$, $C_2 = [(1 - a_{22})/a_{02}]X_0$. It is a matter of elementary algebra to find the equation of the straight line joining this pair of points. But that straight line is the consumption-possibility frontier itself.

However, by using some earlier calculations we can go directly to the equation of the schedule. Equation or inequality (9-6) provides at once a gross-output-possibility frontier:

$$a_{01}X_1 + a_{02}X_2 = X_0$$

But by (9-8) the gross outputs can be expressed as linear functions of the final demands:

$$X_1 = A_{11}C_1 + A_{12}C_2$$
$$X_2 = A_{21}C_1 + A_{22}C_2$$

Hence we can substitute for X_1 and X_2 to get

$$a_{01}(A_{11}C_1 + A_{12}C_2) + a_{02}(A_{21}C_1 + A_{22}C_2) = X_0$$

or, collecting like terms,

$$(a_{01}A_{11} + a_{02}A_{21})C_1 + (a_{01}A_{12} + a_{02}A_{22})C_2 = X_0 \qquad (9\text{-}13)$$

which we may write, introducing two new coefficients A_{01} and A_{02},

$$A_{01}C_1 + A_{02}C_2 = X_0 \qquad (9\text{-}14)$$

Here, explicitly, is the consumption-possibility frontier. The final demands satisfying $A_{01}C_1 + A_{02}C_2 \leq X_0$ are all producible; if the strict inequality holds, not all the labor available is being used, and the point lies under the frontier.

The coefficients A_{01} and A_{02} have a simple interpretation, much like the other capital A's. It can be seen from (9-13) that A_{01} is the direct labor input *not* into a unit of C_1 but into the gross direct and indirect X_1 and X_2 needed to support a unit of C_1. In other words, A_{01} represents the total *direct and indirect* labor embodied in a unit of final consumption of Commodity 1, and A_{02} is the same for a unit of final consumption of Commodity 2. The schedule in (9-14) simply says that only those bills of final demand are producible and efficient which require X_0 units of labor to support them.

A consumption-possibility schedule [Eq. (9-14)], drawn in Fig. 9-4, can be thought of as a social transformation curve. If it is desired to consume only C_1, an amount X_0/A_{01} can be produced, given the available resources and technology. If it is desired to give up some C_1 in favor of C_2, such substitutions are possible along the transformation curve. Because the frontier is a straight line, substitution of C_2 for C_1 takes place at constant costs. The marginal rate of substitution (MRS) is a constant, namely $-(dC_2/dC_1) = A_{01}/A_{02}$. Giving up 1 unit of C_1 liberates (directly and indirectly) A_{01} units of labor. To get 1 more unit of C_2 requires A_{02} units of labor. By giving up 1 unit of C_1, society can therefore procure for itself A_{01}/A_{02} units of C_2. The straight-line constant-cost nature of the transformation curve reflects not only the linearity of the technology, but also the presence of only *one* primary factor and the absence of joint production.[1] One might think that constant costs are fundamentally a consequence of the absence of substitute processes in the Leontief scheme. It will shortly be shown that this is not the case. Even if there were alternative ways of producing the various commodities, with different input ratios, as long as we assume constant returns to scale, one primary factor, and no joint production, we can deduce that the marginal rate of transformation must be constant.

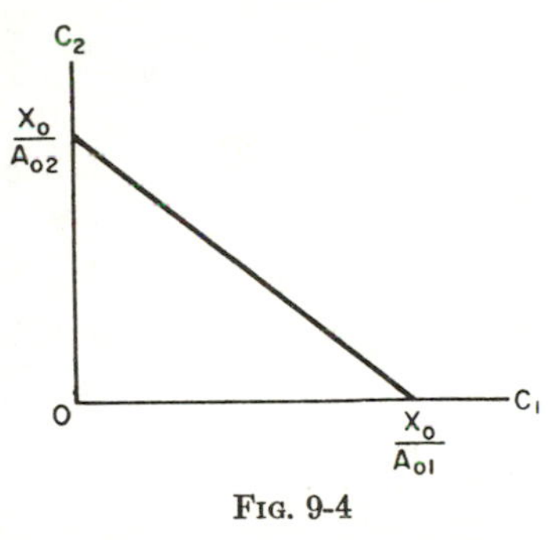

FIG. 9-4

9-4-1. Numerical Example. Let us return to our earlier numerical example and find its consumption-possibility frontier. We have already computed that

$$A_{11} = 1.61, A_{12} = 2.84, A_{21} = 0.31, A_{22} = 1.75$$

We know from Table 9-3 that $a_{01} = 0.04$, $a_{02} = 0.33$. We can calculate, then, that

$$A_{01} = (0.04)(1.61) + (0.33)(0.31) = 0.169$$
$$A_{02} = (0.04)(2.84) + (0.33)(1.75) = 0.691$$

[1] Compare with the constant-cost opportunity-cost curves of the classical theory of international trade.

Therefore the equation of the transformation schedule is

$$0.17C_1 + 0.69C_2 = 50$$

and the MRS is $0.17/0.69 = 0.25$.

As a check, we can find the equation of the line by the alternative method. The point P_1 (compare Fig. 9-2) has coordinates $C_1 = 0.9$, $C_2 = -0.16$. $\bar{P}_1$ has coordinates $C_1 = (0.9/0.04)50 = 1{,}125$, $C_2 = -(0.16/0.04)50 = -200$. The point P_2 has coordinates $C_1 = -1.46$, $C_2 = 0.83$. Thus $\bar{P}_2$ is at $C_1 = -(1.46/0.33)50 = -219$, $C_2 = (0.83/0.33)50 = 125$.

The equation of the line through $\bar{P}_1$ and $\bar{P}_2$ is

$$\frac{C_2 - (-200)}{125 - (-200)} = \frac{C_1 - 1{,}125}{-219 - 1{,}125}$$

$$\frac{C_2 + 200}{325} = \frac{C_1 - 1{,}125}{-1{,}344}$$

This becomes

$$325C_1 + 1{,}344C_2 = 98{,}825$$

Divide through on both sides by 1,976.5, and the equation becomes

$$0.17C_1 + 0.68C_2 = 50$$

which is close enough to what we found by the first method. In particular, we verify that the MRS is 0.25.

9-5. A THEOREM ON SUBSTITUTION

It is a remarkable implication of the Leontief system that even if there were available several different processes for each industry, only one of them would ever be observed. The economy would always behave as if it knew only one set of input ratios for each commodity. This substitution theorem (for references, see Chap. 10) is sometimes misunderstood. It does not mean that *changes* in technological information will not result in changes in observed input ratios. It does mean that with given technology there is one preferred set of input ratios which will continue to be preferred no matter what the desired bill of final consumption happens to be. It does not mean that changes in relative prices will not induce changes in proportions. In fact, part of the point of the theorem is that in a Leontief technology relative prices *can't change*.

The reason is essentially that there is *by assumption* only one fixed factor, only one true social cost. Relative prices of commodities will depend only on their direct and indirect labor content.[1] A change in wage rates,

[1] Compare the marginal rate of substitution deduced above, and see the next section for details.

for example, will simply increase the prices of all commodities in the same proportion, leaving relative prices unchanged. Since there is only one thing to be economized, labor, it is perhaps plausible that one set of activities should turn out to be the most economical of labor, regardless of what final goods are desired.

Let us imagine an expanded technological matrix like Table 9-6; for each industry there are several different columns, representing different processes or methods of production. For example, 1 unit of Commodity 1

TABLE 9-6

	Inputs to Industry 1				Inputs to Industry 2				Final consumption
Commodity 1...	$a_{11}^{(1)}$	$a_{11}^{(2)}$	...	$a_{11}^{(h)}$	$a_{12}^{(1)}$	$a_{12}^{(2)}$	...	$a_{12}^{(k)}$	C_1
Commodity 2...	$a_{21}^{(1)}$	$a_{21}^{(2)}$	...	$a_{21}^{(h)}$	$a_{22}^{(1)}$	$a_{22}^{(2)}$	...	$a_{22}^{(k)}$	C_2
Labor services...	$a_{01}^{(1)}$	$a_{01}^{(2)}$	...	$a_{01}^{(h)}$	$a_{02}^{(1)}$	$a_{02}^{(2)}$	...	$a_{02}^{(k)}$	

can be produced with inputs of $a_{11}^{(1)}$ of Commodity 1, $a_{21}^{(1)}$ of Commodity 2 and $a_{01}^{(1)}$ of labor, or with inputs of $a_{11}^{(2)}$, $a_{21}^{(2)}$, and $a_{01}^{(2)}$, etc. Moreover, under our usual constant returns to scale and additivity assumptions, each industry can operate any subset of its processes simultaneously, and the inputs and outputs can be calculated separately for each process and then combined.

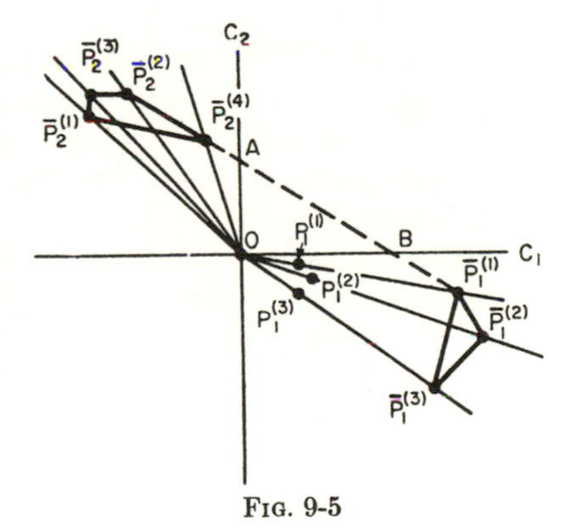

FIG. 9-5

Figure 9-5 is like Fig. 9-3, except that there is a different point $P_1^{(1)}$, $P_1^{(2)}$, . . . , $P_1^{(h)}$ for *each* of the processes known to Industry 1, and similarly, there is a different point $P_2^{(1)}, P_2^{(2)}, \ldots, P_2^{(k)}$ for each process in Industry 2. So also we get several rays for each industry, and on each ray we get a point $\bar{P}_1^{(1)}, \ldots, \bar{P}_1^{(h)}, \bar{P}_2^{(1)}, \ldots, \bar{P}_2^{(k)}$. These $h + k$ points show the net results if the whole supply of labor were to be applied to just one of the known processes. Now in Fig. 9-3 we remarked that if you split the available labor between two processes, you get net results represented by a point on the line joining the corresponding barred points. Exactly the same proposition holds in Fig. 9-5; moreover, we can think of combining two processes in the same industry, and we can think of combining three or more points by taking a weighted average of the barred points.[1] In this way we generate the totality of all weighted averages (with nonnegative weights) of the pure processes. The corresponding set of points in Fig. 9-5 is called the "convex hull" of the points $\bar{P}_1^{(1)}, \ldots, \bar{P}_1^{(3)}, \ldots, \bar{P}_2^{(4)}$. Any point of the convex hull (the irregular-shaped set $O\bar{P}_2^{(1)}\bar{P}_2^{(3)}\bar{P}_2^{(2)}\bar{P}_2^{(4)}AB\bar{P}_1^{(1)}\bar{P}_1^{(2)}\bar{P}_1^{(3)}O$) represents a bill of final consumption producible by some combined operations of the seven known processes—eight, counting the origin. Clearly the consumption-possibility frontier is the line segment AB, a straight line as before. Note well that all points on the frontier AB are obtained as weighted averages of a single process in Industry 1 and a single process in Industry 2. In

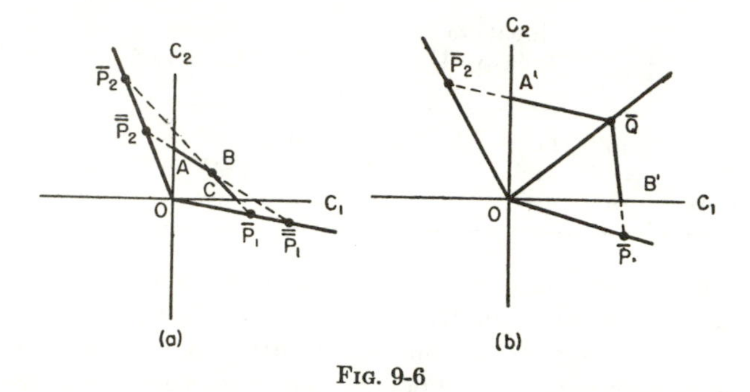

FIG. 9-6

Industry 1, the process $P_1^{(1)}$ is preferred. For Industry 2, the situation is a bit different. There are two preferred processes, $P_2^{(2)}$ and $P_2^{(4)}$; and any mixture of these two is equally good. But it is still true that Industry 2 can light on one set of input proportions once and for all, and need never change, regardless of what bill of goods is desired. This proves the substitution theorem.

Figure 9-6*a* shows what would happen if there were a second fixed factor, say, land. On each process ray we must distinguish two points, $\bar{P}_1$ and $\bar{\bar{P}}_1$ on one ray, and $\bar{P}_2$ and $\bar{\bar{P}}_2$ on the other. $\bar{P}_i$ represents net outputs if all the labor is allocated to the ith process; $\bar{\bar{P}}_i$ represents net outputs if all the land is allocated to the process. In the usual way, we find a consumption frontier for *each* fixed factor taken separately. But only those net outputs are attainable which lie on or under *both* frontiers. Hence the final consumption-possibility schedule is the broken line ABC. The marginal rate of substitution changes suddenly at the vertex point B. Thus the presence of two factors destroys the constant-cost characteristic even when there is only one process per industry, a fortiori when there are several.

[1] We can get away with combinations of two processes only, but every combination of two processes represents a new "process" which can now be mixed with still other processes, etc.

In Fig. 9-6*b* we return to the case of a single fixed factor, but along with $\bar{P}_1$ and $\bar{P}_2$ we introduce a third process, represented by the ray through $\bar{Q}$. This is a joint-production process; it uses labor as an input and produces positive net output of both commodities. It is easy enough to see, by our previous line of reasoning, that the consumption-possibility frontier is given by $A'\bar{Q}B'$. Net outputs on $A'\bar{Q}$ are produced by a blend of P_2 and Q, those on $\bar{Q}B'$ are produced by a blend of Q and P_1.

Other proofs of the substitution theorem appear in the next chapter.

9-6. PRICES IN THE LEONTIEF SYSTEM

As economists we already know enough about the structure of an input-output system to guess with considerable confidence what the accompanying competitive price relationships must be. The (constant) marginal rate of substitution was shown to be A_{01}/A_{02}; this must determine the relative price of the two commodities:

$$\frac{p_1}{p_2} = \frac{A_{01}}{A_{02}} \tag{9-15}$$

Next, we have interpreted A_{01} as the total labor content of 1 unit of final output of Commodity 1. If we designate the wage rate by w, this tells us that

$$\begin{aligned} p_1 &= A_{01}w = (a_{01}A_{11} + a_{02}A_{21})w \\ p_2 &= A_{02}w = (a_{01}A_{12} + a_{02}A_{22})w \end{aligned} \tag{9-16}$$

since labor is the only cost-generating element in the system. This is as far as we can hope to go; a *real* system like Leontief's can only hope to determine relative prices. The absolute level of prices is completely indeterminate. It would be natural here to choose labor as the numeraire.

These price relationships can be verified in another, equally intuitive, way. In a long-run competitive equilibrium we may put prices equal to unit costs. To be more exact, we should say that we may put prices *at most* equal to unit costs; if a commodity is produced at all, the equality must hold, but price may fall short of unit cost for a commodity not being produced—in fact, this is why it is not being produced. The inputs into 1 unit of Commodity 1 are a_{11} units of Commodity 1, a_{21} units of Commodity 2, and a_{01} units of labor; the unit cost is then $a_{11}p_1 + a_{21}p_2 + a_{01}w$. Thus we can write as a condition of equilibrium

$$\begin{aligned} p_1 &\leq a_{11}p_1 + a_{21}p_2 + a_{01}w \\ p_2 &\leq a_{12}p_1 + a_{22}p_2 + a_{02}w \end{aligned}$$

or, after rearranging,

$$\begin{aligned} (1 - a_{11})p_1 - a_{21}p_2 &\leq a_{01}w \\ -a_{12}p_1 + (1 - a_{22})p_2 &\leq a_{02}w \end{aligned} \tag{9-17}$$

In our system we are sure that both commodities will be produced, and hence we can insert both equality signs and solve the resulting pair of linear equations. We get

$$p_1 = \frac{a_{01}(1 - a_{22})}{(1 - a_{11})(1 - a_{22}) - a_{12}a_{21}} w + \frac{a_{02}a_{21}}{(1 - a_{11})(1 - a_{22}) - a_{12}a_{21}} w$$

From (9-13) and (9-10) this will be recognized as just the assertion that $p_1 = A_{01}w$, verifying the first half of (9-16). That $p_2 = A_{02}w$ follows similarly, and hence, by division, the fact that the price ratio equals the marginal rate of substitution.

We can now highlight the fact that a Leontief system is an extremely simplified linear-programming model. Return to our old commodity-accounting inequalities (9-5):

$$\begin{aligned} (1 - a_{11})X_1 - a_{12}X_2 &\geq C_1 \\ -a_{21}X_1 + (1 - a_{22})X_2 &\geq C_2 \end{aligned} \tag{9-5}$$

Comparing these with the price-cost inequalities of (9-17), we observe that the coefficients on the left-hand sides are transposes of each other—columns become rows, and rows become columns. Moreover, the direction of the inequalities has been reversed. This suggests viewing (9-5) and (9-17) as the constraints in dual linear programs (the variables—prices and gross outputs—are necessarily nonnegative):

1. Subject to (9-5), minimize $wa_{01}X_1 + wa_{02}X_2$.
2. Subject to (9-17), maximize $p_1C_1 + p_2C_2$.

In other words, subject to the production of a specified bill of goods, choose gross outputs to minimize total labor costs; and subject to the price-cost inequalities, choose prices to maximize the value of net output.[1]

A glance back at Fig. 9-1 will show that the intersection point L is the only solution to the minimum problem. At L, *both* gross outputs are *smaller* than at any other point satisfying (9-5). Hence *any* positively weighted sum such as total labor cost will be minimized at L. We leave it to the reader as an exercise to draw the corresponding diagram for the constraints of (9-17), and to show that again there is an intersection point P at which *both* prices are greater than at any other point satisfying (9-17). Hence *any* positively weighted sum, such as value of the given net output, will be maximized at P. This characteristic—that there is a *unique* minimum point L regardless of the labor inputs a_{01}, and a_{02}, and a *unique* maximum point P regardless of the final demands C_1 and C_2—is what makes the Leontief system so easy to handle.

[1] The duality theorem reminds us that no gross output will be overproduced unless the corresponding commodity is a free good; and price equals cost unless the corresponding gross output is zero.

Finally, we have a theorem from linear programming that states that the minimum value of the total labor cost just equals the maximum value of the form to be maximized. We deduce then that at the intersection points L and P, when all equalities hold in (9-5) and (9-17),

$$wa_{01}X_1 + wa_{02}X_2 = p_1C_1 + p_2C_2 \tag{9-18}$$

We can interpret this equation as saying that the total value of net output is just imputed as wages to the scarce factor labor. Alternatively, we have in (9-18) the basic identity of income analysis: that net product is the same whether measured as factor costs or market values.

10

The Statical Leontief System (Continued)[1]

Let $(X_1, X_2, \ldots, X_n)$ stand for the totals of producible outputs, and let X_0 stand for the total of a primary nonproduced good (such as "labor").

Let $(C_1, C_2, \ldots, C_n)$ stand for the total final consumption of each of the produced goods. By convention $C_0 = 0$.

Let x_{ij} stand for the amount of *input* of the *i*th good used in the production of the *j*th good's *output*. Consequently x_{0j} represents the labor allocated to the production of the *j*th good's output.

10-1. REAL OR NONPRICE RELATIONS

The total of any good, such as X_i, is allocated as final consumption C_i, or as intermediate inputs $x_{i1}, x_{i2}, \ldots, x_{in}$, so that[2]

$$X_i = x_{i1} + x_{i2} + \cdots + x_{in} + C_i \qquad i = 0, 1, 2, \ldots, n \qquad (10\text{-}1)$$

These represent pure bookkeeping relations, and it will be noted that the primary good X_0, or labor, is subject to a similar relation.

Ruling out joint production and assuming classical constant returns to scale, we may write the production function relating the output X_j to its inputs x_{ij}, as follows:

$$X_j = F^j(x_{0j}, x_{1j}, \ldots, x_{nj}) \qquad j = 1, 2, \ldots, n \qquad (10\text{-}2)$$

where F^j is a homogeneous function of the first degree.

Leontief makes the special fixed-coefficient assumption that *each* input

[1] This chapter, dealing somewhat more technically with the material of the previous one, can be skipped without loss of continuity.

[2] Strictly speaking, $X_i \geq \Sigma x_{ij} + C_i$, but, provided X_i is a "scarce" good, we can assume it to be fully employed and omit the sign $\geq$.

x_{ij} is required in fixed proportion to output X_j, so that[1]

$$x_{ij} = a_{ij}X_j \qquad j = 1, \ldots, n; i = 0, 1, \qquad , n \qquad (10\text{-}2a)$$

where, by definition,

$$\begin{bmatrix} a_{01} & a_{02} & a_{03} & \cdots & a_{0n} \\ \cdots & \cdots & \cdots & \cdots & \cdots \\ a_{11} & a_{12} & a_{13} & \cdots & a_{1n} \\ a_{21} & a_{22} & a_{23} & \cdots & a_{2n} \\ \cdots & \cdots & \cdots & \cdots & \cdots \\ a_{n1} & a_{n2} & a_{n3} & \cdots & a_{nn} \end{bmatrix} = \begin{bmatrix} a_{0j} \\ \cdot\,\cdot \\ a_{ij} \end{bmatrix} = \begin{bmatrix} a_0 \\ \cdot \\ a \end{bmatrix} \qquad (10\text{-}2b)$$

represent the given nonnegative technological coefficients showing the requirements of the ith input needed to produce a single unit of the jth output. Dimensionally, each a_{ij} represents input units required per output unit: it is illegitimate to add $a_{ij} + a_{ik}$ or $a_{ij} + a_{kj}$; note too that doubling the size of the output unit j will double a_{ij}, but doubling the size of the input unit i will halve a_{ij}, for obvious reasons.[2]

Combining (10-1) and (10-2a), we have, for every unknown variable,

$$\begin{aligned} X_i &= a_{i1}X_1 + a_{i2}X_2 + \cdots + a_{in}X_n + C_i \\ &= \sum_{j=1}^{n} a_{ij}X_j + C_i \qquad i = 0, 1, 2, \ldots, n \end{aligned} \qquad (10\text{-}3)$$

where by convention $C_0 = 0$.

If we are given $(C_1, \ldots, C_n)$, Eq. (10-3) represents $n + 1$ linear equations in the $n + 1$ unknown X's. Disregarding the first equation, which defines X_0, we see that the last n linear equations completely determine $(X_1, \ldots, X_n)$ in terms of the final C's. Elementary algebra tells us that, by successive substitution or elimination, we can finally express each produced X_i linearly in terms of the C's.[3] Without going into detail,

[1] Strictly speaking, Leontief writes (10-2) in the special form

$$X_j = F^j = \min\left(\frac{x_{0j}}{a_{0j}}, \frac{x_{1j}}{a_{1j}}, \ldots, \frac{x_{nj}}{a_{nj}}\right)$$

but as long as each good is scarce, the minimum will equal the maximum. Nothing will be wasted, and (10-2) can be replaced by (10-2a). Leontief sometimes sets each $a_{ii} = 0$ by "netting out" intraindustry transactions, but we do not do so.

[2] If the size of units of the ith good is changed in the ratio k_i, then using bars to represent the new values of any variable gives us $\bar{X}_i = k_iX_i$, $\bar{a}_{ij} = k_ia_{ij}/k_j$. Later it will be shown that prices are given by $\bar{P}_i = P_i/k_i$, and the later-defined A's transform like the a's.

[3] See the later numerical and algebraic analysis. Mathematically, Eq. (10-3) can be written in matrix form, $(\mathbf{I} - \mathbf{a})\mathbf{X} = \mathbf{C}$. The solution is $\mathbf{X} = (\mathbf{I} - \mathbf{a})^{-1}\mathbf{C}$, so that the A's in (10-4) are the elements of $(\mathbf{I} - \mathbf{a})^{-1}$. On the existence of this inverse, see below.

we may finally note that this gives us

$$\begin{aligned} X_i &= A_{i1}C_1 + A_{i2}C_2 + \cdots + A_{in}C_n \\ &= \sum_1^n A_{ik}C_k \qquad i = 1, 2, \ldots, n \end{aligned} \tag{10-4}$$

where A_{ik} represents the total X_i needed to produce 1 unit of C_k alone. These A's depend only on the nonlabor a's.

The form of (10-4) is quite remarkable in that it is *linear*. This means that the total of any output needed to produce an assigned target of consumption goods can be built up by adding the separate outputs needed to produce each item of the target. Thus, to produce 100 consumable hats and 500 consumable shirts, we first work out 100 times the requirements for 1 hat alone, and then add to this 500 times the requirements for 1 shirt alone. The A's represent the total production requirements, direct and indirect, of each good needed for each single unit of consumption.

It is not immediately obvious that a relationship of the form (10-4) holds for labor. But recall from the first equation of (10-3) that

$$\begin{aligned} X_0 &= a_{01}X_1 + \cdots + a_{0n}X_n + 0 \\ &= a_{01}(A_{11}C_1 + A_{12}C_2 + \cdots + A_{1n}C_n) + \cdots \\ &\qquad + a_{0n}(A_{n1}C_1 + \cdots + A_{nn}C_n) \\ &= (a_{01}A_{11} + a_{02}A_{21} + \cdots + a_{0n}A_{n1})C_1 + \cdots \\ &\qquad + (a_{01}A_{1n} + \cdots + a_{0n}A_{nn})C_n \end{aligned}$$

The expressions in parentheses are seen to represent the total increases in labor needed in every industry to produce 1 extra unit of the consumption good in question. Again note that our final answer can be built up by a simple *superposition* of independent linear terms. Hence, a relation like that of (10-4) holds for X_0 too, provided we define $(A_{01}, A_{02}, \ldots, A_{0n})$ so that

$$\begin{aligned} A_{01} &= a_{01}A_{11} + a_{02}A_{21} + \cdots + a_{0n}A_{n1} \\ &\cdots\cdots\cdots\cdots\cdots\cdots \\ A_{0n} &= a_{01}A_{1n} + a_{02}A_{2n} + \cdots + a_{0n}A_{nn} \end{aligned} \tag{10-5}$$

where any A_{0j} represents the total labor needed (in all industries) to produce 1 extra net unit of consumable good j. This gives us

$$X_0 = A_{01}C_1 + A_{02}C_2 + \cdots + A_{0n}C_n \tag{10-6}$$

This last relation is the final all-important production-possibility menu for the Leontief economy. For any preassigned "bill of goods"

$(C_1, \ldots, C_n)$, it gives us the needed total labor X_0. Alternatively, for any given amount of total labor, we see how we can linearly convert—at constant costs—one good into any other.

Once again we note the surprisingly simple juxtaposition of linear terms. Schematically, the requirements for any basket of consumption goods can be depicted as follows:

$$
\begin{array}{llll}
C_1 \longrightarrow & X_0 = A_{01}C_1, & X_1 = A_{11}C_1, & \ldots, X_n = A_{n1}C_1 \\
C_2 \longrightarrow & X_0 = A_{02}C_2, & X_1 = A_{12}C_2, & \ldots, X_n = A_{n2}C_2 \\
\cdots & \cdots & \cdots & \cdots \\
\text{Total requirements:} & X_0 = \Sigma \text{ of above}, & X_1 = \Sigma \text{ of above}, & \ldots, X_n = \Sigma \text{ of above}
\end{array}
$$

There are no diminishing returns of one good for labor in the Leontief model, or of one good in terms of another. With labor the only primary factor, the changes in factor proportions underlying the classical laws of variable proportions cannot take place. A simple statical Leontief model would never be able to explain why food prices rise relative to other prices in periods of prosperity; it fails to recognize our *in*ability to produce land or new capital capacity at constant costs. On this score it errs in the one direction of being overly optimistic as to convertibility of goods from war to peace. This optimistic bias must be set off against its pessimistic bias in ruling out technical substitutions of one factor for another.

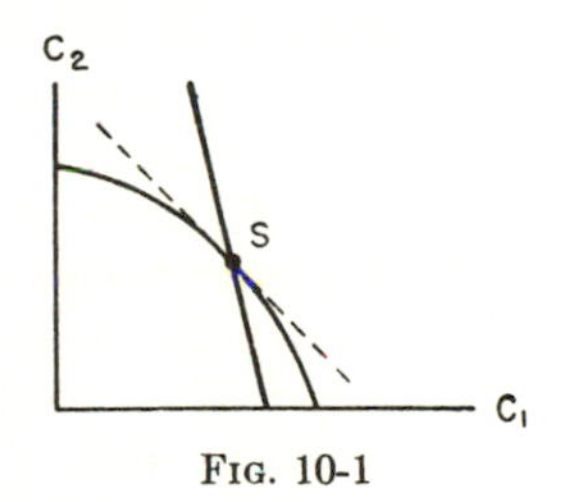

FIG. 10-1

Imagine an economy whose production-possibility menu corresponding to (10-6) has the usual convex shape—either because of variable proportions and diminishing returns in production, or the existence of several primary factors, or both. This is illustrated in Fig. 10-1. Now suppose a Leontief statistician observes point S and tries to squeeze what he measures into the Leontief framework. If he knows that competition prevails and therefore that price ratios measure marginal rates of transformation, he may make his computations in such a way as to yield the tangent line through S as his linear-production-possibility frontier. Because of the assumed convexity of the true frontier, it can be seen that the net result is necessarily an optimistic bias. If, on the other hand, he simply measures the interindustry flows (our x_{ij}) and proceeds as in (10-2*a*), then in all likelihood the resulting frontier will *cut* the true frontier (see Fig. 10-1 again) and the bias will be optimistic over one range and pessimistic over the other. In this case, the computed prices (see Sec. 10-2) would not agree with observed *competitive* prices.

10-2. COST AND PRICE RELATIONS

The coefficient A_{0j} represents the total cost in terms of labor of the jth consumption good C_j. This total labor cost exceeds the direct-labor cost a_{0j} by the amount of "indirect labor" needed to produce the intermediate products x_{ij}, needed in turn to produce the jth consumption good.

The total labor "congealed" in the jth good cannot—as in the simplest Austrian structure of production in which each good is produced only by direct labor and the output of the "previous stage of production"—be simply decomposed into the sum of labor embodied in all the previous stages of production. In such a simple Austrian model we could write $A_{0,17}$ as the sum of direct labor in the seventeenth stage $a_{0,17}$ and the total congealed labor of the raw materials used from the sixteenth stage $A_{0,16}a_{16,17}$.

In a general Leontief model, everything is needed to produce everything. There are no early and late stages, no previous stages, no higher or lower stages. Coal is needed to produce fertilizer; fertilizer is needed to produce coal. Neither is prior to the other. In fact coal may be needed in order to produce coal. These non-Austrian "whirlpools," or circular interdependences, make it impossible ever to decompose total labor of j, A_{0j}, into direct-labor requirements of a finite number of previous stages. Rather we should have to decompose fertilizer into its direct labor plus the previous direct labor involved in coal plus the previous direct labor involved in the fertilizer needed to produce the coal plus an infinite chain of such previous stages. It can be proved that the infinite chain is a dwindling one as we go into the hypothetical remote past, so the sum of all past direct-labor requirements will in fact rigorously add up to each A_{0j}.[1]

However, we have already seen how the A_{0j}'s are defined by solving the simultaneous equations of (10-3) alone. No infinite series need be summed. The only point at issue is the proper interpretation of any A_{0j} as the "total labor requirement or cost of one unit of consumption good j." This interpretation is rigorously provided by (10-5).

To drive home the point and relate total labor cost in a Leontief model to that in an Austrian model, we ought to be able to prove that any A_{0j} is in fact exactly equal to its direct labor cost a_{0j}, plus the total labor costs of each and every intermediate good x_{ij} used in its production. Thus we

[1] H. T. N. Gaitskell, "Notes on the Period of Production," *Zeitschrift für Nationaloekonomie*, 9:215–44 (1938), gives references to the voluminous German literature on whirlpool structures of production; even a finite average period of production can be defined for an infinite series, *which must converge* if consumption and output are all positive.

should have

$$A_{0j} = a_{0j} + A_{01}a_{1j} + A_{02}a_{2j} + \cdots + A_{0n}a_{nj} \qquad j = 1, 2, \ldots, n \tag{10-7}$$

There are n linear equations to determine the n A_{0j}'s. Except in a crypto-Austrian case, where each stage depends upon definable "previous" stages only, these equations must be solved *simultaneously*. Simultaneity is the mathematical economist's way of cutting through circular interdependence and avoiding all infinite-series multiplier chains.

How do we know that the A_{0j}'s as defined by (10-7) are really identical with our previous definition of them in (10-5)? This is a purely algebraic question that admits of a simple affirmative answer, confirming mathematically our intuitive expectations.[1]

This completes a brief survey of the Leontief system. Note that we have purposely dealt with it *in natura*, using no money relations and no competitive markets. So simple a system could be run by fiats, using punch-card machines rather than markets. However, there is no reason why such a system could not be organized by means of perfectly competitive markets, which give prices $(P_0,P_1,P_2, \ldots ,P_n)$ to labor and to each of the n goods.

Under perfectly competitive statical conditions, the equilibrium price for each producible good must be exactly equal to its unit cost of production. The latter consists of the costs per unit of each and every needed intermediate good, plus direct-labor cost. The cost per unit for the jth good of the needed ith input would be P_ia_{ij}, and the direct-labor cost would be the wage times needed labor, or P_0a_{0j}. Thus, for each of the n produced goods, we have the following market conditions:

$$P_j = P_0a_{0j} + P_1a_{1j} + P_2a_{2j} + \cdots + P_na_{nj} \qquad j = 1, 2, \ldots, n \tag{10-8}$$

Now it is obvious that the absolute level of prices plays no role in the Leontief model as we have described it. We cannot hope to solve for determinate prices of all $n + 1$ variables. Instead we may designate any one price as *numeraire* and solve for the remaining n prices in relation to it. In the Leontief system, it is natural to use "wage units," placing $P_0 = 1$, or, what is the same thing, solving for $(P_1/P_0, \ldots ,P_n/P_0)$. Dividing the equations of (10-8) all through by P_0, we have n linear equations to determine our n unknown price ratios.

If we look closely at (10-8), we see that these linear equations have *exactly the same "a" coefficients* as did Eqs. (10-7). This reaffirms what we

[1] The proof follows from the definition of the A's in (10-4) in relation to the a's of (10-3). In matrix terms, (10-5) tells us that $\mathbf{A}_0' = \mathbf{a}_0'(\mathbf{I} - \mathbf{a})^{-1}$. But (10-7) can be written $\mathbf{A}_0'(\mathbf{I} - \mathbf{a}) = \mathbf{a}_0'$, whose solution is just $\mathbf{A}_0' = \mathbf{a}_0'(\mathbf{I} - \mathbf{a})^{-1}$.

have already suspected: The A_{0j} coefficient denoting the total labor costs of the jth good will be exactly the same thing as P_j/P_0, the competitively determined price ratio of that good relative to the wage rate.[1] Note also from (10-6) that the competitive price ratios correspond exactly to the slopes, or marginal rates of transformation, of the production-possibility schedule.

Compare Eqs. (10-8) and our original (10-3). You will note that many of the a's are involved. In fact both involve $(a_{11}, a_{12}, \ldots, a_{n1}, a_{n2}, \ldots, a_{nn})$, the square array of nonlabor a's, but with an important difference: the rows of (10-3) become columns in (10-8); that is, a_{ij} is transposed to become a_{ji}; also, the constant coefficients of (10-3) are consumption goods—often called the "open-end" items of the model—while the constant coefficients of (10-7) are direct-labor coefficients, labor being the sole open-end primary factor.

To summarize, there is a simple "duality" relation between quantities and prices in the Leontief system: transpose the a's of the quantity problem and you get the price problem; similarly transpose the a's of the price problem and you get the quantity problem.

Such dualities are common in mathematics. In particular, they turned up earlier in connection with the valuation problem of the competitive firm, and they will occur again in connection with game theory. To illustrate one of the macroeconomic applications of duality principles, let us show how we can derive one of the basic relations of all national-income accounting.

National income or product can always be looked at in two ways: as the value of a flow of final *outputs* or as the total cost of factor *inputs*. In most accounting systems these become tautologically equal by virtue of a residual definition of "profits" as a factor payment. In a statical competitive system it is an equilibrium condition, not an accounting definition, that equilibrium profits are zero. So for such systems we arrive at a more meaningful equality between the two ways of looking at net income.

The total value of final output in the Leontief system is not hard to evaluate. It definitely is not $P_1X_1 + \cdots + P_nX_n$, since much of each X_i is intermediate goods used up in production. Only consumption C_i counts as final output; so our total is $P_1C_1 + \cdots + P_nC_n$.

The total cost of inputs in the Leontief system is not given by the total cost of all needed materials plus the total labor cost. This would involve double counting, since cost of intermediate goods is itself "decomposable," as we have seen, into labor costs. In a statical time-saturated, one-

[1] We can calculate price ratios directly by solving (10-8). Alternatively, if we have already solved (10-3) for (10-4), which is a much more formidable computation job, we can calculate our price ratios indirectly from (10-5).

primary-factor Leontief system, all "value added" is measured by labor cost alone, so that the second way of measuring product is by $P_0(x_{01} + x_{02} + \cdots + x_{0n})$, or more simply, P_0X_0, the wage bill.

The fundamental identity we seek must then be

$$P_1C_1 + P_2C_2 + \cdots + P_nC_n \equiv P_0X_0 \tag{10-9}$$

Will this hold? Yes. To verify this *algebraic* fact most simply, recall our successful identification of P_j/P_0 with A_{0j}; and then note that our constant-cost production-possibility menu for society, previously established in (10-6), is exactly equivalent to the desired national-income identity.

This income identity is so basic as to appear almost trivial. But involved in it is the whole "adding-up or exhaustion-of-the-product problem," so intimately tied up with the assumption of constant returns to scale and Euler's theorem on homogeneous functions.[1]

10-3. QUANTITATIVE MEASUREMENT OF A LEONTIEF MODEL

So far we have dealt with physical quantities of outputs: of X_i's, C_i's, and x_{ij}'s. The a_{ij}'s consequently are definite, physically measured inputs/outputs. Leontief, however, dealt originally in dollar values only, and most statisticians have since followed him in this. To the superficial eye, the whole subject appears to be more a branch of money national-income accounting than the structure of physical production. Let us therefore try to relate the statistical measurements of dollar flows to our model.

Leontief begins by measuring value flows: $\bar{X}_i$, which is P_iX_i, rather than P_i and X_i separately; $\bar{C}_i$, which is P_iC_i; $\bar{x}_{ij}$, which is P_ix_{ij}, etc. These value quantities can be arranged in a tableau, as shown in Table 10-1.

The sum of the rows shows the total value that has been sold or allocated to consumption and all industrial uses. It is a little dangerous to add the columns at this stage; but if we are willing to accept the fact that all profits are zero and all capital accumulation and other complications can be ignored, then we may risk doing so. The sum of any column would then, by definition, be the same as the sum of the corresponding row, and these are included in Table 10-1 in parentheses. Note that the sum of all the items in the table equals the sum of either set of marginal totals. This dollar sum is a very gross total, involving double counting

[1] If marginal input requirements had deviated from the a's depicting average input requirements, competitive equilibrium would have been impossible and our model would have to be described in quite different nonlinear terms.

A less obvious duality relation may be mentioned: The increase in P_j/P_0 resulting from a unit increase in X_i's direct labor cost exactly equals the increase in total output X_i needed for a unit increase in consumption C_j; both are equal to A_{ij}.

of intermediate goods. The net national product in our static model is given by the first-row or column sum. The dotted lines separate the open-end consumption and labor part of our system from the inner industrial part.

TABLE 10-1

Sales of: \ Purchases by:	Households	Industry 1	Industry 2	. . .	Industry n	Total sales
Labor	0	$\bar{x}_{01}$	$\bar{x}_{02}$	. . .	$\bar{x}_{0n}$	$\bar{X}_0 = \Sigma\bar{x}_{0j}$
Industry 1	$\bar{C}_1$	$\bar{x}_{11}$	$\bar{x}_{12}$	. . .	$\bar{x}_{1n}$	$\bar{X}_1 = \Sigma\bar{x}_{1j}$
Industry 2	$\bar{C}_2$	$\bar{x}_{21}$	$\bar{x}_{22}$	. . .	$\bar{x}_{2n}$	$\bar{X}_2 = \Sigma\bar{x}_{2j}$
	. . .	. . .	. . .	. . .	. . .	. . . .
Industry n	$\bar{C}_n$	$\bar{x}_{n1}$	$\bar{x}_{n2}$	. . .	$\bar{x}_{nn}$	$\bar{X}_n = \Sigma\bar{x}_{nj}$
Total purchases	$(\bar{X}_0)$	$(\bar{X}_1)$	$(\bar{X}_2)$	. . .	$(\bar{X}_n)$	

Now nobody can stop us from forming a new table by dividing the elements of each column by the total of the column. In our new table, $\bar{x}_{ij}$ is replaced by $\bar{x}_{ij}/\bar{X}_j$, which we may call $\bar{a}_{ij}$. These new elements will be the fraction of total revenue (= cost) of the jth industry that goes to buy the ith input. The total of all such fractions, including direct-labor costs, must of course add up to unity, as shown by the marginal totals at the bottom of each column of Table 10-2.

TABLE 10-2. TECHNOLOGICAL-VALUE COEFFICIENTS PER DOLLAR OF OUTPUT

$\bar{a}_{01}$	$\bar{a}_{02}$	$\bar{a}_{03}$	. . .	$\bar{a}_{0n}$
$\bar{a}_{11}$	$\bar{a}_{12}$	$\bar{a}_{13}$	. . .	$\bar{a}_{1n}$
. . .	. . .	. . .	. . .	. . .
$\bar{a}_{n1}$	$\bar{a}_{n2}$	$\bar{a}_{n3}$	. . .	$\bar{a}_{nn}$
$1 = \Sigma\bar{a}_{i1}$	$1 = \Sigma\bar{a}_{i2}$	$1 = \Sigma\bar{a}_{i3}$	. . .	$1 = \Sigma\bar{a}_{in}$

In Table 10-2, we have omitted the consumption column, since $\bar{a}_{10}$, $\bar{a}_{20}$, . . . , $\bar{a}_{n0}$ would represent the percentages of total income spent by consumers on different goods. These would be psychological budgetary facts rather than technological facts, and in his later exposition Leontief has tended to work with open-end systems that exclude these and thus avoid Malthusian complications. Even had we included them, it would make no sense to add the elements of any row: the result would be $\bar{x}_{i0}/\bar{X}_0 + \bar{x}_{i1}/\bar{X}_1 + \cdots + \bar{x}_{in}/\bar{X}_n$; because of the different column divisors, this is a meaningless total, about which little could be said.

The resemblance between our $\bar{a}_{ij}$ coefficients and the original physical a_{ij} coefficients of (10-2b) is striking. The $\bar{a}$'s are simply a's defined in special units. We can write down all the remaining relations between the barred quantities, regarding these as simply one choice of units for all our variables. The reader may verify that the following relations do hold:

$$\bar{X}_i = \sum_{j=1}^{n} \bar{a}_{ij}\bar{X}_j + \bar{C}_i \qquad i = 0, 1, \ldots, n \tag{10-3a}$$

$$\bar{X}_i = \sum_{j=1}^{n} \bar{A}_{ij}\bar{C}_j \qquad i = 1, \ldots, n \tag{10-4a}$$

However, because the columns of $\bar{a}_{ij}$ add up to exactly 1, we can see that the solution of the price equations

$$\bar{P}_j = \bar{P}_0\bar{a}_{0j} + \sum_{i=1}^{n} \bar{P}_i\bar{a}_{ij} \tag{10-7a}$$

is

$$1 = 1 \cdot \bar{a}_{0j} + 1 \cdot \bar{a}_{1j} + \cdots + 1 \cdot \bar{a}_{nj}$$

that is, $\bar{P}_0 = \bar{P}_1 = \cdots = \bar{P}_n = 1$.

This shows us that the barred variables have the following interpretation: Leontief is working with dimensional units of outputs and inputs defined by "a dollar's worth." Thus if eggs are 50 cents a dozen, Leontief is using 2 dozen eggs as his physical unit in which $\bar{X}_{\text{eggs}}$ is expressed. The prices of these artificial units—$\bar{P}_{\text{eggs}}$, not P_{eggs}—are all exactly 1, by convention. The interested reader can verify that Eqs. (10-5), when put in terms of barred a's, will yield unit values for any new $\bar{A}_{0j}$, and that Eq. (10-6) becomes the rather-trivial-looking relation

$$\text{Total wages} = \bar{X}_0 = 1 \cdot \bar{C}_1 + 1 \cdot \bar{C}_2 + \cdots + 1 \cdot \bar{C}_n = \text{total final consumption} \tag{10-6a}$$

As Leontief[1] has shown, the use of value data alone gives rise to certain paradoxes. It is as if we are looking at the pale shadows on a screen of a system and it turns out that two systems with the same shadow may be quite different indeed.

Thus, two systems with the same $\bar{a}$ pattern might differ in the following ways: (1) one system might use dollars and the other francs (or dimes), which is only a trivial difference; (2) one might be characterized by double the price level of the other, with all else unchanged; (3) one might be a tenth the scale of the other (e.g., the State of New York versus the United States); (4) one might have an entirely different taste pattern for consumption goods, with a resulting difference in the allocation of labor to industries and in the mix of outputs.

[1] Wassily W. Leontief, *The Structure of American Economy, 1919–1939*, 2d ed., p. 64, Oxford University Press, New York, 1951.

None of the above differences is significant. More important are the following: (5) one system might be opulent and the other poverty-stricken by virtue of lower labor productivity in every line; (6) one system might impart an entirely different physical meaning to "a dollar's worth of coal," so that it requires more physical coal in every use but in compensation is able to produce that much more physical coal with the same inputs of the other items—so long as the physical labor requirement for a physical unit of coal consumption or, more importantly, of housewarming remains the same, this is of no consequence.

The usefulness of the simple Leontief system as a predictive device rests upon two conditions: (1) It depends on the degree to which each industry (such as j) continues to expend its money among labor and other industries in the same fractions regardless of the changes going on (from year to year, from place to place, from new discovery to new discovery, etc.). (2) Since what consumers want is physical-consumption goods rather than dollars per se, it depends too upon the degree to which a given physical pattern of consumption goods can be predictably converted into physical production.

In general, this means that *physical* a_{ij}'s must be predictable or constant, and not the percentage expenditure $\bar{a}_{ij}$'s, alone. A crucial test between any two periods—such as 1939 and 1948—would be to compare ${}_{39}\bar{a}_{ij\,39}P_j/{}_{39}P_i$ with the similar magnitude for 1948.[1]

10-4. CONSOLIDATION AND AGGREGATION

In terms of value units it is easy to see what happens when we consolidate two or more industries and treat them as if they were a single industry. For example, let us consolidate together all the last industries from m to n: call this M and leave all other industries unchanged. To economize on writing bars, let the barred x's be written as y's. The results are given in Table 10-3.

Since we have not followed Leontief's practice of netting out all intra-industry items, there is no difficulty connected with the definition of y_{MM} and no need to eliminate items of the type y_{mm}.

[1] Note that the needed P's for 1939 and 1948 are not 1's but must be *actual* prices or price indexes, derived from outside of the Leontief accounting data. As a matter of fact, a purist would consider *any* change in relative prices as a refutation of the Leontief system. Since agreement is never perfect, and since $n^2 + n$ comparisons are numerous, we might have to compromise by comparing actual total physical requirements and production for a given year with totals computed from the other year's coefficients and decide whether the results are better than could be accomplished by other methods.

TABLE 10-3

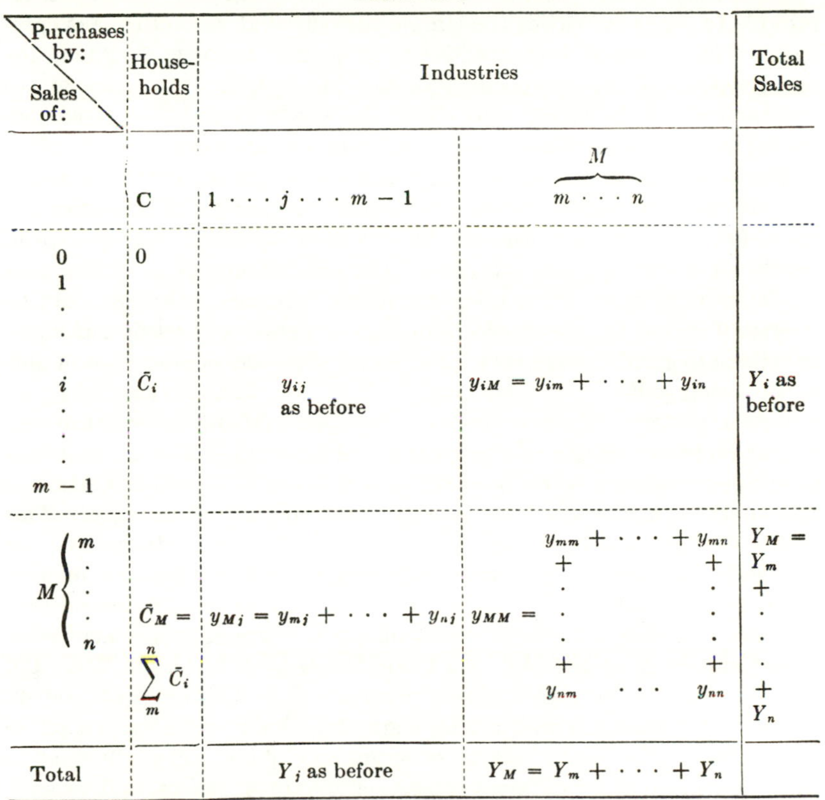

Purchases by: / Sales of:	Households	Industries		Total Sales
	C	$1 \cdots j \cdots m-1$	$\overbrace{m \cdots n}^{M}$	
0	0			
1				
$\vdots$				
i	$\bar{C}_i$	y_{ij} as before	$y_{iM} = y_{im} + \cdots + y_{in}$	Y_i as before
$\vdots$				
$m-1$				
$M \begin{cases} m \\ \vdots \\ n \end{cases}$	$\bar{C}_M = \sum_m^n \bar{C}_i$	$y_{Mj} = y_{mj} + \cdots + y_{nj}$	$y_{MM} = \begin{matrix} y_{mm} & + \cdots + & y_{mn} \\ + & & + \\ \vdots & & \vdots \\ + & & + \\ y_{nm} & \cdots & y_{nn} \end{matrix}$	$Y_M = Y_m + \cdots + Y_n$
Total		Y_j as before	$Y_M = Y_m + \cdots + Y_n$	

We can compute new $\bar{a}$'s by the usual formulas

$$\bar{a}_{Mj} = \frac{y_{Mj}}{Y_j} = \frac{\sum_{i=m}^{n} Y_{ij}}{Y_j} = \sum_{i=m}^{n} \bar{a}_{ij} \quad j = 1, \ldots, m-1$$

$$\bar{a}_{iM} = \frac{y_{iM}}{Y_M} = \frac{\sum_{j=m}^{n} y_{ij}}{\sum_{m}^{n} Y_k} \qquad i = 0, 1, \ldots, m-1, M \tag{10-10}$$

$$= \sum_{j=m}^{n} \frac{y_{ij}}{Y_j} \frac{Y_j}{\sum_{m}^{n} Y_k} = \sum_{j=m}^{n} \bar{a}_{ij}\omega_j$$

The first of these relations tell us that the composite industry must provide to other industries the simple sum of what its constituents provide. The second of these relations tell us that the new $\bar{a}$'s giving the requirements of the composite industry are weighted averages of the requirements of the constituent parts, the weights being the proportionate importance of each constituent industry's production.

The new consolidated coefficients $\bar{a}_{iM}$ are no longer invariant constants. Instead their numerical values will depend upon some of the unknown variables of the problem whose values we seek: specifically, they depend on the proportionate importance of the constituent parts of the consolidated industry.[1] What does this mean? It means that we shall be in error if we use the observed $\bar{a}_{iM}$ of a given period and set up our linear equations to grind out answers for a period when the composition of output has changed.

How important will our errors be? Equations (10-10) show us the conditions under which the resulting error will be small. Any $\bar{a}_{iM}$ will be a close approximation to the needed correct value if (1) all the separate $\bar{a}_{ij}$'s of which it is a weighted average are nearly equal in value. This means that we should try to classify industries so that those that we aggregate together are industries requiring the same types and relative quantities of inputs for their production. Autos and walking shoes serve the same purpose but do not meet our test; autos and military tanks serve different purposes, but they have similar production functions and can therefore be lumped together without serious error. (Of course, intraindustry planners will not be content to know the total values of their sales to other industries: in programming ahead, they will want to know how much of the different constituents—trucks, tanks, bulldozers—other industries and consumers will be wanting. If your interest is in detail, no aggregation can satisfy you. A saving consideration is the fact, not explicitly relevant to a simple statical Leontief system but important in practice, that constituents with similar production functions can probably use each other's capital capacity. Consequently, predicting the future aggregate may suffice for present planning.)

Equations (10-10) show us that a given $\bar{a}_{iM}$ will err very little, provided that (2) production of all the constituent parts of the aggregate invariably changes in about the same proportion so that the weights ω_j remain constant; or failing this, the weight changes should be uncorrelated with the differences in the separate $\bar{a}_{ij}$'s.

When is this condition of unchanged relative production likely to be

[1] That is to say, if the "product mix" in the aggregated industry should change, the aggregate input coefficients would change too. This would occur even if all the "true" underlying unaggregated input coefficients were stable. Since all of what we habitually call "industries" are aggregates to some extent, this is a practical difficulty.

met? One probable case is that of different stages of a given vertical process of production. Spinning and weaving can be usefully consolidated into one industry because they go up and down together. Or to put the matter more technically, the fundamental A coefficients defined by Eqs. (10-4) are likely to have the row for spinning which is exactly proportionate to the row for weaving.[1] (The intraindustry programmer is quite content to know aggregate demand for all output of the vertical stages since he can easily infer the needs for the constituent vertical stages.)

Our first type of aggregation was possible when different industries had similar input needs (cars and tanks, etc.). We may summarize our second type of aggregation by saying that it is possible when different industries are *needed* in the same proportion by other industries (e.g., the need for nuts and bolts, or for spinning and weaving).

10-4-1. Macroanalysis versus Microanalysis. Before leaving the question of consolidation and aggregation, let us notice a third justification for the use of macro rather than micro magnitudes. Let us suppose that in fact the n arbitrarily prescribable consumption items empirically move together, rather than freely and independently. If every point of the economy had an income elasticity of exactly 1, all C's would move in fixed proportions: in fact, we should have one degree of freedom rather than n. Or to take a less special case, suppose all the C's increased in some definite pattern depending upon the consumers' Engel's curves and determinate shifts in income distribution, taxes, etc. Again we would have a single degree of freedom (albeit not a completely linear system).

[1] Strictly speaking, this proportionality holds for all elements except the diagonal set of A's referring to the "own requirements of industries aggregated," resulting from specifying as consumption goods spun thread and woven cloth. However, we would rarely specify as a final consumption item an "unfinished" good such as unwoven thread, so this qualification is unimportant—except to the mathematician who raises his eyebrows at purple cows and inverse matrices with proportional rows.

The general mathematical theorem underlying these two types of aggregation seems to be as follows: If we can find a block of industries such that the partitioned matrix

$$\mathbf{a} = \left[\begin{array}{c|c} \mathbf{E} & \mathbf{B} \\ \hline \mathbf{C} & \mathbf{D} \end{array}\right]$$

has $\mathbf{B}$ of rank 1 (with proportional columns), then we aggregate $\mathbf{D}$ for the first reason; if instead $\mathbf{C}$ is of rank 1 (with proportional rows), we aggregate $\mathbf{D}$ for the second reason. In each case the matrix $(\mathbf{I} - \mathbf{a})^{-1} = \mathbf{A}$ partitions into

$$\left[\begin{array}{c|c} \mathbf{A}_{\mathrm{I\,I}} & \mathbf{A}_{\mathrm{I\,II}} \\ \hline \mathbf{A}_{\mathrm{II\,I}} & \mathbf{A}_{\mathrm{II\,II}} \end{array}\right]$$

with the respective off-diagonal matrices ($\mathbf{A}_{\mathrm{I\,II}}$ in the first case and $\mathbf{A}_{\mathrm{II\,I}}$ in the second) consisting of proportional rows.

In such a case, the crude methods of correlating each item with national income would give a satisfactory prediction of all requirements.[1] Such methods are commonly used; perhaps we should be surprised that the results are not worse than they are. However, if we are considering a shift from a peace economy to a war economy, the shift of final demand is such as to belie the hypothesis of a single degree of freedom. An ideal Leontief system can then hope to do better by showing how more steel and less car washers will be required—even at the same level of employment and "aggregate income."

It may turn out to be the case, however, that the C's can be adequately described by a few degrees of freedom m, where m is greater than 1 but a good deal less than n. Thus, if every economy could be regarded as a quantitative mixture of a peace and war economy, we should have $m = 2$ degrees of freedom. To the extent that we can approximate reality by $m < n$ degrees of freedom, our $n \times n$ requirements matrix $[A_{ij}]$ can have its n columns combined into a fewer number of columns, giving us finally a new rectangular n-row–m-column requirement matrix.[2] Specifically, this means that aluminum requirements can be estimated once you specify the degree to which the economy is to be war, high-consumption, autarchic, etc., for only a few descriptive variables.

But how do we get quantitative knowledge of the new A's? If we had the full Leontief detail, we could take weighted sums of the separate columns of A's. But the whole point of this discussion is to see how far we can go without having detailed industrial breakdowns. In that case we must hope that history or nature has presented us with statistical data covering identifiable differences in the m factors determining the consumption composition: by multiple regression or other techniques, we can hope to estimate the quantitative variation in any given X_i as each $(G_1, \ldots, G_m)$ varies. Given enough sufficiently varied observations, we could take the regression of each of n X's on the collection of m G's. The mn regression coefficients would be our estimates of the elements of the $n \times m$ matrix $[\mathbf{Ar}]$ of the last footnote. A limiting case is when the C_j really are independent, in which case, by taking the regression of each X_i on all the C_j, we estimate the n^2 elements of the matrix

$$[A_{ij}] = (\mathbf{I} - \mathbf{a})^{-1}$$

Having done this we could actually estimate the input coefficients them-

[1] If we identify the single degree of freedom with a general variable G_1, we can hope by regression methods to observe for each X_i terms of the form $\Sigma A_{ik} r_{k1}$, where the r's are constants depending on G_1 alone.

[2] In the simplest linear case in which $\mathbf{C} = \mathbf{rG}$, $\mathbf{r}$ being an $n \times m$ matrix, we get $\mathbf{X} = \mathbf{AC} = (\mathbf{Ar})\mathbf{G}$ and $[\mathbf{Ar}]$ is the new $n \times m$ matrix. If $\mathbf{C}$ depends nonlinearly on $\mathbf{G}$, the r's will be functions of the G's rather than constants.

selves by $\mathbf{I} - [A_{ij}]^{-1}$. But remember that this approach would require that the C's vary widely and independently and that we have more observations than C's. The first condition might require a fine commodity breakdown, which would in turn require, via the second condition, an extremely large number of observations.

Since nature rarely performs the ideal controlled experiments that we are interested in, such estimation will often be crude or impossible. In that case we may be thrown back on hypothetical thought experiments, from the imagined results of which rough estimates are formed. This is what is meant by "using judgment." The results can be very good or very bad, depending upon how good the judgment turns out to be. (It may be mentioned that one possible aid to good judgment in determining, let us say, what war rather than peace would mean for 1960 requirements is the use of a Leontief system worked out in detail for some other time or place. There may be enough rough correspondence and continuity in the structure of somewhat similar economies to make such information better than useless.)

This third general way of avoiding detailed breakdowns of industries has formally little relation to the problem of consolidating industries. In the absence of the special a's needed for the first two methods, we can think of ourselves as aggregating or consolidating the C's into simple patterns. But this does not require or permit us to consolidate the industries into a new $m \times m$ pattern, which is then to be run through the usual mill of Leontief algebra.[1]

10-5. LEONTIEF'S CLOSED-END SYSTEM

During World War II and since, Leontief has worked with an "open-end system" rather than his earlier "closed system."[2] Thus far we have followed him in this because the closed system is much harder to understand. In it we become Malthusian and speak of the consumption by people (e.g., of food) as really being much like the consumption by machines or horses. Just as fertilizer is needed to produce corn, silk shirts are thought of as being needed to "produce" the labor of households. Of course, to Malthus and others who believed in a minimum-of-subsistence theory of wages, the consumption items were as physiologically necessary as is hay to a horse or oil to a diesel engine. Originally, Leontief adopted the convention of regarding consumption (even

[1] Dantzig, Koopmans, Morgenstern, Hoffenberg and Evans, Leontief, and many others have been studying the problem of aggregation; and all who work with statistical data are implicitly giving solutions every day of their lives.

[2] Cf. the new chapters in his second edition of *The Structure of American Economy, 1919–1939* (Oxford University Press, New York, 1951), with the old.

of luxuries) as the input requirements of the household output of labor. Consequently, C_i becomes $a_{i0}X_0$, where a_{i0} is treated just like any other a_{ij} constant.

Actually, the ratios a_{i0} are determined by people's psychological propensities and habits with respect to spending extra income, as Leontief recognized. A completely self-contained system, with no degrees of freedom left, must, in order to exist, satisfy special balance relations, a commodity-by-commodity Say's law. Hence the new a_{i0}'s must be dependent on all the old technical a_{ij}'s.[1] The full significance of a closed system can only be appreciated in connection with the questions of (1) dynamic growth of a system that plows back or accumulates part of its consumable output and, (2) the capability of a system to stand still and reproduce itself at the same time that households are experiencing specified levels of consumption. This must be deferred to the later discussion of von Neumann and other dynamic models.[2]

The deeper problem raised by the closed system has nothing to do with the convention adopted for consumption. It has to do with the question of what we wish to predict. If we wish to predict changes in the level of employment brought about by changes in any autonomous variable—such as foreign investment, tax rates, or investment—we may have to recognize that the C's of the open-end Leontief model are not constants that will stay at prescribed levels, but rather are variables related to changes in employment X_0. In that case, all the familiar elements of the income-multiplier analysis must apply, and marginal propensities to consume—which we may designate as proportional to a_{i0} or to any other set of constants—can be shown to enter into the final results *much as if* we gave them the earlier Leontief-Malthus interpretation. So the closed system is of real interest.

Instead of trying to show that consumers' psychological propensities are like industries' technological input requirements, let us reverse the analogy. We can think of a Leontief dollar tableau as giving the break-

[1] Leontief expressed this by the necessary vanishing of the augmented determinant

$$\Delta = \left|\begin{array}{c|cccc} 1 & -a_{01} & \cdots & & -a_{0n} \\ \hline -a_{10} & 1-a_{11} & \cdots & & -a_{1n} \\ \cdot & \cdot & & & \cdot \\ \cdot & \cdot & & & \cdot \\ \cdot & \cdot & & & \cdot \\ -a_{n0} & -a_{n1} & \cdots & & 1-a_{nn} \end{array}\right|$$

and the resulting indeterminacy of scale of the homogeneous system. This is a pure convention and bookkeeping tautology as applied to any table such as Table 10-1, where saving and algebraic growth are ignored.

[2] See Chaps. 11 and 12.

down in dollar expenditure patterns of all industries; and it is natural to adjoin to it the column showing the percentage breakdown (marginal or average) of the household consumption dollar.[1]

If the C's are variables (or if parts of the C's are "induced" terms) we may conveniently label them with the unused symbols $\bar{x}_{i0}$ and put them on the left-hand side of (10-3*a*). But that will leave zeros on the right-hand side and a zero solution to the equations for a stable multiplier system.[2] To register a change in the system we must put on the right-hand side some "autonomous multiplicand" terms $(E_0, E_1, \ldots, E_n)$, e.g., a dollar of government or private spending on shoes, etc. Then we solve the system for the resulting changes in X's and in labor. Mathematically we have the linear relations like (10-3*a*) and (10-4*b*):

$$\begin{aligned} \bar{X}_i - \sum_0^n \bar{a}_{ij}\bar{X}_j &= E_i \qquad i = 0, 1, \ldots, n \\ \bar{X}_i &= \sum_0^n \bar{\alpha}_{ij}E_j \qquad i = 0, 1, \ldots, n \end{aligned} \tag{10-11}$$

where the $\bar{\alpha}$'s differ from the A's of (10-4*a*) by virtue of the fact that the zeroth rows and columns have been adjoined to the system.

How do the $\bar{\alpha}$'s, defined when the $\bar{C}$'s are variable, differ from the $\bar{A}$'s defined for fixed right-hand constant $\bar{C}$ coefficients? This is an important formal question. If we know the answer to it, we shall be able to predict the effect of modifying (10-3*a*) and (10-4*a*) in the opposite direction; i.e., we shall know what the effect on the A's would be if we were to hold some variable such as X_1 strictly constant in solving the remaining $n - 1$ equations.[3] (Thus, X_1 might be exportable food, held constant by rationing and foreign-exchange controls.)

The relation between A's with an excluded variable to α's with that variable included can be simply stated in terms of determinants and of inverse matrices. Since the presence or absence of bars does not matter,

[1] There is one important difference. Zero-profit industries spend all their money. But if a multiplier income analysis is not to explode, some portion of consumer dollars must be not-spent; i.e., $\bar{a}_{10} + \bar{a}_{20} + \cdots + \bar{a}_{n0} = (1 -$ positive "leakage," or saving terms), for empirical reasons and for "stability."

[2] In the strict closed system of Malthusian type, a nonzero solution of indeterminate scale exists by virtue of the relation

$$\bar{a}_{0j} + \bar{a}_{1j} + \cdots + \bar{a}_{nj} \equiv 1 \qquad \text{for } j = 0, 1, \ldots, n$$

Incidentally, the system is unlike Malthus' system in its neglect of land and primary nonproducible inorganic factors.

[3] Harlan M. Smith in chap. 6 of T. C. Koopmans (ed.), *Activity Analysis of Production and Allocation*, John Wiley & Sons, Inc., New York, 1951, discusses this type of question in detail, as does Leontief in part 4 of his second edition.

they are eliminated. The result is as follows, and will be more easily remembered if we agree to write A_{ij} as $\alpha_{ij \cdot 00}$.

$$\begin{aligned} A_{ij} &= \alpha_{ij \cdot 00} = \alpha_{ij} - \frac{\alpha_{i0}\alpha_{0j}}{\alpha_{00}} \\ \alpha_{ij} &= \alpha_{ij \cdot 00} + \frac{\alpha_{i0}\alpha_{0j}}{\alpha_{00}} \qquad i, j, = 1, 2, \ldots, n \end{aligned} \tag{10-12}$$

The second of these equivalent forms is easiest to grasp intuitively. If we hold X_0 constant and do not let it give rise to secondary induced spending, we shall get a smaller multiplier than if we let X_0 grow and induce further growth in all the X's. By how much additional will any X_i grow if we let X_0 grow? By the amount that X_0 will grow, α_{0j}, times the growth in i associated with each unit increase in X_0, which is α_{i0}/α_{00}.[1]

Much argumentation has taken place as to (1) what should be autonomous and what induced in an income model, and (2) what should be "multiplicand" and what "multiplier." The answer must depend upon the question asked, e.g., what change do you assume takes place, and the factual hypotheses concerning empirical invariances.

10-6. SUBSTITUTABILITY IN LEONTIEF SYSTEMS

Since Leontief works with so-called fixed coefficients of production, a_{ij}, it is usually thought that he must rule out the possibility of substitution,

[1] In partitioned form we can write the following relationship for the enlarged closed system:

$$\left[\begin{array}{c|c} 1 & -a_{01} \quad -a_{02} \quad \cdots \quad -a_{0n} \\ \hline -a_{10} & \\ -a_{20} & \\ \vdots & \mathbf{I} - \mathbf{a} \\ -a_{n0} & \end{array}\right] \left[\begin{array}{ccccc} \alpha_{00} & \alpha_{01} & \alpha_{02} & \cdots & \alpha_{0n} \\ \hline \alpha_{10} & \alpha_{11} & \alpha_{12} & \cdots & \alpha_{1n} \\ \alpha_{20} & \alpha_{21} & \alpha_{22} & \cdots & \alpha_{2n} \\ \vdots & \vdots & & & \vdots \\ \alpha_{n0} & \alpha_{n1} & \alpha_{n2} & \cdots & \alpha_{nn} \end{array}\right] = \left[\begin{array}{c|cccc} 1 & 0 & 0 & \cdots & 0 \\ \hline 0 & 1 & 0 & \cdots & 0 \\ 0 & 0 & 1 & \cdots & 0 \\ \vdots & \vdots & \vdots & & \vdots \\ 0 & 0 & 0 & \cdots & 1 \end{array}\right]$$

or

$$\begin{bmatrix} 1 & -\mathbf{a}_0 \\ -\mathbf{a}^0 & \mathbf{I} - \mathbf{a} \end{bmatrix} \begin{bmatrix} \alpha_{00} & \mathbf{P}' \\ \mathbf{Q} & \mathbf{R} \end{bmatrix} = \begin{bmatrix} 1 & \mathbf{0}' \\ \mathbf{0} & \mathbf{I} \end{bmatrix}$$

Equations (10-12) follow from carrying out the matrix multiplication in partitioned form and from straightforward calculation using the fact that the $\mathbf{A}$ matrix is the inverse of $\mathbf{I} - \mathbf{a}$.

as assumed in the classical Clark-Wicksteed-Walras theory of production and general equilibrium. However, it can be shown[1] that in a one-primary-factor Leontief system all the Leontief theory is compatible with the general case of substitutability. Even if substitution is *physically* possible, it will be ruled out on *economic* grounds.

The conclusion and reasoning can be briefly sketched. But first it may be useful to indicate in an intuitive way why all this is so. Suppose we raise wages in a Leontief system. What will happen to employment in any line of activity? An economist will be tempted to answer: If substitutability is possible, other factors (such as machinery, etc.) will

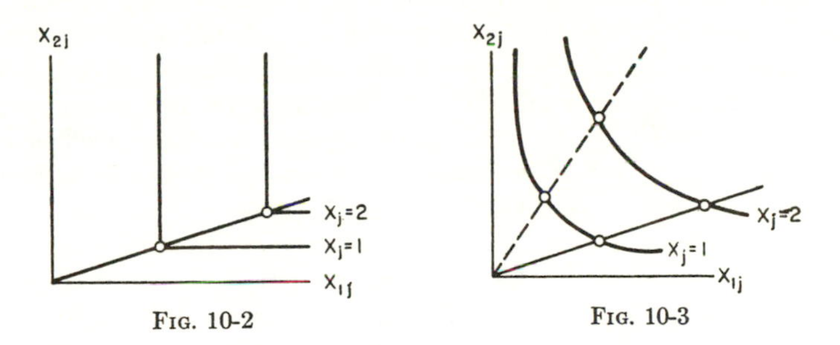

Fig. 10-2 Fig. 10-3

be substituted for labor and employment will drop. Remember, however, that everything is congealed labor in a Leontief system. Hence, when you raise wages you are also raising the cost of machines by the same proportion. Even if technical substitutability is possible, there will be no actual substitution because there will be no change in the relative prices of any factors. The same Leontief a's will continue to be observed.

Arguments like the above are to be found in wage discussions of a century ago. As applied to the real world, they are not conclusive, since an increase in money wages may or may not mean a change in real wages, and may or may not mean a change in labor costs relative to the cost of short-run fixed resources, and may or may not be followed by a change in interest rates and in the dynamic temporal pattern of production. But as applied to a simple Leontief system, which abstracts from time and technological change and which assumes that everything but labor is currently reproducible at constant costs, the argument is ironclad.

Potential substitutability need never give rise to actual substitution because there can never be any relative price changes no matter how extreme are the changes in consumption.

Figure 10-2 shows the production function for the Leontief fixed coefficients case. Figure 10-3 shows the general classical case of a production

[1] See Koopmans (ed.), *op. cit.*, chap 7 and chap. 10, p. 171; also chaps. 8 and 9 by Koopmans and Arrow.

function subject to the usual law of constant returns to scale and diminishing returns to changes in proportions. Knowledge of a single contour, perhaps that corresponding to unit output, will tell us how any output can be produced. The unit-output contour for Industry j can directly give us one implicit relation between the technical coefficients $(a_{0j},a_{1j}, \ldots ,a_{nj})$. Instead of having a single jth column in the Leontief matrix, we have in the jth column a basket of possible technologies, a menu, so to speak, from which we can choose. In Fig. 10-3, we can pick a's as given by the solid ray through the origin, or if we wish we can use a's as given along the dotted ray.

We shall show the following: If the circled points along the solid ray are observed for one set of consumption C's, then no matter how we change the C's we shall never observe any other techniques (such as indicated along the broken line). This means that we shall never be able, from the observed facts alone, to infer whether the fixed coefficients of Fig. 10-2 are really true or whether the variable substitutable coefficients of Fig. 10-3 are true. Nor does it matter.[1]

The cited *Activity Analysis of Production and Allocation* contains a mathematical proof of the above assertions based on marginal-productivity analysis.[2] This need not be repeated here. Within the framework of Leontief algebra, we can see why an alternative set of coefficients, such as $(a_{01}^*,a_{11}^*, \ldots ,a_{n1}^*)$ in Fig. 10-3, can never turn out to be distinctly preferable to the a's observed in the initial situation $(a_{01}, a_{11}, \ldots ,a_{n1})$.

To see this let us examine our price equations (10-8). Using a's rather than a^*'s, we have determined our $(P_1/P_0, \ldots ,P_n/P_0)$. Now let us replace our a's in the first industry's cost-of-production equation by the alternative a^*'s. Let us try the old P's for all the other goods and see whether our new cost of production is greater or less than the old. What must the answer be? Certainly in the original setup, with the original C's, one or the other is optimal; for definiteness, let us suppose that the a's gave a lower P_1/P_0 than did the a^*'s, so that we definitely rejected the

[1] However, if we were to change some of the other a's by invention, or if we were to set one of the other factors as a primary factor, not reproducible at constant costs in the period under discussion, then there would be a difference in what we would observe in the two alternative technological situations of Figs. 10-2 and 10-3.

[2] Equations (5), p. 144, represent the crucial conditions. These show that the *proportions* of the inputs in *all* industries are fully determined by the marginal-productivity equilibrium conditions, *independently of all scales of consumption and production.* (Note that a_{ij} in the present treatment has the meaning of a_{ji} in earlier treatments.) See chapters by Koopmans and Arrow for a proof based on linear-programming concepts rather than on marginal-productivity partial derivatives. (Strictly speaking, there may be cases of indifference where substitution could take place, but need not do so.)

latter. (Here "we" means the "impersonal invisible hand of perfect competition," which can be expected to reach optimal configurations in this ideal setup of constant-returns-to-scale and statical conditions.)

Now, let us change the C's drastically. Again we can make the same computation for Eqs. (10-8). But notice that these price, or unit-cost-of-production, equations are unchanged by any change in C's. There are no extensive scale quantities, such as C or X, in these equations. Whatever was definitely true in the first place remains definitely true. Again the a^*'s must be rejected as too costly; and if any entrepreneur is too stupid to see this, the competitive law of survival of the fittest will eliminate him. It follows by induction that the same a's must be observed for every industry, regardless of the change in the bill of goods.[1]

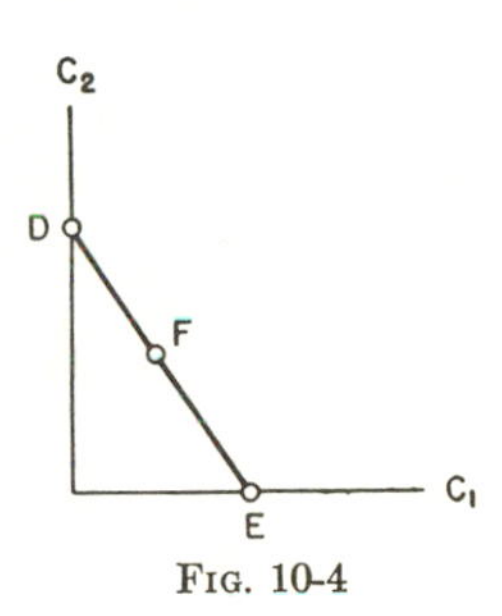

FIG. 10-4

Figure 10-4 gives a picturization of the nonsubstitution theorem for the two-industry case. Suppose only C_2 were to be prescribed with $C_1 = 0$, then we could pick best methods for every industry so as to yield the most C_2 for a given unit of labor. The point D shows the maximum of C_2 producible by our first set of methods. Similarly, if we want only C_1, we can pick a best set of methods

[1] As mentioned before, there is the possibility of "ties," which requires an inessential qualification. A deeper problem will bother the careful reader. In the above rejection of the a^*'s, we used the old a prices for all the other goods. What if we had used the new prices for all the other goods resulting from choice of the a^*'s? Would we get the same answer? Or might it now turn out that the a^*'s appeared to give better P_1/P_0 than the a's? If this last should be the case, it would be catastrophic to our proof; it would then be possible for the a's to be both better and worse than the a^*'s, depending upon our choice of other P's.

Fortunately, there can be no contradiction. We can use the last $n - 1$ equations of (10-8) to solve for the $n - 1$ variables $(P_2/P_0, \ldots, P_n/P_0)$ as linear functions of the parameter P_1/P_0. Every coefficient in these functions will be independent of the choice of a^* or a coefficients in the first equation and will actually turn out to be positive coefficients for any observable Leontief system, as can easily be shown from our algebraic discussions of the properties of such systems (see pp. 253–257). Now write the first equation in the a^* or a form and substitute for the remaining P's the above linear expressions. This will give us final linear expressions in P_1/P_0 alone. If the a^* expression is definitely worse than the a expression, in the sense of giving higher P_1/P_0 and hence higher remaining P's, then it must continue to be so regardless of the C's, which nowhere enter (10-8). This proves our theorem for the case in which only Industry 1 has a choice between a and a^*. If every industry j has a choice between a column a_i and a_i^*, we must repeat the argument and show that for the definitely optimal set $(a_1, \ldots, a_n)$ *every P* will be less than for any set involving one or more a^*'s. In terms of linear programming, we can express all this by saying that the dual problem in the P's has activities that "dominate" other activities, regardless of the C's.

or a best Leontief matrix so as to reach a maximum of C_1, as shown at E. The second set of methods might conceivably be different from the first; but what we can prove is that the first set of methods will necessarily get us to the point E.

Now it is clear that we can end up anywhere on the straight line between D and E. How? By dividing up our unit of labor between the above-described sets of methods in a 50:50 ratio, 60:40, 90:10, 10:90, etc. At any such intermediate point, say F, we are using both sets of methods simultaneously. If we wish we can define a third Leontief matrix, which is a blend of the two polar matrices, the weighting being that given by the relation of F to D and E. From Eq. (10-6) of our previous analysis, we know that this third simple Leontief system has a straight-line consumption-possibility schedule going through F. If it didn't coincide with DE, there would be a contradiction to the statement that D and E are each optimal. Therefore it must coincide. Hence we have proved that a single simple Leontief system, the third one mentioned, can without any substitution do as well as anyone can do with substitution, which completes the proof. (It may be noted that the two coexisting systems at F must yield exactly the same prices or they could not continue to coexist.)

This discussion should clarify the reasons why substitution occurred in the linear-programming models of earlier chapters but not in the static input-output system. There are two important differences (in both respects input-output is a special case of the linear-programming setup): absence of joint production and the presence of only one primary factor in the Leontief system. If either assumption is given up, the economic possibility of substitution arises. We can think of a Leontief matrix as giving one or more activities for each industry. If the industries are bound together by flows of intermediate inputs, then we can expect at least one activity from each industry to be in operation. The question is which one, and will the choice change if the final demands change? In the case of the competitive firm the answer was "yes." In the Leontief case, because of the differences mentioned, the answer is "no." Because there is only one scarce factor to be economized, the choice of activities is independent of final demand. If there were two or more scarce factors, activities would have to be chosen to economize most on the one whose fixed supply is most burdened by the desired consumption mix (e.g., land, if agricultural consumption should be heavily stressed). Joint production would have a similar effect—some activities might be ruled out because they produce commodities in proportions too different from final demands. With a shift in demand these activities might become profitable.

10-7. EMPIRICAL-ALGEBRAIC PROPERTIES OF A LEONTIEF SYSTEM

Using matrix terminology, we can readily summarize the Leontief system as follows:

$$\mathbf{X} = \mathbf{aX} + \mathbf{C} \quad \text{or} \quad (\mathbf{I} - \mathbf{a})\mathbf{X} = \mathbf{C} \tag{10-3}$$

$$\mathbf{X} = \mathbf{AC} \quad \text{where } \mathbf{A} = (\mathbf{I} - \mathbf{a})^{-1} \tag{10-4}$$

$$X_0 = \mathbf{a}_0'\mathbf{X} = \mathbf{a}_0'\mathbf{AC} = \mathbf{a}_0'(\mathbf{I} - \mathbf{a})^{-1}\mathbf{C} = \mathbf{A}_0'\mathbf{C} \quad \text{(10-5) and (10-6)}$$

where $\mathbf{A}_0'$ is the row matrix or row vector $[A_{01}, \ldots, A_{0n}]$.

$$\left(\frac{\mathbf{P}}{P_0}\right)' = \left(\frac{\mathbf{P}}{P_0}\right)'\mathbf{a} + \mathbf{a}_0 \quad \text{or} \quad \left(\frac{\mathbf{P}}{P_0}\right)'(\mathbf{I} - \mathbf{a}) = \mathbf{a}_0'$$

$$= \mathbf{a}_0'(\mathbf{I} - \mathbf{a})^{-1} = \mathbf{A}_0' \tag{10-7}$$

Corresponding to the elementary formula for a convergent geometric series $1 + r + r^2 + \cdots = 1/(1 - r)$, we can find "multiplier expansions" to approximate our unknown X's and P's without ever solving simultaneous equations. Thus,

$$\mathbf{X} = (\mathbf{I} - \mathbf{a})^{-1}\mathbf{C} = (\mathbf{I} + \mathbf{a} + \mathbf{a}^2 + \cdots)\mathbf{C}$$

$$= \mathbf{C} + \mathbf{aC} + \mathbf{a}(\mathbf{aC}) + \mathbf{a}(\mathbf{a}^2\mathbf{C}) + \cdots \tag{10-13}$$

gives us the Cornfield-Leontief multiplier process: according to this, we first compute the output requirements of the new **C** itself; then we compute the first-round direct requirements to produce the **C**, which gives us **aC**; then we compute the second-round direct requirements to produce the first-round items; etc. We thus build up a growing total until the terms in the dwindling infinite chain dwindle to negligible proportions. For Leontief a's, the process can be shown to be convergent, and rather rapidly so.

To arrive at final P's as the sum of direct labor congealed in an infinite number of previous stages, we write down the Gaitskell multiplier chain:

$$\left(\frac{\mathbf{P}}{P_0}\right)' = \mathbf{a}_0'(\mathbf{I} - \mathbf{a})^{-1} = \mathbf{a}_0'(\mathbf{I} + \mathbf{a} + \mathbf{a}^2 + \cdots)$$

$$= \mathbf{a}_0' + (\mathbf{a}_0')\mathbf{a} + (\mathbf{a}_0'\mathbf{a})\mathbf{a} + (\mathbf{a}_0'\mathbf{a}^2)\mathbf{a} + \cdots \tag{10-14}$$

Here we interpret total labor cost of a good as the sum of its initial direct-labor cost, plus the direct-labor costs of the inputs it directly uses on the first round, plus the direct-labor costs of the second-round factors needed to produce the first-round factors, etc., until the terms of the infinite series become negligible. This series will be convergent; it is simply another way of looking at the previous multiplier series. Needless to say, the rounds of which we speak do not take place in calendar time, with the second round following the first, as in Robertson-Metzler dynamical Keynesian systems. Artificial computational time is involved, and if we insist on giving a calendar-time interpretation we must think of the Gaitskell process as going *backward* in time and the Cornfield process as showing how much production must be started many periods back if we are to meet the new consumption targets today.[1]

Intuitively, we should expect a decrease in *any* nonlabor requirement a_{ij} to lower all the X's and a decrease in any a_{0j} to lower the X_0 required. We should also expect such technological improvements to lower all (P/P_0)'s. This turns out to be the case, since it is easily verified that

$$\frac{\partial A_{ij}}{\partial a_{rs}} = A_{ir}A_{sj} \geq 0$$

To see why this is so, we need only reason that if the direct requirement a_{rs} increases infinitesimally, the total requirement of r must increase by enough to produce A_{sj}, and therefore the total requirement of i must increase by a factor $A_{ir}A_{sj}$. All such derivatives are nonnegative. It follows that decreasing the requirements of any well-behaved Leontief system leaves it well behaved. But increasing requirements will eventually render it incapable of producing any positive net consumption at all and will give rise to divergent multipliers, to negative values of variables, and to other anomalies that could never be observed in nature by a Leontief statistician.

How do we know that a Leontief system will be well behaved in the sense of having convergent multipliers, unique solutions, and nonnegative consumption, inputs, and outputs? It turns out that our system is of the beautifully simple type, studied by the mathematicians Minkowski, Frobenius, and Markoff, with well-behaved properties, which can be simply analyzed. It further turns out that the Leontief system has close affinities to Keynesian multicountry income models and to the Hicksian model of stability of multiple markets.

10-8. INDECOMPOSABLE AND DECOMPOSABLE GROUPS OF INDUSTRIES

Before stating a general theorem on observable Leontief systems, we must note some possible arrangements of industries. (1) Every industry

[1] Jerome Cornfield's BLS memo seems never to have been published. But see W. W. Leontief, "Interindustrial Relationships," in *Proceedings of a Symposium on Large Scale Digital Calculating Machinery*, Harvard University Press, Cambridge, Mass., 1948; F. Waugh, "Inversion of the Leontief Matrix by Power Series," *Econometrica*, **18**:142 (April, 1950); L. A. Metzler, "Stability of Multiple Markets: The Hicks Conditions," *Econometrica*, **13**:277 (October, 1945). R. M. Goodwin, J. S. Chipman, R. M. Solow, and others have recently published in this field.

might directly use some positive input of *every* other industry. Failing this, (2) every industry might *indirectly* use some positive input of every other industry, if not buying directly from it, at least buying from intermediary industries which buy directly or indirectly from it—the chain of intermediary industries consisting of 1, 2, . . . , up to $n - 1$ industries.

If sales could be calculated to the last dollar, it is probable that any actual economy would have the above so-called "indecomposable" property, in which all pairs of industries are interlocked directly or indirectly in a two-way fashion.

This, however, is in contrast with the simple Austrian structure of production, in which an industry will directly or indirectly sell to another but not buy from it. Thus, a group of industries may have the property that some of its pairs are only in a one-way (direct or indirect) connection: Frobenius, the mathematician to whom most of these concepts are due, called this a "decomposable" group.[1]

According to Air Force and BLS computations, we can renumber American industries so that the structure is almost triangular, so that Industry 20 buys from no industry with index greater than 20, etc. Thus, treating small a's as zero, we could approximate the facts by a decomposable model.

Strictly speaking, though, zeros and small positive numbers are not the same thing. And the concepts of decomposability and indecomposability are purely qualitative and nonquantitative: they depend only on the pattern of positive and zero a's, and not at all on the size of the positive a's. To the empirical purist, a theorem stated for indecomposable systems may therefore be of greatest interest—which is a blessing, since such a theorem is most briefly stated. It can then be generalized to include decomposable systems as well.

In all that follows we shall be considering an observed or observable Leontief system with every x_{ij} nonnegative. The system is indecomposable so that each pair of industries is directly or indirectly in a two-way connection. By "observable" we mean a system which is productive in the sense that it can produce positive net outputs. An unproductive system in this sense could produce net output only by running down preexisting stocks.

THEOREM. *Any indecomposable observable Leontief system has each and every one of the following strictly equivalent properties; i.e., each one implies all the rest.*

[1] An extreme case of decomposability is that in which one subgroup of industries is *entirely* independent of another subgroup, with zero buying and selling between industries from the separate subgroups. This is called a "completely decomposable group," and we may obviously handle it by treating *its separate parts* entirely separately.

1a. *At least one a_{0j} is positive.*

b. All price ratios P_j/P_0 are positive.

2. *At least one price ratio P_j/P_0 is positive.*

3a. *At least one bill of final demand is producible.*

b. Any bill of final demand is producible, provided only that there is sufficient labor available.

4a. *There is at least one set of measurement units in which no row sum (column sum) is greater than unity and at least one row sum (column sum) is less than unity.*

b. In the special units in which each $P_j/P_0 = 1$, $\bar{a}_{1j} + \bar{a}_{2j} + \cdots + \bar{a}_{nj} \leq 1$, with the inequality holding for at least one index j.

c. In the special units in which each $X_j = 1$, $\hat{a}_{i1} + \hat{a}_{i2} + \cdots + \hat{a}_{in} \leq 1$, with the inequality holding for at least one index i.

5a. *The characteristic roots of the matrix $[a_{ij}]$, which are invariant to any change in units, are all less than 1 in absolute value, so that the "multiplier series" $\mathbf{I} + \mathbf{a} + \mathbf{a}^2 + \cdots$ converges to $(\mathbf{I} - \mathbf{a})^{-1} = (\mathbf{A}_{ij})$.*

b. $\mathbf{I} + \bar{\mathbf{a}} + \bar{\mathbf{a}}^2 + \cdots = (\mathbf{I} - \bar{\mathbf{a}})^{-1}$.

c. $\mathbf{I} + \hat{\mathbf{a}} + \hat{\mathbf{a}}^2 + \cdots = (\mathbf{I} - \hat{\mathbf{a}})^{-1}$.

d. $\mathbf{a}^\infty = \bar{\mathbf{a}}^\infty = \hat{\mathbf{a}}^\infty = 0$.

6. *All the elements of $(\mathbf{I} - \mathbf{a})^{-1}$ are positive.*

7. *$(\mathbf{I} - \theta\mathbf{a})$ is nonsingular for all θ, $0 \leq \theta \leq 1$.*

8. *Decreasing any requirement coefficient a_{ij} decreases every element of $\mathbf{A}$.*

9a. *$\mathbf{I} - \mathbf{a}$ is Hicksian; i.e., all its principal minors are positive.*

b. All the indecomposable subsystems of a have all the properties of $\mathbf{a}$.

If we start not from an a matrix but from an observed tableau $\bar{x}_{ij}$ and define the usual $\bar{a}_{ij}$, the above equivalences still hold for these coefficients.

Note that if knowledge of a's comes not from an observed tableau but rather from engineering estimates of technologies of separate industries, the result might *not* be a well-behaved viable Leontief system capable of steadily producing consumption goods. Thus, until the crucial experiments were made, no one knew whether a self-sustaining chain reaction of atomic fission was possible. Similarly, processes that require us to use up finite stocks of exhaustible materials may not be capable of a sustained steady-state reaction.

We can now reconcile our formulation with Leontief's practice of "netting out" all intrafirm and intraindustry transactions so as to make all his diagonal terms $a_{ii} = 0$. So long as we stick to statics, this is pure convention and either procedure is perfectly adequate. If 10 per cent of coal output is used within the mines itself, Leontief works with an output Z, which is only $1 - 0.1$ of our X. Now if 18 units of fertilizer is needed to produce 1 of our units of coal, then 18 units is needed to produce 0.9 of a Leontief unit; for a full Leontief unit, therefore, 20 units is needed, which shows how his a's must be related to ours so as to lead to the same substantive results.

Mathematically, Leontief outputs Z_i are given by

$$Z_i = (1 - a_{ii})X_i$$

and (10-3) is written

$$Z_i = C_i + \sum_j \left(\frac{a_{ij}}{1 - a_{jj}}\right)(1 - a_{jj})X_j = C_i + \sum_j \left(\frac{a_{ij}}{1 - a_{jj}}\right) Z_j$$

where the j is summed over all variables except i and where the fractional expressions are the Leontief a's, with the diagonal ones zero. Note that our analysis has shown that every well-behaved Leontief system will have positive $1 - a_{ii}$, so the transformations are always possible.

When we leave the realm of statics and assume a time interval between inputs and output, the problem ceases to be one of convention. That corn is needed today to produce next year's corn may be literally as true as that fertilizer is needed today to produce next year's corn. The proper a_{ii}'s then become questions of brute fact, and for this reason we have adhered to the more general formulation.

A literary explanation of the meaning of the above equivalences is omitted here at this time. References to the mathematical and economic literature can be found in a paper by Solow.[1]

That 9*a* implies 6 was known to Minkowski and in the economic literature was early proved by Mosak in connection with Hicks-type multiple markets; the converse is due to Hawkins and Simon. Metzler, in connection with Keynesian income models, proved the relation between 4*b* and 5*b* and many similar results. Solow integrated the related theorems and showed how the Frobenius concepts of "indecomposability" and "acyclicity" affect the inequalities and dynamic iterations.

10-8-1. Decomposable Systems. For completeness, we briefly treat the decomposable case. Even though they agree with intuition, the results are complex to state, and this section may be skipped without great loss.

If the industries are not all in two-way communication with each other, it may happen that we can find distinct subgroups of industries which are never in direct or indirect contact with each other either as suppliers or demanders of inputs. Such "completely separable" subgroups can clearly be treated completely independently of each other, and of course there are all zero repercussions in one subgroup resulting from a change in consumption or input coefficients in another independent subgroup

[1] *Econometrica*, **21**: 29–46 (January, 1952).

In matrix terms we can, in the case of our completely separable subgroups, renumber industries so that

$$\mathbf{a} = \begin{bmatrix} \mathbf{a}_1 & 0 & 0 & \dots & 0 \\ 0 & \mathbf{a}_2 & 0 & \dots & 0 \\ \dots & \dots & \dots & \dots & \dots \\ 0 & 0 & 0 & \dots & \mathbf{a}_m \end{bmatrix} \qquad \mathbf{A} = \begin{bmatrix} \mathbf{A}_1 & 0 & \dots & 0 \\ 0 & \mathbf{A}_2 & \dots & 0 \\ \dots & \dots & \dots & \dots \\ 0 & 0 & \dots & \mathbf{A}_m \end{bmatrix}$$

where each $\mathbf{a}_k$ and $\mathbf{A}_k = (\mathbf{I} - \mathbf{a}_k)^{-1}$ is square and consists of nonnegative elements.

The completely separable case is very simple. More interesting is the situation within each $\mathbf{a}_k$, to which all our remaining remarks apply. We can concentrate on $\mathbf{a}_1$ and suppose that it has no industries which cannot be split further into completely separable subsystems. It follows that every industry is, directly or indirectly, in some kind of connection with every other industry. Thus, it might be the case that Industry 1 both buys from and sells to Industry 2; and Industry 1 might sell to Industry 3 but not buy from it; similarly, Industry 4 might buy from Industry 1 but not sell to it; Industry 5, on the other hand, might neither buy nor sell from Industry 1, but might be indirectly linked to Industry 1 by virtue of the fact that it does have transactions with either Industry 2 or 3 or 4.

Of course, the nature of the linkages between the industries might be such that we do have the already-described case of an indecomposable system—with each industry in direct or indirect two-way contact with every other industry. The interesting intermediate case for present discussion is that in which (1) there is always some linkage between every pair of industries so that complete separability is ruled out, but (2) not all the linkages are two-way.

In this case, we can rearrange our industries so that our $\mathbf{a}_1$, which is now called $\mathbf{a}$ for simplicity, can be written as

$$\mathbf{a} = \begin{bmatrix} \mathbf{h}_1 & * & \dots & * \\ 0 & \mathbf{h}_2 & \dots & * \\ \dots & \dots & \dots & \dots \\ 0 & 0 & \dots & \mathbf{h}_m \end{bmatrix}$$

where each diagonal $\mathbf{h}_i$ is an indecomposable submatrix, where below the diagonals are only zeros, and where above the diagonals are blocks of nonnegative elements with at least one positive element in each column. There may be some zeros above the diagonal, but not all can be zero or the system would be completely separable.

The industries in the first subgroup $\mathbf{h}_1$ buy only from each other; at least one does sell to some other industries. The industries in the second subgroup h_2 buy and sell to each other; at least one buys from an industry in the first block, and at least one sells to the remaining industries. The industries in the third subblock buy and sell to each other; at least one such industry buys from the "earlier" blocks 1 or 2, and at least one sells to a "later" block. Note that the position of the first block is unique: in a decomposable but not separable system, only one set of industries can fail to buy from any other set of industries. However, the relative positions of $\mathbf{h}_2$ and $\mathbf{h}_3$, or of the remaining $\mathbf{h}_i$'s, may be arbitrary, as the following example illustrates:

$$\begin{bmatrix} 0.1 & 0.1 & 0.1 & 0 & 0.1 \\ 0.1 & 0.1 & 0 & 0 & 0 \\ 0 & 0 & 0.1 & 0.1 & 0 \\ 0 & 0 & 0.1 & 0.1 & 0 \\ 0 & 0 & 0 & 0 & 0.1 \end{bmatrix} \quad \text{or} \quad \begin{bmatrix} 0.1 & 0.1 & 0.1 & 0.1 & 0 \\ 0.1 & 0.1 & 0 & 0 & 0 \\ 0 & 0 & 0.1 & 0 & 0 \\ 0 & 0 & 0 & 0.1 & 0.1 \\ 0 & 0 & 0 & 0.1 & 0.1 \end{bmatrix}$$

Since no earlier block of industries ever requires any input from later industries, it follows that an increase in an earlier C will not cause an increase in a later X. Hence, the A matrix must have all zeros below the diagonal.

It is clear that an increased C for any industry in any subgroup will have to increase X's in *all* other industries of the same subgroup. This is because all members of each subgroup are in two-way connection with each other. It follows that the A's in the diagonal blocks are all distinctly positive.

It is also clear that every industry requires some amount of input from at least one of the first subgroup's industries. It follows that an increase in *any* C will require some extra output from one of the first industries. And since each of the first group of industries requires (directly or indirectly) all of the other, it follows that an increase in *any* C whatsoever will increase every X in the first subgroup. Hence *all* the A's in the first row of blocks are definitely positive.

Beyond this, we can say nothing definite about the sign of any other A. Unfortunately, we cannot be sure that the $\mathbf{h}_2$ industries are really "earlier" than the $\mathbf{h}_3$ industries. Our previous numerical example shows that $\mathbf{h}_2$ may really be "nonlater" than $\mathbf{h}_3$; and at the same time $\mathbf{h}_3$ may be "nonlater" than $\mathbf{h}_2$. In such ambiguous cases, an increase in a C in either group will have zero repercussions in the other group. Thus, the above-diagonal A's may be zero rather than definitely positive; in no observable Leontief system can an A be actually negative.

We can make one definite statement about the above-diagonal A's. In any particular block above the diagonal, there must either be all zeros or all positive numbers; no mixtures of zeros and positive numbers is admissible. For suppose that an increased C of any industry, say $C_{j'}$, of the jth block, were to increase the X of at least one industry in $\mathbf{h}_k$. Call it $X_{k'}$. Then all the industries in $\mathbf{h}_k$, which by definition are in two-way connection with each other, must be needed for the production of $X_{k'}$ and so all totals must rise. This shows that all the A's in the column pertaining to the industry whose C has increased must be of the same algebraic sign.

But exactly the same sign as the above must hold for all changes induced by a C of any other industry in the jth block. This is because such a change necessarily increases *all the X's* in that jth block, including X_j. But an increase in X_j will have the same qualitative effect on $X_{k'}$ as did an increase in $C_{j'}$, which proves that every element in the (k,j) block is positive.

We may summarize the form of $\mathbf{A}$ corresponding to the earlier given general decomposable but nonseparable $\mathbf{a}$. Thus,

$$\mathbf{A} = \begin{bmatrix} + & + & \cdots & + \\ 0 & + & \cdots & + \text{ or } 0 \\ \hline & & & \\ 0 & 0 & \cdots & + \end{bmatrix}$$

where plus or zero in a block above the diagonal means all pluses or all zeros in that block.

One final remark. In Leontief's open-end system, with labor or X_0 the sole primary factor, the $(n+1)^2$ matrix of the system of Eq. (10-3) can be written

$$\left[\begin{array}{c|c} 1 & -\mathbf{a}_0' \\ \hline \mathbf{0} & \mathbf{I} - \mathbf{a} \end{array}\right]$$

Inclusive of labor the system is decomposable, since labor, being non-producible, is definitely "earlier" than all the other inputs.

But if we rule out the Land of Cockaigne, where candy grew on trees and was available without even a need for plucking, then labor will not be completely separable from the rest of the system. This ensures that no goods are free and that all wage costs A_{0j} are distinctly positive.

10-9. A NUMERICAL EXAMPLE

Consider a simple Leontief system consisting of labor and two industries, X_1 manufacturing and X_2 agriculture. Labor is measured in man-hours, manufacturing in its physical units, and agriculture in its physical units. The coefficients of production are given by

$$\begin{bmatrix} \mathbf{a}_0 \\ \hdashline \mathbf{a} \end{bmatrix} = \begin{bmatrix} a_{01} & a_{02} \\ \hdashline a_{11} & a_{12} \\ a_{21} & a_{22} \end{bmatrix} = \begin{bmatrix} 4 & 100 \\ \hdashline 0.1 & 40 \\ 0.01 & 0 \end{bmatrix}$$

Equations (10-3) then give us

$$(1 - 0.1)X_1 - 40X_2 = C_1$$
$$-0.01X_1 + (1 - 0)X_2 = C_2 \qquad \text{or} \qquad X_2 = 0.01X_1 + C_2$$

and using the last equation to substitute in the first gives

$$0.9X_1 = C_1 + 40(0.01X_1 + C_2) = C_1 + 40C_2 + 0.4X_1$$
$$X_1 = 2C_1 + 80C_2$$

and hence

$$X_2 = 0.01(2C + 80C_2) + C_2$$
$$= 0.02C_1 + 1.8C_2$$

Thus, the fundamental A_{ij} coefficients of Eqs. (10-4) are

$$\begin{bmatrix} A_{11} & A_{12} \\ A_{21} & A_{22} \end{bmatrix} = \begin{bmatrix} 2 & 80 \\ 0.02 & 1.8 \end{bmatrix}$$

and the reader can verify that solving Eqs. (10-3) for each C_i separately (with all other C's zero) will in fact lead to each of the columns of this **A** matrix. It is easy to verify that **A** is the so-called "inverse matrix" to the matrix with a's subtracted from 1's in the diagonal (i.e., $\mathbf{I} - \mathbf{a}$).

From Eqs. (10-5) we find

$$A_{01} = 4(2) + 100(0.02) = 10$$
$$A_{02} = 4(80) + 100(1.8) = 500$$

and it is easy to identify these with the price ratios $(P_1/P_0, P_2/P_0)$ satisfying (10-8):

$$10 = 4 + 10(0.1) + 500(0.01)$$
$$500 = 100 + 10(40) + 500(0)$$

The production-possibility schedule for society is given by (10-6):

$$X_0 = 10C_1 + 500C_2$$

If we arbitrarily assume wages to be at \$2 per man-hour and specify definite consumption totals, $C_1 = 50$ million units and $C_2 = 2$ million units, then we can easily calculate the P's and X's, and so arrive at the national accounts giving values. It is easy to verify that for such an economy the Leontief statistician would record the *tableau économique* given in Table 10-4 in dollar terms.

A useful exercise for the reader would be to work backward from this table as the statistician must do. (1) Work out the $\bar{a}$ coefficients and use them to determine the new table when all consumption dollars are spent on each good alone, or on any arbitrary *dollar* combination of the two consumption goods. (2) Being told that $P_0 = \$2$, $P_1 = \$20$, and $P_2 = \$1{,}000$, work out all the physical quantities and the original unbarred a's and all the A's. Show that a wartime shift from agriculture to industry would not affect relative prices.

As a last useful exercise, given the a's, we can use the Cornfield-Leontief multiplier method to compute the requirements for $C_1 = 50$ and $C_2 = 2$. Instead of working in dollars, as the BLS would largely do, we work in physical quantities to facilitate comparison of the approximate method with the exact simultaneous solution of Eqs. (10-3), which yields $X_1 = 260$ and $X_2 = 4.6$. Table 10-5 shows the successive approximations, or rounds. Note that after five rounds X_1 sums up to 231.6835 out of 260 units, X_2 sums up to 3.95435 out of 4.6 units, and X_0 sums up to 1,342.169 out of 1,500 units.

TABLE 10-4. VALUES, IN MILLIONS OF DOLLARS

Seller \ Buying industry	Consumption	Industry	Agriculture	Total sales
Labor	0	2,080	920	3,000
Industry	1,000	520	3,680	5,200
Agriculture	2,000	2,600	0	4,600
Total purchases	3,000	5,200	4,600	

An alternative way of computing labor requirements X_0 would be to approximate to A_{01} and A_{02} by the Gaitskell multiplier chain and then to apply the result to C_1 and C_2. If many different C's were to be prescribed, and if we were interested *only* in labor, this would be a preferable procedure. However, if we were interested in all the X's for many different C's, it would be best to compute by the Cornfield iteration all the A's. Thus we work out a table such as Table 10-5 for $(C_1,C_2) = (1,0)$ and for $(C_1,C_2) = (0,1)$; the extension to the n variable case is obvious.[1]

Need the multiplier chains converge? Looking at the five rounds of X_1, you might have your doubts. But algebraic analysis elsewhere shows that there must be convergence. This can be seen arithmetically as follows: (1) note that convergence for one set of units means convergence for all sets; (2) note that using value units and $\bar{a}$'s gives columns $\bar{a}_{11} + \bar{a}_{21} = 1 - \bar{a}_{01}$ and $\bar{a}_{12} + \bar{a}_{22} = 1 - \bar{a}_{02}$. These right-hand terms represent a positive percentage of "leakage" at each stage: just as sav-

[1] Note that we can easily compute all requirements for any one C_k by iterating **a** against the single-column vector $[0, 0, \ldots, \delta_{ii} = 1, 0, \ldots, 0]$ and summing the resulting multiplier chain. It is not so obvious, but we can get any one industry's response $(A_{k1}, \ldots, A_{kn})$ to all the C's by repeatedly postmultiplying the single-row vector $[0, 0 \ldots, \delta_{kk} = 1, 0, \ldots, 0]$ by **a** and summing the resulting Gaitskell chain. This follows from the duality mentioned earlier.

ing-hoarding leakages cause convergence of the Keynes-Kahn multiplier chain, so does the above leakage force the Gaitskell multiplier chain to converge. This shows that the Cornfield multiplier must also converge, a fact which can also be proved by showing that use of output units (so that $\hat{X}_i \equiv 1$) gives new a's whose rows add up to 1 minus the positive percentage of consumption to total output.[1]

A survey of methods of numerical computation of input-output systems can be briefly given. The elementary high-school methods of

TABLE 10-5

	Rounds					
	0	1	2	3	4	5
		(.1) 50 + (40) 2	(.1) 85 + (40) .5	(.1) 28.5 + (40) .85	(.1) 36.85 + (40) .285	(.1) 15.085 + (40) .3685
X_1	50	85	28.5	36.85	15.085	16.2485
		(.01) 50 + (0) 2	(.01) 85 + (0) .5	(.01) 28.5 + (0) .85	(.01) 36.85 + (0) .285	(.01) 15.085 + (0) .3685
X_2	2	.5	.85	.285	.3685	.15085
	(4) 50 + (100) 2	(4) 85 + (100) .5	(4) 28.5 + (100) .85	(4) 36.85 + (100) .285	(4) 15.085 + (100) .3685	(4) 16.2485 + (100) .15085
X_0	400	390	199	175.9	97.19	80.079

NOTE: The numbers in parentheses are iterated again and again. As given here, we sum matrix terms of the form $\mathbf{T}(n) = \mathbf{a}\mathbf{T}(n-1) = \mathbf{a}^n\mathbf{C}$. Any numerical error will persist. A self-correcting variant is to use an iteration defined by

$$\mathbf{S}(n) = \mathbf{C} + \mathbf{a}\mathbf{S}(n-1)$$

which will converge in a self-correcting way for any initial conditions $\mathbf{S}(0)$. If we select the latter equal to $\mathbf{C}$, then $\mathbf{T}(n) = \mathbf{S}(n) - \mathbf{S}(n-1)$ and $\mathbf{S}(n) = \sum_0^n \mathbf{T}(i)$. Often better initial guesses than $\mathbf{C}$ can be made, since the latter is known to be too low.

[1] Specifically, we define new units $\hat{X}_i = k_iX_i$, $\hat{P}_i = P_i/k_i$, $\hat{C}_i = k_iC_i$, $\hat{a}_{ij} = k_ia_{ij}/k_j$, where $k_i = 1/X_i$, so that $\hat{X}_i \equiv 1$ and $\hat{a}_{i1} + \cdots + \hat{a}_{in} = 1 - \hat{C}_i \leq 1$ for $1 \leq i \leq n$. In the iterations it follows that $\sum_1^n T_i(n) = \sum_1^n (1 - \hat{C}_i)T_i(n-1) < \sum_1^n T_i(n-1)$, provided that one $C > 0$. Since the sum of nonnegative T's goes to zero, so must each component; and it is clear that the convergence is exponential.

"cross elimination" or "substitution" are optimal numerical methods. (They are often today referred to under the names of Gauss, Doolittle, Chio, Aitken, Crout, Dwyer, and others.) Usually a systematic ritual is best, in which the last (or first) equation in (10-3) is used to eliminate X_n (or X_1) from all the remaining equations; then the last of the remaining $n - 1$ equations is used to eliminate X_{n-1}, etc., until we finally solve for X_1; and then we solve for X_2 in terms of the known X_1, etc. We proceed with the "back solution" until finally we solve for X_n in terms of the remaining variables. All this may be most quickly done for given numerical C's; alternatively, at slightly greater length, it can be done for literal C's—or what is the same thing, for columns of zeros and 1, so that the inverse matrix $[A_{ij}]$ is calculated. Similar remarks apply to the calculation of the P's.

The numerical work grows with the cube of n. With a desk calculator, a single experienced computer can invert a 10×10 system in less than one day. For systems of 50×50, or beyond, large-scale digital computers, which use the same high-school methods, would be very useful.[1]

In actual practice, a modified Cornfield or Gaitskell multiplier may in a few rounds give a good approximate answer. Empirically, it seems to be true that we can renumber industries so that the structure is almost "triangular," with almost negligible feedbacks and whirlpools. This means that we can avoid solving simultaneous equations, instead solving recursively and in tandem, for the variables. This yields approximate values for the X's. These can be improved by taking into account a few local whirlpools, either by local multiplier chains or by solving a few simultaneous equations for circularly interlocking groups of industries.[2] The ease of solving a Leontief system and its immunity to accumulation of round-off errors are for empirical reasons much greater than we could in general expect for any $n \times n$ equations.

[1] Leontief used the Wilbur-Kelvin analog computer at Massachusetts Institute of Technology to solve his first 10×10 system. His later 20×20 system was solved by a desk calculator using the standard Gauss-Doolittle ritual. His 35×35 system was inverted by the Harvard Mark I computer. The projected 100×100 and 400×400 systems are to be solved or approximated by use of large-scale electronic calculators.

[2] We renumber industries so that $a_{j+k,j}$ for $k > 0$ turns out to be almost zero. Then the iteration $X_i(t+1) - \sum_{j=1}^{i} a_{ij}X_j(t+1) = C_i + \sum_{j=i+1}^{n} a_{ij}X_j(t)$ will rapidly converge from initial conditions $X_i(0) = C_i$ or from any arbitrary guess. Intuitive guessing or "relaxation" methods can alter and accelerate the approach to a correct solution.

11

Dynamic Aspects of Linear Models

11-1. INTRODUCTION AND OUTLINE

The object of this and the next chapter is to analyze some simple situations in economics that involve optimizing over time. Earlier in this book linear programming was described as a special kind of theory and practice of maximizing this or minimizing that, subject to certain constraints on the decision variables. Up to a point, the fact that some maximization processes extend over time makes no difference at all. For instance, if the process (e.g., the operation of a firm or of a military campaign) is to last for some given finite number of periods of time, no new principles are involved. Most economists are familiar, in one context or another, with the device of treating commodities available at different times or different places as different commodities. The same procedure can be used in this special kind of dynamic optimization problem, together with some self-evident identities concerning inventories carried over from one period to the next. This use of new variables for each time period reduces the problem formally to a standard programming one, and no new methods are needed for its solution.

But time does make a difference in economics—witness years of controversy over the theory of capital. To treat dynamic problems as nothing but special cases of static ones may simply rob us of the insights that a more direct theory might yield. After all, n commodities at each of T dates are not *simply* nT separate commodities. There is a structure: sometimes it is useful to view them as T groups of commodities with date in common, sometimes as n groups with physical characteristics in common. Certain features of the situation (prices and discount rates) make economic sense when applied to the groups and to their ordering in time. The occurrence of physically identical commodities in time leads to a kind of recursive structure that it is useful to exploit. Our immediate preoccupation is with the explicitly dynamic formulation and analysis of optimization and programming problems.

The natural parallel in economic theory is the theory of capital and investment (we speak of "investment programs"). This parallel forces itself upon us when we try to connect up maximization over time with some kind of market mechanism. The shadow price duality relations turn out to involve what can only be interpreted by the economist as discount factors, interest rates, and the like. The programming approach casts new light on some vexing questions of classical and neoclassical capital theory.

Perhaps this is the place to explain the connection between the dynamic models now to be discussed and the static input-output scheme of Leontief. The remarkable thing about the Leontief model of production is that it is "locked." No real optimization needs to take place. Any particular bill of goods can be produced in one and only one way.[1] Professor Leontief has recently made his theory dynamic by introducing stocks of fixed capital and inventories. In so doing he has attempted to maintain its "locked" character. Given proper initial conditions his dynamic system is meant to be determinate; it should generate its own future, still with no explicit choice or optimization at all.

Later on in this chapter it will be indicated that this cannot really be done. No matter how rigid we make our assumptions about fixed coefficients of production or fixed capital-output ratios, the introduction of a time dimension and stocks of capital inevitably "unlocks" the model. It is no longer possible to dispense with explicit choice and optimization. Commodities desired later can be produced now and stored, or resources can instead be devoted to investment in facilities for subsequent production of the commodity. Redundant capital can be held idle, or output proportions can be adjusted so that no capital is redundant. Output proportions must in any case be decided by some rule. In a dynamic Leontief model a choice must be made at every instant of time, deliberately or by a market mechanism. From this point of view the "unlocked" Leontief dynamic model falls into the framework of programming or maximization over time as a special case.

Section 11-2 will discuss the simplest possible linear model of production over time, the case of a single dynamic process of production. We may call this the Ramsey model.[2] The case of a nonlinear dynamic production process will also be discussed, as will certain simple extensions to the case of two or more alternative processes for producing a single commodity, and to the case of two commodities produced essentially independently.

Section 11-3 will cover the dynamic Leontief model. We may describe

[1] Even supposing there are originally many ways of producing each good this comes about through behind-the-scenes optimization. See the "substitution theorem," Chap. 10, pages 248–252.

[2] After Frank Ramsey, the protégé of Keynes who studied it in great detail in "A Mathematical Theory of Saving," *Economic Journal*, **38**:543 (1928).

this as the case of many commodities, each produced by a single dynamic production process, but with all the production processes mutually interrelated and interdependent.

Section 11-4 is a brief exposition of a still more general model of production over time, which we may identify with the name of von Neumann. It allows many commodities, several alternative production processes for each commodity, and even joint production. The same model is taken up again in Chap. 13.

Section 11-5 merely introduces smooth transformation functions of a kind familiar from neoclassical international trade, welfare, and other theory. Later, in the next chapter, we shall investigate the problem of efficient and optimal programs of investment and capital development, and there we shall have occasion to use this smooth production model.

In some ways this chapter can be viewed as a preliminary to the next. Our object here, once we get past the simple one-good case, is to explore production possibilities from one period of time to the next, but no further. Most practical planners and students of economic growth are interested in processes extending over longer periods. The next chapter will take up the nature and structure of optimal production and investment programs extending to arbitrary horizons. But the indispensable building blocks for this theory are the one-period production-possibility schedules of this chapter.

11-2. THE RAMSEY MODEL

11-2-1. A Single Linear Dynamic Process. Consider the simplest possible case where a single commodity—rabbits, trees, gold coins, or simply output—can be used at time t as an input to produce itself as an output at time $t + 1$. The output at any time such as $t + 1$ can be split into two nonnegative parts: consumption, or $C(t + 1)$, and input for the next period's output, to be designated $x(t + 1)$. We make the simplest *linear* assumption that 1 unit of input gives rise in the next period to a units of output, where a is a given technical constant, and that constant returns to scale prevail. Hence total output at time $t + 2$ will be $ax(t + 1)$. Our production function at any time period may be written as

$$C(t + 1) + x(t + 1) \leq ax(t) \qquad x(t),\ x(t + 1),\ C(t + 1) \text{ all } \geq 0 \quad (11\text{-}1)$$

where the left-hand side is total output at $t + 1$, expressed in terms of the quantity of input one period earlier. The inequality is included in recognition of the fact that people might by stupidity or inadvertence get less output than is technically feasible. The linear character of the technology is clearly visible.

If at time $t = 0$ we start out with K units of our goods, at that time our choice between consumption then and provision for the future is given by

$$C(0) + x(0) \leq K$$

and if we are maximizing, the inequality can be dropped.

Our choice for period $t = 1$ is given by

$$C(1) + x(1) \leq ax(0) \leq a[K - C(0)]$$

Similarly, for the second period, our choice is

$$C(2) + x(2) \leq ax(1) \leq a\{a[K - C(0)] - C(1)\}$$

or in general for any period, we can verify the formula

$$C(n) + x(n) \leq a^nK - a^nC(0) - a^{n-1}C(1) - \cdots - aC(n-1) \quad (11\text{-}2)$$

The economist who remembers his formula for present discounted value of an income stream will be able to think of a as 1 + interest rate and of the productive process as a bank that pays this rate of compound interest on all deposits; he will recognize the above formula as saying that the amount in the bank today is the cumulated value of the original deposit minus the cumulated value of all consumption withdrawals. This formula can also be written in the equivalent alternative form,

$$C(0) + \frac{C(1)}{a} + \frac{C(2)}{a^2} + \cdots + \frac{C(n)}{a^n} + \frac{x(n)}{a^n} \leq K \quad (11\text{-}2a)$$

The reader can provide a "present-value" interpretation of (11-2*a*).

There is very little scope for choice in this model. If we make all the inequalities into equalities, relations like (11-2) are optimal ones. They give us the maximum amount of consumption in *any* period for specified initial and terminal amounts of output and for prescribed amounts of consumption in *all* other periods. It gives us the "menu" of consumption possibilities among which we can choose. For example we have for $t = 1$, $C(1) + aC(0) \leq aK - x(1)$. If we specify a value for $x(1)$ (the "capital stock" for subsequent production), we get a feasible set of consumptions and an efficiency frontier of optimal consumptions. The slope of the efficiency frontier (or its negative) is, as usual, the marginal rate of substitution between $C(0)$ and $C(1)$: $MRS_{10} = -\partial C(1)/\partial C(0) = a$. This is illustrated in Fig. 11-1*a*.

We can use the fundamental relation (11-2) to find the feasible and the efficient consumption programs extending over any number of periods. In two dimensions we can fix the terminal capital stock $x(n)$, and all but *two* consumptions $C(i)$ and $C(j)$, $(j > i)$, and draw the possibilities open to us.[1] The fundamental relation becomes $a^{n-i}C(i) + a^{n-j}C(j) \leq$ a constant depending on a, K, $C(0)$, . . . , $C(n)$, $x(n)$.[2] We find in this way that the marginal rate of substitution between any two consumptions is constant, namely,

$$MRS_{ji} = -\frac{\partial C(j)}{\partial C(i)} = a^{j-i} \tag{11-3}$$

This is illustrated in Fig. 11-1*b*. If the system is "productive," $a > 1$, and a unit of consumption foregone now can be transformed into more

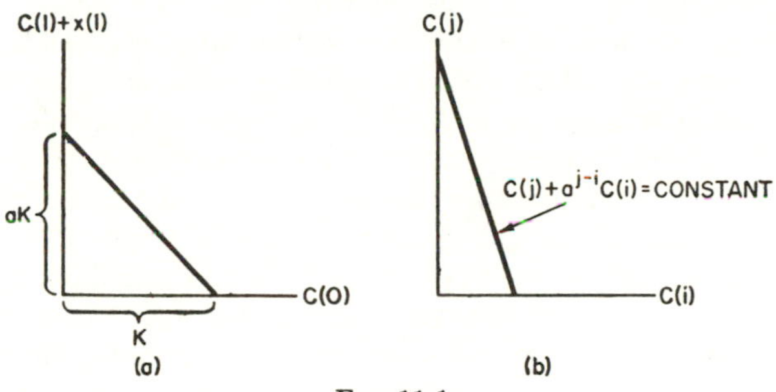

FIG. 11-1

and more consumption as it is postponed until later and later. Geometrically, the efficiency locus gets steeper as $j - i$ increases.

In this case it was trivially easy to get to the optimum. We simply decided never to waste any unconsumed output but to put it all back into the productive process. Can we justify this procedure by means of the statical theory of linear programming? We can. Dynamic processes can be squeezed into the statical framework.

A single productive process will involve, over a long enough period of time, an indefinitely large number of variables. The production relation (11-1) is really as many equations as we have time periods; and there are as many variables $x(0)$, $x(1)$, . . . , $x(n)$, . . . as we care to consider. Let us keep matters simple and consider $t = 0, 1, 2, 3, 4$ only, and let us seek to maximize $C(3)$ with K, $C(0)$, $C(1)$, $C(2)$, $C(4)$, and $x(4)$ all being arbitrarily prescribed. In standard linear-programming terms, we wish to maximize

$$C(3) = B_3x(3) + B_2x(2) + B_1x(1) + B_0x(0)$$
$$= -1 \cdot x(3) + ax(2) + 0 + 0$$

[1] Naturally the other consumptions must be fixed low enough so that (11-2) can be satisfied by *some* $C(i)$, $C(j)$.

[2] This constant is just $a^nK - a^n\bar{C}(0) - a^{n-1}\bar{C}(1) - \cdots - a^{n-i+1}\bar{C}(i-1) - a^{n-i-1}\bar{C}(i+1) - \cdots - a^{n-j+1}\bar{C}(j-1) - a^{n-j-1}\bar{C}(j+1) - \cdots - \bar{C}(n) - \bar{x}(n)$, where the barred variables are to be treated as arbitrary constants.

subject to

$$
\begin{array}{rrrl}
 & & x(0) & \leq K - C(0) \\
 & x(1) & -\,ax(0) & \leq -\,C(1) \\
x(2) & -\,ax(1) & & \leq -\,C(2) \\
-ax(3) & & & \leq -\,C(4) - x(4)
\end{array}
\qquad (11\text{-}4)
$$

[In the maximand we have put $C(3) = -x(3) + ax(2)$, since we know that $C(3) \leq -x(3) + ax(2)$, and for a maximum the equality must hold.] We shall be at the maximum $C(3)$ only if every inequality is discarded. This can be seen in the following way. To maximize $C(3)$ we must make $x(3)$ as small as possible and $x(2)$ as large as possible consistent with the constraints. This enforces equality in the last two constraints, and in addition tells us to make $x(1)$ as large as possible. But this requires making the second constraint an equality and making $x(0)$ as large as possible, i.e., discarding the inequality in the first constraint.

The remarkable thing about this strict-equality property is that it does not change as we vary $C(0)$, $C(1)$, etc. A related singular feature is the fact that if we were to extend our sequence beyond $t = 4$, all the efficient decisions made up to that point would be invariant. Both of these properties are associated with the simple and special form taken by the constraints in such primitive dynamic sequences. The array of coefficients is of the form

$$
\begin{array}{cccccc}
\cdot & \cdot & 0 & 0 & 0 & 1 \\
\cdot & \cdot & 0 & 0 & 1 & -a \\
\cdot & \cdot & 0 & 1 & -a & 0 \\
\cdot & \cdot & 1 & -a & 0 & 0 \\
\cdot & \cdot & \cdot & \cdot & \cdot & \cdot
\end{array}
$$

and it builds downward and to the left without any feedback influences on the earlier inequalities.

An easy but important special case—related to von Neumann's model—arises if all output is used to produce further outputs so that

$$0 = C(0) = C(1) = \cdots = C(t) = \cdots$$

The basic formula (11-2) still holds and shows that an optimal system will grow in geometric progression

$$x(t) = Ka^t \qquad (11\text{-}5)$$

with a growth factor given by

$$\frac{x(t+1)}{x(t)} = a \qquad (11\text{-}6)$$

Previous experience with duality relationships as well as general economic reasoning lead us to expect prices to play a role even in such simple dynamic sequences. Specifically, we should guess that the exchange ratio between the outputs of two periods would be equal to the marginal rate of substitution in production between the outputs of those periods. Let us set $p(0) = 1$, which is just a choice of units. Then (11-3) suggests that

$$-\frac{\partial C(0)}{\partial C(t)} = \frac{p(t)}{p(0)} = a^{-t} \tag{11-7}$$

$$\frac{p(t+1)}{p(t)} = a^{-1} \tag{11-8}$$

One way of verifying that these are true equilibrium relationships is to seek a set of (shadow) prices which will make profits zero on each process used. There is only one process available for each period; in it we invest one unit of $x(t)$ worth $p(t)$ and get in return a units of output worth $p(t+1)a$. Our profit is defined as

$$\pi(t) \leq p(t+1)a - p(t) \tag{11-9}$$

This is what must be zero for *every t*, so that the price relations of (11-8), and thence (11-7), immediately follow.[1]

The formal similarity between (11-5), (11-6) and (11-7), (11-8) is our first dynamic instance of the duality between prices and quantities. Here we have an especially simple case, since there is only one commodity and almost no optimizing or equilibrating to be done.

11-2-2. Digression: A Single Nonlinear Process. In the simple one-commodity world we have so far been exploring, the assumption of constant returns to scale can easily be lifted. Nothing changes except that the dynamic production process becomes nonlinear, with increasing or decreasing returns to scale, or phases of both. The analysis of this more general case will give us some insights which can then in turn reflect more light on the linear model.

Let technology be such that x units of input yield $y = f(x)$ units of output. Then the basic equation (11-1) becomes

$$x(t+1) + C(t+1) \leq f(x_t) \tag{11-10}$$

where $f(x)$ is an increasing function.

Nothing prevents us from proceeding as before. We start with $x(0) + C(0) \leq K$. Hence $x(1) + C(1) \leq f(x_0) \leq f[K - C(0)]$. Still another iteration and $x(2) + C(2) \leq f(x_1) \leq f\{f[K - C(0)] - C(1)\}$. And yet again, $x(3) + C(3) \leq f(x_2) \leq f(f\{f[K - C(0)] - C(1)\} - C(2))$. As

[1] It will be a useful exercise for the reader to formulate the dual linear-programming problem to (11-4) and to similar maximum problems. He should solve the resulting minimum problems (remembering the duality principle) and interpret the results economically.

before, if we optimize, all the equality signs hold. It is easy to see, but hard to write down, how this process can be continued to give us a generalization of Eq. (11-2). Some simple examples and graphs will tell us all we need.

First, let us take $f(x) = \sqrt{x}$, a case of diminishing returns to scale (we can imagine that a second factor, land, is used in production but is available in a fixed nonaugmentable amount). Figure 11-2 is the analog of Fig. 11-1 and shows the familiar convex transformation or substitution curve between consumption at times zero and 1. The marginal rate of substitution of $C(1)$ for $C(0)$ is no longer a constant, but diminishes

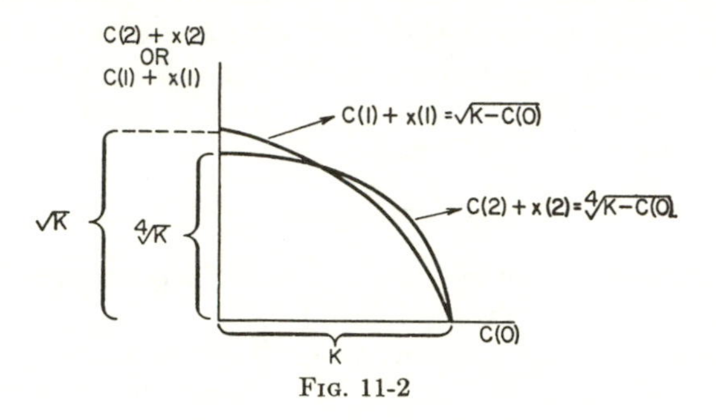

FIG. 11-2

with $C(0)$.[1] On the same graph we can show the terms on which $C(2)$ of the next period can be substituted for original $C(0)$, as of a constant intermediate $C(1)$. For convenience we can set $C(1) = 0$. The new substitution curve shows even more sharply diminishing MRS than the first.[2] There are other less obvious things to notice. For example, the MRS_{20} depends on what particular level we give to $C(1)$, but in the linear case it did not. The curve for $C(2)$ could lie either wholly outside or could cross that for $C(1)$, depending on the size of K, but in the linear case it would be wholly outside if $a > 1$, or wholly inside if $a < 1$, but always one or the other. A nonlinear system can be "productive" or self-sustaining at some levels of output but not at others, as we shall show later.

Finally, we can wonder about the substitution curves for $C(3)$ and also about how they are affected by intervening consumptions. If we take the easy way out and put all intervening consumption equal to zero, a little experiment will show that the general formula is

$$C(n) + x(n) = [K - C(0)]^{1/2^n}$$

[1] Since $C(1) + x(1) = \sqrt{K - C(0)}$, $MRS_{10} = 1/2\sqrt{K - C(0)}$.

[2] Since $C(2) + x(2) = [\sqrt{K - C(0)} - C(1)]^{1/2}$ with $C(1) = 0$,

$$MRS_{20} = 1/\{4[\sqrt{K - C(0)} - C(1)]^{1/2}\sqrt{K - C(0)}\}$$

or in our special case, $1/\{4[K - C(0)]\}^{3/4}$.

A little further experimentation will show that the substitution curve approaches a limit, namely, the boxlike one in Fig. 11-3.[1] This behavior also differs from the earlier linear case; there $C(n)$ must go to infinity ($a > 1$) or zero ($a < 1$). But a diminishing-returns-to-scale economy *can* have a finite equilibrium.

It must not be thought, however, that this *must* happen. Consider for example the production function $y = x + \sqrt{x}$. Then

$$x(1) + C(1) = K - C(0) + \sqrt{K - C(0)}$$

and

$$x(2) + C(2) = x_1 + \sqrt{x_1} = K - C(0) - C(1) + \sqrt{K - C(0)} + [K - C(0) - C(1) + \sqrt{K - C(0)}]^{1/2}$$

These two substitution curves are shown in Fig. 11-4 with, as usual, the intermediate consumption $C(1) = 0$ in the latter curve. Each successive substitution curve between $C(0)$ and $C(1)$, between $C(0)$ and $C(2)$, between $C(0)$ and $C(3)$, etc., lies entirely outside the previous ones.[2] If $C(0)$ does not entirely exhaust the initial stock of capital, future consumption can be increased beyond all bounds as we postpone it indefinitely. But choices are always along a convex substitution curve. And MRS's depend on the levels of *all* intermediate consumption.

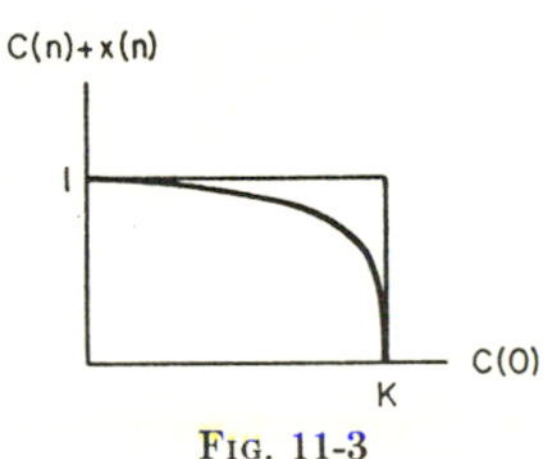

FIG. 11-3

The reader will find it instructive to work out for himself a simple case of increasing returns to scale. He should draw some of the diagrams showing the opposite convexity to the more traditional ones. An easy example to work with is $y = f(x) = x^2$.

There is another way of looking at the nonlinear case. Go back to the basic Eq. (11-10) and eliminate the inequality, with the assumption that we optimize. The equation then can be thought of as a recurrence relation or difference equation: $x(t + 1) + C(t + 1) = f[x(t)]$. If we maintain the earlier device of putting all intermediate consumption equal to zero, we can also absorb $C(t + 1)$ into $x(t + 1)$, and the equation takes

[1] $C(n) + x(n)$ approaches a finite limit, namely, 1. The number 1 plays no really special role. It enters here in this way. Let us ask what stock of capital x has the property that it is *just* self-supporting; i.e., it will produce an output just equal to itself. This x must satisfy $x = f(x) = \sqrt{x}$. Obviously $x = 1$ (discarding the trivial solution $x = 0$). Had we assumed, say, $f(x) = m\sqrt{x}$, we would have had to solve $x = m\sqrt{x}$, which gives $x = m^2$, and this is the limit that $C(n) + x(n)$ would approach.

[2] This is with $C(1) = 0$. Needless to say, $C(1)$ can be made large enough to reduce the choices for $C(0)$ and $C(2)$ wholly inside the initial curve for $C(0)$ and $C(1)$.

the form $x(t + 1) = f[x(t)]$. Thus, as we already know,

$$x(1) = f[K - C(0)]$$
$$x(2) = f\{f[K - C(0)]\}$$

etc. Graphically this is very simple: Fig. 11-5 has x on the horizontal and $f(x)$ on the vertical axis and also contains a 45-degree line. The production function has been drawn with decreasing returns. Anyone who has ever traced out a cobweb cycle or, better still, a Cournot duopoly-reaction path will see immediately how the arrows show the path to the ultimate equilibrium, the input-output which satisfies

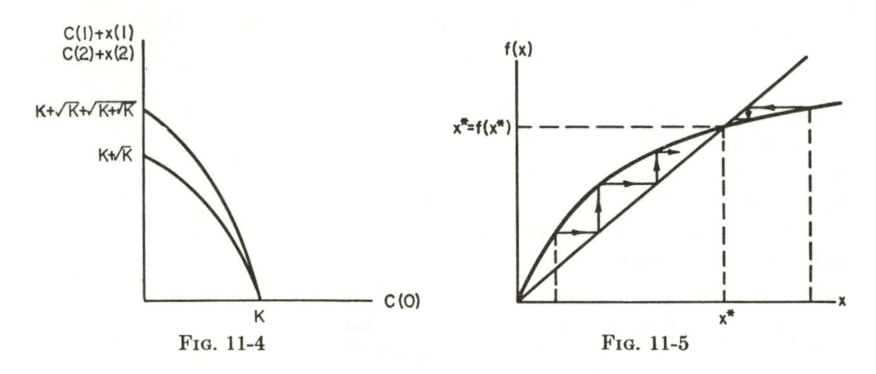

FIG. 11-4 FIG. 11-5

$x^* = f(x^*)$. Note how a diminishing-returns system can have a finite equilibrium, and how this is not because marginal productivity falls to zero (it doesn't), but because average productivity falls below 1. More input will always yield some extra output, but beyond a certain point input will fail to reproduce itself, and so the system decays. Figure 11-6*a* shows the opposite state of affairs for an increasing-returns system. The same kind of equilibrium may exist, but it will be unstable. Once capital stock exceeds a critical level x^* it can reproduce itself and more, and output can expand and capital grow over time. But if the initial capital is below x^*, even total reinvestment will not prevent the stock from dissipating. The linear system treated earlier is shown in Fig. 11-6*b*. If $a > 1$, then any initial capital can expand without limit as (11-5) showed. If $a < 1$, the reader can show how decay is inevitable.

Here we can also treat perhaps the most interesting case: decreasing returns to scale, but indefinitely expansible production. The production function $y = x + \sqrt{x}$ is drawn in the now-familiar way on Fig. 11-6*c*. The difference equation is $x_{t+1} = x_t + \sqrt{x_t}$ and the arrows show how output and capital grow without limit. No decay occurs, because average productivity of capital $y/x = 1 + 1/\sqrt{x}$, while decreasing with the size of the capital stock, never falls to 1. We can get some notion of the rate of growth of capital and output by writing

$$\frac{x_{t+1} - x_t}{x_t} = \frac{1}{\sqrt{x_t}}$$

The relative rate of growth diminishes steadily but never falls to zero. Study of nonlinear models such as this one provides a healthy antidote to the tacit assumption in linear models of the Harrod-Domar type that

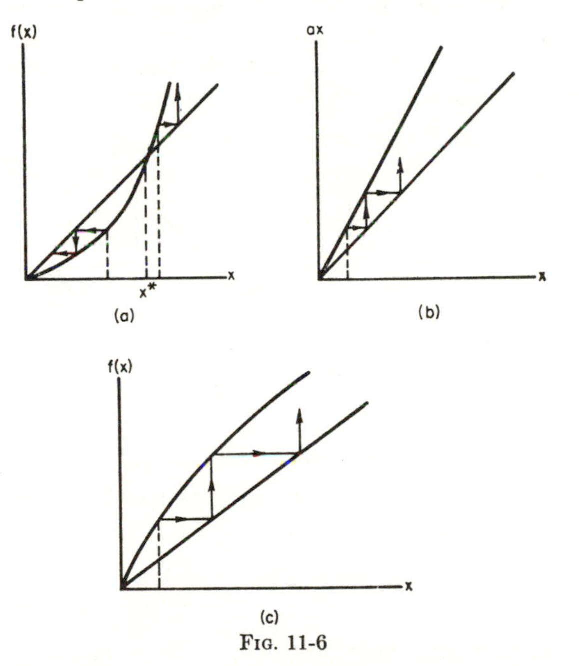

FIG. 11-6

"growth" is geometric growth. In our present case the system grows at about the speed of t^2, or much more slowly than geometrically.

The same device enables us to treat a slightly more general case where consumption $C(t)$ is held at some nonzero constant level through time. The difference equation becomes

$$\begin{aligned} x(t+1) &= f[x(t)] - \bar{C} \\ &= g[x(t)] \end{aligned} \tag{11-11}$$

The new g function is just the old production function lowered by $\bar{C}$ all along the line. Figure 11-7 should by now speak for itself. In both the decreasing-returns case and the linear case, a first phase appears in which, while a small stock of capital can reproduce itself, it cannot simultaneously throw off the required consumption. If it tries to do so it will decay. The nonlinear case may still have an equilibrium x^* which satisfies $x^* = f(x^*) - \bar{C}$.

The shadow price and marginal-rate-of-substitution propositions are less simple in this nonlinear model. The economic theorist will recognize right away that we cannot impose a zero-profit condition once we lose constant returns to scale. Instead all we have is a zero-marginal-profit condition. By creating a second factor to absorb the residual we could again have zero profits. Price ratios will still equal marginal rates of substitution, but the latter now vary with society's consumption-investment choices. In the linear case [see Eqs. (11-7) and (11-8)] the price ratios were independent of these decisions.

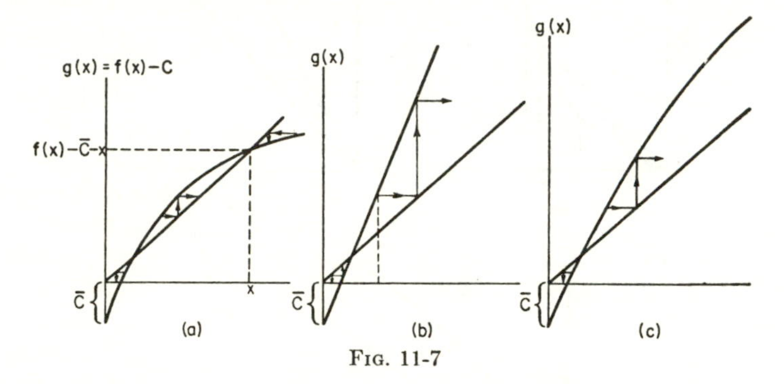

FIG. 11-7

A little calculus and a lot of notation will suffice to calculate any MRS_{ji} from the fundamental relation

$$x(n) + C(n) = f[f(f \cdots f\{f[K - C(0)] - C(1)\} \cdots - C(n-2)) - C(n-1)] \quad (11\text{-}12)$$

Once this is done we shall have, as usual, $MRS_{ji} = p(i)/p(j)$. An example is given above (p. 272, footnote 2) in which an MRS_{20} is explicitly calculated, and it shows how the price ratio varies with the choice of $C(0)$ and $C(1)$. Thus, if we use the square-root production function and assume $C(1) = 0$, then

$$MRS_{20} = \frac{p(0)}{p(2)} = \frac{1}{4[K - C(0)]^{3/4}}$$

The larger the $C(0)$, the lower the relative price $p(2)$, as one would expect. These price ratios and MRS's are, of course, just the (now variable) slopes of the corresponding transformation curves.[1]

One more relation. For adjacent periods we have

$$x(t+1) + C(t+1) = f[x(t)] = f\{f[x(t-1)] - C(t)]\}$$

[1] The reader who can differentiate functions of functions should try his hand at deducing the MRS and relative price relations for $y = x + \sqrt{x}$.

Hence,

$$MRS_{t+1,t} = -\frac{\partial C(t+1)}{\partial C(t)} = f'[x(t)] = \frac{p(t)}{p(t+1)} \tag{11-13}$$

On the other hand, total profit for the operations of period t is

$$p(t+1)f[x(t)] - p(t)x(t) = \pi[x(t)]$$

Setting marginal profit equal to zero we find

$$\frac{d\pi}{dx(t)} = 0 = p(t+1)f'[x(t)] - p(t) \tag{11-14}$$

or $p(t)/p(t+1) = f'[x(t)]$, exactly as in (11-13). We can substitute the $x(t)$ time path in this and find in this way the course of prices over time.

Now we return to linear processes, leaving until the next chapter further discussion of the nonlinear case.

11-2-3. Alternative Processes. The case of a single process is almost trivial from the standpoint of linear programming and maximizing decisions. Scarcely any intelligent choices had to be made. Consider therefore a slightly more interesting case, but still an exceedingly simple one. Suppose that we can allocate the unconsumed output of any period as inputs to either of two productive processes. Let $x_1(t)$ be the input to the first process and let $a_1x_1(t)$ be the maximum output resulting in the following period; likewise, let $x_2(t)$ be the second process input and let $a_2x_2(t)$ be its output in the following period. Note that we have two production functions; but we can add the identical outputs of the processes and allocate the total between consumption $C(t+1)$ and the two inputs for the *following* period's production, to get the relation

$$C(t+1) + x_1(t+1) + x_2(t+1) \leq a_1x_1(t) + a_2x_2(t) \tag{11-15}$$

This is the generalization of our first equation; and as in the case of a single process, we suspect that the inequality signs can be dropped in any optimal process.

But now if we start with initial output

$$C(0) + x_1(0) + x_2(0) = K$$

the resulting process does not give us a unique choice among the consumptions of different periods. Depending upon how we decide to determine $x_1(t)$ and $x_2(t)$ at each stage, we shall get a different menu of consumption choices. How do we resolve this ambiguity? By the same principle as before: *With initial and final outputs being given us, and with all but one consumption item being given us, we must try to maximize that remaining consumption item.*

The solution is intuitively obvious—as obvious as the answer to this question: If two banks offer you different interest rates on your bank balances, how should you invest your money? Only in the bank that gives you the higher interest rate is the obvious answer. Likewise, in this problem, if $a_1 > a_2$, we never allocate any input to the second process; and our solution is exactly as in the case of a single process, but with a subscript 1 on all the earlier a symbols. Thus,

$$\begin{gathered} 0 = x_2(0) = x_2(1) = x_2(2) = \cdots = x_2(t) = \cdots \\ K \geq C(0) + \frac{C(1)}{a_1} + \frac{C(2)}{a_1^2} + \cdots + \frac{C(t)}{a_1^t} + \frac{x_1(t)}{a_1^t} \\ p(t+1) = a_1^{-1}p(t) \end{gathered} \tag{11-16}$$

where the p's are the prices of the output of any period. Note the negative profitability of ever using any $x_2(t)$ as given by

$$\pi_2(t) = p(t+1)a_2 - p(t) = (a_2 - a_1)p(t+1) < 0$$

As in the single process, we get a geometric progression if all consumption is zero and all output is ploughed back into the business. We also get the obvious dual relations

$$\begin{aligned} \max \frac{x(t+1)}{x(t)} &= a_1 = \max(a_1,a_2) \\ \min \frac{p(t+1)}{p(t)} &= a_1^{-1} = \min(a_1^{-1},a_2^{-1}) \end{aligned} \tag{11-17}$$

Note that in this special case, as in the single-process case, the system can go instantly into its maximal rate of growth regardless of how we start it off. This will not be generally true. Note too that three or more independent alternative processes, with specific a_1, a_2, a_3, . . . , would be subject to the same rule: Select the process with the highest a (i.e., the highest "net reproductive rate," or "rate of interest *in natura*").

11-2-4. Transient Unbalance. The two linear cases considered so far have had the special property of always being in a constant relative-price configuration: regardless of the pattern of desired consumption, the same price ratios have prevailed over time; and regardless of the initial endowments of the various commodities, the system can generate itself at a stable geometric rate. In more general systems this will not be the case. An example of joint production will make this clear.

Suppose that a unit of the first commodity, x_1, will reproduce itself by tripling in every period. Suppose that at the same time it produces as a by-product a unit of a second product, x_2. Assume that x_2 can also reproduce itself by doubling every period, so that there are two separate ways of getting x_2. As before, assume that the output of any good may be consumed in any period or can be used as input for the next period's output. Mathematically, our equations are

$$C_1(t+1) + x_1(t+1) \leq 3x_1(t) \qquad (11\text{-}18)$$
$$C_2(t+1) + x_2(t+1) \leq 1x_1(t) + 2x_2(t)$$

How will the system grow if there is no consumption? Suppose we start out with no x_1 and 1 unit of x_2, so that $[x_1(0), x_2(0)] = (0, 1)$. In the next period x_2 will double, so we have $[x_1(1), x_2(1)] = (0, 2)$. This will be followed by $(0, 4), (0, 8), (0, 16), \ldots, (0, 2^t)$, which is obviously a steady geometric progression.

But suppose we had started with one unit of each good, with (1, 1). The x_1 will triple itself and provide $1x_2$ as a by-product. The x_2 unit

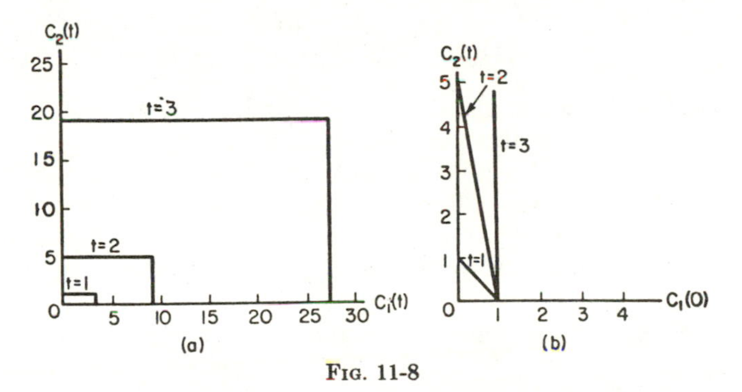

Fig. 11-8

will double itself. Altogether we will have (3, 3). Similarly in the next periods $(9, 9), (27, 27), \ldots, (3^t, 3^t)$. Again we have a steady geometric progression, but with a tripling per period instead of a doubling.

Suppose we start with any amounts of the two goods—say (1, 10). The obvious way to handle this is to decompose it into our previous two cases: into one (1, 1) and nine (0, 1)'s. Does the result grow at a tripling rate or a doubling rate? The answer is: Neither, but somewhere in between; after a long period of time the tripling items become overwhelmingly great compared to the doubling ones, so that in the limit, the strongest geometric progression dominates. The interested reader may easily work out the case (10, 1) and show that x_2 will at first grow at more than a tripling rate.

Let us bring consumption back into the picture. What are our choices between, say, $C_1(1)$ and $C_2(1)$? Or between $C_1(2)$ and $C_2(2)$? Or between $C_1(t)$ and $C_2(t)$? Or between $C_1(0)$ and $C_2(2)$? The answer now depends upon our initial values. In Fig. 11-8*a* we have assumed

$$x_1(0) = K_1 = 1$$
$$x_2(0) = K_2 = 0$$

and have indicated the consumption menus at $t = 0, 1, 2, 3$ (on the postulate that consumption at other dates is zero).

Figure 11-8*b* shows the substitution relations between $C_2(t)$ and $C_1(0)$ under the same conditions. Note that the price ratio,

$$\frac{p_2(t)}{p_1(0)} = \frac{-\partial C_1(0)}{\partial C_2(t)}$$

is steadily dropping in the sequence $1, 5^{-1}, 19^{-1}, \ldots, (3^t - 2^t)^{-1}, \ldots,$ which is no longer a simple geometric sequence, except asymptotically.

Mathematically, we must always have

$$\begin{aligned} x_1(t) &\leq K_1 3^t - C_1(0)3^t - C_1(1)3^{t-1} - \cdots - C_1(t) \\ x_2(t) &\leq K_2 2^t - C_2(0)2^t - \cdots - C_2(t) + [K_1 - C_1(0)](3^t - 2^t) \\ &\quad - C_1(1)(3^{t-1} - 2^{t-1}) - \cdots - 1C_1(t-1) \end{aligned} \tag{11-19}$$

One rather subtle point should be mentioned. The above charts and price formulas hold on the assumption that we are choosing between $C_2(t)$ and $C_1(0)$, not giving a hang about future consumption and inputs. Let us also prescribe some future inputs, $x_2(t+1)$, $x_2(t+2)$, . . . , etc. If these are prescribed low enough, then our formulas may still hold. But if we insist on their being still higher, then the effective menu between $C_1(0)$ and $C_2(t)$ will change. Thus if $x_2(t+k)$ is made high enough to be the limiting factor, then the effective MRS between $C_1(0)$ and $C_2(t)$ will be

$$\left[-\frac{\partial C_1(0)}{\partial C_2(t)}\right]_{t+k} = \frac{2^k}{3^{t+k} - 2^{t+k}} \leq \frac{1}{3^t - 2^t} \tag{11-20}$$

This can be verified by differentiation of the formula for $x_2(t+k)$ in (11-19).

The important thing to notice about this third case is the fact that constant costs no longer hold. Depending upon the patterns of $C_1(t)$ and $C_2(t)$ that we specify, there will be different rates at which the different consumption items can be substituted one for another.

11-2-5. Exceptions to Steady Growth. In every case so far, setting consumption at zero has resulted ultimately in an approach to a state of steady maximal growth. Must this always be the case? The answer is clearly no.

Suppose rabbits obey the production function

$$x_1(t+1) = 100x_1(t)$$

and cheese obeys the production function

$$x_2(t+1) = 2x_2(t)$$

The production functions are entirely independent and lead to two different "own rates of interest" and two different rates of growth. Obviously the price of rabbits must deteriorate over time relative to the price of cheese[1] [see Eq. (11-7)].

In this context we can consider the Malthusian theory of population. Without preventive checks or the positive check of inadequate food, population might double every generation. Malthus argued that food could not grow at so fast a rate so that checks would operate to prevent a doubling of population every generation. Some of his early critics argued that food consisted of animals and plants and that these also grew by nature in a geometric progression. Granting this for the moment, we may still deny that the geometric rate of growth of subsistence need be at the same rate as for humans, namely, a doubling every quarter century or so. There are two possibilities: subsistence may grow, when all the time being drawn on to feed man at the level he is accustomed to, at a faster rate or a slower rate than man himself. It is pretty clear that the needed component of *slowest* growth will set the pace for the whole. If shmoos double every century instead of every quarter century, and man must eat shmoos to live, then positive checks will keep man's growth rate down to that of shmoos. But if man can make subsistence grow so as to double in less than every quarter century, then man will be the bottleneck.

In point of fact, Malthus based his theory on the assumption that inorganic natural resources—"land"—were limited and could not grow in effectiveness in a geometric ratio. The zero own-rate of growth of land would set the pace for organic food and for man, so that a stationary population would be reached in a technologically stationary society.

11-3. GENERALIZED LEONTIEF SYSTEMS

It is time we turned to the more interesting case of several commodities, interdependently used in producing each other. The problem of optimal allocation of resources, so trivially simple in the one-good model, now steps closer to the center of the stage.

11-3-1. The Transformation, or Efficiency Locus. Henceforward we shall operate under the assumption of constant returns to scale, which frees us from worry about the distribution of output among competing

[1] In a sense we might consider

$$x_1(t) = K\ 100^t$$
$$x_2(t) = 0\ \frac{p_1(t)}{p_2(t)} = 0$$

a limiting steady-growth motion.

firms. If we make our time period short enough and admit enough joint products, we can easily handle both fixed and circulating capital: we simply assume that a machine tool produces not only fabricated metal products, but also a one-period-older machine tool.

At any time t, then, there are stocks of each factor or commodity $S_1(t), S_2(t), \ldots, S_n(t)$. If convenient, we may count labor as one of these. There will always be one or more optimal ways of allocating these initial resources among the possible technological processes of the economy. By "optimal" we mean simply that the next period should dispose of *maximal* total amounts of (1) consumption $C_i(t+1)$ of these same

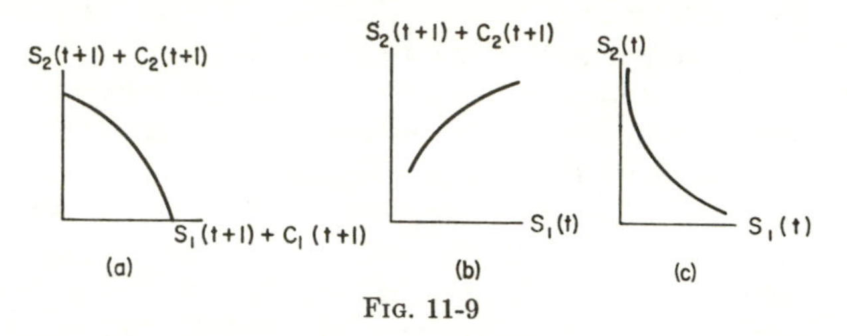

FIG. 11-9

goods plus (2) further stocks $S_i(t+1)$ of each good to be fed back into the economy as inputs. These maximal amounts form a *production-possibility frontier*, or *efficiency locus*. A point on this locus has the properties that it is producible from the initial stocks and that any movement or change to another producible configuration requires us to sacrifice some of one or more goods.

We may write this transformation locus in the form

$$T[S_1(t), \ldots, S_n(t); S_1(t+1) + C_1(t+1), \ldots, S_n(t+1) + C_n(t+1)] = 0 \quad (11\text{-}21)$$

A variant interpretation of this locus is that for given initial stocks and given values for all but one of the $S_i(t+1) + C_i(t+1)$, it gives the maximum consumption plus investment possible in the remaining commodity.

In the one-good case of Sec. 11-2, efficiency simply meant not wasting any input. In the notation used there, we always put $x_{t+1} + C_{t+1} = f(x_t)$. If we write this $x_{t+1} + C_{t+1} - f(x_t) = 0$, we have (11-21) exactly. Thus we are generalizing the Ramsey model.

The transformation locus is subject to constant returns to scale and defines a *convex* frontier: the weighted mean between any two frontier points will be on or "inside" the frontier. This statement, and Figs. 11-9*a–c*, show the usual laws of diminishing returns and diminishing rates of substitution obeyed by this locus. In each figure all variables are held constant except the two whose axes are drawn.

The existence at each t of this efficiency locus is for the moment taken for granted as a familiar piece of economics. Actually, it is one of the tasks of economic theory—either by marginal productivity, or linear programming, or some more general analysis—to deduce the allocation conditions that characterize this locus. We shall shortly have to cover a little of this ground in the special case of the dynamic Leontief model.

From now on, to make life simple, we shall make believe that there are only two goods. Even so, we shall have for each point of time $S_1(t)$, $S_2(t)$, $S_1(t+1)$, $S_2(t+1)$, $C_1(t+1)$, $C_2(t+1)$ to worry about. The reader who yearns for generality will be able to see that all our considerations extend straightforwardly to n goods. Occasional footnotes will point the way.[1]

11-3-2. The Leontief Dynamic System.[2] The dynamic input-output system is a straightforward generalization of the static model of the last chapter. Its distinguishing characteristics are the same: (1) joint products are ruled out; and (2) for each output there is only one possible activity or technological process, with fixed proportions. As in the static model, it then appears that no quantitative problems of optimization could arise (except for the simple problem of whether we should

[1] This is a good place to mention some miscellaneous matters. Many writers, including Ramsey and Leontief, prefer to deal with continuous time rather than discrete periods. In their treatments consumption becomes an instantaneous rate of flow; and net capital formation [what we would call $S_i(t+1) - S_i(t)$] becomes a time derivative or rate of increase $\dot{S}_i = dS_i(t)/dt$. The locus (11-21) would be written as

$$T(S_1, \ldots, S_n; \dot{S}_1 + \dot{C}_1, \ldots, \dot{S}_n + \dot{C}_n) = 0 \qquad \text{(11-21}a\text{)}$$

where the dots on the C's emphasize that they are rates of flow, not stocks. Equation (11-21a) can be rigorously derived from (11-21) by a limiting process, as the length of the discrete time period shrinks to zero. There is no real difference between the discrete and continuous points of view. We choose the former just to avoid the appearance of differential equations. Our model also includes as a special case the von Neumann discrete-time general-equilibrium model, in which there is a finite number of possible activities, each obeying constant returns to scale. When we come to this model in detail, later on, we shall see that it is a straightforward problem in everyday linear programming to go from the von Neumann input and output coefficients to our efficiency locus. As the linear-programming nature of the model suggests, a feature of the von Neumann finite-activities formulation is that the transformation locus becomes a polyhedron with a finite number of flat faces, meeting in a finite number of vertices. The reader can imagine the curves of Fig. 11-9 replaced by broken lines of the same general appearance. As the number of activities gets larger and larger, we can hope in the limit to generate smooth loci of neoclassical type. This case of a finite number of alternative activities is also associated with the name of the late Fred M. Taylor. It is all but explicit in the first two editions of Leon Walras's *Elements of Pure Economics*. In the third and fourth editions, smooth neoclassical marginal productivities make their appearance.

[2] Wassily Leontief, et al., *Studies in the Structure of the American Economy*, Chap. 3, Oxford University Press, New York, 1953.

use all of each input—the problem of inequalities). This is the point of view taken by Leontief himself. Our own approach will be clearer if we begin in general neoclassical terms and then specialize.

Assume that current consumption of any good $C_i(t)$ is produced as the sole product of one industry, the ith. We need a new symbol $X_i(t)$ to represent the total flow of output of the industry in period t. These current flows do not appear in the basic efficiency locus of (11-21), and we shall show later how they can be eliminated and the whole argument conducted in terms of stocks. In the meantime, $X_i(t)$ can be used for three purposes: (1) as consumption for the next period $C_i(t + 1)$, (2) as net addition to the stock of a particular kind of capital good, $S_i(t + 1) - S_i(t)$, or (3) as current flow of materials needed for production in the economy's two (or n) industries, $X_{i1}(t)$, $X_{i2}(t)$, etc. For example, X_{12} is the flow from the first to the second industry, and X_{11} is internally used flow. There is a balance relation:

$$X_i(t) \geq C_i(t + 1) + S_i(t + 1) - S_i(t) + X_{i1}(t) + X_{i2}(t) \qquad i = 1, 2 \tag{11-22}$$

The inequality sign would come into play only if the output of an industry were so great as to surpass any useful purpose, requiring us to dispose (costlessly) of the surplus.

Now how are the current outputs $X_1(t)$ and $X_2(t)$ produced? Leontief assumes a production function relating each single[1] flow of output to two classes of factor inputs: flows of raw materials or current inputs, already denoted as X_{11}, X_{12}, X_{21}, X_{22}; and stocks of capital goods (inventories, machines, buildings, etc.) used in each industry, namely, S_{11}, S_{12}, S_{21}, S_{22}. As with current flows, the first subscript describes the physical nature of the commodity concerned, and the second subscript refers to the industry in which the capital good is employed. Thus $S_{11} + S_{12}$ is the economy's stock of capital in the form of the first commodity, and (S_{11}, S_{21}) (it would be senseless to add these two numbers) describes the capital structure of the industry producing the first commodity.[2] In the simplest model these capital stocks are to be thought of as being present for production purposes, but not as being *used up* by current production. In other words, all stocks of capital are being currently maintained; some of the factors required by each industry may be used not to produce current output, but just to keep the capital items physically maintained.[3]

[1] This is where the postulated absence of joint production comes in, a limitation dropped in the von Neumann model to be discussed later.

[2] Since we are getting along with only two commodities, we let the same physical commodity do duty as both current input and capital good. In general, some commodities may appear only as flows, others only as stocks.

[3] It would be an excellent exercise for the reader to see how this assumption can be relaxed to permit capital to depreciate both through use and from the passage of time.

All this said, we can write each current output as a constant returns-to-scale function of all the relevant inputs:

$$\begin{aligned} X_1(t) &= F^1[X_{11}(t),X_{21}(t),S_{11}(t),S_{21}(t)] \\ X_2(t) &= F^2[X_{12}(t),X_{22}(t),S_{12}(t),S_{22}(t)] \end{aligned} \qquad (11\text{-}23)$$

These production functions may or may not have the smooth marginal productivities of neoclassical theory. Also some inputs may simply not appear in the production function of any particular industry. Some variation in timing is possible without changing the basic structure of the model. For example, in (11-23) we might think of outputs at $t + 1$ being produced by flows at time t, thus eliminating a kind of simultaneity in (11-23) and (11-22) which together imply that somehow flow inputs are produced and used up in the same period.

11-3-3. The Structure of Capital Stock. Before proceeding, we must settle the relation of the elements of the double subscript S_{ij} to each other and the relation of the double subscript to the single subscript totals S_i. Initially we assume that each grade of capital S_i is homogeneous and nonspecific to any industry.[1] Hence the total $S_i(t)$ may be regarded as the sum of the separate allocations of the stock among different industries, i.e.,

$$S_i(t) \geq S_{i1}(t) + S_{i2}(t) \qquad i = 1, 2 \qquad (11\text{-}24)$$

Here an important warning should be sounded. We have written (11-24) so that it can be an inequality, not an equation. If an inequality is in effect, this means that there is excess capacity in terms of the capital good in question. Now an inequality, even if it holds for every t, tells us *nothing* about the rates of change of the two sides. We *cannot* infer from (11-24) that

$$\Delta S_i(t) \geq \Delta S_{i1}(t) + \Delta S_{i2}(t) \qquad (11\text{-}24a)$$

where $\Delta S_i(t)$ means $S_i(t + 1) - S_i(t)$, etc. If the equality holds and continues to hold in (11-24), then (11-24a) must be an equality also. But when (11-24) is an inequality, the two sides are independent and

[1] See Leontief, *op. cit.*, p. 69, for discussion of the case in which capital equipment is specific to an industry and nontransferable. We could extend our model to cover such phenomena completely. We could define *each* specific capital good as a separate commodity with its own producing industry. Then if we like, we can regard the capital as transferable; it just happens that no other industry uses this particular kind of equipment. The price we pay is (1) an increase in the number of industries, (2) many zeros in the production functions, for factors not used in an industry; and (3) if any capital good also serves in consumption, there will be alternative sources of production for what appears to be several specific capital goods but is *really* a single consumption service. In any case, nontransferability introduces no essentially new issues not already implicit in the Leontief system.

their rates of growth are independent, and either rate of growth may exceed the other.

Leontief does find it convenient in his first exposition to assume $\Delta S_i(t)$ to be nonnegative, so that disinvestment of capital is ruled out for the economy as a whole. This introduces the "irreversibility" connected with the existence of a maximum possible rate of disinvestment, as stressed by Goodwin, Hicks, and other modern expositors of business-cycle theories containing a nonlinear acceleration principle. Actually, Leontief is prepared to generalize slightly and go on to handle a maximum rate of disinvestment different from zero. A particularly simple alternative would be the assumption that gross investment must be nonnegative. Then net investment $\Delta S_i(t)$ can be negative so long as it does not numerically exceed the rate of current depreciation of capital. Since we are not explicitly allowing for depreciation, gross and net investment coincide as a matter of definitional convenience.

To summarize: We are assuming that $\Delta S_i(t) = S_i(t+1) - S_i(t) \geq 0$. However, the components of any capital item, $\Delta S_{ij}(t)$, may be of any sign, with capital transferred out of use in any one industry into any other industry, or into excess capacity.

11-3-4. Fixed Coefficients of Production. As things stand, our economic model is grossly incomplete. Obviously, the behavior in time of our system will not be determinate until it has some way of deciding its choices among alternative methods of production and allocations of available resources. This inevitably involves a problem of optimization: Given "society's" tastes, either the "invisible hand," operating through a decentralized competitive pricing system, effectuates all such decisions and makes the outcome determinate; or else some computing intelligence must solve the same problem by methods of linear programming or calculus. Only when enough efficiency conditions (marginal equalities and inequalities) are added to (11-22), (11-23), and (11-24) will our system become determinate.

Leontief tries to avoid *all* such problems of optimization by assuming fixed coefficients of production with only one way of producing each output. We can write down in our terminology these special nonsubstitutability assumptions and see how far they take us toward pinning the system down.

The Leontief production functions require a *fixed* minimum amount, say a_{ij}, of $X_{ij}(t)$ for each unit of $X_j(t)$ *and* a *fixed* minimum amount, say b_{ij}, of $S_{ij}(t)$ for each unit of $X_j(t)$ produced. The word "minimum" must not be forgotten; under our assumption of free disposal, production is limited by the *first* bottleneck reached. The a's are flow coefficients, of dimensionality flow input per unit of flow output. The b's are capital coefficients, of dimensionality stock input per unit of flow output, i.e., commodity/(commodity/time). Hence, like all acceleration coefficients and unlike the a's, the b's depend on the time unit used: expressed in weeks they yield numbers 52 times as large as when expressed in years.

Leontief's special production function can be brought into the framework of (11-23) for each t by saying

$$X_1 = F^1(X_{11},X_{21},S_{11},S_{21}) = \min\left[\frac{X_{11}}{a_{11}},\frac{X_{21}}{a_{21}},\frac{S_{11}}{b_{11}},\frac{S_{21}}{b_{21}}\right]$$
$$X_2 = F^2(X_{12},X_{22},S_{12},S_{22}) = \min\left[\frac{X_{12}}{a_{12}},\frac{X_{22}}{a_{22}},\frac{S_{12}}{b_{12}},\frac{S_{22}}{b_{22}}\right] \qquad (11\text{-}25)$$

where $\min(a,b,c, \ldots ,z)$ means the smallest of the numbers $a, b, \ldots, z$, and we agree to omit from the parentheses any term whose a_{ij} or b_{ij} is zero. Such inputs never limit production. We can't draw the complete isoquant surfaces, because each production function contains four inputs. But if we imagine all inputs but two as held constant, the equal-output contours appear as in Fig. 11-10.

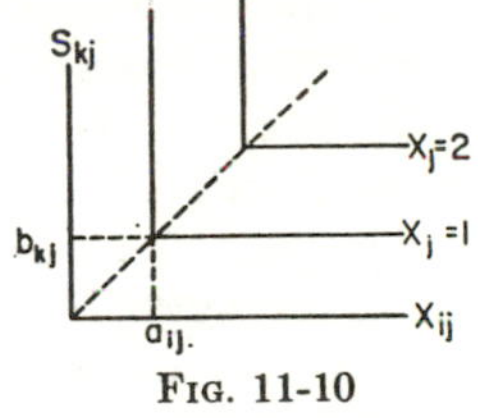

FIG. 11-10

We can read (11-25) to say

$$X_1 \leq \frac{X_{11}}{a_{11}},\ X_1 \leq \frac{X_{21}}{a_{21}},\ X_1 \leq \frac{S_{11}}{b_{11}},\ X_1 \leq \frac{S_{21}}{b_{21}}$$

since if X_1 is equal to the smallest of these ratios, it certainly cannot exceed any of them. A similar reading can be given for X_2. Now we can get the equivalent form:

$$X_{11} \geq a_{11}X_1,\ X_{21} \geq a_{21}X_1,\ S_{11} \geq b_{11}X_1,\ S_{21} \geq b_{21}X_1$$
$$X_{12} \geq a_{12}X_2,\ X_{22} \geq a_{22}X_2,\ S_{12} \geq b_{12}X_2,\ S_{22} \geq b_{22}X_2 \qquad (11\text{-}26)$$

with at least one *equality* holding in each line.[1]

If a flow or stock is not required at all by an industry, the corresponding a_{ij} or b_{ij} coefficients will be zero; but these technically given constants are never negative (which would mean that the "input" is really an output) or infinite. Aside from this, there are no other restrictions on the capital coefficients. Two industries can have identical capital structures; some industries may use no capital; others may have heavy requirements.[2]

The a_{ij} flow coefficients don't get off so easily, however. They must

[1] Had we written an inequality back in (11-23) and (11-25) we could omit this remark about equality. But why worry? We could deduce the same remark as a theorem from the simplest efficiency condition: clearly we must *never* throw away something of *every* input.

[2] Thus the $[b_{ij}]$ matrix can be singular. Even if it is not, we can say nothing about its inverse as to sign or size of elements.

still satisfy the so-called Hawkins-Simon conditions. These were discussed in connection with the statical Leontief system in the previous chapter. We recall the interpretation that we placed there on the Hawkins-Simon conditions: to produce 1 unit of a good must never require, directly or indirectly (as second- or third- or later-round inputs), more than 1 unit of itself.[1] If we were to set all the capital coefficients equal to zero, we could convert the dynamic system into the older statical system. Naturally enough, if the dynamical system is to be viable, capable of reproducing itself and at the same time yielding some consumption flows, the a_{ij} coefficients must at least satisfy all the productivity conditions they satisfied when capital requirements were ignored. Furthermore, it can be shown that no more stringent conditions need to be imposed on the a's, even when there are positive capital requirements. The net yield that provided consumption in the static model is now divided among consumption and gross investment.

An alternative and perhaps more realistic way of formulating the model would be to change the timing a little. Suppose that society has available, at time t, stocks $S_1(t)$ and $S_2(t)$ of the two commodities. Instead of imagining, as we do, that the flow inputs are produced and used up *within* the production period, we could think of the initial stocks as having to provide *both* for the flow inputs *and* for the capital stocks. Instead of (11-24) we would then require that $S_i \geq X_{i1} + X_{i2} + S_{i1} + S_{i2}$. The over-all balance equations (11-22) are unchanged. The main difference between the two formulations is that in this alternative setup, if the Hawkins-Simon conditions fail to be satisfied by the a's, the system simply runs down gradually over time; the initial stocks are eaten into as flow inputs that the system is too unproductive to make good. Much the same effect on the meaning of the Hawkins-Simon conditions results if we make output at time $t + 1$ depend on flow inputs at time t (and provide for flows at $t + 1$).

11-3-5. The Transformation, or Efficiency Locus Again. Before we reduce the Leontief system to the efficiency-locus form of (11-21), let us pull together the various equations and inequalities of the model. All our notation is established. For each t, S_1 and S_2 stand for the total stock of capital in the form of Commodities 1 and 2; $\Delta S_1 = S_1(t + 1) - S_1(t)$ and $\Delta S_2 = S_2(t + 1) - S_2(t)$ are their rates of growth; X_1 and X_2 are total current outputs; and b_{ij} and a_{ij} are the observable nonnegative unit requirements of the jth industry for the ith capital and flow inputs.

[1] Mathematically speaking, for every connected or indecomposable Leontief system, in which every industry directly or indirectly buys from and sells to every other industry, the open-end system will be capable of producing some positive consumption if and only if the determinant (and hence every principal minor) of the matrix $\mathbf{I} - \mathbf{a} = [\delta_{ij} - a_{ij}]$ is positive.

Finally C_1 and C_2 are the open-end final consumptions of the two commodities. The current input flows X_{ij} and the capital-stock breakdowns S_{ij} are now superfluous, for we have expressed the special Leontief production functions in the form (11-26). We can then substitute for X_{ij} in (11-22) and for S_{ij} in (11-24). Note that in both cases the inequalities all flow unambiguously in the same direction. Let us make these substitutions and recall that the X_i (by their very nature) and the ΔS_i (by explicit assumption) are nonnegative. Then the Leontief dynamic system can *for every t* be written:[1]

$$\begin{aligned} X_1 &\geq a_{11}X_1 + a_{12}X_2 + \Delta S_1 + C_1 \\ X_2 &\geq a_{21}X_1 + a_{22}X_2 + \Delta S_2 + C_2 \\ S_1 &\geq b_{11}X_1 + b_{12}X_2 \qquad \Delta S_1, \Delta S_2, X_1, X_2 \text{ all} \geq 0 \\ S_2 &\geq b_{21}X_1 + b_{22}X_2 \end{aligned} \tag{11-27}$$

We might describe (11-27) as a set of difference *in*equations; the dynamic element is introduced by the terms $S_i(t + 1) - S_i(t)$. At this stage we have parted company with Leontief himself.[2] He insists on equality everywhere in (11-27), no excess capacity anywhere, and permits himself to do what we are unable to do in (11-24*a*). Thus he ends up with a system of difference *equations* corresponding to (11-27).[3] This lends an air of determinacy to his presentation which we shall find to be logically unjustified when we return to this point subsequently.

Our objective is to get from (11-27) to the familiar efficiency locus (11-21). This is essentially just what we did in the static Leontief model. For given amounts of the primary factor labor, we there eliminated the intermediate commodity flows and ended up with a production-possibility frontier. This showed the maximal bill-of-goods menu as limited by the available primary factor. In the dynamic model, at any point of time, the available capital stocks play the role of primary factors. They are historically given quantities, and nonaugmentable *for the moment*, although of course the accumulation of capital is exactly the process we are studying. Final demand now includes both consumption flows and net additions to capital stock. We can hope to proceed much as we did statically, by taking $S_1(t)$ and $S_2(t)$ as given, and finding an efficient or maximal set of $(C_1 + \Delta S_1, C_2 + \Delta S_2)$, or $[C_1(t + 1) + S_1(t + 1) - S_1(t),$

[1] For the case of n commodities we have $X_i \geq \sum_{j=1}^{n} a_{ij}X_j + \Delta S_i + C_i$, $S_i \geq \sum_{j=1}^{n} b_{ij}X_j$, $\Delta S_i \geq 0$, $X_i \geq 0$, with $i = 1, 2, \ldots, n$ in every case. In matrix terms with $\mathbf{a} = [a_{ij}]$, $\mathbf{b} = [b_{ij}]$, and $\mathbf{X}(t)$ and $\mathbf{S}(t)$ representing the vectors of X_i and S_i, we have $\mathbf{X} \geq \mathbf{aX} + \Delta\mathbf{S} + \mathbf{C}$, $\mathbf{S} \geq \mathbf{bX}$.

[2] *Op. cit.*, chap. 3, Eqs. (3.1), (3.2), (3.3), (3.5).

[3] Since Leontief works with continuous time he actually deals with differential equations, but the exact correspondence is easy to see.

$C_2(t+1) + S_2(t+1) - S_2(t)]$. Since $S_1(t)$ and $S_2(t)$ are given, we might just as well seek maximal $[C_1(t+1) + S_1(t+1), C_2(t+1) + S_2(t+1)]$.

Indeed, this turns out to be nothing but a standard linear-programming problem for each t. One way to see this is to consider $\bar{S}_1(t)$, $\bar{S}_2(t)$, $\bar{S}_1(t+1)$, $\bar{C}_1(t+1)$ all as given and $X_1(t)$, $X_2(t)$, and $S_2(t+1) + C_2(t+1)$ as variables. The problem is then to maximize

$$0 \cdot X_1(t) + 0 \cdot X_2(t) + 1[S_2(t+1) + C_2(t+1)]$$

subject to

$$\begin{array}{lllll} & (1-a_{11})X_1(t) & - a_{12}X_2(t) & \geq \bar{S}_1(t+1) \\ & & & \quad + \bar{C}_1(t+1) - \bar{S}_1(t) \\ -S_2(t+1) - C_2(t+1) & - a_{21}X_1(t) & + (1-a_{22})X_2(t) & \geq -\bar{S}_2(t) \\ & b_{11}X_1(t) & + b_{12}X_2(t) & \leq \bar{S}_1(t) \\ & b_{21}X_1(t) & + b_{22}X_2(t) & \leq \bar{S}_2(t) \\ S_2(t+1) & & & \geq \bar{S}_2(t) \\ & \multicolumn{2}{c}{X_1(t) \geq 0,\ X_2(t) \geq 0} & \end{array}$$

Of course, for the historically given $S_1(t)$ and $S_2(t)$, there will be some choices of $\bar{C}_1(t+1)$ and $\bar{S}_1(t+1)$ for which the problem will have no solution. With the given capital stocks, the system certainly cannot be capable of expanding indefinitely and providing unlimited consumption in one period of time. But if we run through all admissible choices of $\bar{C}_1(t+1)$ and $\bar{S}_1(t+1)$, we can find for each such choice the maximal producible $S_2(t+1) + C_2(t+1)$. For the given $S_1(t)$ and $S_2(t)$ we can in this way indicate the maximal combinations of $[S_1(t+1) + C_1(t+1), S_2(t+1) + C_2(t+1)]$ that are possible. This collection is nothing but our sought-for efficiency or transformation locus (11-21).

A more symmetrical procedure is to take any pair of nonnegative numbers K_1 and K_2 and find the X_1, X_2, $\Delta S_1 + C_1$, $\Delta S_2 + C_2$ that maximize $K_1(\Delta S_1 + C_1) + K_2(\Delta S_2 + C_2)$ under the constraints[1]

$$\begin{array}{lllll} \Delta S_1 + C_1 & & - (1-a_{11})X_1(t) & + a_{12}X_2(t) & \leq 0 \\ & \Delta S_2 + C_2 & + a_{21}X_1(t) & - (1-a_{22})X_2(t) & \leq 0 \\ & & b_{11}X_1(t) & + b_{12}X_2(t) & \leq S_1(t) \\ & & b_{21}X_1(t) & + b_{22}X_2(t) & \leq S_2(t) \end{array}$$

By varying the constants K_i we trace out the very same locus. We can then impose our conventional constraint that $S_i(t+1) \geq S_i(t)$.

Note that, as expected, the current flows X_{ii} and X_{ij} wash completely out of the efficiency locus. The K's here play the role of guiding prices, or valuations placed on the new stocks. The bigger the ratio K_1/K_2, the more effort will be channeled into increasing S_1, and vice versa. All that

[1] Once more, since $S_1(t)$ and $S_2(t)$ are given, thinking in terms of maximal $\Delta S_i + C_i$ is equivalent to maximizing $C_i(t+1) + S_i(t+1)$, and sometimes neater.

really matters is that the future's capital stock and consumption are limited by the present's capital stock.

We can get a very good idea of the shape of this locus by working directly with (11-27) and doing some manipulating. For example, we can rewrite the first two lines of (11-27) as follows:

$$(1 - a_{11})X_1 - a_{12}X_2 \geq \Delta S_1 + C_1$$
$$-a_{21}X_1 + (1 - a_{22})X_2 \geq \Delta S_2 + C_2$$

Now we multiply the first inequality by a_{21} and the second by $1 - a_{11}$. Both of these are positive numbers (the latter by the Hawkins-Simon conditions), so the inequalities remain valid. This gives us

$$a_{21}(1 - a_{11})X_1 - a_{21}a_{12}X_2 \geq a_{21}(\Delta S_1 + C_1)$$
$$-a_{21}(1 - a_{11})X_1 + (1 - a_{11})(1 - a_{22})X_2 \geq (1 - a_{11})(\Delta S_2 + C_2)$$

Now we add term by term to get

$$[(1 - a_{11})(1 - a_{22}) - a_{21}a_{12}]X_2 \geq a_{21}(\Delta S_1 + C_1) + (1 - a_{11})(\Delta S_2 + C_2)$$

Naturally it is legitimate to add inequalities this way. Finally the coefficient $(1 - a_{11})(1 - a_{22}) - a_{21}a_{12}$ on the left will be recognized as the determinant

$$D = \begin{vmatrix} 1 - a_{11} & -a_{12} \\ -a_{21} & 1 - a_{22} \end{vmatrix}$$

and the Hawkins-Simon condition tells us that this too is a positive number. Hence,

$$X_2 \geq \frac{a_{21}}{D}(\Delta S_1 + C_1) + \frac{(1 - a_{11})}{D}(\Delta S_2 + C_2)$$

Similarly

$$X_1 \geq \frac{1 - a_{22}}{D}(\Delta S_1 + C_1) + \frac{a_{12}}{D}(\Delta S_2 + C_2)$$

or, to simplify,

$$\begin{aligned} X_1 &\geq A_{11}(\Delta S_1 + C_1) + A_{12}(\Delta S_2 + C_2) \\ X_2 &\geq A_{21}(\Delta S_1 + C_1) + A_{22}(\Delta S_2 + C_2) \end{aligned} \tag{11-28}$$

where the A_{ij} coefficients are positive numbers; in fact, these A's are the very same A's that appeared in the static Leontief model (cf. p. 232). We have been working quite statically except that the terms ΔS_i have been added to final demand. Hence A_{ij} is nothing but the total output of Commodity i needed directly and indirectly to support 1 unit of final demand for Commodity j.[1] We could have passed directly from (11-27) to (11-28) via an appeal to the static model.

[1] In the n-commodity case the A_{ij} are the elements of the inverse matrix $(\mathbf{I} - \mathbf{a})^{-1}$. They are known to be nonnegative. Hence from our earlier $(\mathbf{I} - \mathbf{a})\mathbf{X} \geq \Delta\mathbf{S} + \mathbf{C}$ we deduce $\mathbf{X} \geq (\mathbf{I} - \mathbf{a})^{-1}(\Delta\mathbf{S} + \mathbf{C})$, which generalizes (11-28).

Still on the trail of the efficiency locus, let us substitute (11-28) in the second pair of inequalities in (11-27), noting that the direction of all inequalities is preserved. We find, after a little regrouping,

$$S_1 \geq (b_{11}A_{11} + b_{12}A_{21})(\Delta S_1 + C_1) + (b_{11}A_{12} + b_{12}A_{22})(\Delta S_2 + C_2)$$
$$S_2 \geq (b_{21}A_{11} + b_{22}A_{21})(\Delta S_1 + C_1) + (b_{21}A_{12} + b_{22}A_{22})(\Delta S_2 + C_2)$$

which we may shorten to

$$\begin{aligned} S_1 &\geq B_{11}(\Delta S_1 + C_1) + B_{12}(\Delta S_2 + C_2) \\ S_2 &\geq B_{21}(\Delta S_1 + C_1) + B_{22}(\Delta S_2 + C_2) \end{aligned} \tag{11-29}$$

The B's are all positive,[1] since the A's and b's are.[2]

The inequalities (11-29) are the efficiency relations we want. We can rewrite them slightly to say

$$\begin{aligned} B_{11}[S_1(t+1) + C_1(t+1)] + B_{12}[S_2(t+1) + C_2(t+1)] \\ \leq (1 + B_{11})S_1(t) + B_{12}S_2(t) \\ B_{21}[S_1(t+1) + C_1(t+1)] + B_{22}[S_2(t+1) + C_2(t+1)] \\ \leq B_{21}S_1(t) + (1 + B_{22})S_2(t) \end{aligned} \tag{11-30}$$

It is easy enough now to sketch the efficiency locus as in Fig. 11-11.

Each of the equalities in (11-29) is a straight line, and the inequality defines the region *under* the corresponding line. The first part of (11-29), for example, expresses the limitation on final demand (consumption plus net investment) imposed by the fact that there is only a certain amount of capital stock S_1 available to the economy. This leads us directly to the interpretation of the B_{ij}. As can be seen from the original definitions, B_{ij} represents the amount of Capital Stock i needed to support 1 unit of final demand for Commodity j, taking account of the fact that secondary inputs also require capital for their production. Thus the B_{ij} are the

[1] The direct capital coefficients b_{ij} are nonnegative. The A_{ij} coefficients are nonnegative [from the Hawkins-Simon conditions or the series expansion of $(\mathbf{I} - \mathbf{a})^{-1}$, or from the fact that if A_{ij} were negative, an increase in the final demand for Good j would decrease the required output of Good i]. Hence, at the very least, the B_{ij} total (direct and indirect) capital coefficients are nonnegative. Provided that (1) each capital good finds a use in at least one industry; and (2) every industry sells goods directly or indirectly to every other industry, we can make the stronger statement that every B_{ij} is definitely positive.

[2] In n variables,

$$S_i \geq \sum_{j=1}^{n} b_{ij}X_j \geq \sum_{j=1}^{n} b_{ij} \sum_{k=1}^{n} A_{jk}(\Delta S_k + C_k) = \sum_{k=1}^{n} \left(\sum_{j=1}^{n} b_{ij}A_{jk} \right) (\Delta S_k + C_k)$$
$$= \sum_{k=1}^{n} B_{ik}(\Delta S_k + C_k)$$

Matrix notation simplifies this to $\mathbf{S} \geq \mathbf{bX} \geq \mathbf{b}(\mathbf{I} - \mathbf{A})^{-1}(\Delta\mathbf{S} + \mathbf{C}) = \mathbf{B}(\Delta\mathbf{S} + \mathbf{C})$.

summed direct and indirect capital-requirement coefficients. The second line of (11-29) gives the limitation on final demands imposed by the availability of $S_2(t)$. Producible final demands are those that satisfy *both* inequalities, and so we get a polygonal region like that in Fig. 11-11.

The efficiency locus is the solid-line boundary, made up of two "faces" in this simple case. Along one of the faces, S_1 is the bottleneck and the available S_2 is not being completely used. Along the other face, the situation is reversed. At the point L and *only there*, both capital stocks are being fully used and there is no excess capacity. Thus the efficiency locus is defined by (11-29) with the proviso that *at least one* equality shall hold. Of course Fig. 11-11 holds as of the given $S_1(t)$, $S_2(t)$. If these were different, the locus would be different. It would still be made up of a pair of straight lines with the same slopes as those shown in Fig. 11-11 [since the slopes in (11-29) depend only on the B's]; but the lines would be raised or lowered, depending on whether the corresponding S_i were increased or decreased. For example, if both S_i were increased, the efficiency locus would shift uniformly outward. In fact, if both ΔS_i are positive, this is exactly what will happen in period $t + 1$.

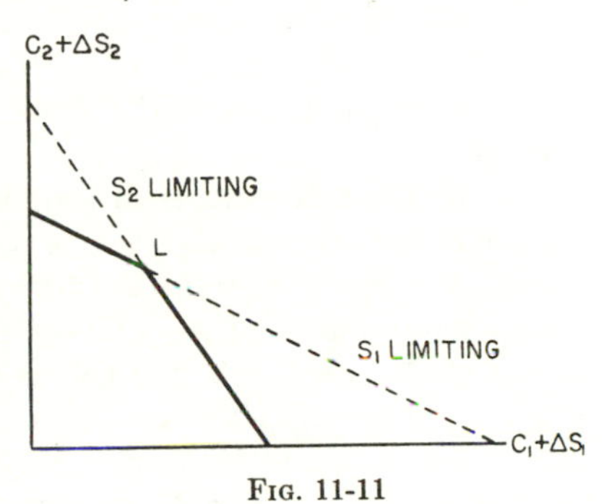

FIG. 11-11

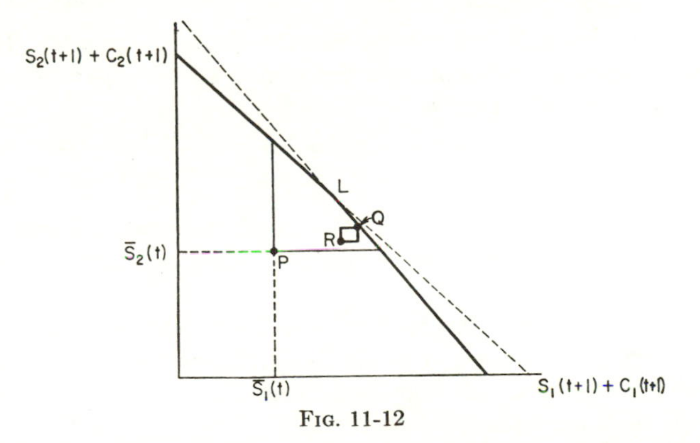

FIG. 11-12

Figure 11-12 is based on (11-30) instead of on (11-29). Along the axes are measured $S_i(t + 1) + C_i(t + 1)$. Figure 11-11 can be derived from Fig. 11-12 just by shifting the origin to the given point $P[\bar{S}_1(t), \bar{S}_2(t)]$. If we hold to the assumption $\Delta S_i \geq 0$, we are limited to that part of the plane lying northeast of the point $[\bar{S}_1(t), \bar{S}_2(t)]$. One advantage of this presentation is that we can see two (and even more) points of time on one diagram. For instance, starting from P, the economy may choose to allocate resources (capital stocks) in such a way as to go to Q on the efficiency locus. From Q we can deduct an allowance for consumption $C_1(t+1)$ and $C_2(t+1)$, and this brings us to the point $R = [S_1(t+1), S_2(t+1)]$, which can be used as an initial point from which to start the whole process over, with a new efficiency locus. The system has progressed, efficiently, from P to R. The extension to *many* periods of time is not so simple as one might think, and will be the subject of the next chapter.

11-3-6. Cost Imputation and the B Coefficients. By definition, B_{12} (for instance) $= b_{11}A_{12} + b_{12}A_{22}$, where the b's are direct capital coefficients and the A's are total (direct and indirect) flow coefficients. It takes a total output of A_{12} units of Commodity 1 to yield 1 unit of final output of Commodity 2, and this in turn requires $b_{11}A_{12}$ units of Capital Stock 1. Similarly, it takes A_{22} units of the second output to yield 1 unit of final demand for Commodity 2, and this requires $b_{12}A_{22}$ units of Capital Stock 1. Thus B_{12} measures the total amount of Capital Stock 1 tied up in the production of 1 unit of net output of Commodity 2. Since in this model the capital stocks are the only scarce resources, the B_{ij} can be interpreted in terms of cost. If the current services of a particular S_i cost (or rent for) r_i dollars per unit of time, and if all other capital were free or were not needed, then the costs of production of Good j would be r_iB_{ij} and in stationary equilibrium this would have to be the price per unit of Good j.

Now what if each S_i had a rent, namely, $(r_1,r_2, \ldots ,r_n)$? Then the total unit costs of flow X_j would be the sum of all the separate costs; and in equilibrium we would have

$$\begin{aligned} p_1 &= r_1B_{11} + r_2B_{21} \\ p_2 &= r_1B_{12} + r_2B_{22} \end{aligned} \tag{11-31}$$

provided that both flow outputs could actually be produced without loss.

Thus the price of each flow is imputed back in total to the scarce factors.

In this connection it is interesting to go back to the second linear-programming problem we used to define the efficiency locus and examine its dual problem (we label the four dual variables p_1, p_2, r_1, r_2 for reasons which will become obvious):

Minimize $r_1S_1(t) + r_2S_2(t)$ subject to

$$\begin{aligned} 1 \cdot p_1 \qquad\qquad\qquad\qquad\qquad\qquad &\geq K_1 \\ 1 \cdot p_2 \qquad\qquad\qquad\qquad &\geq K_2 \\ -(1 - a_{11})p_1 + a_{21}p_2 + b_{11}r_1 + b_{21}r_2 &\geq 0 \\ a_{12}p_1 - (1 - a_{22})p_2 + b_{12}r_1 + b_{22}r_2 &\geq 0 \end{aligned}$$

Now if we interpret the dual variables as flow prices and stock rents we see that we have to choose these shadow prices to minimize the rent on the preexisting capital. In the direct original problem we maximized $K_1(\Delta S_1 + C_1) + K_2(\Delta S_2 + C_2)$, so that the K's were essentially valuations placed on the final demands. The first two dual constraints say that the shadow flow prices must be at least equal to the corresponding valuation K's. If both final demands are positive, we shall have $p_i = K_i$. Now as to the second pair of constraints: If we rewrite them as

$$(1 - a_{11})p_1 - a_{21}p_2 \leq b_{11}r_1 + b_{21}r_2$$
$$-a_{12}p_1 + (1 - a_{22})p_2 \leq b_{12}r_1 + b_{22}r_2$$

and then perform the same kind of arithmetic we used to get from (11-27) to (11-28), the result will be

$$p_1 \leq r_1B_{11} + r_2B_{21}$$
$$p_2 \leq r_1B_{12} + r_2B_{22} \qquad (11\text{-}31a)$$

which is exactly the same as (11-31), except for the inequalities. But if both flows are positive, equality will hold in (11-31a); and correspondingly, if X_1 (for example) were to be zero, there would be no need for the flow price p_1 to equal total factor costs. In fact, price less than cost would reflect the unprofitability of producing any X_1. Moreover, the duality relations tell us that if production takes place at any point other than L in Fig. 11-11, one or the other capital stock will be redundant and its rent r will be zero.

These price and valuation aspects of the theory of capital expansion will come up again in the next chapter.

11-3-7. Balanced Growth in the Leontief System. We return now to the relations (11-27), which summarize the period-to-period production possibilities of the Leontief system. These simultaneous inequalities define the transformation locus (11-2). Because of the special assumption of nonsubstitutability, only simple linear-programming computations were needed to reject inefficient points. Figures 11-11 and 11-12 showed the general appearance of the resulting locus.

Note that the inequalities in (11-27) are as important as the equalities. If our initial stocks $S_1(0)$ and $S_2(0)$ are prescribed arbitrarily, it may be literally impossible to use up both. Indeed, holding S_1 constant and increasing S_2 must ultimately render it redundant, making a free good of its current services. In that case the transformation locus looks as in Fig. 11-13. The transformation locus degenerates to the line AB, to which the stock S_1 limits net output, and there literally is no point L at which all excess capacity is eliminated. We would have $r_2 = 0$ and $p_i = B_{1i}r_1$, $i = 1, 2$.

Of course, the stocks of any historical economy are unlikely to take on completely arbitrary values. So it can certainly happen that all the equalities in (11-27) will hold, with no stock redundant. There will be a whole region of relative values of the initial S's[1] which are *compatible* with zero excess capacities. However, even if some pattern of production *could* use all the stocks, the pattern of open-end demand for the C's and ΔS's might still be such as to create a redundancy of one or more stocks.[2] In fact, this is, in a sense, by far the likeliest outcome. We shall then be on one of the straight-line facets of the locus, and *not* at the vertex point L in Figs. 11-11 and 11-12.

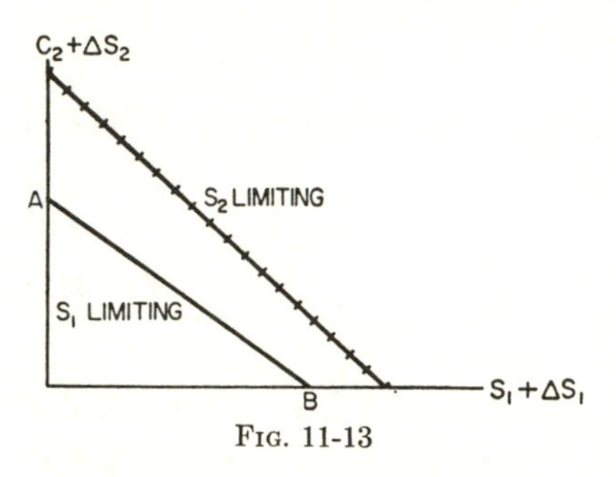

FIG. 11-13

In the interesting Malthus–von Neumann case in which all consumption is zero (a "stock" of labor is "produced" by an input of goods and can be treated like any other S), we can prove that there will always exist one and only one relative configuration of initial stocks that will permit all capacities to grow at the same percentage rate with no excess capacity.[3] Hence in the special Leontief technology there must be such a mode of balanced growth. We can go further and prove that this particular positive rate of balanced growth is the greatest rate of growth of which the system is capable. This means: Suppose there is an initial configuration of stocks such that the system can increase each stock at a relative rate at least equal to λ, say, and some stocks at a greater rate; then λ is smaller than the unique rate of balanced growth described above. Alternatively stated, every initial

[1] Which the reader should find explicitly as an exercise.

[2] Obviously *all* stocks cannot be simultaneously redundant, so in our two-good case one stock at most can be redundant. We might mention here a very singular possibility, namely, that the two capitals should be used (directly and indirectly) in the same proportions in the production of both goods. Then the two lines making up the efficiency locus are parallel. For most initial stocks the two lines never intersect, as in Fig. 11-13. But if the initial stocks are in the right amounts, the two lines coincide and *no* demands induce excess capacity. A similar but more complicated situation can arise with n goods, if the matrix $[B_{ij}]$ is singular.

[3] If we permit excess capacity, we can always obtain steady growth. All we have to do in Fig. 11-12 is draw the ray from the origin through the point P: $[\bar{S}_1(t), \bar{S}_2(t)]$ and extend it until it hits the efficiency frontier. For balanced growth without excess capacity, it is necessary and sufficient that the vertex point L lie on the ray through P. The statement in the text asserts that there is one and only one relative initial configuration for which this occurs.

configuration can give rise to balanced growth with excess capacity. All such rates are smaller than the special rate we have described.[1]

11-3-8. Causal Indeterminacy of the Leontief System. We have proved that every "normal" dynamic Leontief system has one initial relative configuration that can continue to satisfy all the equalities of (11-29), and in fact does so by generating steady, balanced growth. We have also seen that only a very special initial configuration can do this. We must now point out a rather surprising fact: Leontief systems are as often as not such that the slightest disturbance of initial conditions away from the razor's edge of balanced growth will necessarily result in a growth of capitals that will ultimately either (1) violate the requirement that $\Delta S_i \geq 0$ or (2) require us to replace by inequalities one or more of the equalities in (11-29).

A single numerical example will illustrate this result. Suppose $a_{11} = a_{22} = a_{12} = a_{21} = 1/3$ and $b_{11} = b_{22} = 1$, $b_{12} = b_{21} = 0$. Then $B_{11} = B_{22} = 2$, $B_{12} = B_{21} = 1$, and if we insist on equalities, (11-29) becomes

$$2\,\Delta S_1 + \Delta S_2 = S_1$$
$$\Delta S_1 + 2\,\Delta S_2 = S_2$$

If we choose initial conditions $S_1(0) = S_1^0$, $S_2(0) = S_2^0$, these difference equations have a unique solution:

$$S_1(t) = \frac{S_1^0 + S_2^0}{2}\left(\frac{4}{3}\right)^t + \frac{S_1^0 - S_2^0}{2}\,2^t$$
$$S_2(t) = \frac{S_1^0 + S_2^0}{2}\left(\frac{4}{3}\right)^t - \frac{S_1^0 - S_2^0}{2}\,2^t$$

[1] If all equalities are to hold, (11-27) and (11-28) read $(\mathbf{I} - \mathbf{a})\mathbf{X} = \Delta\mathbf{S}$, $\mathbf{bX} = \mathbf{S}$, or $\mathbf{S} = \mathbf{b}(\mathbf{1} - \mathbf{a})^{-1}\Delta\mathbf{S} = \mathbf{B}\Delta\mathbf{S}$. Now a (Frobenius) matrix with positive elements is known to have one and only one positive characteristic root σ^* corresponding to a characteristic vector of all positive elements $(V_1^*, \ldots, V_n^*)$. Balanced growth at rate λ without excess capacity implies $\Delta\mathbf{S} = \lambda\mathbf{S}$; thus $\mathbf{S} = \lambda\mathbf{BS}$ or $\mathbf{BS} = 1/\lambda\mathbf{S}$. Hence the only balanced-growth path without excess capacity is $S_i(t) = V_i^*(1/\sigma^*)^t$. Any initial configuration $\mathbf{S}(t_0)$ proportional to $\mathbf{V}^*$ will continue to grow in these proportions at a relative rate $1/\sigma^*$ and no other initial configuration can perpetuate itself in this way.

That $1/\sigma^*$ is the most rapid rate of growth can be shown as follows: Suppose $\Delta\mathbf{S} \geq \lambda\mathbf{S}$ or equivalently $\mathbf{BS} \leq 1/\lambda\mathbf{S}$, for some positive $\mathbf{S}$ and λ. The inequalities can all be converted into equalities by increasing some elements of $\mathbf{B}$; thus $1/\lambda$ and $\mathbf{S}$ are, respectively, the unique positive characteristic root and vector of a Frobenius matrix whose elements are at least equal to those of $\mathbf{B}$ and are sometimes greater. It is known that increasing an element of such a matrix increases its real root. Hence $1/\lambda > \sigma^*$ or $1/\sigma^* > \lambda$. A more general theorem is stated by Robert M. Solow and Paul A. Samuelson, "Balanced Growth under Constant Returns to Scale," *Econometrica*, **21**:412–424 (July, 1953).

If $S_1^0 = S_2^0$, we are in the maximum rate of balanced growth with $\Delta S_1/S_1 = \Delta S_2/S_2 = 1/3$ per unit time, and $S_1(t) = S_2(t)$ forever.[1] But the important thing to note is that the much more rapidly growing component 2^t appears *with coefficients of opposite signs* in the solutions for $S_1(t)$ and $S_2(t)$. Thus if $S_1^0 \neq S_2^0$, one of the solutions will contain 2^t with a negative sign, and since 2^t gets big faster than $(4/3)^t$ eventually one of the ΔS_i must become negative. For example if $S_1^0 > S_2^0 > 0$, a simple calculation shows that $\Delta S_2(t)$ would finally become negative for

$$t = t^* = \frac{\log\left(\frac{1}{3}\frac{S_1^0 + S_2^0}{S_1^0 - S_2^0}\right)}{\log 3/2}$$

with t^* positive as long as $\Delta S_2(0) > 0$.

After this point of time, what shall be the rules of our system's[2] development? Something must give. We cannot insist that $\Delta S_2 \geq 0$ and that all equalities hold. But from all that has gone before, it is clear that there is no real economic reason to insist on the equalities. The path so defined (as long as a path is defined) is just one among many possible, and efficient, paths. The point of our present discussion is that the dynamic Leontief system is causally ambiguous, and hence there is no simple way of deducing *the* behavior of the system. To repeat for emphasis: Recognition of the prominent role played by the *in*equalities in (11-27) and (11-29) reveals the Leontief dynamic system to be causally indeterminate and incomplete, even if we postulate complete technological nonsubstitutability.

Leontief's own procedure, as we have remarked, is to insist on equalities and determinacy. Numerical examples like ours above indicate that sometimes these "Leontief trajectories" (defined by always moving to the vertex point) arrive at an impasse. They insist that some $\Delta S_i < 0$, which is contrary to assumption. Even if we were to relax the assumption and permit decumulation, this would only postpone trouble, since eventually many Leontief trajectories would lead to some S_i itself being negative and this is clearly nonsensical. Leontief's solution is to maintain determinacy and to make his acceleration coefficients b_{ij} nonlinear. Since he never permits excess capacity he can write

[1] From the symmetry of the a's and b's one would expect balanced growth to require $S_1 = S_2$.

[2] The example $B_{11} = B_{22} = 0.1$, $B_{12} = B_{21} = 1$ shows that there do exist some systems which can develop according to all the equalities. For this system, given any initial positive configuration $S_1^0, S_2^0, \Delta S_1^0, \Delta S_2^0$, the ΔS_i will continue to be positive, and the system will ultimately approach the maximum rate of balanced growth. However, the case in the text is by no means "peculiar" in its choice of B values.

$$\Delta S_1 = b_{11}\,\Delta X_1 + b_{12}\,\Delta X_2$$
$$\Delta S_2 = b_{21}\,\Delta X_1 + b_{22}\,\Delta X_2$$

To avoid contradictions he makes these equations hold only if ΔX_1 and ΔX_2 are ≥ 0. When any ΔX becomes negative, the corresponding b's are put equal to zero (no acceleration on the downswing) with perhaps some allowance for disinvestment via depreciation. Even so, a peculiar kind of impasse may arise at the moment when this switch of regimes is supposed to take place.[1]

From our point of view these "switching troubles" are irrelevant. Figures 11-14*a* and *b* illustrate the causal indeterminacy of the two-good

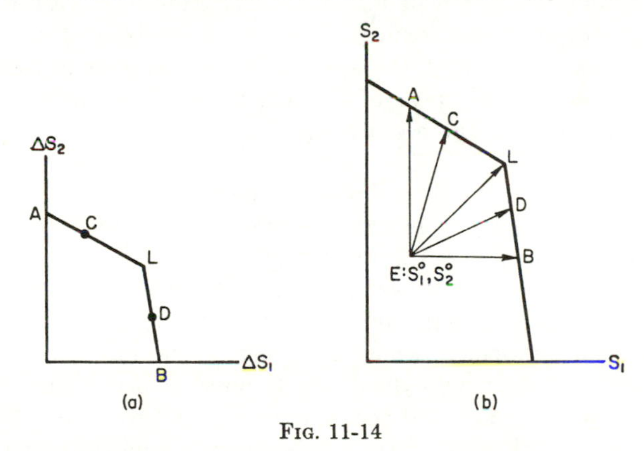

FIG. 11-14

case. Even if the Leontief vertex point L is in the positive quadrant, the economy *must* choose among all the alternative infinity of points on the locus ALB. In Fig. 11-14, from any initial point such as E, the alternative rates of current growth are indicated by the fanning-out arrows. Society must choose at what relative rate it would like to invest in the two commodities. It may choose to increase S_1 and S_2 in the proportions represented by point C or D. A point like C or D (or for that matter A or B) may easily be preferable to L, even though idle capacity may be involved. (Imagine S_1 and S_2 to be military and civilian capital.) A similar need for decision making exists in every subsequent configuration.[2]

[1] It can easily happen that the Leontief trajectories tell a certain ΔX_i to become negative, which means that some b's must become zero. But the *new* trajectories (with some b's zero) tell ΔX_i to become positive (which means that the old rules hold). Then the old rules tell ΔX_i to become negative, etc.

[2] The numerical example in which ΔS_2 became negative illustrates a case in which successive decisions to move to the vertex point L led eventually to a situation in which the constraint lines had shifted (by virtue of the resulting capital accumulation) to a position like that shown in Fig. 11-13, in which the vertex point has dropped below the axis $\Delta S_2 = 0$.

In the last chapter we saw how a static Leontief system dispenses with all choice. There was only one efficient way of producing any given bill of goods. The system was, so to speak, "locked." Leontief, by leaping to the case in which all equalities hold, attempts to achieve the same freedom from the necessity to choose among alternative efficient dynamic plans.[1] We now see that when the system is made dynamic by the introduction of capital stocks, choice becomes inevitable. There are many efficient ways of producing specified consumption goods, differing among themselves in terms of the efficient pattern of capital accumulation implied. Since the historically given initial stocks play the role of primary factors, which can be increased over time by investment, there are too many things to economize. No amount of fixity of flow and capital coefficients can prevent the system from becoming unlocked. We cannot (as Leontief does) speak of *the* path of the system from some given initial configuration. There is no *one* path without further criteria of social choice.

In the next chapter we shall show how such choice criteria can resolve the causal indeterminacy of general systems, subject to technological relations of the kind given in (11-21) or (11-29). Then we shall be in a position to show how a Leontief system's efficient paths of development can be defined. We shall give the name "Leontief trajectories" to the special class of paths or motions in which all the equalities of (11-29) or (11-30) hold, and we shall later show that they do possess special efficiency properties in the process of capital accumulation over time. But they share these properties with an infinity of other motions and take no primacy over these other efficient paths.

11-4. THE VON NEUMANN MODEL

11-4-1. Alternative Production Processes. The Leontief production technology cries out for generalization in two ways. One is to permit joint production so that a single productive activity can have more than one output. The other is to relax the assumption of a fixed ratio of inputs to output. Both of these generalizations can be accomplished without affecting the basically "linear" (constant returns to scale and additivity of the results of separate processes) character of the technology. Actually, the general-equilibrium model of von Neumann,[2] which has both of these generalizations, was first announced in a Princeton lecture in 1931 and was published in German in 1936. We deal here with

[1] Leontief, *op. cit.*, Eq. (3.5), also chooses to write his differential equations in terms of the flows, while we write our difference inequations in terms of the stocks. But this is a matter of convenience.

[2] J. von Neumann, "Über ein ökonomisches Gleichungssystem und eine Verallge-

the second generalization only, retaining the Leontief assumption that each process or activity has only one output. This relieves us of the necessity of keeping track of commodities and activities separately and enables us to identify each industry or activity with the commodity it produces.[1]

Suppose that the total flow X_i in any period can be produced by several Leontief-like production processes. Then $X_i = X_i^{(1)} + X_i^{(2)} + \cdots$, where the superscripts correspond to the various production processes. We can represent by $[a_{1i}^{(1)},a_{2i}^{(1)};b_{1i}^{(1)},b_{2i}^{(1)}]$, $[a_{1i}^{(2)},a_{2i}^{(2)};b_{1i}^{(2)},b_{2i}^{(2)}]$, . . . , the alternative input requirements per unit of output, where the same meaning attaches to the superscripts, and as before, the a's are flow coefficients and the b's are capital coefficients.

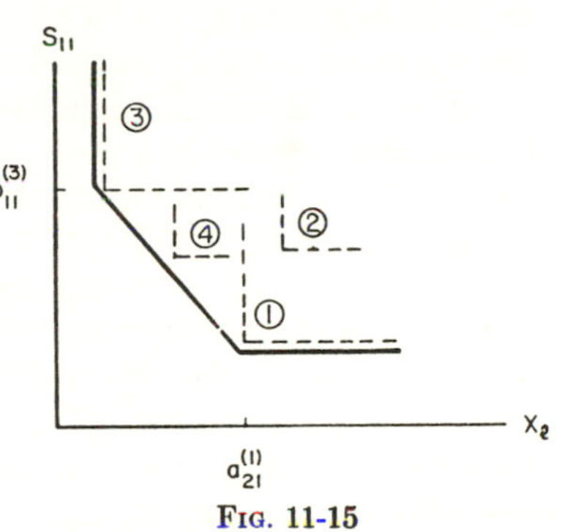

FIG. 11-15

We can't easily draw pictures of the isoquants of this new production function; there are too many inputs. And we can't appeal to the device used earlier in this chapter of holding all but two inputs constant, because each of the alternative processes requires the other inputs to be held constant at different levels. To get some visual idea, however, we can imagine the singular coincidence to have occurred that *all* processes for producing Commodity 1 have *identical* values for two of the four coefficients, say, a_{11} and b_{21}. Then, with those two inputs (X_{11} and S_{21}) held constant at the appropriate level, we can draw, in Fig. 11-15, the unit-output isoquants of four processes producing Commodity 1. Since the separate processes are each simply Leontief-type production activities, we know what their individual isoquants look like (see Fig. 11-10, p. 287), and we could represent on the graph such numbers as $a_{21}^{(3)}$ and $b_{11}^{(2)}$. It is quite obvious that we can forget about Process 2. This is because it requires more of *both* inputs X_{21} and S_{11} (and by assumption

meinerung des Brouwerschen Fixpunktsatzes," *Ergebnisse eines mathematischen Kolloquiums*, (8):73–83 (1935–1936), Franz Deuticke, Leipzig and Vienna, 1937; English translation, "A Model of General Economic Equilibrium," *Review of Economic Studies*, **13**(1):1–9 (1945–1946).

[1] The reader should try his hand at pushing through the exposition of this section with a model that allows joint products. The a_{ij} coefficients can be interpreted as the net input (positive if an input, negative if an output) of Commodity i at the unit level of operation of Activity j. There will be a different number of commodities and activities. This slightly more general model is discussed from a different point of view in Chap. 13.

the same amount of X_{11} and S_{21}) to produce the same output as Process 1 can produce. Process 1 is always better than Process 2 on sheerly technological grounds. For only slightly less obvious reasons, Process 4 can also be disregarded. Process 4 uses less S_{11} to produce 1 unit of output than does Process 3 and less X_{21} than does Process 1. But since we assume constant returns to scale and additivity, by some judicious weighting of Processes 1 and 3 we can get our unit of output of Commodity 1 with inputs anywhere we please along the straight line joining points 1 and 3. And Process 4 is clearly inferior to some such combinations of 1 and 3. If we rule out all such technological inefficiencies, we are left with the heavy line as our combined unit isoquant. Anywhere on the oblique portion of this combined isoquant, Processes 1 and 3 are *both* in use. The isoquant could have two or more facets on each of which two different processes are in use.

11-4-2. The Transformation Locus. At every moment of time, each industry must decide which of its processes or activities to use. This choice involves an optimization problem, which may be solved by linear-programming techniques. The simplest way of viewing this optimization problem is the standard Paretian way: given the existing stocks $S_1(t)$ and $S_2(t)$, the prescribed rates of consumption C_1 and C_2, and one (in general, all but one) of the rates of investment, say ΔS_1, choose the nonnegative activity levels $X_1^{(1)}, X_1^{(2)}, \ldots, X_2^{(1)}, X_2^{(2)}, \ldots$ in such a way as to achieve the highest feasible rate of investment $\Delta S_2 \geq 0$ in the remaining capital stock. Naturally production constraints analogous to (11-27), with account taken of the multiplicity of activities, have to be satisfied. Of course the prescribed consumptions and rates of investment have to be attainable themselves: it would be possible to set some consumptions so high that the program can be met only by living off capital; this possibility is still ruled out. In the familiar neoclassical way we can vary the prescribed rates of investment and trace out a curve or surface analogous to (11-29), which has the well-known Pareto-optimality, or efficiency, property.

The same end result comes to pass if we attach a positive "price," or valuation constant, to each rate of investment and maximize instead a weighted sum of the ΔS_i. This has the advantage of treating all components of final demand symmetrically and leading to an interesting dual problem. By varying the valuation constants we trace out the same efficiency locus as before.

We are thus led to the following problem for each t: Pick nonnegative activity levels and rates of final output $X_1^{(1)} \geq 0$, $X_1^{(2)} \geq 0, \ldots,$ $X_2^{(1)} \geq 0$, $X_2^{(2)} \geq 0, \ldots,$ $C_1 + \Delta S_1 \geq 0$, $C_2 + \Delta S_2 \geq 0$ such that

$$X_1^{(1)} + X_1^{(2)} + \cdots - \sum_{j=1}^{2} a_{1j}^{(1)}X_j^{(1)} - \sum_{j=1}^{2} a_{1j}^{(2)}X_j^{(2)} - \cdots - (C_1 + \Delta S_1) \geq 0$$

$$X_2^{(1)} + X_2^{(2)} + \cdots - \sum_{j=1}^{2} a_{2j}^{(1)}X_j^{(1)} - \sum_{j=1}^{2} a_{2j}^{(2)}X_j^{(2)} - \cdots - (C_2 + \Delta S_2) \geq 0 \qquad (11\text{-}32)$$

$$\sum_{j=1}^{2} b_{1j}^{(1)}X_j^{(1)} + \sum_{j=1}^{2} b_{1j}^{(2)}X_j^{(2)} + \cdots \leq \bar{S}_1$$

$$\sum_{j=1}^{2} b_{2j}^{(1)}X_j^{(1)} + \sum_{j=1}^{2} b_{2j}^{(2)}X_j^{(2)} + \cdots \leq \bar{S}_2$$

and so that $Z = K_1(C_1 + \Delta S_1) + K_2(C_2 + \Delta S_2)$ is at a maximum.

The reader should see that the first set of constraints states that for each of the two commodities the currently produced flow—no matter by which activity it is produced—must at least cover use as current-flow inputs into all the processes of the economy plus consumption plus addition to the capital stocks of the commodity held in the system. The second set of constraints says that the use of capital stocks of each commodity by all the processes of the economy cannot exceed the given stock availabilities.

The solution[1] to this linear-programming problem (there is one for each period of time) will of course depend on the prescribed values of S_1, S_2, K_1, and K_2. The K's are the valuation coefficients (only their ratio really counts) that can be varied through all nonnegative values to generate the whole production-efficiency locus.

What will the efficiency locus of this von Neumann model look like? We can get a good idea by using what we already know about the Leontief system. Suppose we were to select *one* process or activity or set of input coefficients for each commodity. This collection of processes would be a Leontief model, and we know what *its* efficiency locus would look like. There are many such Leontief submodels that we could select from the given von Neumann technology. If there are n_1 processes for producing Commodity 1 and n_2 for producing Commodity 2, then there are n_1n_2 Leontief-type submodels. Let us superimpose the efficiency loci of all such submodels on one graph, as in Fig. 11-16, where there are four altogether. *Any point on or under any one or more of these loci is feasible.*

[1] We can simply omit from problem (11-32) any processes which are known in advance to be technologically inferior to other processes, as explained in connection with Fig. 11-15.

This is because the von Neumann technology can behave like a Leontief technology if it wants to, by suppressing some of its activities. But in general, a von Neumann technology can do better. It can behave like a combination of any two or more of its Leontief subtechnologies by splitting its available resources among them. By constant returns to scale and additivity, this means that any point on the line between two feasible points is also feasible.[1] Filling in this way in Fig. 11-16 we see that the feasible final output points are those on or under the heavy broken line, and the heavy broken line is the efficiency locus we seek.

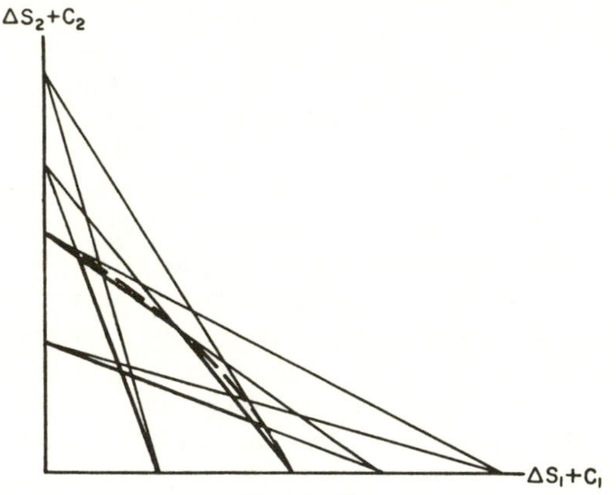

Fig. 11-16

This transformation locus can now have many more corners than it could have in the Leontief nonsubstitution case; but it will almost certainly have fewer than n_1n_2 corners. Evidently the von Neumann model is a halfway house between the single-corner, no-substitution Leontief locus and the smooth, no-corner transformation locus of neoclassical theory.[2]

To relate this feasible set to the linear-programming problem that generates it, we may observe that the "budget lines"

$$K_1(C_1 + \Delta S_1) + K_2(C_2 + \Delta S_2) = \text{constant}$$

form a family of parallel lines, sloping downward to the right. Maximizing such a sum over the feasible set means finding the outermost member of the family which touches the feasible set. By varying the slope of the line $-(K_1/K_2)$ we trace out the northeast frontier of the feasible set, our transformation locus.

[1] That is to say, the set of feasible points is convex.

[2] The rigorous reader may wonder how we know that there are no feasible points outside the heavy line in Fig. 11-16. Such a reader should convince himself as an exercise that every feasible production program in the von Neumann two-good model can be broken down into a linear combination (with positive weights adding to unity) of Leontief-type subtechnologies.

11-4-3. Shadow Prices and the Dual Problem. Exactly as in the Leontief system, the dual problem to (11-32) will determine nonnegative prices p_1 and p_2 of the X_i flows and nonnegative rents r_1 and r_2 of the capital stocks S_i, such that

$$
\begin{array}{l}
p_1, p_2, r_1, r_2 \text{ all} \geq 0 \\
p_1 \leq a_{11}^{(1)}p_1 + a_{21}^{(1)}p_2 + b_{11}^{(1)}r_1 + b_{21}^{(1)}r_2 \\
p_1 \leq a_{11}^{(2)}p_1 + a_{21}^{(2)}p_2 + b_{11}^{(2)}r_1 + b_{21}^{(2)}r_2 \\
\cdots\cdots\cdots\cdots\cdots\cdots\cdots \\
p_2 \leq a_{12}^{(1)}p_1 + a_{22}^{(1)}p_2 + b_{12}^{(1)}r_1 + b_{22}^{(1)}r_2 \\
p_2 \leq a_{12}^{(2)}p_1 + a_{22}^{(2)}p_2 + b_{12}^{(2)}r_1 + b_{22}^{(2)}r_2 \\
p_2 \leq a_{12}^{(3)}p_1 + a_{22}^{(3)}p_2 + b_{12}^{(3)}r_1 + b_{22}^{(3)}r_2 \\
\cdots\cdots\cdots\cdots\cdots\cdots\cdots \\
p_1 \geq K_1 \\
p_2 \geq K_2
\end{array}
\qquad (11\text{-}33)
$$

and so that $Z^* = r_1\bar{S}_1 + r_2\bar{S}_2$ is a minimum.

We can make the now-standard remarks about the shadow prices. (1) The optimal (minimum) value of Z^* will equal the optimal (maximum) value Z of total consumption and capital formation. Thus the entire value of net output is completely imputed back to the scarce capital stocks $\bar{S}_1(t)$ and $\bar{S}_2(t)$. At any nonoptimal configuration, Z^* will exceed Z. (2) No activities have positive profitabilities (which is why optimal imputation exhausts value of net output and nonoptimal imputation exceeds); activities with negative profitabilities, as indicated by strict inequalities in the dual, will be at a zero level in the solution to (11-32); all activities actually used will have zero profitabilities. (3) If any capital stock is redundant, as indicated by an inequality in the original problem, then its services will be a free good in competitive equilibrium, and its rent r_i will be zero. (4) In all "normal" indecomposable economic systems, the X_i flows will never be free goods, and their prices p_i will be positive.[1] (5) No flow price can fall short of the valuation placed on consumption and net investment in that commodity.

11-5. A NEOCLASSICAL MODEL

To conclude this chapter we survey very briefly the way in which the basic transformation locus (11-21) is defined in the case of smooth neoclassical production functions. The explicit distinction between flow and stock inputs is not usually made in standard treatments, but otherwise the technique is familiar and we do not dwell upon it. The marginal-

[1] A p_i can be zero only if X_i is not desired for final demand ($K_i = 0$) *and* if X_i is not an input to any desired commodity or an input to any input to a desired commodity, etc.

productivity calculus and Lagrange multipliers replace the linear-programming setups we have found heretofore.

Our symbols retain the same meaning as before; only now for production functions we write

$$\begin{aligned} X_1 &= F^1(X_{11},X_{21};S_{11},S_{21}) \\ X_2 &= F^2(X_{12},X_{22};S_{12},S_{22}) \end{aligned} \tag{11-34}$$

and F^1 and F^2 possess continuous marginal productivities and obey all the laws of diminishing returns and diminishing marginal rates of substitution. Graphs of sections of these functions have already been given in Fig. 11-9.

Introducing the valuation constants K_1 and K_2, our problem now is to maximize $K_1(C_1 + \Delta S_1) + K_2(C_2 + \Delta S_2)$ subject to the following constraints:

$$\begin{aligned} X_1 &= F^1(X_{11},X_{21};S_{11},S_{21}) \\ X_2 &= F^2(X_{12},X_{22};S_{12},S_{22}) \\ C_1 + \Delta S_1 + X_{11} + X_{12} &= X_1 \\ C_2 + \Delta S_2 + X_{21} + X_{22} &= X_2 \\ S_{11} + S_{12} &= \bar{S}_1 \\ S_{21} + S_{22} &= \bar{S}_2 \end{aligned} \tag{11-35}$$

It would be more exact to write the sign $\leq$ instead of the sign $=$ in each of these constraints, as we have done previously. But the tacit assumption is nearly always made in neoclassical theory that factor proportions are variable over so wide a range that marginal productivities never fall to zero. Similar satiation phenomena are ruled out in consumption. Strong assumptions of this kind will assure us that no commodities will be free and no excess capacity will exist anywhere in the system; i.e., our maxima will be interior maxima, and all equalities hold.[1]

We can substitute for X_1 and X_2 from the second pair of constraints into the first pair, and then introduce Lagrange multipliers λ_1, λ_2, μ_1, μ_2 to form the Lagrangean expression

$$\begin{aligned} L = K_1(C_1 + \Delta S_1) + K_2(C_2 + \Delta S_2) - \lambda_1(C_1 + \Delta S_1 + X_{11} + X_{12} - F^1) \\ - \lambda_2(C_2 + \Delta S_2 + X_{21} + X_{22} - F^2) \\ - \mu_1(S_{11} + S_{12} - \bar{S}_1) - \mu_2(S_{21} + S_{22} - \bar{S}_2) \end{aligned}$$

Now we have to differentiate L partially with respect to the unknowns X_{11}, X_{12}, X_{21}, X_{22}, S_{11}, S_{12}, S_{21}, S_{22}, $C_1 + \Delta S_1$, $C_2 + \Delta S_2$ and set these derivatives equal to zero. This will yield 10 equations. The six con-

[1] These assumptions are often stated by saying that free goods are simply not counted in our list of goods and factors. This, however, is begging the question, since which goods and factors are free is determined by the economic process itself, not exclusively by nature. This problem is taken up again in Chap. 13.

straints provide additional equations to determine X_1, X_2 and the four multipliers.

When we differentiate with respect to $C_i + \Delta S_i$, we get

$$K_1 = \lambda_1,\ K_2 = \lambda_2 \tag{11-36}$$

When we differentiate with respect to X_{ij}, we get

$$\begin{aligned} -\lambda_1\left(1 - \frac{\partial F^1}{\partial X_{11}}\right) &= 0 \\ -\lambda_2\left(1 - \frac{\partial F^2}{\partial X_{22}}\right) &= 0 \\ -\lambda_1 + \lambda_2 \frac{\partial F^2}{\partial X_{12}} &= 0 \\ -\lambda_2 + \lambda_1 \frac{\partial F^1}{\partial X_{21}} &= 0 \end{aligned} \tag{11-36a}$$

When we differentiate with respect to S_{ij}, we get

$$\begin{aligned} \lambda_1 \frac{\partial F^1}{\partial S_{11}} - \mu_1 &= 0 \\ \lambda_1 \frac{\partial F^1}{\partial S_{21}} - \mu_2 &= 0 \\ \lambda_2 \frac{\partial F^2}{\partial S_{12}} - \mu_1 &= 0 \\ \lambda_2 \frac{\partial F^2}{\partial S_{22}} - \mu_2 &= 0 \end{aligned} \tag{11-36b}$$

Remarkably enough we can think of the Lagrange multipliers as shadow prices.[1] λ_1 and λ_2 are our flow prices p_1 and p_2, and μ_1 and μ_2 are our stock rents r_1 and r_2.

Equations (11-36) simply say that at our interior maximum, with something of everything being produced and no idle factors, the flow prices must equal the valuations placed on final demands.

The first two equations of (11-36*a*) say that the marginal flow productivity of a commodity in producing itself must be equal to 1.[2] If it were greater than 1, it would always pay to increase the flow input and get a still larger amount of the same flow as output. If it were less than 1, it would similarly pay to decrease the input.

The second pair of equations in (11-36*a*) and all four equations in (11-36*b*) are straightforward marginal-productivity conditions. The

[1] In fact had we left the inequalities in our constraints, we would be able to say that strict inequalities go along with zero Lagrange multipliers, just as in the duality of linear programming.

[2] Or else the corresponding shadow price must be equal to zero.

former say that value of marginal product of flow input equals price of flow input.[1] The latter say that value of marginal product of stock input must equal rent of stock. It is a useful exercise to see how inequalities could enter into these conditions, along with the duality interrelationships. The interested reader can also go through the usual process of eliminating the multipliers in (11-36*a*) and (11-36*b*) by taking ratios, thus stating the optimality conditions in terms of ratios of marginal productivities or marginal rates of substitution.

Just as before, we can vary the ratio of the valuation constants K_1 and K_2 through all nonnegative values and trace out the whole transformation locus. It gives the nonimprovable combinations of $C_1 + \Delta S_1$ and $C_2 + \Delta S_2$ [or $S_1(t + 1) + C_1(t + 1)$ and $S_2(t + 1) + C_2(t + 1)$] that can be produced from the initial endowments $S_1(t)$ and $S_2(t)$. We see that, when all is said and done, whether we start with fixed proportions, alternative fixed proportions, or neoclassically variable proportions, society must in all cases choose among an optimal menu of final outputs: consumptions plus capital formation. We can rule out the nonoptimal combinations that lie beneath the efficiency locus, but there still remains a basic choice for every period of time.

In the next chapter we shall make use of the efficiency locus developed here to study society's alternatives for consumption and investment programs extending over many periods of time. We shall see that intertemporal efficiency requires *more* than the period-to-period optimality we have studied so far. And we shall find what choices society has to make to render determinate its path of capital accumulation over time.

[1] The first pair in (11-36*a*) could be interpreted this way also.

12

Efficient Programs of Capital Accumulation

12-1. INTRODUCTION

The end result of the previous chapter was to supply us with the instantaneous technological transformation locus (or production-efficiency locus):

$$T[S_1(t), S_2(t); S_1(t+1) + C_1(t+1), S_2(t+1) + C_2(t+1)] = 0 \quad (12\text{-}1)$$

Most of the time we shall want to use this locus in slightly different form, after it has been solved for one of the net outputs, say,

$$S_2(t+1) + C_2(t+1) = F[S_1(t+1) + C_1(t+1); S_1(t), S_2(t)] \quad (12\text{-}2)$$

These loci have the virtue of being able to include as special cases almost any economic theorist's model of capital, e.g., the special point-input, point-output model of Jevons, the Böhm-Bawerk triangular capital model, the various models involving produced durable goods (Evans, Lange, and others).

We know already that this instantaneous, one-period locus requires many efficiency conditions to be satisfied in the background. It might naturally be thought that no more can be required in the way of production-efficiency conditions than that the system be operating optimally at each and every instant of time. Consumers or the market would then decide, via tastes and time preference, how rapidly capital stocks are to grow to increase future consumption possibilities at the expense of current consumption.

Such a view is short-sighted and incomplete. *It overlooks important additional intertemporal production-efficiency conditions which have received little emphasis in the literature of economic theory.* It is the task of this chapter to elucidate these multiperiod requirements for optimality and to indicate some of the dual price and interest implications.

For this analysis we reverse the order of the previous chapter and study the smooth neoclassical production-possibility schedule first. Afterward comes the Leontief no-substitution model. This is to take advantage of the economist's familiarity with marginal-rate-of-substitution and own-rate-of-interest concepts. The Leontief case will then be clearer by analogy, and linear programming provides the needed analytical technique. Throughout this chapter we continue the previous convention of treating the graphically accessible case of two commodities and of counting time in discrete periods. The case of continuously flowing time can be handled by the more sophisticated methods of the calculus of variations in terms of n commodities and capital stocks, but we do not give this extension here.

12-2. INTERTEMPORAL EFFICIENCY CONDITIONS IN THE SMOOTH CASE

The transformation function (12-2) can be thought of as derived from neoclassical production functions in which inputs are smoothly substitutable for each other, obeying the law of constant returns to scale and the generalized law of smoothly diminishing returns as proportions are varied. It represents an efficiency frontier in the sense of giving the maximum obtainable $S_2(t + 1) + C_2(t + 1)$ for specified capital-stock availabilities $S_1(t)$ and $S_2(t)$ and specified carryover $S_1(t + 1) + C_1(t + 1)$. Because the underlying production functions have smooth marginal productivities, so will the transformation function F. Because we banish all scale effects, doubling all the variables in F will just double the left-hand side. F is a homogeneous function of the first degree.[1]

12-2-1. Intertemporal Efficiency Conditions. At any one point of time, F describes the best menu available to society. But life goes on. Whatever stocks $S_1(t + 1)$ and $S_2(t + 1)$ are retained will become the inputs to produce a menu for $t + 2$. We must inquire whether extending the horizon in this way adds anything not already contained in our instantaneous efficiency locus. Imagine initial stocks $[S_1(0), S_2(0)]$ to be given. Imagine consumption $[C_1(1), C_2(1)]$ to be specified at whatever level tastes might have decreed. Then what maximum frontier or best menu of $S_1(2) + C_1(2)$, $S_2(2) + C_2(2)$ can we hope for at the end of two periods? We might just as well take $[C_1(2), C_2(2)]$ as also specified and ask for the maximal frontier of capital stocks $[S_1(2), S_2(2)]$ which we can bequeath to posterity. It is clear, or experiments will soon show, that there are numerous alternative time paths of development which do satisfy the instantaneous relations (12-2) at *each* period of time and which provide the same profile of consumption, but which wind up at the end of Period 2 with different amounts of capital. One such path might

[1] This assumption could be lightened, but then the role of competitive market prices would become ticklish.

easily end up with less of every capital stock than some other path. Obviously, we must regard such a time path as *inefficient, even though it satisfied* (12-2) *at* $t = 0$ *and* $t = 1$. The point is simply this: There are many efficient ways of providing $C_1(1)$ and $C_2(1)$, and each way leaves a different composition of capital stocks $S_1(1)$ and $S_2(1)$. Some of these capital stocks will be quite inappropriate for the subsequent provision of

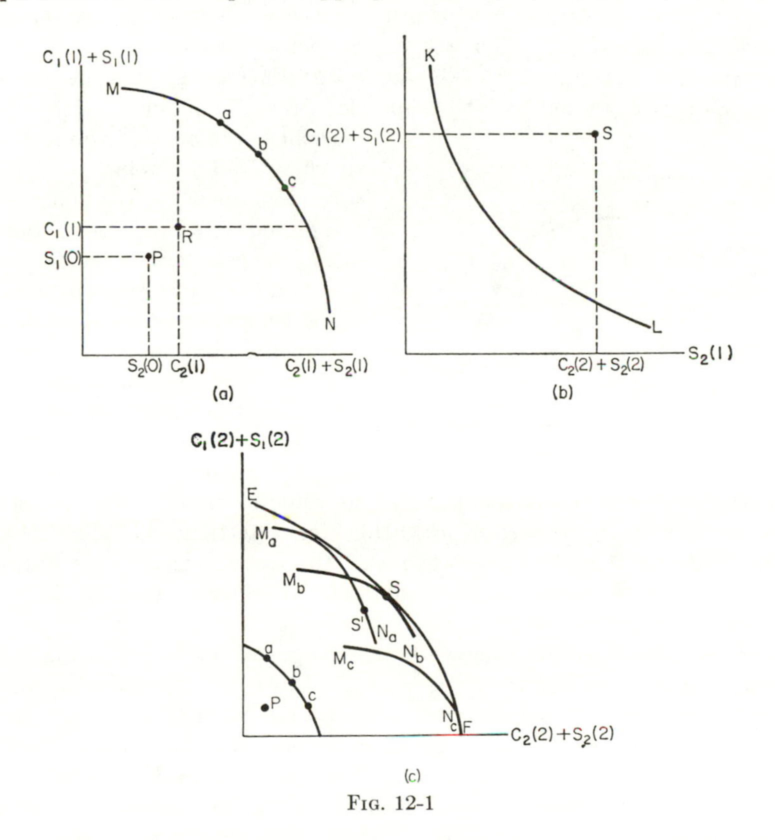

FIG. 12-1

$C_1(2)$ and $C_2(2)$. We must select among the instantaneously efficient time paths only those whose final stocks cannot all be improved upon.

Figure 12-1*a* is familiar from the last chapter. For the initial endowment $[S_1(0), S_2(0)]$ it shows how much is producible at Time 1, over and above the prescribed consumption point R: $[C_1(1), C_2(1)]$. Figure 12-1*b* shows the isoquant, or input aspect, of (12-2). To each point of outputs for Period 2, such as S, there is a concave locus of *minimal* required inputs in the previous period, such as KL.

Now note that each and every point on the instantaneous efficiency frontier MN in Fig. 12-1a will, after Period 1's consumptions have been subtracted off, be regarded as an initial input for the output of Period 2. Hence every point such as a, b, and c on MN generates in Fig. 12-1c a new and different instantaneous efficiency locus such as M_aN_a, M_bN_b, M_cN_c, etc. Intertemporal efficiency means that we want to get as northeast as possible in Fig. 12-1c. Clearly, to get the most of both goods in this sense we must (and can) end up on the *envelope EF* of the separate loci M_aN_a, M_bN_b, etc. The fact that perpetual one-period efficiency can be inefficient over longer periods can now be illustrated. If society wants commodities in Period 2 in the proportions given by point S in Fig. 12-1c, efficiency *requires* that the way station for $t = 1$ be point b. A path to a, and thence to S', violates no one-period efficiency rule, but is inefficient compared with the path to b and S. Only paths leading to the envelope are efficient.

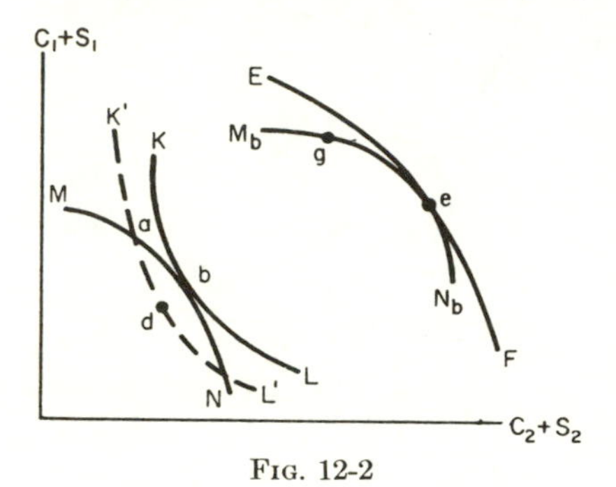

FIG. 12-2

Anyone familiar with modern economic theory could guess the rule that leads to the efficiency envelope. He would suspect that marginal rates of substitution must be proportional in some sense. And he would be right. We shall show presently that a necessary condition for intertemporal efficiency is the following: The MRS between any two goods regarded as outputs of the previous period must equal their MRS as inputs for the next period.

Graphically, this rule is shown by the tangency conditions in Fig. 12-2. Point e is an efficient envelope point. Why? Because the required-inputs isoquant to e, labeled KL, is tangential to MN at b. An inefficient point like g generates a requirements locus $K'L'$ that intersects MN at a. Observe that by moving along MN from a toward b it would have been possible to achieve more of both stocks than at, say, d. But d can produce g. Hence g must be inefficient. This kind of production arbitrage through time is impossible if the MRS tangency conditions hold.

12-2-2. Analytic Formulation. In terms of partial derivatives the proportionality rule can be written

$$\left[\frac{\partial S_2(1)}{\partial S_1(1)}\right]_{C_i(1) \text{ and } S_i(0) \text{ constant}} = \left[\frac{\partial S_2(1)}{\partial S_1(1)}\right]_{C_i(1) \text{ and } S_i(2) \text{ constant}}$$

or in terms of the derivatives of the transformation curve as

$$\frac{\partial F[S_1(1) + C_1(1); S_1(0), S_2(0)]}{\partial S_1(1)} = -\frac{\dfrac{\partial F[S_1(2) + C_1(2); S_1(1), S_2(1)]}{\partial S_1(1)}}{\dfrac{\partial F[S_1(2) + C_1(2); S_1(1), S_2(1)]}{\partial S_2(1)}} \tag{12-3}$$

The left-hand side is the slope of MN, the right-hand side the slope of KL (Fig. 12-2).

To derive this all-important marginal efficiency-envelope condition we have to maximize

$$S_2(2) + C_2(2) = F[S_1(2) + C_1(2); S_1(1), S_2(1)]$$

subject to prescribed $C_1(1)$, $C_2(1)$, $C_1(2)$, $C_2(2)$, $S_1(2)$, $S_1(0)$, $S_2(0)$, and $F[S_1(1); S_1(0), S_2(0)] - S_2(1) - C_2(1) = 0$. The variables in our problem are $S_1(1)$, $S_2(1)$, and $S_2(2)$. The last of these disappears because we can maximize F. We can also get rid of $S_2(1)$ by using the constraint. In abbreviated but unambiguous notation, we have to maximize

$$F[S_1^2 + C_1^2; S_1^1, F(S_1^1; S_1^0, S_2^0) - C_2^1] = f(S_1^1)$$

since $S_1(1)$ is the only variable left. Hence all we have to do is differentiate with respect to S_1^1 and set $f'(S_1^1) = 0$. With a little calculation we get (12-3).[1]

We can strengthen our intuitive grasp of (12-3) by juggling it around to make it read

$$\frac{\partial F[S_1(2) + C_1(2); S_1(1), S_2(1)]}{\partial S_2(1)} = -\frac{\dfrac{\partial F[S_1(2) + C_1(2); S_1(1), S_2(1)]}{\partial S_1(1)}}{\dfrac{\partial F[S_1(1) + C_1(1); S_1(0), S_2(0)]}{\partial S_1(1)}} \tag{12-4}$$

The left-hand side is $\partial S_2(2)/\partial S_2(1)$, a direct *own* rate of interest in terms of Good 2. It shows how much more of Good 2 we could dispose of over this period, had there been a little more of it in the productive "bank" in the previous period. The right-hand side is a little complicated. It tells us how much more of Good 2 we could have in this period had we indirectly sacrificed some $S_2(1)$ to get more $S_1(1)$ (the denominator) and used the latter to produce more $S_2(2)$ (the numerator). The intertemporal efficiency condition (12-4) says that on the margin the direct and indirect processes must yield the same.

[1] The calculus-trained reader can work out the second-order convexity conditions that (12-3) determines a true maximum. We are assuming the maximum to be an interior one, so none of the nondecumulation, nonnegativity conditions come into play.

12-2-3. Many Goods and Many Periods. Without going into detail it can simply be stated that the same envelope rule applies for any number of goods. Any pair of goods must satisfy the rule (as the reader can prove by holding all but two S's constant and going through the previous reasoning). With three goods, this yields two independent conditions.[1] With n goods we select any one as numeraire, pair each of the other $n - 1$ with it, and derive $n - 1$ independent conditions, much like (12-3) or (12-4).

It is much more interesting and important to consider optimal programs extending over more than two future time periods. The two-good

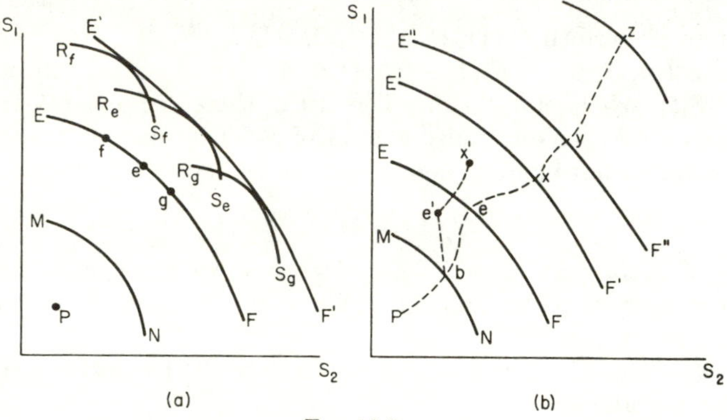

FIG. 12-3

case will provide enough generality. Refer back to Fig. 12-1c. Starting with the initial stocks $[S_1(0), S_2(0)]$ at P, and with all consumption points prescribed, we have the efficiency locus for $t = 1$, and the envelope frontier EF of goods that the system can have left at $t = 2$. Now what can $t = 3$ provide? Each point on EF in Fig. 12-1c or 12-2 can generate a new instantaneous efficiency locus in Period 3, as shown by R_fS_f, R_eS_e, R_gS_g in Fig. 12-3a. The best that society can arrange to do in this period is to reach the envelope $E'F'$, a sort of envelope-to-the-envelope.

The logic is the same for any future period. For $t = 4, 5, \ldots,$ we wish to reach the maximal frontier compatible with technological possibilities, consumption profiles, and initial conditions. Reflection shows that the solution to this problem is given by the succession of envelopes-to-envelopes-to-envelopes, etc. To see this it is enough to note that once we have found the maximal frontier for T periods of time, the frontier for $T + 1$ periods must be the envelope of instantaneous efficiency loci starting from all points of the T envelope. It can never pay to start from

[1] There are three pairs that can be constructed from three goods. But if the rule holds for any two pairs, it must hold for the third.

inside the T envelope, it is impossible to start from outside, and this is the best that can be done from points on it. Figure 12-3b shows the proliferation of envelopes attainable at each subsequent period from the initial point P; it is understood that consumptions $C_i(1)$, $C_i(2)$, . . . , are specified. It will bear repeating that only paths which hop from envelope to envelope to envelope have any claim to efficiency. Once we are off the sequence of envelopes, a uniformly better point can be found, and hence a uniformly better future. Such an optimal path is *Pbexy . . . z* in the diagram. *Pbe'x'* is inefficient.

The relationship between efficiency envelopes and efficient paths should be understood. The envelopes could be defined as the loci of successive terminal points on optimal paths. Conversely, optimal paths are those that go from envelope to envelope. Each optimal path crosses each efficiency envelope once and only once. Each point on an efficiency envelope lies on one optimal path (and only one, if we assume diminishing MRS).

Do we have to find new super-duper envelope rules to cover programs involving more than two periods? Fortunately not, and the above reasoning shows why. Our simple envelope rule (12-3) or (12-4) handles all cases! All we need do is replace 0, 1, 2, by t, $t + 1$, $t + 2$ and require the equation to hold for all t.

12-2-4. Formal Analysis. The mathematical maximum problem involved is easily set down. We are given initial stocks $S_1(0) = S_1^0$, $S_2(0) = S_2^0$; consumptions $C_i(0) = C_i^0$, $C_i(1) = C_i^1$, . . . , $C_i(T) = C_i^T$, $i = 1, 2$; and one terminal stock $S_1(T) = S_1^T$. Subject to

$$F(S_1^t + C_1^t; S_1^{t-1}, S_2^{t-1}) - C_2^t - S_2^t = 0 \qquad t = 1, 2, \ldots, T$$

maximize $S_2^T + C_2^T = F(S_1^T; S_1^{T-1}, S_2^{T-1})$.

The variables in this problem are the intermediate capital stocks S_i^1, S_i^2, . . . , S_i^{T-1}. The S_2^t could be eliminated by the instantaneous efficiency constraint, and the resulting function differentiated with respect to its remaining variables S_1^1, S_1^2, ... But this substitution procedure is lengthy, and here it is best to introduce Lagrange multipliers, one for each time period, and to write the new expression[1] as follows:

$$G = F(S_1^T + C_1^T; S_1^{T-1}, S_2^{T-1}) + \sum_{t=1}^{T-1} \lambda_t [F(S_1^t + C_1^t; S_1^{t-1}, S_2^{t-1}) - S_2^t - C_2^t]$$

$$= F^T + \Sigma\lambda_t[F^t - S_2^t - C_2^t]$$

Now differentiate with respect to S_1^t and S_2^t and set the derivatives

[1] Some readers may prefer to skip this paragraph and just look at the results.

equal to zero:

$$\frac{\partial G}{\partial S_1^t} = \lambda_t \frac{\partial F^t}{\partial S_1^t} + \lambda_{t+1} \frac{\partial F^{t+1}}{\partial S_1^t} = 0 \qquad t = 1, 2, \ldots, T-1$$

$$\frac{\partial G}{\partial S_2^t} = -\lambda_t + \lambda_{t+1} \frac{\partial F^{t+1}}{\partial S_2^t} = 0 \qquad \lambda_T = 1, \text{ by convention}$$

Hence,

$$\lambda_t \frac{\partial F^t}{\partial S_1^t} = -\lambda_{t+1} \frac{\partial F^{t+1}}{\partial S_1^t}$$

$$\lambda_t = \lambda_{t+1} \frac{\partial F^{t+1}}{\partial S_2^t}$$

and dividing one equation by the other we find

$$\frac{\partial F^t}{\partial S_1^t} = -\frac{\partial F^{t+1}/\partial S_1^t}{\partial F^{t+1}/\partial S_2^t}$$

or, written out in full, without abbreviation,

$$\frac{\partial F[S_1(t) + C_1(t); S_1(t-1), S_2(t-1)]}{\partial S_1(t)} = -\frac{\partial F[S_1(t+1) + C_1(t+1); S_1(t), S_2(t)]/\partial S_1(t)}{\partial F[S_1(t+1) + C_1(t+1); S_1(t), S_2(t)]/\partial S_2(t)} \qquad (12\text{-}5)$$

$$t = 1, 2, \ldots, T$$

Comparing this with (12-3), we verify what was said above. The frontier for a T-period program is defined by the original two-period envelope condition (12-3), repeated for each two-period stretch. The interpretation in terms of paths moving always from envelope to envelope has already been given.[1]

It is worth noting explicitly that (12-5), together with the instantaneous locus itself,

$$S_2(t) + C_2(t) = F[S_1(t) + C_1(t); S_1(t-1), S_2(t-1)] \qquad t = 1, \ldots, T \qquad (12\text{-}2a)$$

provides a system of two *difference equations* for the unknown capital accumulation programs $S_1(t)$, $S_2(t)$. Equation (12-5) is of second order—it involves two lags; (12-2a) is of first order. Correspondingly there are three boundary conditions: $S_1(0)$, $S_2(0)$, and $S_1(T)$ are prescribed [as are the $C_i(t)$, which play the role of arbitrary functions]. The dynamic efficiency equations are nonlinear. We shall have to analyze them a bit more closely later.[2]

[1] The secondary conditions for a true maximum are too long-winded to be set down here. The calculus-trained reader should write them out for, say, $T = 3$, and interpret them in terms of generalized diminishing returns.

[2] Instead of beginning with initial stocks as at P in Fig. 12-3 and working outward to successive expanding envelopes, we could have begun with a prescribed terminal

12-2-5. Own-rates of Interest, Flow Prices, and Stock Rents. We have already given a purely "technocratic" interpretation of the efficiency conditions in terms of direct and indirect processes of production over time. It is natural to wonder whether a shadow price formulation is possible, linking up intertemporal efficiency with competitive market behavior. Such is indeed the case.

Let the money price of a unit of Commodity 1, delivered at Time t, be $p_1(t)$ and that of the second commodity $p_2(t)$.[1] We also need the concept of the rent per period of each capital good S_i. Thus $r_1(t)$ is the rent, for the tth period, of the services of 1 unit of S_1, reckoned in money terms. Likewise $r_2(t)$ is the money rent per unit of Stock S_2. These rents are *net* earnings over and above necessary maintenance and replacement expense.

Now consider r_1/p_1, the rent per period of 1 unit of S_1 divided by the price of 1 unit of S_1. This ratio is a pure number, or percentage, per unit time. If $r_1 = 2$ and $p_1 = 20$, then $r_1/p_1 = 1/10$ and we may say that one period's use of S_1 costs $1/10$ unit of S_1. An owner of S_1 can consume one-tenth of his stock annually, and his rental earnings will just suffice to maintain his stock intact. In money terms he can consume $r_1(t)S_1(t)$. If he wishes to devote all his rents to investment in S_1, he can convert money rents of $r_1(t)S_1(t)$ into $[r_1(t)S_1(t)]/[p_1(t)]$ new units of S_1. Hence his capital stock will grow according to the rule

$$S_1(t+1) = S_1(t) + \frac{r_1(t)}{p_1(t)} S_1(t) = \left[1 + \frac{r_1(t)}{p_1(t)}\right] S_1(t) \qquad (12\text{-}6)$$

From this formula it is clear that $r_1(t)/p_1(t)$ behaves like an interest rate; it is in fact the own-rate of interest per period of Good S_1.[2]

The own-rates of different goods need not be equal in equilibrium. In fact, they must not all be equal if relative prices are changing. Any good whose relative price is rising will have a low own-rate; any good whose relative price is falling will have a high own-rate. The fundamen-

---

point $[S_1(T),S_2(T)]$ and worked backward. First there is a locus of minimal requirements at $T - 1$. Each point on this locus in turn generates a locus for $T - 2$, and the inmost envelope of these loci gives the minimal requirements two periods earlier. An excellent exercise would be to carry through the formal analysis in terms of these ever-contracting requirement loci.

[1] Prices are quoted in money terms purely for convenience. There is no cash, and hence no liquidity problem, in our model economy. Money serves only as a unit of account. We could instead have chosen Commodity 1 as numeraire and set $p_1(t) \equiv 1$. Other commodity prices would then be expressed in terms of numeraire.

[2] If S_1 were chosen as numeraire, so that $p_1 \equiv 1$, then the formula would reveal r_1 itself to be an own-rate. The numeraire own-rate is what we think of as *the* rate of interest, "the rent of money." Fisher, Wicksell, Marshall, Thornton, Sraffa, Keynes, Lerner, and others have discussed own-rates.

tal arithmetical relationship between own-rates can be easily worked out. One dollar will buy $1/p_1(t)$ units of S_1. According to (12-6), this will yield money rents of $r_1(t)/p_1(t)$ and the stock itself will have a sale value of $p_1(t+1)/p_1(t)$ in the next period. And the same applies to S_2. Under competition the net advantages of investing in the two stocks must be equal; hence,

$$\frac{r_1(t)}{p_1(t)} + \frac{p_1(t+1)}{p_1(t)} = \frac{r_2(t)}{p_2(t)} + \frac{p_2(t+1)}{p_2(t)} = r_0(t) + 1 \qquad (12\text{-}7)$$

The extreme-right-hand member shows what would happen to a numeraire good; that is, r_0 is essentially a money rate of interest. Equation (12-7) confirms that appreciating goods have low own-rates and depreciating goods have high own-rates. If the equivalence of (12-7) were not realized, an arbitrager could change from one kind of investment to another, thereby tending to make sure profits and also tending to wipe out the discrepancies. For example, if $r_1(t)/p_1(t) + p_1(t+1)/p_1(t)$ were to exceed $r_2(t)/p_2(t) + p_2(t+1)/p_2(t)$, it would clearly pay to convert cash and S_2 into S_1, to hold the S_1 (collecting rents) for one period, and then, if desired, to convert back to cash, or S_2. This would tend to increase $p_1(t)$, decrease $p_2(t)$, and restore the equality. Anyone will be content with a lower rent if he can be sure that his stock will be increasing relatively in value.

12-2-6. Competitive Markets and Dynamic Efficiency. Now let us use what we know about competitive equilibrium to connect up these price relationships with technological characteristics. For stock rents we have the usual value-of-marginal-product equations and for commodity prices we have the usual MRS equations. Thus, $r_2(t)$ must equal the value (at next period's prices) of the marginal product of $S_2(t)$ in the production of $S_2(t+1)$.

$$r_2(t) = \left\{ \frac{\partial F[S_1(t+1);S_1(t),S_2(t)]}{\partial S_2(t)} - 1 \right\} p_2(t+1) \qquad 12\text{-}8)$$

The -1 appears because we don't want to count the initial increment in $S_2(t)$ as part of its own marginal product. Using subscripts 1, 2, and 3 for partial derivatives with respect to the successive arguments, we can simplify the notation so that (12-8) becomes $r_2(t) = (F_3^{t+1} - 1)p_2(t+1)$. Correspondingly, the net marginal product of $S_1(t)$ in producing $S_1(t+1)$ is $(-F_2^t/F_1^t) - 1$, and we get

$$r_1(t) = \left(-\frac{F_2^{t+1}}{F_1^{t+1}} - 1 \right) p_1(t+1) \qquad (12\text{-}8a)$$

Finally, the price ratio $p_1(t)/p_2(t)$ must under competition equal the MRS between $S_1(t)$ and $S_2(t)$ as outputs; hence,[1]

$$\frac{p_1(t)}{p_2(t)} = -F_1{}^t = \frac{\partial S_2(t)}{\partial S_1(t)} \tag{12-8b}$$

$$\frac{p_1(t+1)}{p_2(t+1)} = -F_1{}^{t+1} \tag{12-8c}$$

Equations (12-7) and (12-8*a*–*c*) are the competitive price and own-rate equilibrium conditions for a dynamical capital-accumulating system. We shall now show that *together they imply the purely "technological" intertemporal efficiency conditions* (12-5), which we deduced from entirely nonmarket considerations.

Insert (12-8) and (12-8*a*) in (12-7) to get

$$\left(-\frac{F_2{}^{t+1}}{F_1{}^{t+1}} - 1\right)\frac{p_1(t+1)}{p_1(t)} + \frac{p_1(t+1)}{p_1(t)} = (F_3{}^{t+1} - 1)\frac{p_2(t+1)}{p_2(t)} + \frac{p_2(t+1)}{p_2(t)}$$

The last term on each side cancels off against the -1 in parentheses. In what is left, substitute for $p_1(t)$ and $p_1(t+1)$ from (12-8*b*) and (12-8*c*) and divide out common factors. What is left is

$$-\frac{F_2{}^{t+1}}{F_1{}^t} = F_3{}^{t+1} \qquad \text{or} \qquad \frac{\partial F^t}{\partial S_1{}^t} = -\frac{\partial F^{t+1}/\partial S_1{}^t}{\partial F^{t+1}/\partial S_2{}^t}$$

which is nothing but (12-5). This enables us to assert the following very important "invisible-hand" principle.

If perfectly atomistic competitors cause resources to be channeled into consumption and investment programs so as (1) to maximize their current net profits or in any case to prevent net profits from becoming negative, and (2) to make it a matter of indifference how further increments of investment are scheduled, then an efficient program of capital accumulation will result.

This presumes no uncertainty so that *ex ante* expected prices or rates of change of prices—which each competitor knows but cannot himself affect—will correspond exactly to *ex post* observed prices. Under these strong assumptions of perfect certainty, where the *ex ante* future must

[1] The observant reader will remark that there are two more value-of-marginal-product equations that have not been written down: one for the use of $S_1(t)$ in producing $S_2(t+1)$ and one for the use of $S_2(t)$ in producing $S_1(t+1)$.

Exercise. Formulate these two equations and show that they follow from Eqs. (12-8) to (12-8*c*) and so add no independent information. {Hint: the relevant marginal productivity of $S_1(t)$ is not $\partial F[S_1(t+1);S_1(t),S_2(t)]/\partial S_1(t)$, because a small increment of $S_1(t)$ will *ceteris paribus* increase $S_1(t+1)$, and this side effect on $S_2(t+1)$ must be canceled off. Hence $r_1(t) = (F_2{}^{t+1} - F_1{}^{t+1})p_2(t+1)$.}

Exercise. Using (12-8) and the calculations preceding (12-5), interpret the Lagrangian multipliers λ_t.

agree with the *ex post* past, the whole future pattern of prices is knowable but each small competitor need know with certainty only the present instantaneous rate of change of prices.

A glance back at Fig. 12-3*b* will reduce the relation of prices and intertemporal efficiency to familiar terms. An efficient capital program is one that hops from envelope to envelope, like *Pbexyz*. Now with each such efficient path we can associate exactly one profile of relative prices, that is to say, draw the successive tangent "budget lines" to the envelopes at the successive points of the path. As usual the slopes of these tangent lines will give the relative prices corresponding to this particular path. There are various ways of verifying this. Perhaps the easiest is to recognize first that competitive current-profit maximization will necessarily equalize the price ratio and the instantaneous MRS. This is the content of Eq. (12-8*c*). But Fig. 12-3*a* shows the basic envelope relationship according to which at each point of an efficient path its instantaneous transformation curve is tangent to (i.e., has the same slope as) the envelope. So the slope of the envelope will also come into equality with the price ratio.

A geometrically obvious consequence of this is that along an efficient path, at each point of time the total value of capital stocks $p_1(t)S_1(t) + p_2(t)S_2(t)$ is at a maximum at the corresponding efficiency prices. In other words, if the whole history of future prices were to be announced initially, entrepreneurs would allocate resources in such a way as to maximize the current value of their assets at each point of time. But now we come to a subtle point. Not every time profile of relative prices can lead to consistent behavior in this way. In fact only price profiles which correspond to efficient paths will work. Any other price profile will lead to inconsistency of the following kind: a certain capital program will maximize, say, $V(t_0) = p_1(t_0)S_1(t_0) + p_2(t_0)S_2(t_0)$, for the given prices. But this program will involve $S_1(t_0 - 1)$, $S_2(t_0 - 1)$, and earlier capital stocks which do not maximize $V(t_0 - 1)$, etc., at the earlier given prices. In Fig. 12-3*b*, suppose that prices for $t = 3$ are announced which would maximize $V(3)$ at the point x. Then the corresponding capital program is *Pbex*. But if prices for $t = 2$ other than those determined by the slope at e were announced, then $V(2)$ and $V(3)$ could not *both* be maximized. This inconsistency does not arise if prices corresponding to an efficient path are announced. In this case entrepreneurial short-run and long-run maximization coincide.

But we have just finished proving that *under full competition only the consistent case can arise.* That is the meaning of the invisible-hand principle. The little-appreciated fact is that the arbitrage-induced own-rate relations (12-7) have this effect. They knit successive price ratios together in such a way that only sequences leading to efficient programs can arise. Thus we needn't worry about arbitrarily announced time profiles of future prices. Under competitive assumptions only price ratios obeying (12-7) are eligible. Under such prices, long-run asset-value maximization and current-profit maximization coincide. Economic intuition should tell us this. In the example of the last paragraph, there is a clear differential between a program aimed at maximizing $V(3)$ by production and one aimed at first maximizing $V(2)$ by production and then by trade at current prices converting to capital stocks in proportions best suited to maximizing $V(3)$ in the next period. The function of competition is to wipe out such gaps, and no prices which permit them can endure.

The reader familiar with the theorems of modern welfare economics will not need to be reminded that competition guarantees only that *some* efficient capital program will be followed. There are infinitely many such time paths, fanning out from initial point P in Figs. 12-3*a* and *b*. One goes through each eligible[1] point of MN and continues on so that one path goes through each point of EF and $E'F'$, etc. A *particular* efficient program is picked out by the invisible hand only if one arbitrary bit of information is added, e.g., the price ratio at $t = 1$ (which picks out a point on MN and the corresponding path), or the price-ratio at some horizon date $t = T$ [which picks out the point of budget-line tangency on $E^{(T)}F^{(T)}$ and the path leading to it].

Mathematically, this arbitrariness reflects the fact that the difference equations of intertemporal efficiency (12-2*a*) and (12-5) were shown to be subject to three boundary conditions. Competition ensures that the equations will hold, and history provides two initial conditions. The remaining degree of freedom lets us pick out one more point through which the efficient path must pass.

The truly remarkable thing about the intertemporal invisible hand is that while it results in efficiency over long periods of time, it requires only the most myopic vision on the part of market participants. Just current prices and current rates of change need to be known, and at each moment long-run efficiency is preserved. But for society as a whole there is need for vision at a distance. If, for example, it is desired that at $t = T$ capital stocks should be in some given proportions $S_2(T)/S_1(T)$, only explicit calculation will show what prices $p_1(T)$, $p_2(T)$[2] need to be quoted in order that competition should lead a myopic market inevitably to the appropriate point on the envelope for $t = T$.

One interesting sidelight before we leave the subject of intertemporal

[1] If disinvestment is not allowed, only points on MN which are northeast of P are eligible, and a similar restriction must be made on EF, etc. Observe how this confines eligible paths to a sort of irregular cone emanating from P.

[2] Or $p_1(1)$, $p_2(1)$.

pricing: Consider any efficient capital program and its corresponding profile of prices and own-rates. *At every point of time the value of the capital stock at current efficiency prices, discounted back to the initial time, is a constant,* equal to the initial value. This law of conservation of discounted value of capital (or discounted Net National Product) reflects, as do the grand laws of conservation of energy of physics, the maximizing nature of the path.[1]

12-2-7. Maintainable Consumption Levels. Having deduced the relations that efficiency requires we are now in a position to investigate the behavior of efficient paths under special assumptions about consumption. We might consider (1) the case of zero net capital formation, with all the productive potential of the system going into consumption, or (2) the case of zero consumption (labor treated as just another stock) with all the productive potential going into capital accumulation, or (3) an intermediate case in which some fraction of available resources is used for current consumption and the remainder for net investment. Naturally, the more stringent the assumptions, the more we can say about the resulting paths. In this section we turn to the case 1, in which capital is just being maintained intact.

By substituting $S_i(t+1) = S_i(t)$ into our transformation function (12-2*a*) we get

$$C_2(t+1) = F[S_1(t) + C_1(t+1); S_1(t), S_2(t)] - S_2(t) \qquad (12\text{-}9)$$

This gives a whole frontier of possible consumption combinations that are obtainable from any given endowment of capital $[S_1(t), S_2(t)]$. In fact Eq. (12-9) defines the consumption-possibility schedule of static economic theory. The capital stocks play the role of fixed nonaugmentable resources. This consumption frontier is obtainable at the time t for which the capital stock is prescribed. But more than that, these consumption levels are steadily obtainable *from then on.* This is because of our assumption that capital is maintained intact, neither growing nor diminishing, so that this frontier repeats itself indefinitely.

Figure 12-4 shows the menu of different consumption bundles available for one particular stock of capital. Note the infinity of different consumption possibilities: the ones under the curve are clearly nonoptimal.

In national-income statistics an attempt is made to give a single number which will characterize the consumption frontier. In real life where

[1] The details of the proof are left to the reader who cares. It is to be shown that $[p_1(t+1)S_1(t+1) + p_2(t+1)S_2(t+1)]/[1 + r_0(t)] = p_1(t)S_1(t) + p_2(t)S_2(t)$. Here $r_0(t)$ is the "money" rate of interest and can be substituted for in two different ways from (12-7). The rest is manipulation of (12-8) to (12-8*c*) and (12-5) and use of Euler's theorem on the homogeneous function F. It is simplest to assume all consumption to be zero.

the number of commodities, however aggregated, greatly exceeds two, the national-income statistician is trying to use a single number—real national product—to summarize what is really a surface of $n - 1$ dimensions, an $(n - 1)$-fold infinity of different values.

So far we have been following Leontief in assuming that negative capital formation is physically impossible. The income statistician cannot make such a simplifying assumption; he knows that economies can for a short time speed up their consumption levels at the expense of capital maintenance and replacement. Therefore he cannot accept any observed amounts of consumption and presume that they fairly reflect the economy's real national-product potential. He must make sure that the presupposition of (12-9) is realized, so that $S_i(t) = S_i(t + 1)$; or, if this is not so, he must make appropriate allowances. Actually, in the simple case in which the flow of consumption is *addible* to the stock of capital, the statistician can work with the quantities $C_i(t + 1) + S_i(t + 1)$ or $C_i(t + 1) + \Delta S_i$ and from them compute his measures of real product. All allowances for keeping capital intact will then have been made. His task would be somewhat more difficult, but our analytic task would not, if the transformation curve in (12-2) had been replaced by the more general and perhaps more realistic one:

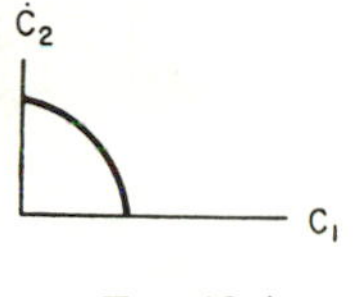

FIG. 12-4

$$S_2(t + 1) = f[S_1(t + 1);C_1(t + 1),C_2(t + 1);S_1(t),S_2(t)]$$

All our previous analysis of efficiency would hold good except that now $\partial f/\partial C_i \neq \partial f/\partial S_i(t + 1)$. However, note that up to now we have had no need of derivatives like $\partial f/\partial C_i$.

From now on we shall drop the assumption that net investment cannot be negative, but we shall retain the assumption that consumption and investment flows are additive and shall at most require that their sum $C_i + \Delta S_i$ be nonnegative.

Equation (12-9) and Fig. 12-4 summarized the stable consumption possibilities of our economic system. But they did so without regard to intertemporal efficiency conditions. As long as we prescribe the arbitrary strait jacket of maintaining *every single* capital stock intact, no discretionary power remains to reject inefficient programs and (12-9) does fairly represent the steady-state consumption possibilities. But it ignores the possibility of changing the capital structure over time while maintaining steady consumption levels.

We can restate the problem in a way that does not ever seem to have been done in the literature of national-income or pure-capital theory. We ask ourselves: Why do we stipulate that capital be maintained intact? We do it because we fear that letting any capital shrink in amount will ultimately jeopardize the maintenance of current consumption levels. If we could be sure that current consumption could be indefinitely maintained, we would not care what specifically is happening to the various capital stocks.

Now it is easy to show that most of the consumption levels shown to be possible in Eq. (12-9) represent definitely inefficient capital programs maintained over time with zero net investment. If we pick at random one of the feasible consumption levels in Fig. 12-4, someone else can show us how to get still more of every consumption good, forever.

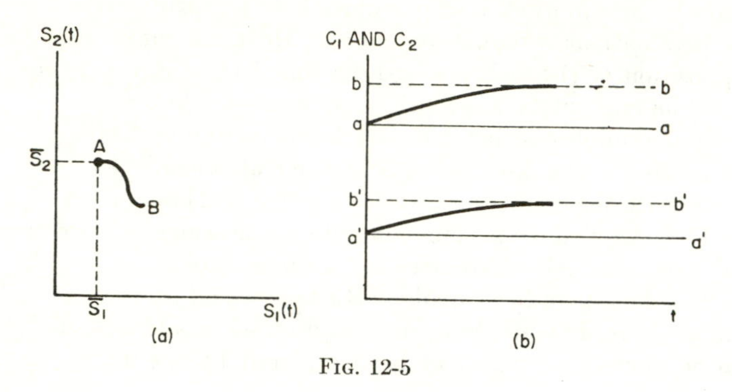

FIG. 12-5

For any given $\bar{S}_1$, $\bar{S}_2$ we can expect only one stationary consumption level to be efficient. For if we are to have $S_i(t) \equiv \bar{S}_i$ and $C_i(t) \equiv \bar{C}_i$, *and* our efficiency conditions (12-5) and (12-2) are to be satisfied, we must have

$$\begin{aligned} 0 &= F_2(\bar{S}_1 + \bar{C}_1;\bar{S}_1,\bar{S}_2) + F_1(\bar{S}_1 + \bar{C}_1;\bar{S}_1,\bar{S}_2)F_3(\bar{S}_1 + \bar{C}_1;\bar{S}_1,\bar{S}_2) \\ &= F(\bar{S}_1 + \bar{C}_1;\bar{S}_1,\bar{S}_2) - \bar{S}_2 - \bar{C}_2 \end{aligned} \tag{12-10}$$

Here are two equations. If we take $\bar{S}_1$ and $\bar{S}_2$ as given, the law of diminishing returns assures us that we can solve these equations for one and only one set of consumptions $\bar{C}_1$, $\bar{C}_2$. Any other prescribed feasible consumption program must be inefficient.

Just what does inefficiency mean in this context? It means that (1) we can drop the assumption that each capital be maintained intact; (2) we can permit some stocks to increase and others to decrease; and (3) we can end up with a program that from now until kingdom come gives us more of every single consumption flow than the prescribed inefficient program.

Figures 12-5*a* and *b* illustrate this. When $S_i(t) \equiv \bar{S}_i$ we stay at A in Fig. 12-5*a* and enjoy steady consumption levels *aa* and *a'a'* in Fig. 12-5*b*. By letting S_1 increase and S_2 decrease to B in Fig. 12-5*a* we are able to enjoy the heavy-line consumption levels. Only the asymptotic levels bb and $b'b'$ corresponding to B are efficient levels satisfying (12-10). From A, some quite different consumption levels are efficient (although, when tastes are consulted, not necessarily *desirable*).

Any particular composition of capital stocks is appropriate for only one composition of steady consumption levels, namely, the consumption levels obtained by finding the efficient path with $\Delta S_i \equiv 0$, which gives (12-10), and then solving these equations for $\bar{C}_1$ and $\bar{C}_2$. If, as is overwhelmingly likely to be the case, the given initial capital structure is not appropriate for the particular desired steady-state consumption program, we can find the capital structure which is appropriate. All we have to do is solve Eq. (12-10) in reverse. By disinvesting in one stock and investing in the other, it will be possible simultaneously to improve consumption of both goods, preserving the desired proportion, and over time to create a capital structure appropriate for the desired composition of consumption. As this is done, physical capital is not maintained intact, but the consumption potential to eternity is maintained and even improved.

In an efficient stationary state, defined by Eq. (12-10), all the own-rates of interest are equal. [This follows from (12-10) in conjunction with Eqs. (12-8) to (12-8*c*).] In turn (12-7) or (12-8*b*) and (12-8*c*) then imply that relative prices are constant. If the economy is capable of capital growth in all its parts—and many economies dependent on exhaustible resources are not—the common own-rate of interest will be positive. Its numerical value will differ depending on the exact taste pattern for consumption flows. Though capable of growth, consumption is so large that the system is stationary. Consequently the earlier figures with expanding frontier envelopes are no longer appropriate. Consumption is here so great as to make the envelope be a single negatively inclined locus passing through the initial $\bar{S}_1$, $\bar{S}_2$ point.

Parenthetically we may remark that positive time preference in the Fisher or Böhm-Bawerk sense would have to be assumed if a maximizing individual or set of individuals is to come into such a stationary (efficient or inefficient) equilibrium.

12-2-8. Closed Consumptionless Systems. So much for the case of steady consumption in a stationary state. Another important and interesting case is that of a closed system with consumption zero and all production plowed back into capital formation.

How can consumption be zero and still keep people and horses alive? In the usual interpretation of Malthus, Marx, von Neumann, and Leontief (who have all studied such systems) it is "extraneous final consumption" that is zero. The minimum of subsistence, needed by horses and men to keep themselves alive and productive and willing to reproduce themselves, is not counted as consumption, but rather is treated as a necessary cost of production or as intermediate goods used up.[1] The stock of horses and of labor become capital stocks like any other.

Formally all we have to do is put $C_i(t) \equiv 0$ in all previous considerations. Nothing is substantively changed, since the consumption flows have been completely arbitrary up to now. We still have the instantaneous efficiency curve

$$S_2(t+1) = F[S_1(t+1); S_1(t), S_2(t)] \tag{12-11}$$

The concept of intertemporal efficiency keeps all its force; there are efficient and inefficient ways of accumulating capital. The intertemporal efficiency conditions are still with us:

$$F_2[S_1(t+1); S_1(t), S_2(t)] + F_3[S_1(t+1); S_1(t), S_2(t)] \times F_1[S_1(t); S_1(t-1), S_2(t-1)] = 0 \tag{12-12}$$

Freed of the arbitrary consumption flows we can study efficient programs, i.e., solutions of (12-11) and (12-12), in some detail.[2]

12-2-9. Balanced Growth. Mention of Malthus irresistibly recalls the notion of geometric growth; and this is all the more interesting since the growth models of Harrod and Domar are essentially very special cases (fixed coefficients and only one commodity) of this model or of the one discussed in footnote 2, this page. We can certainly inquire whether the difference equations of intertemporal efficiency have a steady-growth solution: $S_1(t) = S_1^0 g^t$, $S_2(t) = bS_1(t) = bS_1^0 g^t$. This would imply capi-

[1] Of course the necessary level of internal consumption may be conventional, not physiological.

[2] Before proceeding, we should mention a slightly more general and often-discussed situation. Suppose consumption of each good were always the same fraction of total output: $C_i = k(C_i + \Delta S_i)$ or $C_i(t+1) = (k/1-k)[S_i(t+1) - S_i(t)]$. We could even let k depend on relative prices or relative outputs, independently of scale. Then substituting in the basic production relation (12-2) we get

$$S_2(t+1) = (1-k)F\left[\frac{1}{1-k} S_1(t+1) - \frac{k}{1-k} S_1(t);\ S_1(t),\ S_2(t)\right] + kS_2(t)$$

and since F is homogeneous of first degree we can multiply through the $1-k$ to get

$$S_2(t+1) = F[S_1(t+1) - kS_1(t);\ (1-k)S_1(t);\ (1-k)S_2(t)] + kS_2(t) \tag{12-11a}$$

If $k = 0$, we have the Malthus–von Neumann case of strictly zero consumption, and this reduces to (12-11). As k approaches unity we get closer to the stationary-state case. For $0 < k < 1$ we have strictly proportional consumption flows. An observer who concentrated only on capital stocks could neglect consumption and still describe the facts by (12-11a), which is much like (12-11) and could be treated similarly. Of course the greater the fraction consumed, k, the less "productive" the system would appear to an observer interested only in its growth potential and uninterested in the final consumption it provides. Indeed k can increase so much as to lead to consumption in excess of efficient maintainable levels, and then the system loses its apparent growth potential and begins to live off its shrinking capital.

tal stocks always in the same ratio $S_2/S_1 = b$, both increasing at a rate $g - 1$ per period. If we substitute this into (12-11) we find, remembering that F is homogeneous of first degree,

$$bg = F(g;1,b) \tag{12-13}$$

When we substitute in (12-12) we must remember that the partial derivatives of functions homogeneous of first degree are themselves homogeneous of zero degree, so we can simply cross off common factors of all the variables. This gives

$$F_2(g;1;b) + F_3(g;1,b)F_1(g;1,b) = 0 \tag{12-14}$$

In (12-13) and (12-14) we have two equations in the two unknowns g and b. They are analogous to the characteristic equations associated with linear difference equations with constant coefficients. If any efficient path happens to be of this special balanced-growth type, the b and g will satisfy (12-13) and (12-14); and any numbers g and b (not negative numbers, of course, or we lose all meaning) that satisfy (12-13) and (12-14) determine a steady-growth efficiency path.

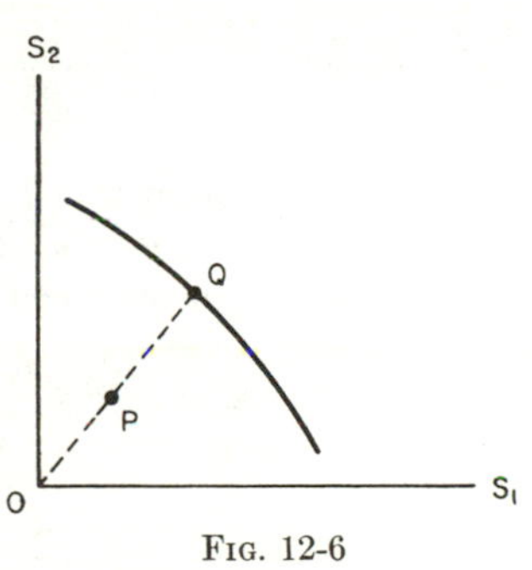

Fig. 12-6

Remember, there are many efficient paths, or solutions to (12-11) and (12-12). Even if we fix the initial capital stocks, there remains a whole one-parameter family of efficient paths, because there is one more arbitrary constant at our disposal. We can use this to prescribe one terminal stock, say $S_1(T)$ or $S_2(T)$, or we can prescribe the terminal composition of capital $S_2(T)/S_1(T)$, or anything else (within reason). Let us take $S_1(0)$ and $S_2(0)$ as fixed and historically given. Then we are currently asking whether any of the many efficient paths into the future are of this especially simple steady-growth type. Even if some are, it would be coincidence indeed if the particular b should come out just equal to our $S_2(0)/S_1(0)$. But balanced growth is an interesting state anyway.

Consider for a moment Eq. (12-13). Its interpretation is simple and is quite independent of any questions of efficiency. Suppose we start with capital stocks in the proportion $S_2/S_1 = b/1$ and wish to preserve this ratio. Then (12-13) says we can, after one period, have g of S_1 and bg of S_2. This is, of course, steady growth for one period. Graphically Fig. 12-6 shows a one-period instantaneous efficiency locus. We start at P, with the slope of OP being b. To preserve this ratio we can go to Q. Then $OQ/OP = g$, the rate by which both stocks can be expanded. If we start with some other ratio b, we wind up with some other g (Fig.

12-6 is, of course, completely independent of scale effects because of constant returns to scale). It is obvious that (12-13) defines g as a single-valued function of b. For every slope OP we can find the corresponding ratio OQ/OP. Note that no question of efficiency has entered, beyond the instantaneous kind.

It is natural to wonder what proportions b result in the largest rate of steady growth. To find out, we have to calculate dg/db and d^2g/db^2 [1] from (12-13):

$$\frac{dg}{db} = \frac{F_3(g;1,b) - g}{b - F_1}$$
$$\frac{d^2g}{db^2} = \frac{F_{11}(dg/db)^2 + 2(F_{13} - 1)(dg/db) + F_{33}}{b - F_1}$$

Since $b > 0$ and $F_1 < 0$, the denominators are positive. We find $dg/db = 0$ when $g = F_3(g;1,b)$. And when $dg/db = 0$, we find

$$\frac{d^2g}{db^2} = F_{33}/(b - F_1) < 0$$

since our assumption that F obeys the law of diminishing returns entails $F_{33} < 0$. What does this add up to?

First, among all balanced-growth expansion paths, *efficient or not*, the fastest attainable balanced expansion rate g^* and its corresponding stock-ratio b^* must satisfy[2]

$$\begin{aligned} g^*b^* &= F(g^*;1,b^*) \\ g^* &= F_3(g^*;1,b^*) \end{aligned} \tag{12-15}$$

Secondly, we can assert[3] that there is only one stock ratio b^* which can yield the maximal rate of growth. We can go slightly further: If we plot g against b, the curve rises to a maximum at b^* and falls away from the maximum forever, on both sides of b^*.

Finally, we have an important connection with efficiency. Euler's theorem says that $g^*b^* = F(g^*;1,b^*) = F_1g^* + F_2 + F_3b^*$. But again from (12-15), $g^* = F_3(g^*;1,b^*)$, and hence,

$$F_1(g^*;1,b^*)F_3(g^*;1,b^*) + F_2(g^*;1,b^*) = 0$$

Look at (12-14). We have proved that the maximal balanced-growth path satisfies (12-14), and of course (like every balanced-growth path)

[1] We assume F to be smooth enough so that these derivatives exist.

[2] For simplicity let us rule out the case in which the maximal growth rate occurs for $S_1 \equiv 0$ or $S_2 \equiv 0$, but the reader should think about it.

[3] We have observed that $d^2g/db^2 < 0$ whenever $dg/db = 0$. Hence every regular stationary point is a maximum. If there were two local maxima, there would have to be a local minimum somewhere between them, which is impossible.

it satisfies (12-13). Hence, *maximal balanced growth is efficient.* If initial stocks should happen to be in proportions $S_2/S_1 = b^*$, then among the possible efficient paths is one which involves steady growth at rate g^*. What is more, it can be shown that *no other balanced-growth path is efficient.* Balanced growth is possible in any proportions, at some rate determined by (12-13), as in Fig. 12-6. But except for the special maximal-growth path, such balanced growth is inefficient. We could find some unbalanced path which would make both stocks grow faster.

To summarize: Malthus-Cassel-Harrod balanced growth is always possible but always inefficient, with one important exception. Every composition of capital has its own possible rate of steady growth. One and only one configuration has the largest possible rate. Call this the von Neumann rate (since he was the first to study this problem, although not in these terms). The von Neumann path is always efficient, whenever initial stocks are in the right proportion. It is the only steady-growth path which is ever efficient. Of course there are many other nonbalanced efficient paths emanating from the same initial conditions.

12-2-10. Balanced Growth in the Very Long Run. States of steady growth have in the past been studied for their own intrinsic interest as a natural generalization of the classical stationary state.[1] From the present point of view we have to ask whether or not steady growth has any maximum or efficiency significance that sets it apart from other programs of capital accumulation. Quite clearly no such importance can be assigned to just any old state of balanced growth. General balanced growth is not even intertemporally efficient, let alone somehow special among efficient paths.

But *maximal* steady growth, growth at the von Neumann rate and in its particular proportions, *is* special. For one thing, it is efficient. In fact, even more is true. In the very long run, maximal balanced growth is in a sense *the* best way for the economy to expand. We must now make this precise.

Suppose we are at Time 0, technocratically planning a capital program for the distant future. We already know how to find the intertemporal efficiency conditions which must govern any such program if society's capital stocks are to be on the outermost frontier attainable at the terminal date T. Given the initial capital stock, society must still choose which of the infinitely many points on the envelope frontier it would like to reach at $t = T$. There are at least three ways that it can do this, illustrated in Fig. 12-7.

[1] Even von Neumann, whose deep analysis turned up properties of steady growth that we do not touch on here, *defines* steady growth as "equilibrium" in his model. The recent habit of aggregating to a one-commodity economy and eliminating all choice effectively masks many of the structural properties of steady growth.

One way is to specify the desired stock $\bar{S}_1(T)$ and then to procure the biggest compatible $S_2(T)$. This defines an efficient path leading from P to Q. Another way is to prescribe the desired ratio of $S_2(T)$ to $S_1(T)$. This prescribes the slope of a radius vector OR, and defines an efficient path leading from P to R, where the radius vector cuts the envelope (the path will not coincide with OR, which in general will not even pass through P). Finally, society could attach relative weights or desirabilities or prices $p_1(T)$ and $p_2(T)$ to the two stocks at the end of the program and choose a path which maximizes $p_1(T)S_1(T) + p_2(T)S_2(T)$, a procedure which emphasizes the "programming" character of the problem.

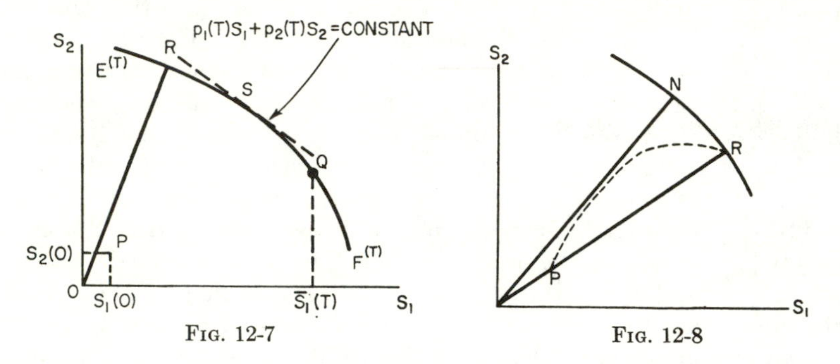

FIG. 12-7

FIG. 12-8

The optimal path would then lead to S, a point of tangency between the envelope and the farthest-out "budget line." Because of the traditional downward-sloping, convex shape of the envelope it doesn't matter which of these three methods we choose. For convenience we use the second, the specification of the terminal stock ratio.

Back to our very-long-run program. Society has decided in what proportions it would like to possess capital stocks at the end of the planning period. The problem is to end up as far out as possible on a specified radius vector, that is, at a point such as R in Fig. 12-8.

Now suppose a terrific coincidence had occurred so that (1) OR passed through P, i.e., the initial stock ratio coincided with the prescribed terminal one; and (2) the slope S_2/S_1 of OR happened to be equal to b^*, the special von Neumann proportions of maximal steady growth. Since this kind of growth is known to be efficient, our problem would already be solved. Society's best bet would be to perform maximal steady growth from P to R, along the path PR.[1]

Such a double coincidence need not detain us. But its analysis suggests the general proposition illustrated in Fig. 12-8: Take *any* initial

[1] Of course it doesn't matter by what method the terminal point is chosen. If initial and terminal points both lie on the radius vector with slope b^*, then steady growth is the solution.

capital structure P and *any* desired terminal structure, like the ray OR. Then if the programming period is very long, the corresponding optimal capital program will be describable as follows: The system first invests so as to alter its capital structure *toward the special von Neumann proportions.* When it has come close to these proportions, it spends most of the programming period performing steady growth at the maximal rate (more precisely, something close to maximal steady growth). The system *expands along or close to the von Neumann ray ON* until the end of the programming period approaches. Then it bends away from ON and invests in such a way as to *alter the capital structure to the desired terminal proportions,* arriving at R as the period ends.

Thus, in this unexpected way, we have found a real normative significance for steady growth—not steady growth in general, but maximal von Neumann growth. It is, in a sense, the single most effective way for the system to grow, so that if we are planning long-run growth, no matter where we start and where we desire to end up, it will pay in the intermediate stages to get into a growth phase of this kind. It is exactly like a turnpike paralleled by a network of minor roads. There is a fastest route between any two points; and if origin and destination are close together and far from the turnpike, the best route may not touch the turnpike. But if origin and destination are far enough apart, it will always pay to get on to the turnpike and cover distance at the best rate of travel, even if this means adding a little mileage at either end. The best intermediate capital configuration is one which will grow most rapidly; even if it is not the desired one, it is temporarily optimal.

12-2-11. Sketch of a Proof.[1] We can fairly briefly verify the proposition stated above, omitting the details. We go back to the difference equations of intertemporal efficiency, (12-11) and (12-12), and we have to show that their solutions have this particular property. First we make a change of variable, which has the advantage of reducing the system to second order and putting the maximal balanced-growth solution in a clear light. Suppose we define

$$\begin{aligned} y_t &= \frac{S_2(t)}{S_1(t)} \\ x_t &= \frac{S_1(t+1)}{S_1(t)} \end{aligned} \tag{12-16}$$

Thus y_t is the capital-stock ratio at time t, and x_t is 1 plus the relative rate of increase of S_1. Now we have to make these substitutions in the basic difference equations (12-11) and (12-12). In (12-11) we divide both sides by $S_1(t)$ and use the fact that F is homogeneous of first degree and that $S_2(t+1)/S_1(t) = [S_2(t+1)S_1(t+1)]/[S_1(t+1)S_1(t)] = y_{t+1}x_t$ to

[1] Some readers may prefer to skip this section.

get

$$y_{t+1}x_t = F(x_t;1,y_t) \tag{12-17}$$

Since the partial derivatives in (12-12) are homogeneous of zero degree, we can divide each argument by the same number and not change anything. Hence

$$F_2(x_t;1,y_t) + F_3(x_t;1,y_t)F_1(x_{t-1};1,y_{t-1}) = 0 \tag{12-18}$$

Observe that (12-17) and (12-18) are a pair of difference equations each of first order. There are two boundary conditions we can set; if we prescribe $S_1(0)$ and $S_2(0)$, the initial stocks, *and* $S_2(T)/S_1(T)$, the terminal-stock ratio, we are also prescribing y_0 and y_T. The three conditions on the S's reduce to only two on y.

Suppose x_t and y_t are both constants. Then according to (12-16), the stock ratio remains fixed over time and S_1 increases at a steady geometric rate. And because S_2/S_1 is fixed, S_2 also increases at the same geometric rate. In other words, a constant solution of (12-17) and (12-18) is a state of steady balanced growth in terms of the S's. Now we already know that the only steady-growth path satisfying the intertemporal efficiency condition is maximal von Neumann growth. But (12-17) and (12-18) came from the efficiency conditions themselves, so we should expect that the only constant solution of (12-17) and (12-18) would be the von Neumann solution:

$$x_t = \frac{S_1(t+1)}{S_1(t)} = g^*$$

$$y_t = \frac{S_2(t)}{S_1(t)} = b^*$$

This is indeed true. If we put $x_t = b$ and $y_t = g$ in (12-17) and (12-18) we get

$$gb = F(g;1,b)$$

$$F_2(g;1,b) + F_3(g;1,b)F_1(g;1,b) = 0$$

which is nothing but (12-13) and (12-14) all over again. These are the equations used to define g^* and b^*. We conclude then that our new equations (12-17) and (12-18) have only one equilibrium or constant solution, namely, $x_t = b^*$, $y_t = g^*$, which implies maximal steady growth for the original system.

The theorem about very-long-run steady growth that we are trying to prove can be stated as follows: As long as T (the terminal date) is large enough, every solution (x_t,y_t) of (12-17) and (12-18) spends a long time with x_t very close to g^* and y_t very close to b^*.

Suppose we take (12-17), write it $y_{t+1} = F(x_t;1,y_t)/x_t$, and expand the right-hand side in a Taylor series around $x_t = g^*$, $y_t = b^*$, keeping only the linear terms. The result is

$$y_{t+1} - b^* = \frac{g^*F_1(g^*;1,b^*) - F(g^*;1,b^*)}{(g^*)^2}(x_t - g^*) + \frac{F_3(g^*;1,b^*)}{g^*}(y_t - b^*)$$

By referring back to (12-15) we find that we can simplify this to

$$y_{t+1} - b^* = \frac{F_1 - b^*}{g^*}(x_t - g^*) + (y_t - b^*) \tag{12-19}$$

$$\Delta(y_t - b^*) = \frac{F_1 - b^*}{g^*}(x_t - g^*) \tag{12-19a}$$

Now we have to linearize (12-18). This is an even dirtier job. Rather than reproduce the calculations here, we leave them to the persistent reader with the following hints: After differentiation, one has to use the fact that the derivatives F_1 and F_3 are homogeneous of zero degree, so that, for example, $F_{11}g^* + F_{12} + F_{13}b^* = 0$; one also has to use the assumption that F is smooth enough so that $F_{ij} = F_{ji}$; and finally one has to use (12-19a). The end result is

$$(x_t - g^*) - (x_{t-1} - g^*) = \frac{F_{33}}{F_{11}}\frac{F_1 - b^*}{g^*}(y_t - b^*) \tag{12-20}$$

or

$$\Delta(x_t - g^*) = \frac{F_{33}}{F_{11}}\frac{F_1 - b^*}{g^*}(y_{t+1} - b^*) \tag{12-20a}$$

From these linear approximations the qualitative properties we are after can be deduced. One way is to observe that if the stock ratio y_t tends to any finite limit, the left-hand side of (12-19a) will tend to zero. Then so must the right-hand side, so x_t tends to g^*. But then the left-hand side of (12-20a) tends to zero, and so must the right-hand side, so y_t tends to b^*.

Another way is to draw a "phase diagram" of (12-19a) and (12-20a) in analogy to the treatment of differential equations. We measure x on the vertical axis and y on the horizontal axis. Any solution of (12-19a) and (12-20a) traces out a path in this plane. These paths turn out to be a family of hyperbolas. The asymptotes are the lines

$$x - g^* = \pm (F_{33}/F_{11})^{1/2}(y - b^*)$$

running through the point $x = g^*$, $y = b^*$. (Because of diminishing returns and diminishing MRS, F_{33} and F_{11} are both negative, so the square root makes sense.)

Such a phase diagram is sketched in Fig. 12-9. It can be verified that the motion is always in the direction of the arrows. For example, $F_1 - b^*$ is negative and F_{33}/F_{11} positive, so that northeast of the point $x = g^*$, $y = b^*$, where $x_t > g^*$ and $y_t > b^*$, (12-19a) tells us that $y_t - b^*$

is decreasing, and (12-20*a*) tells us that $x_t - g^*$ is decreasing. The reader should be certain that he follows this. In Fig. 12-9 we have drawn vertical lines representing one set of boundary conditions y_0 and y_T.

It can be verified that movement along the hyperbolas is slower, the closer the hyperbola lies to the center or singular point (g^*,b^*). So the inner hyperbolas are the slowest, particularly in their inner portions. Now let us concentrate on y_0 and y_T and imagine T getting larger and larger. It will be seen that the only solution starting at $y = y_0$ and ending at $y = y_T$ is one following one of the upper group of nested hyper-

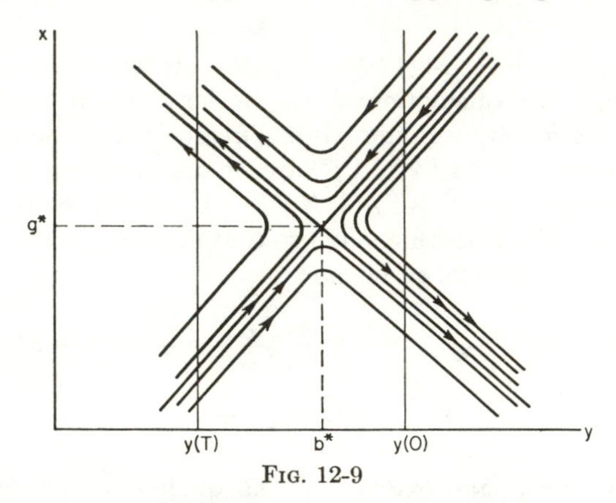

FIG. 12-9

bolas. Further, as T gets larger, the solution passes to lower and lower hyperbolas, which take longer and longer to get from y_0 to y_T. But the lower the hyperbola, the longer it spends near (g^*,b^*). No matter how the boundary conditions are chosen, the same result occurs. Even if y_T is chosen far to the right of y_0, it works. For small T the solution may be along a lower branch of the right-hand hyperbolas or the right-hand branch of the lower hyperbolas. But as T gets longer, there comes a time when only the right-hand hyperbolas will do, and eventually the solution must start at the top and work down one of the right-hand hyperbolas, and it gets forced closer and closer to (g^*,b^*). The reader should practice translating these paths in Fig. 12-9 into paths in a diagram such as Fig. 12-8.

12-2-12. Interest and Prices in Steady Growth. We conclude our analysis of this neoclassical model with the remark that all our earlier price and own-rate considerations carry over to this special case in which all consumption is set at zero. Because of constant returns to scale, all the marginal-product derivatives in (12-8) to (12-8*c*) remain constant during a steady-growth process. Then (12-8*b*) tells us that in such a state relative prices are constant. Therefore if any one price is held absolutely constant, all prices must be constant (for example, if there is a produced numeraire).

Assume that this is so. Also let us restrict ourselves to maximal steady growth, the only efficient kind (and therefore the only kind competition will tolerate). Then comparing (12-8) and (12-15) tells us that $r_2/p_2 = g^* - 1$, or the own-rate of Commodity 2 is equal to the rate of growth of the system. And now a glance at (12-7) shows that the own-rate of Commodity 1 must equal that of Commodity 2 (since prices are not changing), and so it must also equal the rate of growth. And both are equal to the money rate (or numeraire rate) of interest.

We have by no means exhausted the content of our model. But rather than work out all its implications, we use the economic insights it gives to study the dynamic Leontief model from the standpoint of intertemporal efficiency.

12-3. INTERTEMPORAL EFFICIENCY IN LEONTIEF MODELS

We can now drop the assumption of smooth neoclassical production functions with well-behaved marginal productivities. What happens if we replace this model of production with the dynamic Leontief assumptions of fixed flow and capital coefficients, one process for each commodity, and no joint production? The answer is very simple: essentially the same things happen as in the neoclassical case!

There are minor technical differences, of course. The instantaneous efficiency locus, as developed in the preceding chapter, is an angular polygon or polyhedron, with flat faces and occasional sharp vertexes, or corners. But although no longer smooth, it is the same general kind of convex frontier we have been analyzing. The successive outward-expanding envelopes will also behave in much the same way, but they too will have corners. The trouble with the corners is that there no unique marginal rate of substitution is defined. For this reason the MRS and marginal-product proofs offered in the last section fail in the Leontief case. Fortunately linear programming offers a technique which can handle this situation. Apart from this change in analytical techniques, it turns out that there is no qualitative difference between the dynamic Leontief model and the more old-fashioned one we have been studying so exhaustively. Much the same theorems are true about both.

12-3-1. Graphical Formulation—Envelopes. The basic fact with which we start is the instantaneous efficiency locus. This was derived from first principles in the previous chapter, where it was shown that even this basic building block is not a purely technological given, but already involves some unavoidable optimization. In the neoclassical case we could express this locus as a smooth function [as in (12-1)]. Now we have to write it as a set of linear inequalities, which show how each period's net output (consumption plus net capital formation) is limited by the preexisting stocks of capital:

$$\begin{aligned} B_{11}[\Delta S_1(t) + C_1(t)] + B_{12}[\Delta S_2(t) + C_2(t)] &\leqq S_1(t) \\ B_{21}[\Delta S_1(t) + C_1(t)] + B_{22}[\Delta S_2(t) + C_2(t)] &\leqq S_2(t) \end{aligned} \tag{12-21}$$

Taking note of the fact that $\Delta S_i(t) = S_i(t+1) - S_i(t)$, we can rewrite this as

$$\begin{aligned} B_{11}S_1(t+1) + B_{12}S_2(t+1) &\leqq (1 + B_{11})S_1(t) + B_{12}S_2(t) \\ &\qquad - B_{11}C_1(t) - B_{12}C_2(t) \\ B_{21}S_1(t+1) + B_{22}S_2(t+1) &\leqq B_{21}S_1(t) + (1 + B_{22})S_2(t) \\ &\qquad - B_{21}C_1(t) - B_{22}C_2(t) \end{aligned} \tag{12-22}$$

The first inequality expresses the limitation on net output due to the

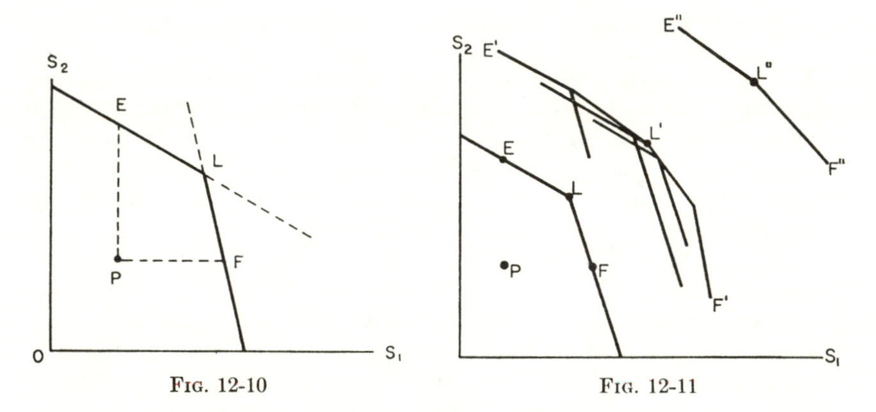

FIG. 12-10

FIG. 12-11

scarcity of $S_1(t)$; the second that due to the limited availability of $S_2(t)$. Figure 12-10 describes this graphically.

The distinction between (12-21) and (12-22) turns on whether we use P or O as an origin. (Also we can deduct the prescribed consumption before drawing the locus, or include it in the locus.)

This ground we have covered before. Now how about capital programs extending over two periods or longer? Just as in the smooth case, each point on the locus *ELF* represents an efficient one-period program. Moreover, each point of *ELF* can be the starting point for the production of period $t + 2$. Analytically this amounts to putting $t + 2$ and $t + 1$ in (12-22), instead of $t + 1$ and t, and inserting the appropriate values for $S_1(t + 1)$ and $S_2(t + 1)$ on the right-hand side. Graphically this gives rise to a whole family of one-period efficiency loci, each coming from a different point on *ELF*. This is illustrated in Fig. 12-11. But of course we must reject any two-period program which is not efficient, i.e., which can be improved upon in the sense that some other program providing the same consumption can end up with more of both capital stocks.

This means that we are interested only in the outward envelope of this family, namely, the frontier $E'L'F'$, and only two-period paths which get from P to the envelope (via ELF) are efficient. Now to consider three-period programs we must draw one-period loci starting from each point of $E'L'F'$ (or two-period envelopes from ELF) and take the envelope of the resulting family. Thus we get $E''L''F''$, and in this way we can continue for any number of periods.

12-3-2. Linear-programming Formulation. The neatest way of finding an efficient capital program extending over T periods is to take as given $S_1(0)$ and $S_2(0)$, to prescribe all the consumptions $C_i(t)$ from $t = 1$ to $t = T$ and to find a program that will maximize $K_1S_1(T) + K_2S_2(t)$ among all feasible programs. As usual, K_1 and K_2 are arbitrary (but nonnegative) valuation constants. By putting the problem this way we seek on the T-period envelope a point of tangency with a "budget line." (If the budget slope happens to coincide with that of a flat face of the envelope, there will be many such points.) As we vary the ratio of the K's (the slope of the budget line), we can trace out the whole efficiency envelope.

We must of course restrict ourselves to feasible programs, i.e., *programs which satisfy* (12-21) *or* (12-22) *at each point of time* ($t = 0, 1, \ldots, T - 1$), and which in addition have all $S_i(t) \geq 0$.[1] Thus we are led to what is easily recognizable as a straightforward (if somewhat grandiose and overblown) linear program (see p. 338). In all this the consumptions $C_1(t)$ and $C_2(t)$ are to be taken as given time sequences, prescribed from outside. Naturally, we shall assume the consumptions to be prescribed in such a way that there is at least one feasible program. One can imagine higher consumptions than the system, with its given technological productivity and limited initial capital, can produce.

In spite of its imposing size (with n commodities and T periods there are nT variables and constraints), this is just another linear program. In fact it is simpler than most, because of the way the constraints build down and to the right. Nothing but zeros appears in (12-23) below the diagonal. This reflects the fact that once, say, $S_i(3)$ are in existence, the earlier history of the capital stocks plays no further part. Their legacy is in the limited availability of $S_1(3)$ and $S_2(3)$, and that is all. This is an exam-

[1] We could also, if we liked, require $S_i(t + 1) \geqq S_i(t)$ (no decumulation of capital). It would make an excellent exercise for the reader to carry through all further considerations with the addition of this constraint, with special attention to the formulation and interpretation of the dual problem.

Maximize

$$0 \cdot S_1(1) + 0 \cdot S_2(1) + \cdots + K_1 S_1(T) + K_2 S_2(T)$$

subject to

$$
\begin{aligned}
&B_{11}S_1(1) + B_{12}S_2(1) \le (1 + B_{11})S_1(0) + B_{12}S_2(0) - B_{11}C_1(1) - B_{12}C_2(1) \\
&B_{21}S_1(1) + B_{22}S_2(1) \le B_{21}S_1(0) + (1 + B_{22})S_2(0) - B_{21}C_1(1) - B_{22}C_2(1) \\
&-(1 + B_{11})S_1(1) - B_{12}S_2(1) + B_{11}S_1(2) + B_{12}S_2(2) \le -B_{11}C_1(2) - B_{12}C_2(2) \\
&-B_{21}S_1(1) - (1 + B_{22})S_2(1) + B_{21}S_1(2) + B_{22}S_2(2) \le -B_{21}C_1(2) - B_{22}C_2(2) \\
&\quad -(1 + B_{11})S_1(2) - B_{12}S_2(2) + B_{11}S_1(3) + B_{12}S_2(3) \le -B_{11}C_1(3) - B_{12}C_2(3) \\
&\quad -B_{21}S_1(2) - (1 + B_{22})S_2(2) + B_{21}S_1(3) + B_{22}S_2(3) \le -B_{21}C_1(3) - B_{22}C_2(3) \\
&\qquad -(1 + B_{11})S_1(3) - B_{12}S_2(3) + B_{11}S_1(4) + B_{12}S_2(4) \le -B_{11}C_1(4) - B_{12}C_2(4) \\
&\qquad -B_{21}S_1(3) - (1 + B_{22})S_2(3) + B_{21}S_1(4) + B_{22}S_2(4) \le -B_{21}C_1(4) - B_{22}C_2(4) \\
&\qquad \cdots\cdots\cdots\cdots\cdots\cdots\cdots\cdots\cdots\cdots \\
&\qquad S_i(t) \ge 0,\ i = 1, 2;\ t = 1, 2, \ldots, T
\end{aligned}
\qquad (12\text{-}23)
$$

Minimize

$$
\begin{aligned}
&u_1(1)[(1 + B_{11})S_1(0) + B_{12}S_2(0)] + u_2(1)[B_{21}S_1(0) + (1 + B_{22})S_2(0)] - u_1(1)[B_{11}C_1(1) + B_{12}C_2(1)] - u_2(1)[B_{21}C_1(1) + B_{22}C_2(1)] \\
&\quad - u_1(2)[B_{11}C_1(2) + B_{12}C_2(2)] - u_2(2)[B_{21}C_1(2) + B_{22}C_2(2)] - \cdots - u_2(T)[B_{21}C_1(T) + B_{22}C_2(T)]
\end{aligned}
$$

subject to

$$
\begin{aligned}
&B_{11}u_1(1) + B_{21}u_2(1) - (1 + B_{11})u_1(2) - B_{21}u_2(2) \ge 0 \\
&B_{12}u_1(1) + B_{22}u_2(1) - B_{12}u_1(2) - (1 + B_{22})u_2(2) \ge 0 \\
&\quad B_{11}u_1(2) + B_{21}u_2(2) - (1 + B_{11})u_1(3) - B_{21}u_2(3) \ge 0 \\
&\quad B_{12}u_1(2) + B_{22}u_2(2) - B_{12}u_1(3) - (1 + B_{22})u_2(3) \ge 0 \\
&\quad \cdots\cdots\cdots\cdots\cdots\cdots \\
&\qquad B_{11}u_1(T-1) + B_{21}u_2(T-1) - (1 + B_{11})u_1(T) - B_{21}u_2(T) \ge 0 \\
&\qquad B_{12}u_1(T-1) + B_{22}u_2(T-1) - B_{12}u_1(T) - (1 + B_{22})u_2(T) \ge 0 \\
&\qquad B_{11}u_1(T) + B_{21}u_2(T) \ge K_1 \\
&\qquad B_{12}u_1(T) + B_{22}u_2(T) \ge K_2 \\
&\qquad u_i(t) \ge 0,\ i = 1, 2;\ t = 1, 2, \ldots, T
\end{aligned}
\qquad (12\text{-}24)
$$

ple of what Dantzig and Jacobs have called "block-triangular" systems in linear programming, and many computational short cuts are available.[1]

For any given K_1 and K_2, we have defined a whole path of capital accumulation $S_i(1)$, $S_i(2)$, . . . , leading up to an optimal $S_1(T)$ and $S_2(T)$. If we run through all possible ratios of the K's, we can trace out in this way not only the efficiency envelope for period T, but also the whole collection of intertemporally efficient capital-accumulation paths fanning out from the initial point $P[S_1(0),S_2(0)]$.

12-3-3. Shadow Prices and the Dual. It will hardly come as a surprise that if we flip the maximum problem (12-23) on its side we get a related dual minimum problem whose variables can be interpreted as competitive shadow prices [see Eq. (12-24), p. 338].

Think of the u's as being shadow prices on the capital stocks at different moments of time. (More precisely they are discounted shadow values.) Then let us compute the shadow profits to be earned by producing 1 unit of S_1 in period $t = 1$. The costs are simply $B_{11}u_1(1) + B_{21}u_2(1)$, the cost of acquiring at $t = 1$ an outfit of capital goods sufficient to produce one unit of S_1. We need not worry about current raw-material costs: the B's are total direct- and indirect-capital coefficients and include in themselves all prime costs resolved back into capital costs. Now in period 2 our revenues are simply $u_1(2)$, the shadow value of the unit of Commodity 1 produced, plus $B_{11}u_1(2) + B_{21}u_2(2)$, the shadow value of the one-period-older outfit of capital, which we still own.[2] Costs minus revenues (all discounted) turns out to be $B_{11}u_1(1) + B_{21}u_2(1) - (1 + B_{11})u_1(2) - B_{21}u_2(2)$. The first constraint in (12-24) simply says that this can't be negative, i.e., that revenues can't exceed costs or that profits can't be positive. We know a bit more: Any time a particular stock is being held in positive amounts, there will actually be zero shadow profits earned in producing the corresponding flow. If the stock is at zero, profits may be negative.[3]

The quantity to be minimized in the dual problem looks complicated but really isn't. What one would expect to find is something like $u_1(0)S_1(0) + u_2(0)S_2(0)$, which would be the imputed shadow value of the initial capital stocks. But we have no shadow prices for $t = 0$. Thus the first two terms in the minimand actually do represent this imputed value, only in terms of the shadow prices for $t = 1$, which is what makes it look so complicated. The negative terms in the minimand

[1] G. B. Dantzig, "Upper Bounds, Secondary Constraints, and Block Triangularity in Linear Programming," *Econometrica*, **23**:174–183 (April, 1955).

[2] Put differently, since all costs are resolved into capital costs and capital does not depreciate, net profit is just sales plus capital gains (or minus losses).

[3] The last pair of constraints in (12-24) has a slightly different interpretation, which should be worked out as an exercise.

are easy to understand: from the imputed value of initial stocks we subtract off the shadow value of the capital tied up in producing the prescribed consumptions. So the minimand amounts to the *net* imputed value of initial capital stock after appropriate allowance for capital set aside to produce consumption rather than capital growth. The duality theorem of linear programming then tells us that in an optimal program this net imputed value just exhausts the maximum "budget line" value of terminal capital stocks $K_1S_1(T) + K_2S_2(T)$. Another way of looking at this is to say that the gross imputed value of initial capital just eats up the value of terminal capital stocks *plus* the shadow value of consumption (or, what is the same thing, the shadow value of the total capital tied up in consumption). In any nonoptimal program, moreover, the net imputed value of initial capital will exceed the value of terminal stocks. This is one of the mechanisms by which competitive markets lead to optimal programs.

Finally, if in an optimal program a capital stock should be in excess supply, so that it is not fully used up in the production of that period, then its corresponding shadow value in the same period will be zero.

12-3-4. The Leontief Trajectories. Both in the general neoclassical case and in the more restricted Leontief model of production, the theory of capital accumulation and growth over time has been developed in terms of a sequence of optimizing choices. In fact this is unavoidable. No matter how restricted we make the technological possibilities of production, once we admit the existence of long-lived capital goods, the necessity of choice appears. There *is* a whole frontier of future possibilities from which society must choose, whether by central decision or decentralized competitive markets (and a historically given distribution of tastes and incomes), whether optimally or inefficiently. In this respect our theory differs radically from Leontief's own and is more complicated. Leontief insists on equalities in (12-23) where we set only inequalities. He assumes that resource allocation is always such that no capital stock is in excess supply, less than fully utilized. In our theory excess capacity may quite possibly happen, accompanied by a zero shadow price.

In terms of Fig. 12-10, Leontief makes his dynamic system *determinate* by simply assuming that from P the economy *must* move to L, and from L (in Fig. 12-11) to L', and so forth. We, on the other hand, simply say that the whole of ELF, and then of $E'L'F'$, is accessible to society (although once a point of ELF is chosen, most of $E'L'F'$ is *no longer* accessible; the long-run goal governs the initial steps). The special capital program which requires all capital to be fully utilized we may nickname the Leontief trajectory, or path.

The first thing to note about the Leontief trajectory is that none may exist. Suppose society moves, as per instructions, from P to L. Then

Leontief says: Draw the one-period efficiency locus from initial point L and move to the corner L'. And then repeat the process from L'. But in this way, as we showed in the previous chapter, there is very likely to come a time when no "next" Leontief corner exists, because to reach it we would have to violate one of the rules of the game, namely, to decumulate capital. And even if we permitted decumulation of capital at any rate, this would only postpone the evil day. Sooner or later, maintenance of all the equalities would require that some capital stock become negative, and here we must certainly stop. The reader of Leontief's pioneer essay will recall how he treats this kind of impasse by so-called "switching rules." From our point of view the impasse arises only because of the artificial rigidity which results from making determinate a system which contains choice in an essential way. Figure 12-12 illustrates the impasse that may befall a Leontief trajectory, namely, at L'' we would have $S_2(3) < S_2(2)$, which is ruled out.

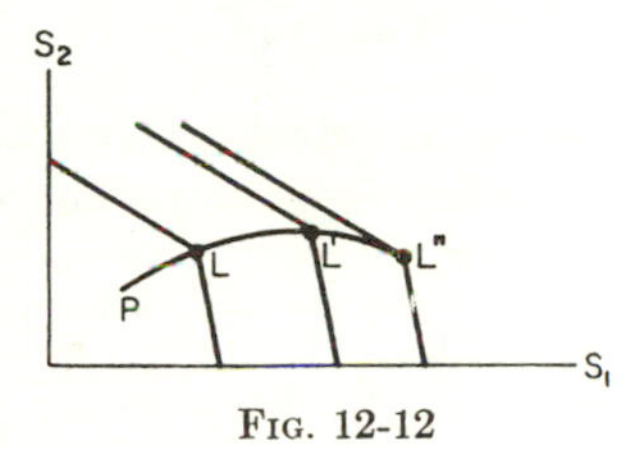

FIG. 12-12

But suppose that over a span of T periods a Leontief trajectory *does* exist. Then it is a theorem (and not an easy one) that *the Leontief trajectory is an efficient path.* In other words, if from given initial conditions it is possible to allocate resources in such a way as to avoid excess capacity, then it must be efficient to do so. Of course, this only says that the Leontief trajectory is one among an infinity of efficient paths, and each of the other efficient paths allows some excess capacity at one time or another. There are some choices of the terminal-stock ratio or some choices of the valuations to be placed on terminal stocks which will make the Leontief trajectory appropriate. For other choices, the Leontief trajectory is just as definitely inappropriate.

There seems to be no simple way of proving the proposition that the Leontief trajectory, if it exists, is an efficient path. We relegate the proof to a footnote which the nonmathematical reader will skip.[1]

[1] In partitioned-matrix terms the linear program (12-23) is as follows: Subject to

$$\begin{bmatrix} \mathbf{B} & \mathbf{0} & \mathbf{0} & \mathbf{0} \ldots \mathbf{0} \\ -(\mathbf{I}+\mathbf{B}) & \mathbf{B} & \mathbf{0} & \mathbf{0} \ldots \mathbf{0} \\ \mathbf{0} & -(\mathbf{I}+\mathbf{B}) & \mathbf{B} & \mathbf{0} \ldots \mathbf{0} \\ \mathbf{0} & \mathbf{0} & -(\mathbf{I}+\mathbf{B}) & \mathbf{B} \ldots \mathbf{0} \\ \cdots & \cdots & \cdots & \cdots \\ \mathbf{0} & \mathbf{0} & \mathbf{0} & \mathbf{0} \ldots \mathbf{B} \end{bmatrix} \begin{bmatrix} \mathbf{S}(1) \\ \mathbf{S}(2) \\ \mathbf{S}(3) \\ \\ \cdots \\ \mathbf{S}(T) \end{bmatrix} \leq \begin{bmatrix} (\mathbf{I}+\mathbf{B})\mathbf{S}(0) \\ \mathbf{0} \\ \mathbf{0} \\ \\ \cdots \\ \mathbf{0} \end{bmatrix} \quad (12\text{-}23a)$$

maximize $\mathbf{K}'\mathbf{S}(T)$.

For simplicity let us assume all consumption to be zero and permit capital to decumulate. Suppose the initial stocks $\mathbf{S}(0)$ to be such that a Leontief trajectory exists, i.e., such that we can find *nonnegative* vectors $[\mathbf{S}(1), \mathbf{S}(2), \ldots, \mathbf{S}(T)]$ satisfying

12-3-5. Stationary States and Maintainable Consumption. We can if we like return to the one-period efficiency locus (12-21) and put each ΔS_1 and ΔS_2 equal to zero. This freezes both capital stocks, and so the same situation will repeat itself period after period. If we require all net investment to be zero, (12-21) becomes

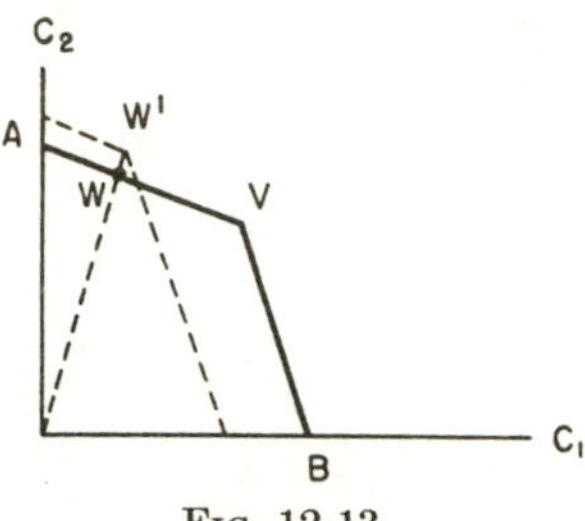

FIG. 12-13

$$\begin{aligned} B_{11}C_1(t) + B_{12}C_2(t) &\le \bar{S}_1 \\ B_{21}C_1(t) + B_{22}C_2(t) &\le \bar{S}_2 \end{aligned} \tag{12-25}$$

where $\bar{S}_1$ and $\bar{S}_2$ are the frozen levels of the two capital stocks. This yields, in Fig. 12-13, a frontier of perpetual consumption levels available to society forever and ever.

(12-23*a*) with all inequalities replaced by equalities. To show that this path is efficient we have to show that there is at least one choice of the valuation vector **K** such that $\mathbf{K}'\mathbf{S}(T)$ is maximized for these particular vectors $\mathbf{S}(t)$. The dual problem to (12-23*a*) is as follows:

Subject to

$$\begin{bmatrix} \mathbf{B}' & -(\mathbf{I}+\mathbf{B})' & & \mathbf{0} & \cdots & \mathbf{0} \\ \mathbf{0} & \mathbf{B}' & -(\mathbf{I}+\mathbf{B})' & \mathbf{0} & \cdots & \mathbf{0} \\ \mathbf{0} & \mathbf{0} & \mathbf{B}' & -(\mathbf{I}+\mathbf{B})' & \cdots & \mathbf{0} \\ \mathbf{0} & \mathbf{0} & \mathbf{0} & \mathbf{B}' & \cdots & \mathbf{0} \\ \cdots & \cdots & \cdots & \cdots & \cdots & \cdots \\ \mathbf{0} & \mathbf{0} & \mathbf{0} & \mathbf{0} & \cdots & \mathbf{B}' \end{bmatrix} \begin{bmatrix} \mathbf{u}(1) \\ \mathbf{u}(2) \\ \\ \\ \cdots \\ \mathbf{u}(T) \end{bmatrix} \ge \begin{bmatrix} \mathbf{0} \\ \mathbf{0} \\ \\ \\ \cdots \\ \mathbf{K} \end{bmatrix} \tag{12-24a}$$

minimize $\mathbf{S}_0'(\mathbf{I}+\mathbf{B})'\mathbf{u}(1)$.

From the duality theorem, if there is a **K** such that the dual problem has a *minimum solution* with all **u**'s *positive*, then the original problem (12-23*a*) will have a solution with all equalities. Let us try all the equalities in (12-24*a*):

$$\mathbf{B}'\mathbf{u}(1) = (\mathbf{I}+\mathbf{B})'\mathbf{u}(2),\ \mathbf{B}'\mathbf{u}(2) = (\mathbf{I}+\mathbf{B})'\mathbf{u}(3),\ \ldots,\ \mathbf{B}'\mathbf{u}(T) = \mathbf{K}$$

There is at least one *positive* solution to this set of equations. For $\mathbf{B}'$ is a matrix of *positive* elements and therefore has a *positive* characteristic root ρ to which corresponds a *positive* characteristic vector $\mathbf{u}^+$. We can set $\mathbf{u}(t) = [\rho/(1+\rho)]^t\mathbf{u}^+$. Then

$$\begin{aligned} \mathbf{B}'\mathbf{u}(t) &= \left(\frac{\rho}{1+\rho}\right)^t \mathbf{B}'\mathbf{u}^+ = \left[\frac{\rho^{t+1}}{(1+\rho)^t}\right]\mathbf{u}^+ = \left(\frac{\rho}{1+\rho}\right)^{t+1}(1+\rho)\mathbf{u}^+ \\ &= \left(\frac{\rho}{1+\rho}\right)^{t+1}(\mathbf{I}+\mathbf{B})'\mathbf{u}^+ = (\mathbf{I}+\mathbf{B})'\mathbf{u}(t+1) \end{aligned}$$

as desired. Then we choose $\mathbf{K} = \mathbf{B}'\mathbf{u}(T)$. It remains to be shown that with this **K**, these **u**'s provide a minimum to the dual problem (12-24*a*). Remember that we also have equality in (12-23*a*). Hence

$$\begin{aligned} \mathbf{K}'\mathbf{S}(T) = \mathbf{u}(T)'\mathbf{B}\mathbf{S}(T) &= \mathbf{u}(T)'(\mathbf{I}+\mathbf{B})\mathbf{S}(T-1) = \mathbf{u}(T-1)'\mathbf{B}\mathbf{S}(T-1) \\ &= \mathbf{u}(T-1)'(\mathbf{I}+\mathbf{B})\mathbf{S}(T-2) = \cdots = \mathbf{u}(1)'(\mathbf{I}+\mathbf{B})\mathbf{S}(0) \end{aligned}$$

using alternatively the equalities of (12-23*a*) and (12-24*a*). Thus for this **K**, these positive **u**'s and these **S**'s, the minimand equals the maximand. Hence we have a solution to the two problems, with positive **u**'s, and the theorem is proved.

That such a stationary state can be inefficient is pretty obvious. Suppose society's tastes are such that it desires to consume C_1 and C_2 in proportions given by W. Since W is not at the vertex, one of the two capital stocks is in excess supply and is not being fully utilized. Suppose this is S_1. Then it would clearly be to society's advantage to run down its stock of S_1 somewhat and simultaneously build up its stock of S_2 somewhat. This could be done according to (12-21) without sacrificing any consumption and, in fact, with an increase in both consumptions. Graphically, the VB part of the frontier will shift parallel to the left as S_1 decumulates, and the AV part will shift parallel upward as S_2 accumulates. The intersection V will move upward and to the left until it falls

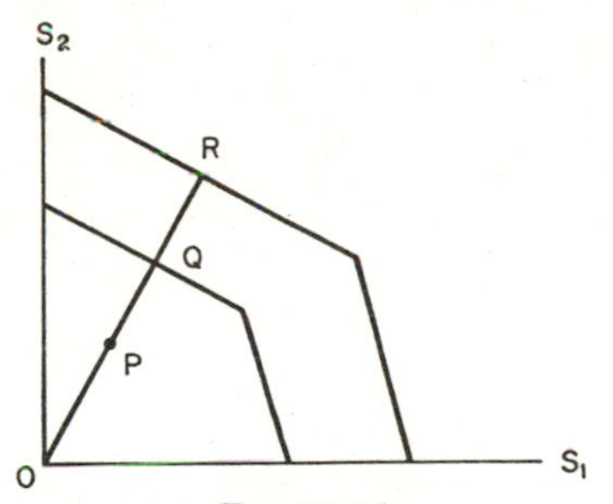

FIG. 12-14

on the ray through W. The final situation is shown by the dotted lines in Fig. 12-13, with consumption at W^1.

The moral is that not every stationary state is efficient, even if the existing capital stock is being efficiently utilized. It is still necessary that the efficiency conditions of (12-23) be satisfied. But in this case the consequences are easy to see.

12-3-6. Closed Growing Systems. Finally we come to the interesting case in which consumption is zero and all the net productivity of the system is plowed back into capital accumulation. Technically this case is covered by what has already been said; since we have treated consumption as arbitrary, we can always set it equal to zero.

The dynamic Leontief technology has essentially the same steady-growth properties as the neoclassical model. Some of this was established in Chap. 11 (see pp. 295–297). A Leontief system can expand in any proportions. In Fig. 12-14 we simply go from P to Q, draw the instantaneous efficiency locus from Q, and go to R, etc.

But in general a path like PQR . . . is inefficient. The envelope for $t = 2$ passes somewhere northeast of R; it would be possible to find an unbalanced expansion path definitely superior to PQR . . . .

There is, however, one (and only one) initial stock ratio which has the special property that its Leontief trajectory is a balanced-growth path; i.e., this initial stock ratio can undergo balanced growth without excess capacity. The situation is shown in Fig. 12-15.

We already know that a Leontief trajectory is efficient. Since $P'Q'R'$ is a Leontief trajectory, we have a case of efficient balanced growth. In fact this is the only balanced-growth path that is efficient. Moreover, we showed in the previous chapter that this special steady-growth rate was faster than any other possible steady-growth rate. Comparing Figs. 12-14 and 12-15, we would have $O'Q'/O'P' > OQ/OP$. This special

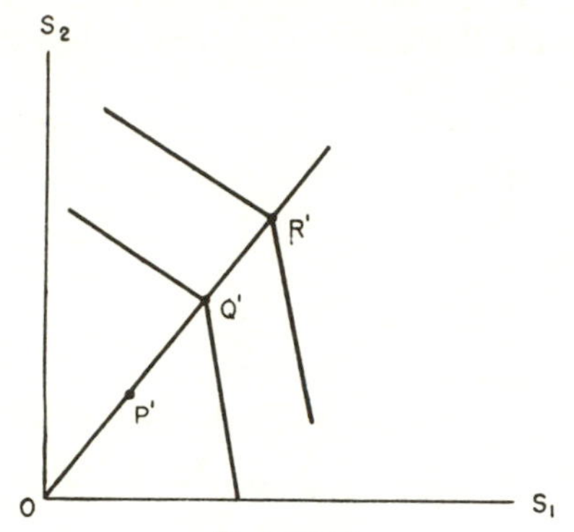

FIG. 12-15

steady-growth path has all the same distinguishing characteristics as the "von Neumann path" that we studied earlier. Indeed the correspondence is quite exact.

The maximal-steady-growth path plays the same role in the Leontief technology as in the neoclassical. If we are optimizing over a sufficiently long run, all efficient paths of capital accumulation have the following property: They first transform the initial capital structure into one approaching the maximal steady-growth proportions; they then spend a long time performing approximate maximal steady growth; and then at the end of the period they bend away and grow into the desired terminal proportions.

More could be said on this subject, but although the terminology and methods of proof would necessarily differ, the economic consequences would be essentially those already described in the corresponding discussion of the neoclassical model of production, particularly with reference to the shadow-price consequences.

The reader familiar with matrix algebra should study in detail the consequences of maximal steady growth for the dual problem. In particular, equality in the dual constraints (as employed in our proof that Leontief trajectories are efficient) yields a special set of shadow prices, proportional to $\mathbf{u}^+$, which represents the competitive equilibrium price configuration for the steadily growing Malthus–Cassel–von Neumann–Harrod economy.[1]

[1] For the mathematical reader only, we sketch a proof that in the very long run all efficient paths approach maximal steady growth. In the notation of the footnote on page 341, it follows from (12-23*a*) and (12-24*a*) that

$$\mathbf{u}(t)'\mathbf{BS}(t) \geq \mathbf{u}(t+1)'(\mathbf{I}+\mathbf{B})\mathbf{S}(t) \geq \mathbf{u}(t+1)'\mathbf{BS}(t+1) \geq 0 \qquad (12\text{-}26)$$

Hence the sequence $\mathbf{u}(t)'\mathbf{BS}(t)$ is nonincreasing and is bounded below; it must approach a limit. In addition we know that $[\mathbf{u}(t)'\mathbf{B} - \mathbf{u}(t+1)'(\mathbf{I}+\mathbf{B})]\mathbf{S}(t)$ tends to zero. Since every component of $\mathbf{S}(t)$ is unbounded and the vector in square brackets is nonnegative, it must be that every component of $\mathbf{u}(t)'\mathbf{B} - \mathbf{u}(t+1)'(\mathbf{I}+\mathbf{B})$ tends to zero. That is, the shadow-price configuration in any very-long-run program must come close to the special configuration $\mathbf{u}^+$. Since we can require $\mathbf{B}'\mathbf{u}(T) = \mathbf{K}$, it can always be arranged that $\mathbf{u}(t)$ is bounded away from zero over the period considered. From (12-26) it also follows that $\mathbf{u}(t+1)'[(\mathbf{I}+\mathbf{B})\mathbf{S}(t) - \mathbf{BS}(t+1)]$ tends to zero. It can be shown that $(\mathbf{I}+\mathbf{B})\mathbf{S}(t) - \mathbf{BS}(t+1)$ tends to zero componentwise. The solutions of the difference equation $\mathbf{BS}(t+1) = (\mathbf{I}+\mathbf{B})\mathbf{S}(t)$ depend on the characteristic roots of $\mathbf{B}$. If ρ is a root of $\mathbf{B}$, then $[(1+\rho)/\rho]^t$ is a fundamental solution of the difference equation. The maximal ρ of $\mathbf{B}$ need not correspond to the maximal $(1+\rho)/\rho$ of the difference equation. But if any other root of the difference equation is dominant and appears with positive weight in a solution, some component of the solution must eventually become negative. Hence only the maximal-steady-growth solution can subsist for the very long run.

13

Linear Programming and the Theory of General Equilibrium

The linear model of production has other uses besides its obvious one as a practical way of computing solutions to practical maximum problems. It doubles as a useful theoretical tool, a convenient way of idealizing the production and profit-maximizing side of a model designed for answering abstract economic questions. The purpose of this chapter is to exhibit linear programming in this role, describing production possibilities in the theory of general economic equilibrium. Section 13-1 will discuss some preliminary and general connections between linear programming and general equilibrium. Section 13-2 will set out in detail a specific model of a competitive economy, which we may call the Walras-Cassel model. The following section shows what modifications in the original model are suggested by the natural requirement that it should possess an economically meaningful equilibrium solution, and in Sec. 13-4 we show how a proof that such a solution exists can be carried out using the theory of linear programming. Section 13-5 contains a brief reference to some recent literature and relates the earlier considerations to more traditional models of production. Finally, the chapter concludes with a similarly quick survey of the dynamical (but again linear) general-equilibrium system of von Neumann.

13-1. EQUILIBRIUM THEORY AND LINEAR ECONOMICS

The systematic study of linear models of production, under names like input-output, linear programming, and activity analysis, is a fairly recent development. But this way of idealizing society's production process has a long and respectable past in the history of economic doctrine.

One of the examples used in Chap. 2 to introduce the concepts of linear programming was the Ricardian theory of comparative cost. Textbooks of international trade still devote detailed analysis to the case of constant costs, where society's production-transformation schedule is a straight line in the case of two commodities or a plane or hyperplane in the case of three or more.

For some purposes it may be convenient to ignore the fact that goods are made from other goods.[1] Then society can be thought of as having a set of production processes or activities, one for each commodity. Each activity converts a unit of labor into a given amount of a particular commodity. These particularly simple activities can be expanded or contracted with constant returns to scale and do not interact with each other except by using up the common resource.[2] In short, the Ricardian model of production is a member of the linear-programming family, old in years and relatively lacking in complications.

International trade provides one of the first examples of general-equilibrium analysis in economics, and in this spirit Frank Graham generalized the Ricardian model to the case of many countries and a simplified world-demand schedule. In this more difficult situation the nature of the world-trading equilibrium and even its very existence are far from obvious. But these problems have lately received a more exact working-over, and the formal tools are exactly those of linear programming.[3]

But the constant-cost model is, after all, only a special case, even within the general linear framework. There is no reason why we should not consider society's production potential as subject to two or three or more resource limitations. When we do this, as in Chaps. 6 and 11, we get a case intermediate between the straight-line constant-cost schedule and the smooth convex transformation curve of neoclassical theory. We get a transformation curve which is a polygon, made up of several flat linear stretches, changing direction at corner vertexes. It is easy to see how this comes about. Each single resource limitation, taken by itself, determines a straight-line constant-cost transformation schedule. But all limitations must be satisfied simultaneously; hence we must always look for the most restrictive limitation. The flat portions of the produc-

[1] Where there is only one scarce and nonaugmentable resource such as labor, the example of the Leontief system (described in Chap. 9) shows how we can cut through all the circularities and interdependencies created by the existence of intermediate goods. The end result is a "total" direct- and indirect-labor input coefficient for each final commodity, and the production-possibility frontier is determined by the condition that the amount of labor used up must not exceed the amount available. Final goods are convertible into each other at constant costs, ultimately via the transfer of labor from one set of occupations to another.

[2] That we need have only one activity for each commodity follows from the presence of only one scarce resource and the absence of joint production, as shown in Chap. 9.

[3] Frank Graham, *The Theory of International Values*, Princeton, N.J., 1948; Lionel McKenzie, "On Equilibrium in Graham's Model of World Trade," *Econometrica*, **22**:147–161 (April, 1954). See also Thomson M. Whitin, "Classical Theory, Graham's Theory, and Linear Programming in International Trade," *Quarterly Journal of Economics*, **67**:520–544 (November, 1953).

tion-possibility curve occur at output proportions for which a single resource constraint is binding. At a vertex or edge there is a change-over; a new constraint becomes binding and the old one ceases to be. Of course some constraints are always redundant; these correspond to resources which are absolutely abundant and will be free goods no matter what the pattern of tastes and output.

This is the model of production used by Walras in the first and second editions of the *Éléments d'économie politique pure.*[1] The famous coefficients of fabrication are defined as "the respective quantities of each of the productive services which enter into the production of a unit of each of the products." As this definition indicates, Walras for the most part finesses the problem of intermediate goods and operates as if production transformed ultimate resources directly into final commodities. But he mentions (p. 233) the existence of produced means of production and shows how they could be accommodated into his scheme by what amounts to the solution of a set of simultaneous linear equations of the Leontief type.[2] We shall return to this question later on when we take up the Walrasian system in detail.

Even while he worked with a fixed-coefficient model Walras had a more general situation in mind. He mentions in the second edition that he takes the coefficients of fabrication as given only for convenience, that of course technical substitution is possible, that the coefficients depend on the prices of the factors of production. In fact, in the third edition he goes over to variable coefficients of fabrication, to be determined by a cost-minimization process as unknowns of the general-equilibrium system.

Now there are two things to be said about this. The first is that by continuing to speak of coefficients of fabrication (inputs per unit of output), variable as factor proportions change but fixed as output changes, Walras maintains the assumption of constant returns to scale. In fact, the assumption is a vital one if he is to be able to talk about the outputs of various commodities without worrying about the allocation of output among firms.[3] Secondly, the mere possibility of substitution of one input for another, of alternative processes of production, to be selected according to some criterion of minimum cost or maximum profit, does not at all take us beyond the scope of the linear-programming model. In fact, in Chaps. 2 and 6, the choice of which of the available production processes are actually to be used is the essence of linear programming. Given the basic assumptions of constant returns to scale and the additivity of the

[1] See, for instance, the Jaffé translation, George Allen & Unwin, Ltd., London, 1954.

[2] *Ibid.*, Chap. 9.

[3] Walras does not miss even this fine point. If all firms in an industry have the same production coefficients, the allocation of output among firms is indeterminate, but unimportant.

various processes, the production-possibility frontier for society is still the broken-line convex polygon we have come across before. It is only when we insist on infinitesimal substitution, on continuously varying marginal rates of transformation, on sensitivity of factor proportions to all price variations no matter how slight, that we have to give up the polygonal frontier for the neoclassical smooth curve.[1]

Cassel,[2] the popularizer of the Walrasian system, went unequivocally back to the fixed-coefficient model and also introduced some further simplifications, to which we shall return.

Before going on to formulate a simple general-equilibrium system in the linear-programming vein, it is worth wondering what kinds of questions can usefully be asked of such a model. Or more precisely, what questions can be asked that cannot be answered by less ambitious models? It seems apparent that a system which leaves many supply and demand functions (or the utility and production functions which lie one step further back) almost completely unspecified as to shape can yield only incomplete results. If we ask the Walrasian equations what will happen to the price of Commodity A if the supply of Factor T shifts to the right, the answer we get is literally the disappointing "That depends"—depends on the shape of just about every schedule appearing in the equations of the system. To learn any more would require drastic simplifying assumptions or numerical knowledge of the schedules or, more likely, both.[3]

Actually, in connection with abstract Walrasian systems, the main question that seems to have been studied in the literature has to do with the *existence* of an equilibrium solution to the collection of equations and with the *uniqueness* of the equilibrium if it exists. If we add some dynamical assumptions describing the response of price and quantities to disequilibrium situations, then we can also study the *stability* of possible equilibria.

Walras was aware that these were legitimate and important questions and believed himself to have done much to settle them. But his remarks on this score were far from rigorous, and satisfactory treatments were given only much later. For example, Walras dismissed the question of existence of an equilibrium (and occasionally also its uniqueness) by show-

[1] In a sense these two possibilities are equally general. Any linear-programming polygon can be closely approximated by a smooth convex curve, and reciprocally, any smooth transformation curve can be closely approximated by a polygon with a finite number of vertexes; i.e., smooth production functions can be replaced by a finite number of alternative linear processes.

[2] G. Cassel, *Theory of Social Economy*, rev. ed., pp. 137–155, Harcourt, Brace and Company, Inc., New York, 1932.

[3] One of the advantages often claimed for the linear-activity-analysis formulation is that it facilitates the collection and organization of empirical information.

ing carefully that his system contained exactly as many equations as unknowns to be determined. But, contrary to what generations of beginning economic theorists have been led to believe, equality of the number of equations and the number of unknowns is neither necessary nor sufficient for the existence of a solution (let alone a unique solution) to a system of equations.

It takes only a little reflection to see this. The example $x^2 + y^2 = 0$ shows that one equation in two unknowns can have a unique solution, namely, $x = 0$, $y = 0$. The example $x^2 + y^2 = \pm 1$ shows that one equation in two unknowns can have either infinitely many solutions or none at all. The system $xy = 10$, $x + y = 1$, two equations in two unknowns, has no solution in real numbers (all that matters here). The system $x - y = 4$, $xz = 0$, $z \log (y + 5) = 0$, three equations in three unknowns, has infinitely many solutions. Evidently a detailed analysis of a system of equations is required before we can pass judgment on the existence of solutions and their number. Equation counting is not enough.

In the case of economic equations this care is all the more necessary because there are some additional, usually unspoken, restrictions on the solutions. To be meaningful, numbers which are to serve as prices and quantities must be nonnegative. Even if we are willing to relax this restriction for prices of some commodities, it can hardly be evaded when we deal with the quantities and with the prices and rents of productive services. Now there is never any guarantee, just from equation counting, that *if* one or more solutions exist, they will contain only nonnegative numbers where nonnegative numbers are required for economic sense.[1] The reader can easily construct examples to illustrate this. The fact that linear-programming analysis takes explicit account of nonnegativity and inequality restrictions hints that these methods will prove useful in this field.

One might ask why all this bother about the existence of an equilibrium solution to the Walrasian equations. If there is anything clear about the real economic world, it is that it exists, it functions. The dynamic super-equations that *really* in some Laplacian way describe the economic system *must* have a solution. But to reason this way is to miss the point. In the first place, it is not so clear that the ever-changing, imperfect, oligopolistic world has a statically timeless, frictionless, perfectly com-

[1] In his pioneering presentation of a marginal-utility theory of consumer choice, Walras was extremely fastidious in pointing out that the usual marginal equivalences need hold only for commodities which the consumer actually purchases in positive amounts or is on the verge of purchasing. For commodities well below the margin only certain inequalities need hold. Many later writers tended to neglect this subtlety. It is a little surprising that Walras should completely overlook the quite analogous point about inequalities when it came to the production part of his system.

petitive equilibrium. In the second place, we can't blithely attribute properties of the real world to an abstract model. It is the *model* we are analyzing, not the world. We wish to use the model or parts of it for studying real economies. It is important to know whether this collection of supply-and-demand relations really captures what is important about economic systems. One test is provided by the existence problem. Just because no *real* existence problem can occur, a system of equations whose assumptions do not guarantee the existence of a solution may fail to be a useful idealization of reality. This may be a minimal test, but it is a test with some cutting power. It will subsequently turn out that the simple Walras-Cassel model *does* have to be modified to assure the existence of solutions.[1] The study of inconsistent systems may indeed be useful precisely because they direct our attention to the sources of inconsistency and display them in a new light. And on the technical side, the pursuit of the existence problem leads to new analytical tools and focuses attention on neglected aspects of the structure of general-equilibrium systems.[2]

13-2. THE WALRAS-CASSEL MODEL

In this section we shall formulate a simple version of the Walras-Cassel general-equilibrium system and look into some of its properties from the activity-analysis point of view. Then in the next two sections we shall show how the theory of linear programming can be used to obtain a proof of the existence of a competitive equilibrium.

Consider an economy with n commodities and m resources or factors of production. Let r_i be the amount of the ith resource supplied and let x_j be the amount of the jth commodity produced. Technical production possibilities are characterized by mn fixed numbers a_{ij}, representing the physical amount of the ith resource used up in the manufacture of a unit of the jth commodity. Thus a_{ij} is an input coefficient[3] of the Walras-Leontief type.

[1] Many of the most important results of the new welfare economics have to do with the optimal character of competitive equilibria and the deadweight burden on society associated with departures. Theorems like this are worked out in the general context of the Walrasian system. Naturally, then, it matters whether there really is a competitive equilibrium *in this model.*

[2] We shall not discuss the stability of general equilibrium in this chapter, except possibly in passing. Suffice it here to say that the famous *tâtonnements* by which Walras tries to describe the dynamical approach to equilibrium fall far short of achieving this purpose. After a long argument the main question is simply begged (*op. cit.*, 2d ed., pp. 236–249, especially 249). Only later with the work of Hicks, Lange, Metzler, Samuelson, and others was a satisfactory approach evolved.

[3] We shall show shortly how these definitions can be modified to allow for intermediate goods.

Producers of the first commodity then demand $a_{i1}x_1$ units of Resource i, producers of the second commodity will demand $a_{i2}x_2$, and the total demand for the ith resource is then $a_{i1}x_1 + a_{i2}x_2 + \cdots + a_{in}x_n$. Putting supply equal to demand for each resource, we get m equations:

$$\begin{aligned} a_{11}x_1 + a_{12}x_2 + \cdots + a_{1n}x_n &= r_1 \\ a_{21}x_1 + a_{22}x_2 + \cdots + a_{2n}x_n &= r_2 \\ \cdots\cdots\cdots\cdots\cdots\cdots\cdots& \\ a_{m1}x_1 + a_{m2}x_2 + \cdots + a_{mn}x_n &= r_m \end{aligned} \tag{13-1}$$

Now we need price variables, $m + n$, in all. Let $p_1, \ldots, p_n$ be the prices of the commodities and $v_1, \ldots, v_m$ the prices or rents of the services of the m resources or factors. The market-demand equations for the commodities can then be written

$$\begin{aligned} x_1 &= F_1(p_1, \ldots, p_n; v_1, \ldots, v_m) \\ x_2 &= F_2(p_1, \ldots, p_n; v_1, \ldots, v_m) \\ \cdots\cdots&\cdots\cdots\cdots\cdots\cdots\cdots \\ x_n &= F_n(p_1, \ldots, p_n; v_1, \ldots, v_m) \end{aligned} \tag{13-2}$$

The factor prices appear in the demand functions to allow for changes in demand induced by shifts in the level and distribution of income.[1] Doubling or halving all commodity and factor prices will leave each household's real position the same, and so will not affect individual or market demands. Technically, all the demand functions F_i are homogeneous of zero degree.

Since we are dealing with long-run competitive equilibrium, another set of conditions states that the price of each commodity must equal its unit costs. Since intermediate goods are washed out, unit costs consist of payments for resources. Per unit of Commodity 1, for instance, resource requirements are a_{11} units of the first resource, a_{21} units of the second, etc. Thus we get n equations:

[1] Some writers on general equilibrium, e.g., Cassel and Wald, have omitted this detail. They also often "inverted" the demand functions and wrote

$$p_i = f_i(x_1, \ldots, x_n)$$

This simplifies the mathematics, but is quite illegitimate. Entirely apart from the classical indeterminacy of absolute prices (which can easily be allowed for in Wald's formulation and which actually was taken explicitly into account by him), this version says that any configuration of market demands can be brought about by one and only one set of prices. Economic theory says no such thing. It is educational that some economists should think that there is no mathematical difference between these two versions and mathematicians should think that there is no economic difference and that both should be wrong.

$$
\begin{aligned}
a_{11}v_1 + a_{21}v_2 + \cdots + a_{m1}v_m &= p_1 \\
a_{12}v_1 + a_{22}v_2 + \cdots + a_{m2}v_m &= p_2 \\
\cdots\cdots\cdots\cdots\cdots\cdots\cdots \\
a_{1n}v_1 + a_{2n}v_2 + \cdots + a_{mn}v_m &= p_n
\end{aligned}
\qquad (13\text{-}3)
$$

All that is needed now to round out the Walras-Cassel system is some consideration of the supply of resources. Quite generally we can suppose that the offer of resources depends on the market prices of all the productive services and in addition on the prices of final commodities. Our last set of equations is

$$
\begin{aligned}
r_1 &= G_1(p_1, \ldots, p_n; v_1, \ldots, v_m) \\
r_2 &= G_2(p_1, \ldots, p_n; v_1, \ldots, v_m) \\
\cdots&\cdots\cdots\cdots\cdots\cdots\cdots \\
r_m &= G_m(p_1, \ldots, p_n; v_1, \ldots, v_m)
\end{aligned}
\qquad (13\text{-}4)
$$

m in number.

Like the demand functions (13-2) these supply functions[1] have to be homogeneous of zero degree. Even more can be said. Each household's demands and supplies are subject to a budget constraint which says that outlays on goods equals income from factor services. Since this is true for each household separately, it is true for the aggregate. Hence the market supply and demand functions are not independent. They satisfy an identity, $\sum_1^n p_j x_j \equiv \sum_1^m v_i r_i$, or

$$\sum_1^n p_j F_j \equiv \sum_1^m v_i G_i$$

an identity in the commodity and factor prices. This has been christened Walras' law. Actually, because of the constant-returns-to-scale nature of the technology we can deduce this exhaustion-of-product relation as an *equilibrium* condition from (13-1) and (13-3).[2]

Sometimes it is expositionally convenient to do as Cassel did and assume that the factor supplies r_i are constants given by nature. That is

[1] We take as given once and for all the individual household's endowment of owned resources. K. Arrow and G. Debreu, "Existence of an Equilibrium for a Competitive Economy," *Econometrica*, **22**:265–290 (July, 1954), take explicit account of each household's initial inventory, and even of the way firms distribute their earnings to households. Their complete microeconomic model is much more general than ours, at the cost of much mathematical complication.

[2] Multiply the first equation of (13-1) by v_1, the second by v_2, etc., and add the m equations together. On the left we get the sum of all possible terms like $a_{ij}v_i x_j$, and on the right-hand side we get total factor payments $\Sigma v_i r_i$. Now multiply the first equation of (13-3) by x_1, the second by x_2, etc., and add the n resulting equations. The left-hand side is again the sum of all terms $a_{ij}v_i x_j$ and the right-hand side is $\Sigma p_j x_j$. Thus the value of output is exactly imputed to the scarce resources.

equivalent to saying that the supply functions (13-4) are perfectly inelastic with respect to all the prices. No additional difficulty of principle occurs when we let factor supplies be elastic, but often no essentially new point of economic interest is illuminated either, and mathematical difficulties are created to boot. When convenient, we shall treat the simpler case of given resource amounts first and then show how the argument can be extended to the general situation.

Summing up, in (13-1) to (13-3) we have $2n + m$ equations for the $2n + m$ unknowns x_j, p_j, and v_i (with the r_i given). If we add the supply relations (13-4), the number of equations is increased to $2n + 2m$, and we acquire the m additional unknowns r_i. At this point we have to do something about the absolute price-relative price dichotomy. Walras' law tells us that (13-2) and (13-4) really contain only $m + n - 1$ independent equations. The demand and supply functions are such that if all but one of these $m + n$ equations are satisfied, the last one must be. But even as we lose one equation we lose one unknown. For suppose we have found a solution and we proceed to multiply all the prices (p's and v's) by the same constant. Nothing changes in (13-2) and (13-4) because the demand and supply functions are homogeneous of zero degree. Nothing changes in (13-1) because the prices do not enter. And (13-3) also continues to hold, since both sides will simply be multiplied by the same constant. As every economist knows, this "real" system determines only relative prices; the absolute price level is at our disposal. The Walrasian way of handling this is to choose one commodity, say the first, as numeraire. We can then arbitrarily set $p_1 = 1$ and thus reduce the number of unknowns by one. In effect we solve for relative prices, relative to the price of the numeraire. Walras even omits the demand equation for x_1 from (13-2) and determines x_1 by the equation

$$1 \cdot x_1 + p_2x_2 + \cdots + p_nx_n = v_1r_1 + v_2r_2 + \cdots + v_mx_m$$

We shall not do this. Instead it will prove convenient to leave the equations as they stand and to keep in mind that we are free to subject the "absolute" prices to one reasonable condition. For instance, instead of setting $p_1 = 1$, it may be more helpful to set

$$p_1 + p_2 + \cdots + p_n + v_1 + v_2 + \cdots + v_m = 1$$

This is a condition without any apparent economic meaning, but it is clear that if we like, we can find an absolute price level that will satisfy this condition. In any case, the addition of one such "normalizing" condition preserves the equality of the number of independent equations and the number of unknowns. By now the reader should realize that we cannot simply conclude from this that the Walras-Cassel system has a solution, still less that it has exactly one solution, or that it has an economically meaningful solution with all the prices and quantities taking on *nonnegative* values.

13-2-1. Intermediate Goods and Alternative Processes. Before making a closer analysis, let us digress to see how intermediate goods and alternative production processes fit into the framework of the system (13-1) to (13-3) or (13-1) to (13-4). It is particularly easy to handle intermediate goods in the simple case of Leontief-type production functions: only one process per commodity and no joint production. For then we can proceed exactly by way of the Leontief recipe as given in Chap. 9 or 10. Define new variables y_i, the final output of Commodity i, the net national-product contribution of the ith commodity; that is to say, net out the intermediate-good component from x_i, leaving the bill of goods only. The y_i are the appropriate variables to appear in the demand functions. In Chap. 9 it was shown that in the case of the Leontief system total output x_i can be expressed as a linear combination of final demands. We know there are nonnegative constants A_{ij} such that[1]

$$x_i = A_{i1}y_1 + A_{i2}y_2 + \cdots + A_{in}y_n \qquad i = 1, 2, \ldots, n$$

Now Eqs. (13-1) still stand, for resource use depends on total, not final, output. But in (13-1) we can substitute for each x_i its linear expression in terms of the y's. The first equation of (13-1) becomes

$$\begin{aligned} a_{11}(A_{11}y_1 + \cdots + A_{1n}y_n) + a_{12}(A_{21}y_1 + \cdots + A_{2n}y_n) + \cdots \\ + a_{1n}(A_{n1}y_1 + \cdots + A_{nn}y_n) = (a_{11}A_{11} + a_{12}A_{21} + \cdots \\ + a_{1n}A_{n1})y_1 + (a_{11}A_{12} + a_{12}A_{22} + \cdots + a_{1n}A_{n2})y_2 + \cdots \\ + (a_{11}A_{1n} + \cdots + a_{1n}A_{nn})y_n = c_{11}y_1 + c_{12}y_2 + \cdots + c_{1n}y_n = r_1 \end{aligned}$$

The same goes for every equation of (13-1). We get a new set of m equations just like (13-1) with the a coefficients replaced by new c coefficients which can be computed from the a's and the Leontief A's. We have, namely, $c_{ij} = a_{i1}A_{1j} + a_{i2}A_{2j} + \cdots + a_{in}A_{nj}$. Obviously the c's, like the a's and A's, are nonnegative. The cost-covering equations (13-3) must now show that the price of each commodity is just equal to its unit resource cost, including the resource cost of the intermediate commodities consumed in its manufacture. The reader can easily convince himself that to accomplish this it is only necessary again to replace the a's by the new c's. The end result is that we have new equations just like (13-1) to (13-4), with new but qualitatively similar coefficients, and the total outputs x replaced by the final outputs y everywhere, even in the demand functions. Thus the existence of intermediate goods does not alter the model at all.[2]

[1] See Chap. 9, where A_{ij} is interpreted as the total direct and indirect input of Commodity i per unit of final output of Commodity j.

[2] This process of elimination is just what Walras sketches in a rudimentary way: *op. cit.*, p. 241.

When we come to the general linear model of production, things are not so easy. We have three complications to allow for: (1) intermediate goods, (2) joint production—more than one output to a single process, and (3) the existence of alternative production processes. Of these, only the last really makes a difference, but it makes enough of a difference to prevent us from reducing this general case to *exactly* the Walras-Cassel equations (13-1) to (13-4). It is possible to formulate the general linear case so that it comes out *very much like* the Walras-Cassel model, and just about everything we shall subsequently say about the latter carries over directly. We shall not take time here to discuss the general case, because to do so would involve a lot of extra notation for which we have no real need. Instead we can take a couple of paragraphs to see why the Walras-Cassel formulation is not quite general enough to encompass the triply complicated linear model.

The reason is fairly simple. Because of the last two complications mentioned, we have to distinguish between processes and commodities. A process is a fixed-proportion, divisible, constant-returns-to-scale method for turning resources and some commodities into other commodities. We could introduce new numbers, say k_{ij}, to represent the *output* of Commodity i from Process j operated at unit level. If i is a net input to Process j, k_{ij} will be negative. We can let our old x_j represent the level of operation of the jth process—there are n processes altogether. Let us say there are s commodities (not resources). Generally there will be many more processes than commodities, and this is the nub of the matter. Equations (13-1) will still hold: process levels determine the demand for resources. But (13-2) needs to be modified. There is no final demand for activities as such. Final demand is for commodities. So we need symbols to represent commodity outputs, and these would appear as the left-hand variables in (13-2), as functions of commodity and factor prices. Then (13-3) would need a slight modification: in competitive equilibrium the resource cost of a process will have to equal the "net value added" of the process: value of commodity outputs minus value of commodity inputs.

It almost seems as if we have succeeded in doing what was just said to be impossible—putting the general linear case in Walras-Cassel form. But we haven't, and the tip-off is that different variables appear in (13-1) and (13-2). The former contains activity levels, the latter commodity outputs. It is easy enough to compute commodity outputs from activity levels—add up the process outputs of any given commodity. But it is impossible straightforwardly to compute process levels from commodity outputs. There are more processes than commodities, and the same bundle of commodities can be produced in many ways. Which way it *will* be produced depends on a cost-minimization process which is not yet written into our equations. The closest we can come is to express each commodity output in terms of all the process levels and put this expression on the left-hand side of (13-2), which would now read

$$\sum_{j=1}^{n} k_{ij}x_j = F_i(p_1, \ldots, p_s; v_1, \ldots, v_m) \qquad i = 1, \ldots, s$$

Certainly this is enough like the Walras-Cassel equation (13-2) to make it plausible that nothing terrifically new is involved [the modification to (13-3) is even less vital]. This is not to blink at the fact that the more general model is genuinely more general. It forces the issue of *choice* among activities. But still, if we can handle the simpler model, we are more than halfway to mastering the more complicated one.[1]

13-3. EXISTENCE OF SOLUTIONS

With an eye to the existence of solutions let us consider the system consisting of the market equations for factors (13-1), the demand functions for final goods (13-2), and the price-equals-unit-cost equations (13-3). We delete the factor-supply functions (13-4) temporarily and consider the factor amounts as given quantities. In this simple case, one possible obstacle to the existence of a solution sticks out at first glance. The subsystem (13-1) consists of m linear equations, one for each resource, in n unknown commodity outputs. Now it is possible to lay down fairly simple rules which determine when m linear equations in n unknowns possess a solution, and if so, how many.[2] One of these rules states: If m, the number of resources and equations, exceeds n, the number of final goods and unknowns, then Eqs. (13-1) will possess no solution unless the coefficients a_{ij} and the factor supplies r_k stand in a very special relationship to each other. Since the a's are supposed to represent technological constants and the r's the existing, inelastically supplied, factor supplies, there is no reason to expect any such good fortune. We must conclude that if $m > n$, the Walras-Cassel system as we

[1] The interested reader will learn much from trying to write out in detail the equations of the general case. Remember the distinctions between activities and commodities, between commodity prices and value added (how do you compute the latter from the former and the k's?). For hints see Sec. 13-6, and refer back to Chap 6. The basic reference on the general linear model of production is Tjalling Koopmans, Chap. 3 in *Activity Analysis of Production and Allocation*, John Wiley & Sons, Inc., New York, 1951. The mathematics is quite stiff. For a simpler exposition see John Chipman, "Linear Programming," *Review of Economics and Statistics*, **35**:101–117 (May, 1953).

[2] See Appendix B. We are not speaking now of the sterner requirement that no x can be negative. First we must guarantee that there is some solution. For if no solution exists to the subsystem (13-1), no solution can exist for the whole system.

have written it will in general have no equilibrium solution.[1] Suppose there is only one commodity, a unit of which is produced by 1 unit of labor and 1 unit of land. If the available supplies are 2 labor and 1 land, how can Eqs. (13-1) be satisfied and all of both factors used?

There are ways around this obstacle. In the first place, the constancy of factor supplies was an assumption of convenience, and if it leads to difficulty, it must be abandoned. This converts $r_1, \ldots, r_m$ into variables and removes the immediate problem. It is no sure cure, however, as we shall later see. To be realistic we would have to admit that some factors *are* inelastic in supply and that the quantities supplied of others can't be made to vary with complete freedom (for example, there may be a minimum below which supply will not fall even at zero price). It is quite conceivable that some valid factor-supply functions would still leave (13-1) incapable of solution. Actually we shall later evolve a different kind of modification of (13-1) which *will* guarantee a solution even in the case of given factor quantities.

Another possible way out of this impasse is to give up the fixed-coefficient assumption and permit the a_{ij} to vary as functions of factor prices. Actually this would add very little *economic* generality to the model. Remember that with minor alterations the model could be interpreted as already involving alternative production processes for the same products, i.e., variable coefficients. As remarked in the preceding section we must then think of n not as the number of final goods but as the number of processes. This makes it less likely that we shall find m bigger than n, so that the problem we are now discussing is less likely to arise. By going over to the familiar smooth production function we in effect create an infinity of processes, so that m can never be bigger than n. The outward sign of this would be the appearance of factor prices as new variables in (13-1). As before, this kills the problem in its present form, but is not by itself sufficient to ensure the existence of meaningful solutions. In the example at the top of this page we could let the ratio of labor to land in production vary as freely as it likes between 0.5 and 1.99. It would still be impossible to satisfy (13-1). In any case, we want to pursue the implications of the activity-analysis model of production, and so we shall continue to treat the a's as constants. There will be a little more to say about smooth transformation functions subsequently.

Taking Eqs. (13-1) literally, let us ask exactly what they state, in economic terms. They require that the demand for each resource should just equal the given constant supply. This amounts to full employment of each factor. In effect, the solution of the equations would be something like the intersection of a derived demand curve with a perfectly

[1] This point was raised by H. von Stackelberg, "Zwei kritische Bemerkungen zur Preistheorie Gustav Cassels," *Zeitschrift für Nationalökonomie*, June, 1933, p. 464.

inelastic supply curve. Why do we insist on this equality of supply and demand as a necessary characteristic of competitive equilibrium? Because if demand were to exceed supply, the price of the resource would rise; if supply were to exceed demand, the price would fall. In any case, one can hardly imagine a set of outputs using up more of a resource than is available; that is a physical impossibility. But the reverse is not a *physical* impossibility. Some of the available amounts of a particular resource *could* be left unused, unemployed. But then the price of that resource would fall. So it would, but there is a limit to how far a price can fall. It can't fall to less than zero.[1] And if the price falls to zero and the supply still exceeds the demand at a zero price, there *will* be some of the resource unemployed. There is nothing internally inconsistent about this as a description of competitive equilibrium. What we have described is nothing but the process by which some resources become free goods. They cease to be "scarce," in this sense: no price can be obtained for their services.

If we describe this condition as being one of unemployment ("nonemployment," or "redundancy," would be a better word), the reader must not think of it as something which "ought" to be eliminated without further thought. This kind of nonemployment reflects *real* production and taste conditions; it signals that the intrinsic economic value of the redundant factor has fallen to zero. We do not necessarily think it a calamity if a resource becomes a free good. Nonemployment of labor is calamitous for social and ethical reasons. Recent writers on the economics of overpopulated and underdeveloped areas suggest that in some cases it is a fair statement that the marginal productivity of agricultural labor is zero; i.e., that labor could be withdrawn from agriculture with no consequent decrease in output. This is a case of absolute redundancy, and only the obvious social and sociological reasons explain why positive wages are paid. In more developed countries it is sometimes tempting to attribute "structural" unemployment to this kind of redundancy. This must be done with care; if labor alone, with no other inputs, can produce any single commodity of value (shaves and haircuts, domestic and personal service, etc.), then labor cannot be redundant.

Considerations like this don't contradict the "law of supply and demand" in any sense. We usually visualize price determination as in Fig. 13-1*a*. But the situation of Fig. 13-1*b* is by no means inadmissible. The paradox is resolved if we realize that the supply curve is not simply SS', but OSS', and price is determined definitely at zero. Quantity is a little ambiguous, but if we are talking about a productive resource, quan-

[1] It is true that one can sometimes think of certain goods (or "bads") as having a negative price. But ultimate factors of production like land or labor are hardly likely to be offered for a negative return. Only positive or zero prices are thinkable.

tity will be determined on the demand curve at OQ. In the diagram the supply curve is not perfectly inelastic, but of course it might be.

It was early pointed out by Zeuthen and Neisser[1] that the market determines which goods shall be free and which scarce. There is no external badge of intrinsic scarcity or abundance. It depends on the structure of demand, on the availability of complementary factors, on production relations. Thus the list of resources appearing in the Walras-Cassel model must be all-inclusive. One of the jobs of the general-equi-

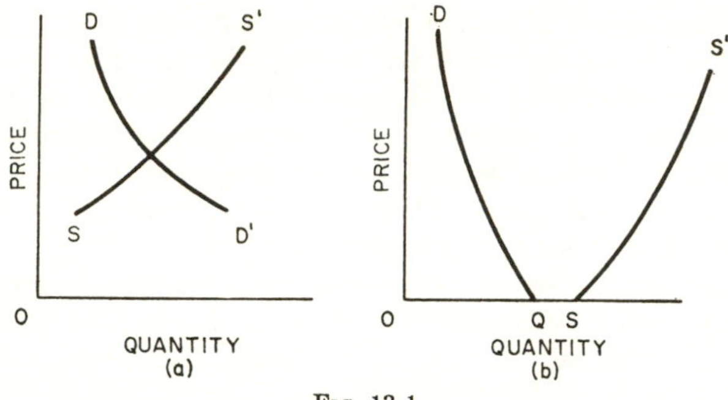

FIG. 13-1

librium system is to decide, in terms of the given data, which resources will be free and which will command a positive price.

Zeuthen suggested one way of incorporating this process into the model. Equations (13-1) have to be modified to read

$$\begin{array}{c} a_{11}x_1 + a_{12}x_2 + \cdots + a_{1n}x_n \leq r_1 \\ a_{21}x_1 + a_{22}x_2 + \cdots + a_{2n}x_n \leq r_2 \\ \cdots\cdots\cdots\cdots\cdots\cdots \\ a_{m1}x_1 + a_{m2}x_2 + \cdots + a_{mn}x_n \leq r_m \end{array} \qquad (13\text{-}1a)$$

with the further condition that *if the strict inequality holds in any line of* (13-1a), *i.e., if any resource—say the kth—is less than fully employed, then its price* v_k *must be zero*.

One thing we note right away is that as far as the inequalities (13-1a) are concerned, there will always exist x's which satisfy them, regardless of how many resources there are and how many final goods. Since the

[1] F. Zeuthen, "Das Prinzip der Knappheit, technische Kombination und ökonomische Qualität," *Zeitschrift für Nationalökonomie*, October, 1932, p. 2; H. Neisser, "Lohnhöhe und Beschäftigungsgrad in Marktgleichgewicht," *Weltwirtschaftliches Archiv*, October, 1932, p. 422. See also W. L. Valk, *Production, Pricing and Employment in the Stationary State*, DeErven F. Bohn N.V. and P. S. King, Haarlem and London, 1937.

r's are all[1] positive and the a's nonnegative, there will in fact always be an infinity of sets of x's which satisfy all the inequalities. So our initial objection to the system has been eliminated.

Perhaps a small numerical example will help clarify what has been done. Let $a_{11} = 1$, $a_{12} = 2$, $a_{21} = 3$, $a_{22} = 4$, $r_1 = 2$, $r_2 = 20$. There are thus two final goods and two resources. Suppose both goods have independent unit-elastic demand curves (independent of factor prices): $p_1 = 8/x_1$, $p_2 = 10/x_2$. Then the three systems of equations read

$$x_1 + 2x_2 = 2, \quad x_1 = \frac{8}{p_1}, \; v_1 + 3v_2 = p_1$$

$$3x_1 + 4x_2 = 20, \; x_2 = \frac{10}{p_2}, \; 2v_1 + 4v_2 = p_2$$

Inspection shows that Commodity 1 uses the two resources in the ratio 3:1, and Commodity 2 uses them in the ratio 4:2. But the supplies of the factors are in the ratio 10:1. It will not surprise us to learn that no way of juggling the two outputs will ever fully employ both factors. An even simpler case to visualize would be one in which the production of the two commodities requires the two resources in *exactly the same* ratio, say 2:1. Then no matter what the outputs turn out to be, the resources will always be demanded in just the ratio 2:1. If the independently given resource supplies are in any other ratio but 2:1, it is a manifest impossibility for both to be fully employed. The best that can be done in this respect is for production of the two commodities to expand (in proportions dictated by demand considerations) until one resource is fully used. The abundant resource will suffer nonemployment, and its price under our assumptions will fall to zero. In general, the only factor demand ratios that can come about under *any* pattern of production are those *between* the largest and smallest ratios in which single processes demand the factors. Making the factor supplies elastic might clear up the situation, or it might not. The reader should experiment until he has constructed plausible supply functions which still fail to yield a solution.

In the numerical example, solution of the general-equilibrium system strikes a snag with the very first set of equations. If we solve the two linear equations in x_1 and x_2, we obtain $x_1 = 16$, $x_2 = -7$, which makes no economic sense. The commodity which uses the abundant factor less intensively and the scarce factor more intensively has a negative output. The equations are trying to run this production *in reverse* to procure

[1] If any r is zero, we can set at zero the output of any commodity which actually requires that resource (i.e., if $r_j = 0$ and $a_{jk} > 0$, then $x_k = 0$). We are left with the commodities which don't use the absent resource, so we can throw away the corresponding line in (13-1a). Hence we might as well imagine all r's positive.

some of the scarce factor to be used with the released abundant resource *and* the nonemployed part of the redundant resource. But production doesn't work that way. One can't transform sausage into pigs by running the grinder in reverse, and so we must reject this "solution."

We can give several graphical descriptions of this difficulty. In Fig. 13-2*a* the straight lines representing the two equations are drawn. The "solution" is the point *P* where the lines intersect; but this point is not in the nonnegative quadrant which alone has economic meaning. Since resource use cannot exceed, but can fall short of, the availability limits,

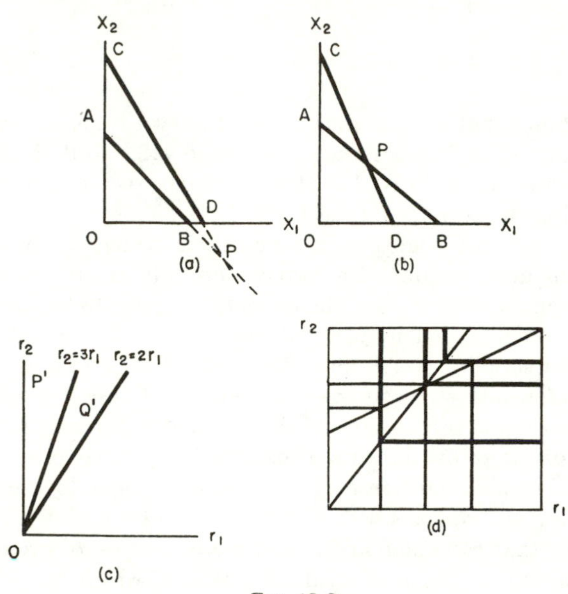

FIG. 13-2

it is always the *inmost* line, closest to the origin, that binds. The physically producible outputs are those in the triangle *OAB*. Were the two lines to have the appearance shown in Fig. 13-2*b*, then the point *P* would be a meaningful solution and it would be *technologically* possible to employ both factors fully. This is not the whole story; there is no guarantee that demand conditions are such as to make *P* an equilibrium. The possible outputs are those in the polygon *OAPD*. If output should settle anywhere on *AP*, one resource would be less than fully employed. On the segment *PD*, the other resource would be less than fully employed. In the constellation of Fig. 13-2*b*, which resource, if either, suffers nonemployment depends on *real* taste and demand conditions. In Fig. 13-2*a* one resource is "absolutely" in excess. Another, "dual," graphical depiction appears in Fig. 13-2*c*. Here the variables are the resource amounts demanded. The resource requirements for a single process are represented by rays from the origin; resources are used in fixed ratios, but the absolute amount expands with the process intensity. The two rays determine a "cone" between them. Any point in the cone and only such a point represents factor demands achievable by these processes operated at nonnegative levels. If the factor supplies are represented by a point such as Q', inside the cone, then full employment is physically achievable; if the factor supplies are represented by a point such as P', outside the cone, full employment of both resources is not achievable.

Still another graphical representation is the Edgeworth box diagram in Fig. 13-2*d*. The (right-angled) isoquants for the two products are drawn from the southwest and northeast corners. In the usual case, the locus of efficient resource allocations, the "contract curve," is given by points of tangency between the two families of smoothly curved isoquants. In the present case the contract curve becomes a zone or area. Any point *between* the rays through the isoquant corners represents an unimprovable pair of outputs. But except at the one point where the rays cross, one factor is redundant. This intersection point corresponds to P in Fig. 13-2*b*. (What does it correspond to in Fig. 13-2*c*? *Hint*: How would the resource supplies appear in Fig. 13-2*c*?) Anywhere outside the contract zone both factors are redundant; this is clearly inefficient. If the dimensions of the box were to change, or if the slopes of the rays were different, it would be possible for there to be no intersection point in Fig. 13-2*c*. Then the same resource is always redundant.

Let us return to the numerical example. We would like to solve the first set of equations for x_1 and x_2, insert these numbers in the demand functions to get p_1 and p_2, and use these prices in the final set of equations which could be solved for v_1 and v_2. But we are blocked at the start by our inability to get meaningful x's. Note that since the second factor is obviously the excessively abundant one our amended rules tell us that we may have $3x_1 + 4x_2 < 20$ and $v_2 = 0$. But if $v_2 = 0$, the third pair of equations becomes $v_1 = p_1$, $2v_1 = p_2$; that is, $p_2 = 2p_1$. Make this substitution in the demand functions: $p_1 = 8/x_1$, $2p_1 = 10/x_2$. We deduce that $8/x_1 = 5/x_2$, or $5x_1 - 8x_2 = 0$. Combine this with the remaining equation of the first set: $x_1 + 2x_2 = 2$. The reader can discover that these two equations have the solution $x_1 = 8/9$, $x_2 = 5/9$. Therefore, $p_1 = 9$, $p_2 = 18$, $v_1 = 9$. We have a solution to the general equilibrium system, namely, $x_1 = 8/9$, $x_2 = 5/9$, $p_1 = 9$, $p_2 = 18$, $v_1 = 9$, $v_2 = 0$. The first resource is fully employed. Of the second resource an amount equal to $3 \times 8/9 + 4 \times 5/9 = 44/9$ is used; the rest is nonemployed, and the rent of this factor has fallen to zero. It is, in fact, a free good, sand in the Sahara. The modification of the original system has provided a solution where none existed before. In this case the reader can check that the solution was unique at each stage, and hence unique over-all.

It was earlier remarked that the kind of nonemployment dealt with here is purely structural. It is not caused by a shortage of effective demand; in fact no expansion (or even redirection) of demand could eliminate it. The unemployed factor is simply redundant. However, this is not the only kind of nonemployment possible in the model. Unemployment can occur even in the situation described by Fig. 13-2*b*; this kind of unemployment is not technologically necessary but reflects real taste or demand conditions, and hence is as "structural" as the other. Consider the following numerical example:

$$x_1 + 2x_2 = 10 \qquad x_1 = \frac{10}{p_1} \qquad v_1 + 4v_2 = p_1$$

$$4x_1 + 5x_2 = 30 \qquad x_2 = \frac{1}{p_2} \qquad 2v_1 + 5v_2 = p_2$$

Note that the production of Good 1 uses factors in the ratio 4:1; the second good uses them in the ratio 5:2; and the resources are supplied in the ratio 3:1. Thus over-all full employment is possible. Indeed if we solve the first two equations we find $x_1 = 10/3$, $x_2 = 10/3$. With these outputs, both factors are used up. We have cleared the hurdle that stopped us last time. From the demand functions we find $p_1 = 3$, $p_2 = 3/10$. So far, so good. Now inserting these values in the final set of equations, we find $v_1 = -46/10$, $v_2 = 19/10$. There is a unique "full-employment" pair of outputs. These outputs will be cleared off the market at a certain pair of prices. But these prices can be made to cover costs only if the first resource commands a negative wage. Looked at from the other side, the demand for the first factor intersects the vertical supply curve only at a negative price, as in Fig. 13-1. This solution has no economic meaning. But we can resort to our modified setup. Suppose we set the price of the first factor at its minimum value, namely, $v_1 = 0$. Then the last equations show that $4v_2 = p_1$, $5v_2 = p_2$, hence $p_1 = 4/5\, p_2$. From the demand equations, $x_1/x_2 = 50/4$, or

$$4x_1 - 50x_2 = 0$$

Combine this with the equation of the scarce factor, $4x_1 + 5x_2 = 30$, and we find $x_1 = 375/54$, $x_2 = 30/54$. The total amount of the first resource used up is $375/54 + 60/54 = 435/54$. There are $105/54$ unemployed units of the first factor. Summing up, our modified general-equilibrium system has the solution $x_1 = 375/54$, $x_2 = 30/54$, $p_1 = 540/375$, $p_2 = 54/30$, $v_1 = 0$, $v_2 = 135/375$. This kind of nonemployment would disappear were demand relationships to change. The reader can verify for himself that if the demand functions were altered to $x_1 = 375/54\, p_1$,

$x_2 = \frac{5}{6} p_2$, the system would possess a solution with both factors fully employed.[1] But there is no economic reason for "tampering" with demand functions; people's tastes are given facts. If only people liked elephant meat more, African jungle land would bear high rents.

The possibility, indeed the likelihood, of structural unemployment such as we have been discussing was noticed by writers already cited, like Neisser, Stackelberg, Valk, and others. Some, like Stackelberg, took the position that this possibility reflected mainly the excessive rigidity of the Cassel system. In particular the assumptions of (1) inelastic factor supplies and (2) fixed technical coefficients violate the usual neoclassical picture and seem to be largely responsible for the appearance of redundant factors. It is quite apparent that *elastic* factor-supply functions could work against the occurrence of unemployment. If a fall in factor return reduces the supply of the factor, the extent of redundancy diminishes. Some of the unemployment is converted into "voluntary" unemployment. It is equally apparent that the insertion of factor-supply schedules like (13-4) cannot without further ado assure the existence of solutions to the system. Even on the simplest level, a factor price may be reduced to zero without cutting sufficiently the quantity offered. Once this zero-price point is reached, no further price reduction is possible. No doubt one could put strong enough conditions on the factor-supply schedules to rule out involuntary unemployment. But these would be too strict to be satisfying—some factors are inelastic in supply, and backward-rising supply curves can offer more trouble—so that even with supply schedules added, (13-1) will have to be replaced by (13-1a) if a solution is always to exist.

The case of variable coefficients of production is more difficult to discuss without getting involved in deep questions which are not really of direct concern to us here. Certainly the notion that it may be technically impossible to employ the given quantities of all factors seems to be inextricably bound to the idea of fixed, or at least limited, coefficients. If a fall in the price of an initially redundant factor decreases the supply and at the same time induces all industries to use the factor at higher intensity, both blades of the scissors are working to wipe out the margin of unemployment. If productive technique is such that for each factor there is always a desired commodity in which its marginal product can never fall to zero, no matter how intensively the factor is used, then

[1] It is vital to realize that this kind of unemployment is still not "Keynesian" unemployment. For even in our modified model with (13-1a) replacing (13-1), our previous proof that all income is spent, $\Sigma p_j x_j = \Sigma r_i v_i$, still holds. This can be established from the proof given in footnote 2, page 353. It only needs to be noticed that every line of (13-1a) that is a strict inequality gets multiplied by the corresponding v, which must be zero. So the inequality does not disturb the final result.

structural redundancy of this kind is ruled out. Thus, even in the linear-programming case of fixed coefficients there are situations in which structural unemployment of the kind illustrated in Fig. 13-2 can't occur. For example, suppose for each factor we can find a never-free good into which the factor is the *sole* input. Then in Fig. 13-2*c* there are rays running along the positive r_1 and r_2 axes and the cone between them is the whole quadrant. *Any* factor endowment ratio will then be fully employed.

This remark points up the fact that what happens when coefficients are variable is qualitatively no different from the nonemployment that occurs in the linear-programming case. In the former model, factor returns fall until factor intensity becomes high enough to absorb the total offer. In general, as factor intensity increases, the marginal productivity of the factor falls. In the fixed-coefficient case we may regard the appearance of unemployment together with a zero wage as indication that the marginal productivity of the factor has fallen to zero. And at near-zero wages, the demand for most factors probably becomes quite elastic;[1] i.e., in many cases there exists for any factor one or more low-productivity employments in which it is the only input. For labor, the standard example is personal service of a kind that requires almost no equipment. When the wage falls to zero (or thereabouts)[2] employment like this can take up all the slack in the factor supply. Thus the linear-programming model produces essentially the same result as the standard neoclassical setup. What appears as unemployment in one appears as low productivity in the other. And with the further inclusion of one-factor fringe employments (of low value productivity) even the latter phenomenon can be brought under the purview of linear programming.[3] That some such qualitative parallel should exist is not surprising. If a set of empirical facts can be idealized by either of the two theories of production, it can also be idealized by the other.

13-4. RIGOROUS PROOF OF EXISTENCE OF SOLUTIONS

The first rigorous study of the Walras-Cassel equilibrium conditions was made by Abraham Wald in 1935.[4] Wald investigated the existence

[1] In Fig. 13-1*b* we can think of the demand curve as following the course *DQS* and the supply curve as running along *OSS'*. Equilibrium could be anywhere in *QS* depending on the nature of the fringe employments.

[2] No economist should have any difficulty visualizing what modifications have to be made in the argument if there is a (biological, sociological, or legal) minimum wage.

[3] All the above ignores questions of Keynesian effective demand. In real life, when money demand falls it may be difficult or impossible to have factor prices fall correspondingly, and the result is involuntary unemployment. In the fixed-coefficients case, unemployment may be high.

[4] Wald's results were first presented in the *Proceedings* of Karl Menger's mathematical seminar in Vienna (vol. 6, 1935, and vol. 7, 1936). They were also summar-

and uniqueness of solution to the system consisting of (13-1a), (13-2), and (13-3), with factor supplies given.[1] Apart from some assumptions which are too obvious to require discussion,[2] Wald's only further restrictions are on the demand functions in (13-2). These restrictions are not all intuitively satisfying, and we shall find it convenient to alter them.

Since Wald (as remarked earlier) wrote his demand functions in the economically illegitimate form $p_i = F_i(x_1, \ldots, x_n)$, we cannot follow his procedure. Some of the assumptions we must make about the nature of our correct demand functions (13-2) will, however, be parallel to his, and some will not. For example, Wald required *his* demand functions to be defined for *all* positive quantities $x_1, \ldots, x_n$. An economist can't accept this condition; there may be some bundles of commodities that the market will not take at any positive price. If, for instance, x_1 and x_2 are perfect complements, only bundles containing them in proper proportions are eligible. We do not have to make this assumption. Instead we assume that *our* demand functions are defined for *all* configurations of commodity and factor prices.[3] This condition is quite unexceptionable. The market must give a definite response to any wage-price situation. (Wald also omitted the factor prices from his demand functions. This could be justified if market demand depended only on total income but not on its distribution, or by some equivalent assumption. But since it is little extra effort to include the factor prices, we do so.)

Next we follow Wald in requiring our demand functions to be continuous. This is a minimal assumption of mathematical regularity which can hardly be dispensed with. Without it one could hardly hope to prove the existence of a competitive equilibrium (compare the case of the simple Marshallian cross when the demand curve can have "holes," i.e., fail to be continuous).

We can dispense with another economically unreasonable assumption that Wald made. He postulated that no matter how high the price of a commodity became, there was still some positive demand. In other words, it would take an infinite price (whatever that means) to choke the

---

ized in a paper which has since been translated and published in *Econometrica*, **19**:368 (October, 1951), under the title "Some Systems of Equations of Mathematical Economics."

[1] The inequalities of (13-1a) should always be thought of as including the additional prescription that if strict inequality holds for any factor, the wage of that factor must be zero.

[2] Specifically: (1) each resource is present in a positive amount; (2) all input coefficients a_{ij} are nonnegative; (3) for each j, at least one a_{ij} is positive (i.e., every commodity requires at least one input in its production).

[3] Except, perhaps, for the meaningless case of all commodity prices equal to zero, which hardly need detain us. Recall also that we are free to impose one arbitrary restriction on absolute prices, e.g., $\Sigma p_j + \Sigma v_i = 1$.

demand down to zero. Exactly the opposite assertion would seem to be more plausible, that there is a price high enough to eliminate the demand for any commodity. Wald seemed to need this condition only to ensure that the equilibrium outputs x_i are all positive. In our further considerations we shall have no need to prevent one or more outputs from falling to zero, and so we can afford to leave this whole question open.

Finally, Wald made a rather peculiar assumption. He required the demand functions to satisfy what later came to be called the (weak) axiom of revealed preference. To be precise, if p_i^1 and x_i^1 are prices and quantities satisfying (13-2) and p_i^2 and x_i^2 are another such set, then $\Sigma p_i^1 x_i^1 \geq \Sigma p_i^1 x_i^2$ must imply $\Sigma p_i^2 x_i^2 < \Sigma p_i^2 x_i^1$. In words, if at the prices p^1 commodity bundle x^2 costs no more than x^1, and the consumer chooses x^1, then bundle x^1 has been revealed to be superior to x^2. Therefore, at prices p^2, when x^2 is actually bought it must cost less than x^1 or the preference previously revealed would be contradicted. Why is this assumption peculiar? Because the demand functions (13-2) are market demand functions, not individual demand functions. "Rationality" cannot be required of market demand functions because changes in prices normally change the distribution of income. With a changed income distribution, *different* "preferences" will be revealed. In other words, Wald really assumed that there is essentially only one rational consumer. But the only use made of this overly strong assumption is to show that the competitive equilibrium is *unique*. This is, as every economist who has studied international trade knows, an overly strong conclusion, and so we can dispense with this assumption too.

Wald's theorem is as follows. The assumptions just discussed, together with the three assumptions mentioned in footnote 2 on page 367, imply that the system (13-1a), (13-2), and (13-3) possesses an economically meaningful solution in the variables x, p, and v. This solution is unique in x and p: only one set of outputs and commodity prices is consistent with equilibrium. If in addition we can rule out such phenomena as two resources being used in the same proportions in *every* production process, then the equilibrium factor prices v are also unique. Evidently some such qualification is needed: if the production of every commodity requires one man and one horse, then clearly only the return to the combined man-horse team is determinable.[1]

Wald's proof of this theorem is extremely intricate and opaque. Nevertheless, it is a beautiful achievement, perhaps the most difficult piece of rigorous economics up to that time. In the next few pages we shall outline a short and relatively transparent proof, which uses as tools the duality theorem of linear programming, and one purely mathematical

[1] The exact statement is: If the matrix $[a_{ij}]$ appearing in (13-1a) is of rank m, the solution is unique in the factor prices.

result, the fixed-point theorem of Kakutani, which will be described when it is needed. Actually we shall solve a slightly different problem from Wald's. We retain the three assumptions of footnote 2 on page 367. In addition, we need only the stated assumptions that market demand is defined for all price configurations and is continuous. At the end we shall also show how the revealed preference assumption can be made to imply the absence of multiple equilibria.

Now look back for a moment at the price-equals-unit-cost equations (13-3). Why must they hold? Because if price exceeds unit cost, output will increase and factor prices be bid up, thus lowering price and increasing unit cost. Because if unit cost exceeds price, output will decline and so will factor prices, thus raising price and lowering unit cost. But there is a natural limit to this last process. Output can fall as far as zero, but not lower. Suppose output is down to zero and unit cost still exceeds price. There is nothing wrong with this picture at all. It is exactly what one would expect of commodities not being produced—that price should not cover unit costs at any positive output. It is no accident that silk diapers are so rare. We are led to the following change in the general equilibrium system: replace the n equations (13-3) by n inequalities:

$$\begin{aligned} a_{11}v_1 + a_{21}v_2 + \cdots + a_{m1}v_m &\geq p_1 \\ a_{12}v_1 + a_{22}v_2 + \cdots + a_{m2}v_m &\geq p_2 \\ \cdots\cdots\cdots\cdots\cdots\cdots\cdots \\ a_{1n}v_1 + a_{2n}v_2 + \cdots + a_{mn}v_m &\geq p_n \end{aligned} \qquad (13\text{-}3a)$$

with the provision that *if inequality holds in one or more lines of* (13-3a), *the corresponding outputs x must be zero.*

We shall show that, subject to the assumptions stated above, the general-equilibrium system consisting of (13-1a), (13-2), and (13-3a) possesses an economically meaningful, nonnegative solution in the variables x, p, and v. (If in addition the demand functions satisfy the axiom of revealed preference, the solution is unique in x and p.) Then we shall widen the system to include factor-supply functions (13-4) and sketch how our theorem extends to this case.

Suppose we rewrite (13-1a) for ready reference:

$$\begin{aligned} a_{11}x_1 + a_{12}x_2 + \cdots + a_{1n}x_n &\leq r_1 \\ a_{21}x_1 + a_{22}x_2 + \cdots + a_{2n}x_n &\leq r_2 \\ \cdots\cdots\cdots\cdots\cdots\cdots\cdots \\ a_{m1}x_1 + a_{m2}x_2 + \cdots + a_{mn}x_n &\leq r_m \end{aligned} \qquad (13\text{-}1a)$$

If we recollect that the variables x and v have to be nonnegative, (13-1a) and (13-3a) look very much like the constraints of linear-programming problems. In fact, they look like the constraints of *dual* linear-programming problems, since the same coefficients appear in both sets of inequalities, except that they are transposed. To go one step further, we can even write down the dual linear-programming problems concerned. One is as follows:

Maximize $p_1x_1 + p_2x_2 + \cdots + p_nx_n$ (= value of output) subject to (13-1a) and $x_i \geq 0$.

The dual problem is as follows:

Minimize $r_1v_1 + r_2v_2 + \cdots + r_mv_m$ (= total factor income) subject to (13-3a) and $v_j \geq 0$.

Now we can use some results from the theory of linear programming. One fundamental theorem says that a linear-programming problem has a solution if both it and its dual are feasible (cf. Chap. 4, p. 104). But this condition is satisfied here: there *are* nonnegative x's and v's satisfying (13-1a) and (13-3a). Whatever the positive r's and nonnegative p's, one can always find a set of x's so small that (13-1a) is satisfied and a set of v's so large that (13-3a) is satisfied. Therefore the pair of dual-programming problems possess solutions. It is also known[1] that whenever we solve a pair of dual problems, the maximum value of the linear form being maximized (here, $\sum_1^n p_jx_j$) equals the minimum value of the form being minimized (here, $\sum_1^m r_iv_i$). Thus if we can find an equilibrium solution of the general equilibrium system which is *also* a solution of the pair of dual programming problems, we have another way of showing that total expenditure equals total factor returns. The converse is also true: if total expenditure equals total returns in a solution of the general equilibrium system, then the linear-programming problems are also solved. But we already know that the demand and supply functions are such as to force this equality. Hence the logic of linear programming has inevitably alerted us to the following vital but little-appreciated fact. *Hidden in every competitive general-equilibrium system is a maximum problem for value of output and a minimum problem for factor returns.*

We are not through exploiting linear-programming theory. The duality theorem (Sec. 4-15) also tells us that whenever a constraint in (13-1a) or (13-3a) is satisfied with strict inequality, the corresponding dual variable is zero. But this is exactly what we want; this corresponds to the qualifications that economic reasoning has led us to append to the inequalities (13-1a) and (13-3a).[2]

[1] See Chap. 4, page 104.

[2] Moreover we can deduce the following interesting facts. Suppose $n > m$ (more goods than factors). Then barring singular cases, whatever the p's turn out to be,

Let us see where we are. We seek a solution to the system of equations and inequalities (13-1a), (13-2), and (13-3a). Our unknowns are n x's, n p's, and m v's. We have been led to formulate a pair of dual linear programs in which x and v appear as unknowns and p as givens. We have shown that for *any* set of nonnegative p's, the dual problems possess solutions. If we pick a set of prices p_j arbitrarily and solve the dual problems, the resulting x's and v's are such that (13-1a), (13-3a), and accompanying qualifications are satisfied. We have almost solved our problem. The gap is that if we choose p's at random and get x's and v's from the linear-programming problems, we have no guarantee that this particular set of p's, x's, and v's will satisfy the demand relations (13-2). To put it differently, we can start with any p's and get a set of x's satisfying (13-1a) and v's satisfying (13-3a). But the p's and v's when inserted in the demand functions determine a second set of commodity outputs, or x's. If this second set of x's should coincide with the x's we have already found (13-2) would also be satisfied and our general equilibrium system would have a solution. Suppose this coincidence does not occur. Then we repeat the process with another arbitrary initial set of p's. As we thus run over all possible sets of p's, will there always be at least one set of p's for which this coincidence will occur? The answer will be yes, there must always be at least one such, and we shall thus succeed in proving the existence of an equilibrium solution. But to complete the proof we have to make use of the fixed-point theorem of Kakutani which we must now describe.

Fixed-point theorems are useful mathematical devices for proving that solutions to some kinds of equations actually exist. Think of a function as a way of associating with each member or point of a set some other member of the same set. (The points in question can be numbers, in which case we have a function of a single variable, or vectors of n numbers, in which case the "function" is really n functions of the same n variables, associating with each n-tuple another n-tuple.) A fixed point of such a function is a member of the set which is associated with (or "mapped into") itself by this function. The simplest fixed-point theorems can be proved in a minute. For example, let x range over the interval from 0 to 1 *inclusive*. Let $f(x)$ be a *continuous* function whose values are also between 0 and 1; i.e., f maps the interval $0 \leq x \leq 1$ into itself. Then it follows that f has a fixed point; i.e., there is an x^* such that $x^* = f(x^*)$. To prove this graphically, draw the square whose side is the unit interval on the x axis, as shown in Fig. 13-3. In the usual way, $f(x)$

---

the pattern of outputs established will have m goods being produced and $n - m$ not being produced. If $m > n$ (more factors than goods), there will be in equilibrium only n positive v's, and the remaining $m - n$ factor services will be free goods. For the underlying theorem in linear programming, see Sec. 4-15.

is represented by a continuous curve which lies in this square. Draw the diagonal $y = x$. Any time the curve $y = f(x)$ crosses the diagonal $y = x$, we have $x = f(x)$ and a fixed point. The reader can see that it is impossible to draw a continuous curve which begins on the vertical $x = 0$, ends on the vertical $x = 1$, and fails to touch the diagonal (including its end points).

More analytically, consider $f(0)$ and $f(1)$. If $f(0) = 0$ or $f(1) = 1$ we have a fixed point. Otherwise $f(0) > 0$ and $f(1) < 1$. Now let $g(x) = f(x) - x$, and g is a continuous function. We have $g(0) > 0$, $g(1) < 0$. A continuous function can't skip any values, so that $g(x)$ in getting from positive to negative values must pass through zero. But $g(x) = 0$ means $x = f(x)$. The crucial point here is the continuity of f. Without it, of course, the theorem is not true.

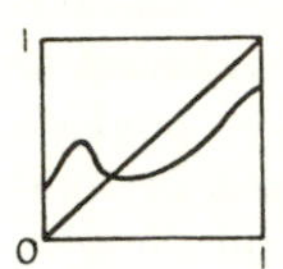

Fig. 13-3

More complicated fixed-point theorems, like Kakutani's, are not so intuitive. The reader can try to visualize a *continuous* mapping of, say, a circular disk into itself and look for the fixed point or points. For example, a rotation of the disk has the center as fixed point; a simple folding-over has a whole diameter full of fixed points.[1]

Suppose S is a bounded closed convex set of points y. With each point y of S we associate a closed convex subset of S, say H_y. The sets H_y must be related to the points y in a smooth way (technically the mapping $y \rightarrow H_y$ must be upper semicontinuous). Then the Kakutani theorem asserts that at least one of the H_y will contain the y with which it is associated.[2]

We use this theorem in the following way. We impose, as we may, the condition that $\Sigma p_j + \Sigma v_i = 1$. The set of nonnegative p's and v's satisfying this condition is our set S. It is closed, bounded, and convex. With each (p,v) point of S we associate an x according to the demand functions (13-2); i.e., p and v determine the outputs that the market will

[1] An enjoyable discussion of the two-dimensional fixed-point theorem is to be found in R. Courant and Herbert Robbins, *What Is Mathematics?* Oxford University Press, New York, 1941.

[2] "Bounded" means just what it says: there is some sphere or cube in which all of S is contained, which means that every point of S is within a finite distance of the origin. "Closed" means, roughly speaking, that S includes all its boundary points. The force of this is that if we take any sequence of points of S which tends to some limit point, then that limit point will also be in S. "Convexity" is discussed at some length in Chap. 14. A mapping or function which associates with each y a set H_y is upper semicontinuous if the following is true: take any sequence of points y tending to a point z; take any sequence of points, one from each of the corresponding image sets H_y, and suppose that this sequence tends to a limit point w; then w is a point of H_z. This last is a rather technical concept; it is meant to rule out situations in which, as you run along neighboring y's, the sets H_y contract *all of a sudden.*

demand at those prices. This market basket of outputs may not be producible, i.e., may not satisfy (13-1a). If it does not, decrease all the x's *proportionally* until (13-1a) is *just* satisfied. Alternatively these outputs may be producible but not efficient; i.e., they may satisfy (13-1a) with all the strict inequalities holding. In that case it would be feasible to increase all outputs simultaneously. Let us do so, increasing them *proportionally* until at least one equality holds, so that we can't increase them any more. The net result is this: with each (p,v) of S we have associated an efficient[1] set of outputs. First we found

$$x_1=F_1(p_1, \ldots ,p_n;v_1, \ldots ,v_m), \ldots , x_n=F_n(p_1, \ldots ,p_n;v_1, \ldots ,v_m)$$

Then from $(x_1, \ldots ,x_n)$ we found a new but proportional set of outputs $(kx_1, \ldots ,kx_n)$ where the constant k was chosen to make these new outputs efficient. Of course the new outputs will no longer satisfy the demand functions, but we shall worry about that later.

Now we know that an *efficient* set of outputs has the following very important property: for *some* set of nonnegative p's, it maximizes $\Sigma p_i x_i$ among all feasible outputs.[2] If x happens to fall on a flat face of the efficiency frontier, it will maximize value of output for only one particular price configuration. If x falls on a corner or an edge, there will be a whole set of price configurations for which this is so, but these price configurations will form what is technically called a "convex polyhedral cone," with accent on the *convex*.

So with our initial (p,v) we have associated an x, then an efficient x, and then a set of p's (perhaps only one) for which the efficient x is a solution of the linear-programming problem (13-1a). Now we can insert in turn each of the p's of this set into the linear-programming problem (13-3a) and solve the latter. Hence with each of the p's we associate the one or more v's which minimize $\Sigma r_j v_j$ subject to (13-3a). At the end of this tedious process we can say: with each initial (p,v) of S we have associated a whole collection of (p,v)'s. As a final step we can change each (p,v) of the collection proportionally until its members add up to 1, that is, until it belongs to S. This final set we shall call H_{pv}.

Now each H_{pv} is a closed and convex[3] subset of S. Furthermore, all the conditions of the Kakutani theorem are satisfied: the mapping of the initial (p,v) into its H_{pv} is upper semicontinuous because (1) the demand functions are continuous; (2) the mapping of x onto an efficient x is con-

[1] Not quite efficient, but the exception is a subtlety and we ignore it.

[2] This proposition (really a foundation stone of welfare economics) will be explored in more detail in the next chapter. But it is a common fact of economic theory that we can trace out the maximal production-possibility frontier by maximizing $\Sigma p_i x_i$ and then varying the prices.

[3] That it is convex can be shown from the fact that by construction we always have $\Sigma p_i x_i = \Sigma r_j v_j$.

tinuous; and (3) the final mapping of the efficient x into H_{pv} is upper semicontinuous. The theorem then tells us that there is at least one (p,v)—call it $(\bar{p},\bar{v})$—that is contained in its own $H_{\bar{p}\bar{v}}$. For this special $(\bar{p},\bar{v})$ we know by construction that (13-3a) is satisfied. Also we have its associated efficient x, say $\bar{x}$, for which (13-1a) is satisfied. We must now close the last gap and show that for $\bar{x}$, $\bar{p}$, and $\bar{v}$ the demand functions (13-2) are satisfied. But this is easy. Take the x that *does* satisfy the demand equations. The demand functions are such that $\Sigma \bar{p}_i x_i = \Sigma r_j \bar{v}_j$. But from the linear-programming nature of (13-1a) and (13-3a) we know that for the efficient $\bar{x}$ also, $\Sigma \bar{p}_i \bar{x}_i = \Sigma r_j \bar{v}_j$. By construction $\bar{x}$ is proportional to x, $\bar{x}_i = kx_i$. It can only be then that $k = 1$ and $\bar{x} = x$. The demand functions are satisfied, and a competitive general equilibrium x, p, v must exist.[1] Home is the wanderer.

This has been a long-drawn-out mathematical excursion. However, there does not seem to be any easy answer to the complicated question of the consistency of the Walras-Cassel system. Ask a complicated question and you get a complicated answer!

We still have to say something about the full general-equilibrium system with elastic supply functions (13-4). Fortunately, this causes almost no additional pain (perhaps because of the numbness induced by the "simple" case!). All through the proof we took the r's as given numbers. Resource supplies were perfectly inelastic. If they are not so, we make one amendment to the proof. We start as before with an initial (p,v). Then along with an x from the demand functions we get an r from the supply functions. And after that we proceed word for word as before.

Finally, suppose that there are two solutions p^1, x^1 and p^2, x^2. From the linear-programming problem hidden in (13-1a) we deduce that $\Sigma p_i^1 x_i^1 \geq \Sigma p_i^1 x_i^2$, because x^1 maximizes $\Sigma p^1 x^1$ for feasible x. Similarly $\Sigma p^2 x^2 \geq \Sigma p^2 x^1$. Looking back at page 368, we see that this violates the axiom of revealed preference. Hence if we accept this assumption,[2] the solution is unique in p and x.

[1] The use of the Kakutani theorem to prove the existence of an equilibrium is McKenzie's idea. See his study of Graham's international-trade model, referred to in footnote on page 347. Our treatment is similar to his, but somewhat simpler (if the reader can believe such a thing). The simplification comes about through the explicit use of linear programming. See also H. W. Kuhn, "On a Theorem of Wald," Chap. 16 in Kuhn and A. W. Tucker (eds.), *Linear Inequalities and Related Systems*, Annals of Mathematics Studies No. 38, Princeton University Press, Princeton, N.J., 1946.

[2] The logic here should be familiar to economists. There is no reason why a set of *market* demand curves should be of a kind that might have been deduced from an indifference map. Distributional shifts are likely to remove that kind of consistency that comes from utility maximization.

If there are 2 or more persons, the assumption that total market demand satisfies the "weak axiom" implies that their indifference surfaces (if they have transitive

One last word. We have shown en route that the equilibrium values p, x, v, and r have certain maximum and minimum significance. This will come in handy when we later turn to some problems in welfare economics.

13-5. COMPARISON WITH THE NEOCLASSICAL MODEL

In the last couple of sections we have formulated a static general-equilibrium system, discussed some of its properties, and even proved the existence of a competitive solution. The production end of this system was of the linear-programming type, and as a result no talk of marginal productivities emerged. Before leaving this subject perhaps we might briefly write down a more orthodox neoclassical system with smooth marginal productivities, to see how it differs from the linear-programming model and how it is analogous.

To concentrate on essentials, we can stick to the elementary case in which there are no whirlpools of goods made from intermediate goods. Commodities are made directly from primary factors. The reader can convince himself that intermediate goods can be handled by the same methods used in our previous linear model.[1] Also, we shall assume that all industry production functions have constant returns to scale. Without this assumption we would have to worry about the allocation of output among firms. And incidentally, if we imagine all firms in an industry as having identical U-shaped average-cost curves, variations in long-run equilibrium industry output will take place by variation of the number of identical firms, all producing at minimum average cost, and the industry production function *will* have constant returns to scale. This much granted, we can write down the n industry *production functions* (a concept that the reader will probably view as a long-lost friend):

$$\begin{aligned} x_1 &= X^1(r_{11},r_{21}, \ldots ,r_{m1}) \\ x_2 &= X^2(r_{12},r_{22}, \ldots ,r_{m2}) \\ &\cdots\cdots\cdots\cdots\cdots \\ x_n &= X^n(r_{1n},r_{2n}, \ldots ,r_{mn}) \end{aligned} \qquad (13\text{-}5)$$

---

preferences) must have unitary income elasticities for all goods; and then the market demand can be thought of as coming from a homogeneous ordinal utility indicator $U(x_i, \ldots ,x_n) \equiv \lambda^{-1}U(\lambda x_{ij}, \ldots ,\lambda x_n)$ for all positive λ. The equilibrium solution would be at the unique point of contact of the convex production-possibility frontier defined by (13-1a) and the highest-attainable strongly convex indifference surface. It would be exactly as if the market consisted of a single rational consumer.

[1] See, for example, P. A. Samuelson, "Prices of Factors and Goods in General Equilibrium," *Review of Economic Studies*, **21**(1):17.

The new symbols r_{ij} represent the physical quantity of the ith resource actually used by the jth industry (no joint production here). In the spirit of our earlier work, we could put $\leq$ signs in (13-5) and add a convention about zero prices, but instead we stick to the orthodox practice of supposing all production to be efficient in the sense that equality holds and no factors are free.

There were no "production functions" in our linear model, but there were processes or activities, ways of transforming resources into commodities, which performed the same role. The main difference is easily seen graphically. Draw the isoquants, or equal-output curves, of a typical production function in (13-5). Because of the constant returns to scale we need only draw one contour line, say the one for one unit of output; all the rest are simply radial enlargements of this. The usual assumptions give a smooth concave curve as in Fig. 13-4*a*. If there is only one process for

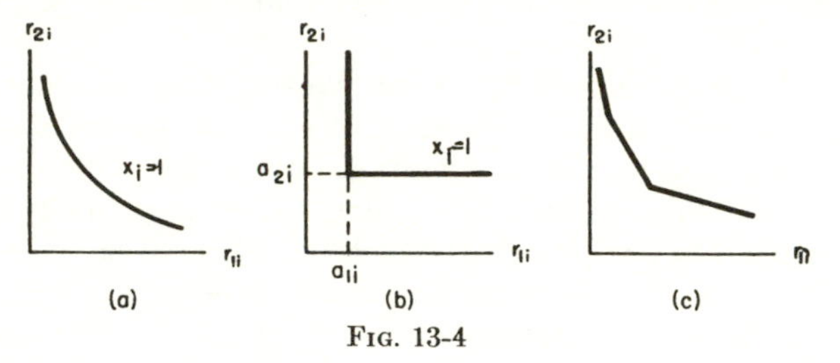

FIG. 13-4

each commodity, we get the now familiar right-angled isoquant of Fig. 13-4*b*. In our general linear model with alternative processes we get Fig. 13-4*c*. Each vertex represents a different process; the flat parts of the isoquant represent concurrent operation of two adjacent processes.

While we are at it, we can find the equation of the unit isoquant pictured in Fig. 13-4*a*. Because of constant returns to scale, knowing this is as good as knowing the whole production function, and moreover we shall be able to get rid of the r_{ij} and replace them with more familiar variables. Take the first production function of (13-5), divide both sides by x_1, and remember that with constant returns to scale if you multiply or divide output by any number, you simply multiply or divide all inputs by the same number. Therefore

$$1 = \frac{x_1}{x_1} = \frac{1}{x_1} X^1(r_{11}, r_{21}, \ . \ . \ . \ , r_{m1}) = X^1\left(\frac{r_{11}}{x_1}, \frac{r_{21}}{x_1}, \ . \ . \ . \ , \frac{r_{m1}}{x_1}\right)$$

Now a quantity like r_{21}/x_1 is the first industry's input of Resource 2 per unit of its own output of Commodity 1. It is just an input coefficient—what we previously called a_{21}. The difference now is that a_{21} is neither a known constant nor one of a finite set of known constants; it is continuously variable, an unknown of the system, something we must determine. Still, we can rewrite each of the production functions in terms of the a's to give us an alternative set of equations with the same content as before:

$$\begin{aligned} 1 &= X^1(a_{11},a_{21}, \ldots ,a_{m1}) \\ 1 &= X^2(a_{12},a_{22}, \ldots ,a_{m2}) \\ &\cdots\cdots\cdots\cdots \\ 1 &= X^n(a_{1n},a_{2n}, \ldots ,a_{mn}) \end{aligned} \tag{13-5a}$$

Previously production decisions consisted of choosing activity or process levels. Here entrepreneurs have to decide on outputs and input coefficients. There are mn input coefficients, and under competition they will be determined by a cost-minimization process. And this leads by the familiar reasoning to equilibrium conditions of the following form: The value of the marginal product of a factor must be the same in every industry and equal to the price of the factor. Let $\partial X^i/\partial r_{ji} = X_j^i$ be the marginal physical productivity of factor j in Industry i (a quantity that depends only on factor *proportions* and not on absolute levels of output or input). Then we can get a new set of $m \times n$ equations (or inequalities):

$$v_j \geq p_i X_j^i \qquad j = 1, \ldots, m; i = 1, \ldots, n \tag{13-6}$$

The inequality has been slipped in, because equality need hold only if Factor j is actually used in Industry i. But there are no potato farms at Times Square and no surgeons used to dig ditches. To (13-6) we add the proviso that if the inequality holds, the corresponding r_{ji} must be zero. There is one inequation of (13-6) for every factor-commodity pair. We can replace all inequalities by equalities only if every factor is used in every industry.

There appears to be nothing in our linear system of the earlier part of this chapter to correspond to the marginal productivity inequations (13-6). But actually there is. To see this, let us ask first whether our smooth orthodox system oughtn't to include a requirement that price equal unit cost for all commodities actually produced in equilibrium. Evidently this competitive equilibrium condition is as applicable here as it was in the linear model's inequalities (13-3a). So it is; but we can show that the marginal-productivity conditions (13-6) already *imply* that price equals unit cost. Take all the equations of (13-6) corresponding to a single industry (represented by an index i). They are

$$v_1 \geq p_i X_1^i, v_2 \geq p_i X_2^i, \ldots, v_m \geq p_i X_m^i$$

Multiply both sides of the first of these by a_{1i}, both sides of the second by a_{2i}, and so forth, and then add the results. We get

$$a_{1i}v_1 + a_{2i}v_2 + \cdots + a_{mi}v_m = p_i(a_{1i}X_1^i + a_{2i}X_2^i + \cdots + a_{mi}X_m^i) \tag{13-7}$$

We can drop the inequality sign because wherever there is an inequality the corresponding a_{ji} is zero; and the inequality is converted into an equality after multiplication. Now the expression in parentheses on the right is exactly equal to $X^i(a_{1i}, \ldots, a_{mi})$. This follows from Euler's theorem and the fact that the production functions are homogeneous of first degree. For in parentheses we have all the partial derivatives or marginal products of $X^i(a_{1i}, \ldots, a_{mi})$, each multiplied by the corresponding factor amount for unit output. The adding-up theorem says that with constant returns to scale this will exactly exhaust the output X^i. But according to (13-5*a*), $X^i = 1$; that is, we are operating along the unit isoquant. Inserting all this in (13-7) we finally get

$$a_{1i}v_1 + a_{2i}v_2 + \cdots + a_{mi}v_m = p_i$$

and we get this for every industry which actually produces any positive output. [If $x_k = 0$, then Eq. (13-5*a*), and in fact the whole notion of an input coefficient a_{jk}, loses all sense.] But compare this with (13-3*a*); it is the same thing. It states that price equals unit cost for each positively produced good. It is easy to show that unit cost may exceed price for unproduced goods.

Thus in the smooth model there is no need to require separately that price equal unit cost; this is already implied by the minimum-cost marginal-productivity conditions and the homogeneity of the production function. If production functions did not have constant returns to scale, competition would require that we add as a separate equilibrium condition that price equals unit cost.

This reasoning leads us to guess that the price-unit-cost relations in the linear model perform the same functions as the marginal-product relations do in the smooth model. And of course this is right. In the linear model we have processes, not production functions. We can choose units so that operation of the *i*th process at unit level produces one unit of Commodity *i*. Then the increment of value produced by unit operation of the process is just p_i. This is like a "value of marginal product" and corresponds to the right-hand side of (13-6). What is the factor cost of a unit level of operation of the *i*th process? Answer: $a_{1i}v_1 + a_{2i}v_2 + \cdots + a_{mi}v_m$. This corresponds to the left-hand side of (13-6). The requirement that the value produced by a unit increment in Process *i* should not exceed the factor cost, and should equal it if the process is operated, leads to

$$a_{1i}v_1 + a_{2i}v_2 + \cdots + a_{mi}v_m \geq p_i$$

and of course this is just our price-unit-cost relation (13-3*a*) again. In the linear model this relation coalesces completely with the marginal-productivity relations.[1] In the smooth-constant-returns model the latter implies the former, but the two conditions can be thought of separately.

So far our smooth neoclassical model consists of (13-5a) and (13-6). The rest of it corresponds exactly to the linear model. We still need (13-1a), which says that supply equals demand for each factor, or else for some factors supply exceeds demand and then the factor price is zero. We can rewrite (13-1a) briefly as

$$a_{i1}x_1 + a_{i2}x_2 + \cdots + a_{in}x_n \leq r_i \qquad i = 1, \ldots, m \qquad (13\text{-}1a)$$

The only difference here is that all the a's are variables of the system. Also we need demand-for-goods and supply-of-factors equations, and here too we can simply take over what we used in the linear model. We have our generalized demand functions

$$x_i = F_i(p_1, \ldots, p_n; v_1, \ldots, v_m) \qquad i = 1, \ldots, n \qquad (13\text{-}2a)$$

and supply functions

$$r_j = G_j(p_1, \ldots, p_n; v_1, \ldots, v_m) \qquad j = 1, \ldots, m \qquad (13\text{-}4)$$

Counting up, we have mn unknown a's, n unknown outputs, x, n unknown prices, p, m unknown factor supplies r, and an equal number of unknown factor prices v—a total of $mn + 2n + 2m$ unknowns. On the other side we have the n production functions (13-5a), mn marginal-productivity relations (13-6), m factor-market clearing relations (13-1a) and the $n + m$ commodity-demand and factor-supply equations (13-2a) and (13-4)—also $mn + 2n + 2m$ in total number. It is a commonplace that a system like this determines only relative prices. Mathematically, the demand functions (13-2) and the supply functions (13-4) are homogeneous of zero degree in the p's and v's. And changing v's and p's in proportion has no effect on (13-6). Economically, doubling all prices, including factor prices, would change nothing real in the system. As before we could choose a commodity, say the first, as numeraire and put $p_1 = 1$. This reduces the number of unknowns by one. But Walras's law and some of our own earlier reasoning tell us that total sales must identically equal total-factor incomes: the demand and supply equations for goods and factors are not independent. Therefore the demand function for x_1 can be eliminated from the system, and we are still even in equations and unknowns.

We didn't do this in our discussion of the linear model, although we could have. Instead of pinning down $p_1 = 1$, we there in effect pinned

[1] For some related discussion, the reader should glance back at our earlier discussion of duality and the competitive firm (Chap. 7, pp. 174–178).

down the value of a peculiar market basket of goods and factors, namely, 1 unit of each. We put

$$p_1 \cdot 1 + \cdots + p_n \cdot 1 + v_1 \cdot 1 + \cdots + v_m \cdot 1 = 1$$

It is shameful but true that this was solely for mathematical convenience. Of course, once we have found an equilibrium we can adjust absolute prices so that this arbitrary condition is no longer satisfied. We could even adjust prices to make $p_1 = 1$ if we so desire (provided only that $p_1 \neq 0$ in the equilibrium point).

Of course the counting of equations and unknowns in the smooth model can't guarantee the existence of solutions any more than it could in the linear model. All the same difficulties arise, including the inherent non-negativity of the unknowns. But as the close similarity of the two models would suggest, the same methods and approach we used in the

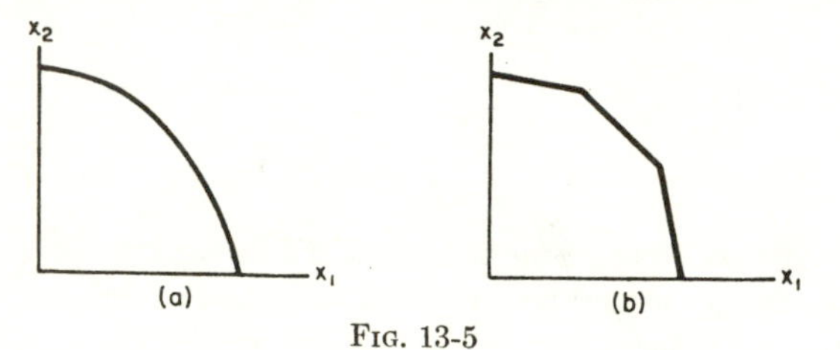

FIG. 13-5

linear model will also provide a proof of the existence of a competitive equilibrium in the smooth model. We must of course make the usual assumptions of generalized diminishing returns.

In fact the parallel between the two models carries all the way through. Although it is a little harder to see in the smooth case, the same nice duality between prices and outputs holds there also, and with it the very important fact that for any set of resource supplies and prices, competition acts to maximize the total value of output and to minimize the total cost of inputs and only succeeds in making these two totals equal. The importance of this will appear in the next chapter on welfare economics. When we get right down to it, the only real difference between the two models is the one illustrated in Figs. 13-4*a* and *c*. Or in another familiar diagram, for given resource amounts, production possibilities in the smooth model are represented by a transformation (or opportunity-cost) curve familiar mainly from international-trade theory, as in Fig. 13-5*a*. The linear model leads to something like Fig. 13-5*b*. The essential geometrical similarity is the convexity in both cases of the set of feasible outputs. It is easy to see that by having more and more processes, and thus more and more shorter line segments in Fig. 13-5*b*, the linear model and the smooth model come closer and closer to identity.

In all the considerations of this chapter the reader has probably noticed that the demand side has been slighted. The market-demand functions and market-supply functions were introduced without preliminary. There was no mention of individual households and their utility-maximizing activities. Even on the production side we began at once with global-industry production possibilities, slurring completely over the profit-maximizing activities of individual entrepreneurs. The expositional reasons for this will be only too clear to the reader.

For completeness's sake it should be mentioned that it is possible to begin at the beginning and build up a complete-equilibrium system, either linear or smooth or, even more generally, mixed—part smooth, part linear. One can start with a list of individual households, their labor potentialities, their tastes for work, leisure, and commodities, and their ownership of other resources. These households are supposed to take all prices as given and to maximize their satisfaction in the standard way. This leads as usual to individual (and, by addition, to market) supply and demand functions. One can start with a list of firms, each with given techniques (i.e., production possibilities), subject to the usual restrictions. The firms are supposed to take all prices as given and to maximize their profits in the standard way. Markets are supposed to be cleared. There is no need to enter upon the details here. Naturally the investigation of such a system is more complicated than the work we have done here. There are more variables, and more is determined; commodity and factor prices once set, the fortunes of firms and the distribution of incomes among households follow. The essential fact is that, given reasonable assumptions, the existence of a competitive equilibrium can be proved, and by methods which are essentially similar to the ones we have used, although necessarily more complicated.[1]

13-6. VON NEUMANN'S MODEL OF EQUILIBRIUM GROWTH

The celebrated general-equilibrium model of von Neumann[2] is in some ways very similar to the systems so far discussed. Most especially, it deals with production in exactly the linear-programming way, with alter-

[1] For details the reader is referred to the paper by Arrow and Debreu, *op. cit.*

[2] The paper was first read at Princeton in 1932 and published six years later in the *Proceedings* of Karl Menger's mathematics seminar in Vienna. An English translation under the title "A Model of General Equilibrium" appeared in the *Review of Economic Studies*, **13**(1):1–9, with a commentary by D. G. Champernowne. The interested reader should study chap. 4 by N. Georgescu-Roegen, in T. C. Koopmans (ed.), *Activity Analysis of Production and Allocation*, which gives alternative proofs and further results in the same vein.

native processes and intermediate goods (in fact nothing but intermediate goods), leading up to linear inequalities. In fact, the von Neumann paper may be considered the first explicit formulation of the linear model of production with this degree of generality. There are also some radical differences from the earlier systems of this chapter. There is no consumer-demand side to the model; there are no primary factors; production is explicitly made to take time; and "equilibrium" is defined in a very special way. In some respects von Neumann's work is more closely akin to the dynamic capital models investigated in Chap. 12. Indeed, some of the explicitly dynamical aspects have already been discussed in that context. We shall use this last section of this chapter to look a little further into the structure of the von Neumann model; this can be done briefly, mainly as an exercise in some of the concepts developed in earlier sections. This exposition may be of some interest because the original paper is extremely forbidding to the nonmathematician. We shall use an adaptation of our earlier notation rather than von Neumann's own.

Let us repeat more precisely the basic assumptions mentioned rather casually above. In the first place, we are now dealing with a closed system, a pure production model. It is "closed" in the sense that there is no final demand and no fixed factor. We can think of labor as being "produced" by households with consumption goods as inputs (in fixed proportions and fixed amounts per unit of output of labor). Actually there can be more than one way of producing labor, just as there can for other commodities. The important thing is that there is no autonomous demand for commodities and no resources which cannot be produced like other goods. Thus "land" can play no role in this model, unless we imagine the system as occupying a small part of a large undeveloped continent; then land can be had on the same kind of terms (constant costs) as other commodities, by clearing forests. But as soon as the frontier disappears and all available land is in use, the von Neumann model will no longer serve as an idealization. Since von Neumann's concern is mainly with growth, it is apparent that fixed factors would be an embarrassment, would in fact limit us ultimately to zero growth, the stationary state. Of course there is no reason why we should not consider this case of zero growth as a limiting case easily comprehended by the von Neumann model. (It would be interesting to consider the "production" of technical knowledge as a possible economic activity.) Another assumption is that each process of production takes exactly one unit of time. This can be arranged even if different processes really have different duration, by introducing fictitious intermediate stages in the longer processes. One of the advantages of the von Neumann model is that it can handle capital goods without fuss and bother. A nondepreciating capital good simply enters both as input and as output in the corresponding process. If the capital good depreciates 3 per cent per unit of time, 1 unit of the good may appear as input, and 0.97 unit as output.

The essential fact about the system is that goods are produced from other goods by processes or activities of the fixed-coefficient kind we have been studying. There can be alternative processes, intermediate goods, and joint production. All this is describable in a notation much like our earlier one. Let a_{ij} be the *input* of Commodity i per unit level of operation of Process j. Let k_{ij} be the *output* of Commodity i per unit level of operation of Process j. The a's and k's are inherently either positive or zero. There are, as at the end of Sec. 13-2, n processes and s commodities (remember there are no primary factors). The process numbered 1 is a way of converting a_{11} units of the first commodity, a_{21} units of the second, up to a_{s1} units of the last, into k_{11} units of the first good, k_{21} units of the second, etc. The same commodity may appear both as input and output. There are n such processes.[1] Outputs appear at 1 unit of time after inputs are used.

Since this is a closed productive system the outputs of one period are the inputs of the next. And while there are no literally fixed resources, time has the effect of instantaneously fixing the supply of inputs. At any given moment of time, production is limited by the *currently available* inputs, i.e., by the previous period's outputs. The reader is familiar with this idea from the earlier chapters on dynamic models; it is an old notion in economics and used to be applied to the provision of subsistence for the labor force.

Let us represent by $x_j(t)$ the intensity at which the jth process is operated in Period t. The a's and k's were known numbers. The x's are going to be unknowns of our problem; they are nonnegative by definition. At the close of Period t the economic system has available the outputs of Period t's productive activity—no more, no less. Inventories can be allowed for by inventing a process which has 1 unit of a commodity for input and 1 unit of the same commodity for output (storage costs and spoilage can be entered as inputs also).[2] Thus the available supply of

[1] Von Neumann makes the additional assumption that for every pair (i,j), $a_{ij} + k_{ij} > 0$. In words, every commodity appears in every process, either as input or output. The point of this is to ensure that the economic system does not split up into two or more groups of commodities which have no interrelations. In that case we are dealing with two or more *different* economic systems, not one. This condition has been lightened in two interesting recent papers on the von Neumann model. See J. Kemeny, O. Morgenstern, and G. Thompson, "A Generalization of the von Neumann Model of an Expanding Economy," *Econometrica*, **24**(2):115–135 (April, 1956); and D. Gale, "The Closed Linear Model of Production," Chap. 18 in H. W. Kuhn and A. W. Tucker (eds.), *op. cit.*

[2] But see the preceding footnote for some difficulties which will have to be adjusted.

Commodity 1 is $k_{11}x_1(t) + k_{12}x_2(t) + \cdots + k_{1n}x_n(t)$, the output of Commodity 1 from each process, added up over all processes; and similarly for every other commodity. For the ith, the available supply is $\sum_{j=1}^{n} k_{ij}x_j(t)$.

Now the available supplies limit the process intensities available in Period $t + 1$. They play the role of primary factors; in fact, so far as Period $t + 1$ is concerned, they are fixed in amount. Now suppose in Period $t + 1$ activity levels are $x_1(t + 1), \ldots, x_n(t + 1)$. How much of Commodity 1 is required for this to be a feasible set of production plans? The answer is $a_{11}x_1(t + 1) + a_{12}x_2(t + 1) + \cdots + a_{1n}x_n(t + 1)$. For this to be physically possible, the required input has to be no more than the available supply. Treating each other commodity in the same way we get

$$\begin{aligned} a_{11}x_1(t + 1) + \cdots + a_{1n}x_n(t + 1) &\leq k_{11}x_1(t) + \cdots + k_{1n}x_n(t) \\ \cdots\cdots\cdots\cdots\cdots\cdots & \cdots\cdots\cdots\cdots\cdots\cdots \\ a_{s1}x_1(t + 1) + \cdots + a_{sn}x_n(t + 1) &\leq k_{s1}x_1(t) + \cdots + k_{sn}x_n(t) \end{aligned} \tag{13-8}$$

These inequalities correspond exactly to (13-1a).[1] And just as in (13-1a), we have to add the condition, reflecting the competitive nature of the economy, that any commodity for which an inequality holds is absolutely redundant and must have a zero price.

Now let us turn to the profit side of the production process. Because of constant returns to scale and additivity we can do this globally and need not worry about individual firms. Let $\pi_1(t), \ldots, \pi_s(t)$ be the prices of the s commodities at Time t, and let us fix our attention on a particular process, say the hth. If operated at unit level in Period t, its inputs are $a_{1h}, a_{2h}, \ldots, a_{sh}$, and so its unit costs amount to $a_{1h}\pi_1(t) + a_{2h}\pi_2(t) + \cdots + a_{sh}\pi_s(t)$. One period later outputs appear: $k_{1h}, k_{2h}, \ldots, k_{sh}$, with a unit revenue of $k_{1h}\pi_1(t + 1) + k_{2h}\pi_2(t + 1) + \cdots + k_{sh}\pi_s(t + 1)$. But because of the lapse of time, what we have to compare with costs is not revenue, but the discounted present value of the revenue. If the market rate of interest in Period t is ρ_t, we would divide the unit revenue by $1 + \rho_t$. Corresponding to our earlier equations and inequalities (13-3a), we can write down at once that in a competitive economy equilibrium requires that discounted revenue from unit operation of a process cannot exceed unit costs (or else the process would

[1] Similar inequalities are to be found in our discussion of the Leontief dynamic model, Chap. 11, pp. 284 and 292.

expand until input prices rise and output prices fall enough to wipe out the profit). Thus

$$\begin{aligned} (1 + \rho_t)[a_{11}\pi_1(t) + \cdots + a_{s1}\pi_s(t)] &\geq k_{11}\pi_1(t+1) + \cdots \\ &\qquad + k_{s1}\pi_s(t+1) \\ \cdots\cdots\cdots\cdots\cdots\cdots\cdots\cdots\cdots\cdots\cdots\cdots \\ (1 + \rho_t)[a_{1n}\pi_1(t) + \cdots + a_{sn}\pi_s(t)] &\geq k_{1n}\pi_1(t+1) + \cdots \\ &\qquad + k_{sn}\pi_s(t+1) \end{aligned} \tag{13-9}$$

Of course, we have to add that if the *inequality* holds for any process, that process operates at a loss and its level of operation, its x, must be zero. This is the usual competitive zero-profit condition.

In a sense this is all there is to the von Neumann model. There are no demand functions to lead us into difficulty in proving the existence of a solution. In fact, there are embarrassingly many possible evolutions for a system described by (13-8) and (13-9). In earlier chapters we saw that Professor Leontief's capital model had recourse to a full-utilization-of-capacity assumption in order to get some kind of determinateness or narrowing down of possibilities in a similar situation. Subsequently we replaced that condition with an efficiency or optimization-over-time requirement. Earlier in this chapter, final demand functions provided a way of limiting the possible outcomes and actually left us with a problem of proving all the various conditions to be consistent. Von Neumann takes a quite different way of limiting the range. He defines equilibrium in a very special way. "Equilibrium" is a state of steady, balanced growth in which all process intensities remain in the same proportion and simply get multiplied by a common constant α every unit of time. This is a generalization of the "stationary state" of Ricardo, Mill, and the classicals. If $\alpha = 1$, we have a stationary state; if $\alpha > 1$, we have growth; if $\alpha < 1$, we have balanced shrinkage. On the price side we shall define equilibrium to mean constant prices and interest rate.

Let us translate these definitions into Eqs. (13-8) and (13-9). We can replace $x_j(t+1)$ in (13-8) by $\alpha x_j(t)$, and in fact we can drop the t altogether since it will be the same on both sides and hence will play no real role. In (13-9) we simply drop the time indication everywhere. This leads to

$$\begin{aligned} \alpha(a_{11}x_1 + \cdots + a_{1n}x_n) &\leq k_{11}x_1 + \cdots + k_{1n}x_n \\ \cdots\cdots\cdots\cdots\cdots\cdots\cdots\cdots \\ \alpha(a_{s1}x_1 + \cdots + a_{sn}x_n) &\leq k_{s1}x_1 + \cdots + k_{sn}x_n \end{aligned} \tag{13-8a}$$

and

$$\begin{aligned} (1 + \rho)(a_{11}\pi_1 + \cdots + a_{s1}\pi_s) &\geq k_{11}\pi_1 + \cdots + k_{s1}\pi_s \\ \cdots\cdots\cdots\cdots\cdots\cdots\cdots\cdots \\ (1 + \rho)(a_{1n}\pi_1 + \cdots + a_{sn}\pi_s) &\geq k_{1n}\pi_1 + \cdots + k_{sn}\pi_s \end{aligned} \tag{13-9a}$$

The dual nature of (13-8a) and (13-9a) should now be expected by the reader. Remember that whenever a strict inequality appears in (13-8a) or (13-9a), the corresponding dual variable must be zero. There are, as unknowns, n x's, s π's, α, and ρ. But since only relative prices and relative intensities matter, there are really only $s + n$ unknowns altogether; the x's, π's, and α have to be nonnegative. ρ may be negative, but $1 + \rho$ must be nonnegative.[1]

The family resemblance to our earlier system is visible. The main difference is the appearance of the factors α and $1 + \rho$. These make demonstrating the existence of an equilibrium a problem of almost, but not quite, the same order of difficulty and magnitude as it was for the Walras-Cassel system. It is interesting that von Neumann's pioneering proof[2] depends on his prior establishment of a fixed-point theorem of essentially the same kind as the one that we used in Sec. 13-4.

We can do no more here than summarize a few of the properties that von Neumann proves the system (13-8a) and (13-9a) to possess. First of all, there must be at least one equality holding in each system, since otherwise *all* the dual variables would be zero, a case which we rule out as uninteresting. Then remembering the interpretation of α, we see that it must be the growth factor of the *slowest-growing* commodity. By definition, (13-8a) says that every commodity grows by a factor of at least α (or at a rate of at least $\alpha - 1$). We have just indicated that (13-8a) contains at least one equality; thus at least one commodity grows at exactly the factor α, and this commodity or commodities must be the slowest-growing one. Any faster-growing commodity is (or becomes) a free good.

For similar reasons there will always be one or more equalities in (13-9a). Now (13-9a) can be interpreted as saying that for each process *actually used*, total revenue must exactly equal $1 + \rho$ times total cost. The factor $1 + \rho$ makes it true that each process used earns just enough to cover costs, including interest on the "capital" committed to production for one period. By reasoning like that in the previous paragraph we conclude that $1 + \rho$ is the ratio of revenues to costs for the most profitable processes. Any less profitable process is unused.

Multiply each line of (13-8a) by the corresponding π, and note that this has the effect of converting inequality into equality, since wherever there is inequality, the multiplying π is zero. Now add all the right sides and

[1] It is worth noting that von Neumann's original paper exploited intensively the duality properties of (13-8a) and (13-9a) and the relation to maximum, minimum, and saddle-point problems.

[2] A game-theoretic proof is given in G. L. Thompson, "On the Solution of a Game-Theoretic Problem," and a different, very elementary proof is given by D. Gale in the paper cited above, both in H. W. Kuhn and A. W. Tucker (eds.), *op. cit.*

all the left sides. Then α can be expressed as a ratio of two expressions ("bilinear forms"), in fact as the ratio of value of output to value of input. Now do exactly the same thing to (13-9*a*), and it happens that $1 + \rho$ can be expressed as exactly the *same* ratio. Thus it is a theorem that *if* a solution exists (obeying the dual restriction), *then* $\alpha = 1 + \rho$. The rate of growth equals the rate of interest. The economics of this result is clear. In equilibrium, used processes have zero profits and must therefore pay out interest just equal to the percentage by which the value of output exceeds the value of input. What is true of each used process must be true of the whole system.[1]

It is fairly evident on inspection that (13-8*a*) will have solutions for sufficiently small α, but not if α gets too large. Conversely; (13-9*a*) will have solutions for sufficiently large $1 + \rho$, but not if $1 + \rho$ gets too small. We have also just seen that in a solution $\alpha = 1 + \rho$. Thus there are two possibilities: either the range of α values for which (13-8*a*) is soluble overlaps the range of $1 + \rho$ values for which (13-9*a*) is soluble, or it does not. In the latter case there is no solution to the system; in the former case there is at least one. Von Neumann proved that the former situation always obtains. He proved more, namely, that the overlap consists of a *single point*. Hence the common solution value is the maximum value in the α range *and* the minimum of the $1 + \rho$ range.

Hence the equilibrium values of α and $1 + \rho$ have extremely interesting interpretations. Forget about prices for a moment, concentrate on (13-8*a*), and ask: What is the largest value of α for which (13-8*a*) has any solution at all? That is, What is the largest rate of growth such that the system is physically capable of expanding every output by at least this rate? This maximal rate of growth is the equilibrium rate of growth, α. This is intuitive: if the system were in equilibrium at a less than maximal rate of growth and any set of prices, it would pay entrepreneurs to move over into the higher rate of growth, using different processes (for *all* commodities), make profits at the going prices, and show that it is not an equilibrium.

[1] This ratio of revenues to costs for the whole economy (and its analog for single processes) is an extremely important concept. Much of von Neumann's theory can be phrased in terms of it. For the system as a whole this ratio depends on activity levels and prices. One can interpret competition as requiring that as of *given* prices, entrepreneurs choose *activity levels* to make this ratio a maximum; *and* that the market "chooses" prices in such a way as to minimize this ratio as of given activity levels. It will follow that all used activities have the same revenue/cost ratio (or else entrepreneurs would get rid of the ones with lower ratios). There is a dual statement that the reader should formulate. Since interest ("waiting") is the only social cost in this system, one can interpret the market as minimizing social cost. Georgescu-Roegen, *op. cit.*, treats the von Neumann model from this point of view. The reader should study his stimulating paper.

Now look at (13-9a) and ask: What is the lowest rate of interest at which a profitless system of prices is at all possible, simply forgetting about process levels? It is conceivable that this rate of interest may be negative. But it is clear that no profitless system of prices is possible if $1 + \rho < 0$: an interest rate of minus 101 per cent! In that case (13-9a) can have no solution, so the lowest rate of interest for which a set of zero-profit prices exists must still leave $1 + \rho \geq 0$. In any case, this lowest rate of interest is the equilibrium rate of interest, whose existence has been assured by von Neumann's proof.

Remember that in equilibrium, $1 + \rho = \alpha$; rate of interest equals rate of growth. From the two paragraphs above it is clear that however many solutions there are in x and π (and there may be more than one), there can be *only one* solution in α and $1 + \rho$. For if there were two α's, how could both be the largest technically possible rate of growth? So the rate of growth *cum* rate of interest of a von Neumann system is uniquely determined even if the price-process intensity pattern is not.[1]

EXERCISES

13-1. Consider a system with two commodities and three factors and no intermediate goods. There is only one process of production for each commodity, with inputs as given in the following table.

Input of Factor per Unit of Output

	Factors		
	1	2	3
Commodity 1	1	½	4
Commodity 2	1	2	¼
Factor amounts	4	6	10

The factors are available in amounts as given in the last line of the table. Two demand functions are as follows:

$$p_1 = x_2 - x_1$$
$$p_2 = \sqrt{36x_3 - x_2^2}$$

The third demand is, of course, implicitly specified, with $p_3 = 1$.

Find a competitive equilibrium for this system.

[1] The reader who is slightly familiar with game theory and knows what a saddle-point is will recognize that the ratio of value of output to value of input referred to above (p. 387) has a saddle-point at a von Neumann solution. It is maximized with respect to process levels and minimized with respect to prices. At the saddle-point this ratio is equal to α and to $1 + \rho$.

13-2. Do the same if the supply of factor 1 should be 2 units; 6 units; 16 units. Could you interpret the four (r_1,v_1) points thus found as points on the demand curve for the first factor?

13-3. In Exercise 13-1, suppose that the system had a second process for producing the first commodity, with inputs 1, 1, 1 of the three factors per unit of output. Find a competitive equilibrium.

13-4. For each of the solutions found so far, compute the value of total sales and verify that it equals total factor incomes. In each case, adjust the demand functions in such a way that total sales should be $100; in such a way that $p_1 = 1$.

13-5. Formulate the static Leontief system of Chap. 9 as a general-equilibrium system with one primary factor. Add demand functions. Use the methods of this chapter to obtain some of the results of Chap. 9. In particular, study the "nonsubstitution" theorem of Sec. 9-5 by our present linear-programming methods.

13-6. Indicate how intermediate goods could be included in the smooth-variable-coefficients–general-equilibrium system.

14

Linear Programming and Welfare Economics

14-1. INTRODUCTION AND OUTLINE

This chapter will be a continuation of the preceding one in at least three ways. For one thing, it is similarly designed to indicate how the linear-programming model of production provides a convenient and intuitive framework for the discussion of some well-worn and venerable questions of economic theory. By exercising the model in a familiar context the reader can acquire some "feel" for the way it works and how it corresponds to the textbook economics of marginal correspondences.

Secondly, without being very precise we can say that the main content of modern welfare economics has to do with certain normative aspects of competitive equilibrium. Before we come to welfare economics proper, then, it is only caution first to convince ourselves that the concept of a competitive equilibrium is internally consistent. In the previous chapter it was shown that under essentially the same assumptions as are used in welfare economics, a competitive equilibrium does always have to exist for the model we are talking about. We are not working in a vacuum.

Finally, it will be recalled that in the course of proving the existence of a competitive equilibrium for a linear-programming economy we showed how such a system can always be thought of as maximizing and minimizing certain quantities. In fact, once this fact was recognized, the main part of the proof was done. It will now turn out that to prove the main theorem of modern welfare economics, practically all we have to do is go back and pick up some of the pieces of this same proposition and reassemble them in slightly different form. The proposition that a competitive economic system (under our assumptions about production) is a mechanism for maximizing certain value sums contains the key to its normative welfare properties.

The plan of this chapter is as follows. Most of it—the next five sections—is devoted to an analysis of productive efficiency.[1]

[1] Although this is an old notion, it has been most thoroughly analyzed and exploited

This concept has come up several times in earlier discussions, particularly in the chapters on dynamic programming. In the context of welfare economics, efficiency is a vitally important concept, so we study it in some detail. It turns out that the precise place where linear programming and welfare theory come together is in this notion of efficiency. We look at it first in general terms, then in a particularly simple and satisfying mathematical (or geometrical) way, and finally we show how efficiency is related both to solutions of linear-programming problems and to the market price mechanism of a competitive economic system.

Efficiency is a property of *production*. It is of immense importance to welfare theory, but it leaves completely untouched everything to do with consumer satisfactions. In Sec. 14-7 we introduce the concept of Pareto-optimality, which closes this gap, and in Sec. 14-8 we sketch how the basic theorem of welfare economics can be proved, making good use of the efficiency properties already worked out.

14-2. EFFICIENT PRODUCTION PATTERNS

The distinction between efficient and inefficient patterns of output has been made fairly casually at several points in this book. It is pretty hard to discuss either the allocation of resources in general or linear programming in particular without stumbling explicitly or implicitly on the concept of efficient points. For example, think of a linear-programming problem where the thing to be maximized, the objective function, depends *positively* on all the variables. Then all the choice variables represent *goods*—more of them is better than less. The simple comparative cost setup is like this; so is the problem of the individual firm. In such a situation, if you can increase one or more of the choice variables without decreasing any of them, the value of the objective function has to rise, so it couldn't have been at a maximum to start with. Therefore, we can just rule out of consideration any feasible values for the choice variables which can be "dominated" by some other feasible choice (i.e., such that it is possible to increase some choice variables without decreasing others and without violating any constraints). Choices or points which can be dominated are called inefficient points. Points which cannot be dominated by any other feasible points are called efficient[1] points.

---

by T. C. Koopmans. The reader should look at his two expository papers, "Efficient Allocation of Resources," *Econometrica*, **19**:455–465 (October, 1951) (Cowles Commission Paper, New Series, no. 52), and "Activity Analysis and its Applications," *American Economic Review*, **43**:406–414 (May, 1953) (Cowles Commission Paper, New Series, no. 75).

[1] For a minimum problem with positive coefficients in the objective function we could just turn these definitions upside down, since *decreasing* the choice variables becomes desirable.

Resource allocation problems are mostly of this kind. We try to maximize a price-weighted sum or we try to maximize a military potential which depends positively on the weapons available. Hence, in this context, efficient production patterns will play a special role.

14-2-1. Precise Definition. For convenience and continuity we shall carry on most of our discussion in terms of the Walras-Cassel model we analyzed so exhaustively in the previous chapter. With it in mind we can give a precise definition of efficiency. Suppose first that resource endowments are fixed. Then a pattern of outputs $(x_1, \ldots, x_n)$ is efficient if (1) it is actually feasible or producible [i.e., satisfies inequalities (13-1*a*), p. 360]; and (2) if there is no other producible pattern of outputs $(y_1, \ldots, y_n)$ which is such that $y_1 \geq x_1, y_2 \geq x_2, \ldots, y_n \geq x_n$, with strict inequality holding at least once (or else y and x are the same outputs). If we work with the more general model in which factor endowments are not fixed, we need a slightly different definition. We then describe an input-output pattern $(x_1, \ldots, x_n; r_1, \ldots, r_m)$ as efficient if (1) it is a feasible pattern [if it satisfies (13-1*a*)]; and (2) if there is no other feasible pattern $(y_1, \ldots, y_n; s_1, \ldots, s_m)$ which is such that $y_1 \geq x_1, \ldots, y_n \geq x_n$, with at least one strict inequality, and $s_1 \leq r_1, \ldots, s_m \leq r_m$. In general, a pattern is efficient if there is no way of increasing some outputs without either decreasing some other outputs or increasing some resource inputs.

In the simple Walras-Cassel model there are only final goods and nonproduced factors. We have simplified by omitting the fact of intermediate goods. In real life and in slightly more complicated models there are *pure* intermediate goods which are not desired for themselves at all but only for further processing. Such, for example, is pig iron. Pure intermediate goods require somewhat different treatment:[1] they do not (like final goods) directly provide social utility, nor are they (like primary factors) direct social costs. When there is a question of efficiency we compare input-output configurations only as to their final outputs and primary inputs. Whether the steel is produced with more pig iron and less scrap or vice versa is immaterial. What matters is the output of finished steel and the input of ore, labor, etc.

The only part of the general-equilibrium system we had to refer to in defining efficiency was the resource-use relations (13-1*a*), but not the demand and supply functions, nor the price-unit cost relations. This reflects the fact that efficiency is a purely technological concept, having to do only with production. Welfare questions, on the other hand, are usually framed in terms of consumer satisfactions. We shall come to this in Secs. 14-7 and 14-8. Meanwhile, we can make the obvious remark

[1] See T. C. Koopmans (ed.), *Activity Analysis of Production and Allocation*, Chap. 3, John Wiley & Sons, Inc., New York, 1951.

that inefficient production patterns can never be good from the welfare point of view. Society may have to choose among the efficient points by some other criteria, but the inefficient points can be eliminated without a second thought. By the very definition, an inefficient point can be improved upon or dominated by some other allocation of resources which will yield more of some outputs without yielding less of any output or requiring more of any input. As long as any household is nonsatiated with the extra commodities produced, the efficient situation is certainly better than the inefficient one.[1]

14-2-2. Geometrical Discussion. Let us draw some pictures for the case of two commodities and three factors. Even this would require five dimensions, so we shall again take the factor amounts as given. Then in Fig. 14-1*a* we draw the constraints due to each of the three resources

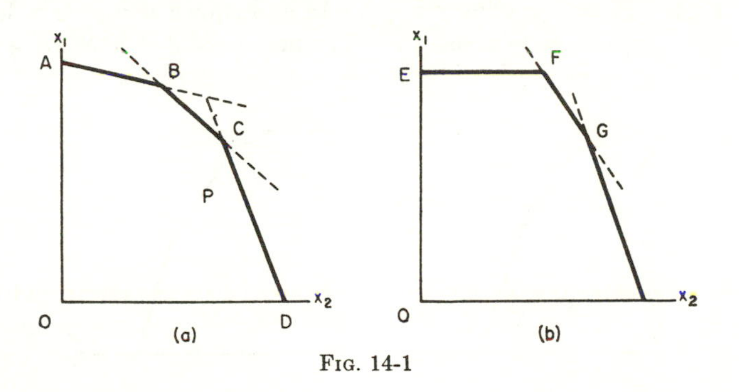

FIG. 14-1

in turn. An output point just uses up a resource if it lies on the corresponding line; it leaves some of the resource unused if it lies between the line and the origin. Since all three resource limitations must be satisfied, the collection of producible outputs is the whole polygon $OABCD$. Note that for each of the flat portions of the frontier $ABCD$, only one resource limitation is binding, the others being in surplus. At the vertexes B and C two resource limitations are binding and the third is free.[2]

In Fig. 14-1*a* the efficient outputs are those along the frontier $ABCD$. A movement which increases one or both outputs while decreasing neither is a movement in the northeast direction (including due north and due east). Any northeastward movement from an efficient point takes us outside the feasible set. From a point like P, however, a feasible north-

[1] We are dodging the possibility that under the institutional rules a "bad" distribution of income may be associated with the efficient production pattern. Lump-sum taxes and subsidies could get around the problem.

[2] *Exercise.* Draw a situation with two goods and three resources in which for some output all three resource limits hold.

east movement is available. Hence P is inefficient. It can't be stressed too much that we cannot say that B is "better" than P just because B is efficient and P is not. What we can say is that there are *some* efficient points (C is an example) that *are* unambiguously better than P.

Figure 14-1*b* differs from 14-1*a* only in having a horizontal portion *EF* in the boundary of the feasible set. This means that the resource whose limited supply is binding along *EF* is not used in the production of x_2 at all, but only in x_1. Points on the segment *EF* are clearly not efficient, except F itself. The efficient set is the rest of the frontier *FGH*.

Now we saw in Chap. 13 that the relations (13-1*a*) could be thought of as the constraints in a linear-programming problem. The outputs producible with the given resources are just the feasible x's in this problem. Among these feasible x's we have to find the one that maximizes $p_1x_1 + p_2x_2$. Figure 14-2*a* reproduces Fig. 14-1*a* and shows also several lines of the family $p_1x_1 + p_2x_2 =$ constant. The members of this family are parallel lines, and the higher the constant, the further out the line; that is, $k_1 < k_2 < k_3 < k_4$. To solve the linear program is to find the outermost "budget line" which has a point in common with the feasible set. In Fig. 14-2*a*, this obviously occurs at C.

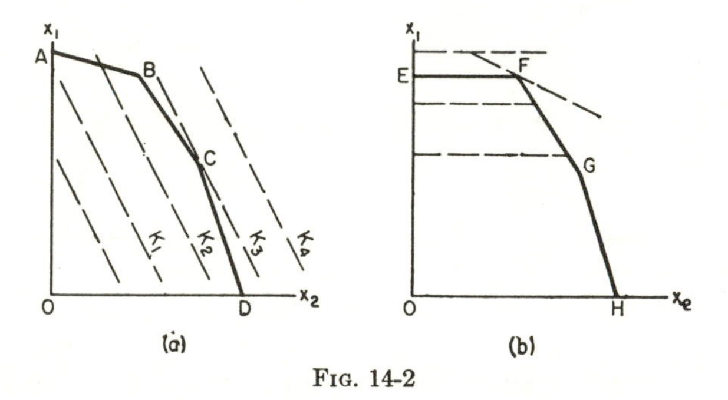

Fig. 14-2

Now suppose we change the values of p_1 and p_2 (only the ratio p_2/p_1 counts) until the budget line is very steep. Then the maximum will occur at D. (If we put $p_1 = 0$, so that the budget line is vertical, we are maximizing p_2x_2, so naturally we make x_2 as large as we can.) In fact, there is a whole range of values of p_2/p_1 for which the maximum occurs at D. But as we pass through flatter and flatter budget lines (i.e., value x_1 more and x_2 less), eventually the outermost budget line will coincide with CD. In this case there is no unique maximum solution—any one of the infinity of points from C to D will do—but the value of $p_1x_1 + p_2x_2$ at the maximum is of course unique. Again there is a whole range of budget-line slopes for which the maximum is attained at C, and then finally a budget line coincides with BC. Finally, for very flat budget slopes the maximum is at A.

Note what has happened. By varying the slope of the budget line (the coefficients p_1 and p_2 of the objective function) and in each case solving the corresponding maximum problem, *we have traced out the set of efficient outputs.* Not only does every maximum occur at an efficient point, but every efficient point is the maximum solution for some budget line. We have a kind of correspondence between efficient points and price configurations. In this correspondence the corners A, B, C, D play a special role—each of them corresponds to a whole range of price configurations. The flat faces AB, BC, CD correspond each to a single price configuration.

We must be careful to see exactly what is special in this setup. The essential thing is that the objective function, the value of output, has no negative coefficients. We are maximizing something that gets bigger (or at least that doesn't get smaller) when we increase a choice variable. In these circumstances the connection between efficiency and linear programs is intuitively clear.

To sum up: Suppose we have a linear program with a positively weighted objective function. Then by varying the weights through all possible values and solving the linear programs, we trace out the whole efficient set. The economic significance of this result is clear. If we can find a set of economic institutions which will always work so as to maximize a positive-weighted value sum of outputs, then willy-nilly it will always bring about an efficient output configuration, without anyone taking any thought. Conversely, if a superbrain is instructed to bring about a particular efficient point, a super-enough brain can always figure out what p's to set in the market place, and the market will automatically maximize $\Sigma p_i x_i$ and march to the desired efficient point.[1] Here is a very close connection between linear programming and economic optimization.

Figure 14-2*b*, like 14-1*b*, illustrates a minor subtlety. If we maximize $p_1 x_1 + 0 \cdot x_2$ (a horizontal budget line), the maximum is attained anywhere on EF. But we know that only F is efficient. The trouble is that if we attach zero weight to x_2 nothing stops us from getting inefficient with respect to x_2. In this special situation, and only then, maximizing by linear programming gets us some inefficient points (along with the efficient point F). On the other hand, if we tilt the objective function a bit from the horizontal, the maximum stays at F, but *not* at the other (inefficient) points of EF. If we maximize a *strictly* positive-weighted sum, we get only efficient points.

[1] Not quite. If a point on a flat face is desired, the best we can do by this method is to get on the face. A little extra nudging may be necessary to pick out the desired point.

For some of us, even three-dimensional geometry puts a strain on the imagination. Nevertheless, it is probably worth trying to visualize the three-commodity situation. Each resource constraint is represented by a plane. The feasible set is a polyhedron whose outermost boundary is the efficient set. This boundary is made up of faces (pieces of single resource planes), edges (where two resource planes come together), and sharp corners (where three resource planes intersect). Budget planes replace budget lines. Everything is a little more complicated, but nothing is really any different.

14-3. SOME SIMPLE MATHEMATICS: A DIGRESSION

In this section we shall prepare for a slightly more rigorous analysis of efficiency by at least naming a few mathematical concepts which make the task easier. The word "easier" is used here advisedly. The mathematics has the advantage of being intuitive, not to say obvious.

Everybody knows that if we draw x_1 and x_2 axes, a single linear equation $a_1x_1 + a_2x_2 = c$ represents a straight line. It is less well known but just as elementary that a straight line divides the plane into two parts: a part in which $a_1x_1 + a_2x_2 \geq c$ and a part in which $a_1x_1 + a_2x_2 \leq c$. These parts are called "half spaces." The straight line itself belongs to both half spaces. If a_1 and a_2 as well as c are both positive, it is especially easy to tell which half space is which. The sign $\geq$ goes with the half space away from the origin. If a_1 and a_2 are of different signs, it may take a little experimentation.

Three and more dimensions are a little harder to visualize, but the same thing goes with minor changes. A single linear equation,

$$a_1x_1 + a_2x_2 + a_3x_3 = c$$

describes a plane and divides the whole three-dimensional space into the half spaces $a_1x_1 + a_2x_2 + a_3x_3 \geq c$ and $a_1x_1 + a_2x_2 + a_3x_3 \leq c$. If we change c we get a new plane parallel to the first, and above it if we increase c, below it if we decrease c.

The relevance of all this to our present subject is clear. *Each* resource limitation confines the feasible outputs to a half space. (*On* the line or plane the resource is entirely used up. *Inside* the half space some of the resource is left idle.) The common-sense condition that outputs not be negative also determines a half space: $x_1 \geq 0$ is just the special case with $a_1 = 1$, a_2, a_3, and c all zero. Thus the whole collection of producible outputs is the common part (the mathematical term is "intersection") of *all* these half spaces.

Another basic geometrical notion is that of *convexity*. A set of points in the x_1, x_2 plane is convex if the following is true: If we take any two points in the set and draw the straight line between them, the whole line also lies in the set. In Fig. 14-3, A, B, and C are convex; D, E, and F are not.

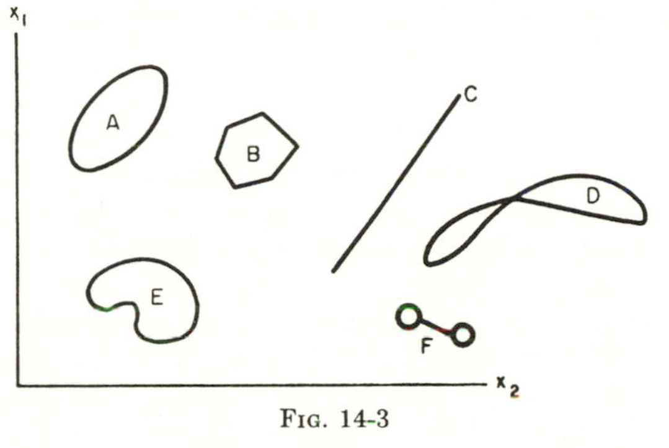

FIG. 14-3

We need some sort of algebraic notation for the points on the straight line joining two given points. This is easily come by. If we have two points with coordinates (x_1,x_2) and (x'_1,x'_2) respectively, all points on the line between them are *weighted averages* of the two end points. Thus all points on the line segment have coordinates $[\alpha x_1 + (1 - \alpha)x'_1,\ \alpha x_2 + (1 - \alpha)x'_2]$. Here α has to be a number between 0 and 1—otherwise we get points which are on the line all right, but not *between* the two given points. If α is zero, we get (x'_1,x'_2) itself. If α is close to zero we get a point on the segment close to (x'_1,x'_2). As α increases, we get points closer and closer to (x_1,x_2); and finally, when α reaches 1, we get (x_1,x_2) itself. We can now paraphrase our definition of convexity: A set in the (x_1,x_2) plane is convex if whenever (x'_1,x'_2) and (x''_1,x''_2) are points of the set, so are all points $[\alpha x'_1 + (1 - \alpha)x''_1,\ \alpha x'_2 + (1 - \alpha)x''_2]$ for $0 \leq \alpha \leq 1$.

We can now prove that in the Walras-Cassel model, the *set of feasible outputs is convex*. The feasible set is the collection of outputs lying in all the half spaces of (13-1a):

$$\begin{aligned} a_{11}x_1 + a_{12}x_2 + \cdots + a_{1n}x_n &\leq r_1 \\ a_{21}x_1 + a_{22}x_2 + \cdots + a_{2n}x_n &\leq r_2 \end{aligned}$$

and so forth. Now suppose we have two *feasible* output points, $x' = (x'_1,x'_2, \ldots ,x'_n)$ and $x'' = (x''_1,x''_2, \ldots ,x''_n)$. Since they are feasible, if we substitute them in (13-1a), the inequalities will all be satisfied. What happens if we choose an arbitrary α between 0 and 1 and substitute a point on the line between x' and x'' into (13-1a)? Take the first inequality. Its left-hand side becomes

$$a_{11}[\alpha x'_1 + (1 - \alpha)x''_1] + a_{12}[\alpha x'_2 + (1 - \alpha)x''_2] + \cdots + a_{1n}[\alpha x'_n + (1 - \alpha)x''_n]$$

Collecting the α and $1 - \alpha$ terms separately, we get

$$\alpha(a_{11}x_1' + a_{12}x_2' + \cdots + a_{1n}x_n') + (1 - \alpha)(a_{11}x_1'' + a_{12}x_2'' + \cdots + a_{1n}x_n'')$$

Each of the terms in brackets is $\leq r_1$. Therefore the whole expression is $\leq \alpha r_1 + (1 - \alpha)r_1$, which is just r_1. Hence the first inequality is satisfied, and so are all the rest. We chose arbitrary feasible x' and x'' and showed that any point on the line between them is also feasible. Hence the set of feasible outputs is convex. Of course we know this geometrically, from the way the feasible set is built up out of the resource-limitation lines or planes.

Convex sets have a very simple and very important property. Imagine any convex set and some point *not* in the set, such as the set K and the point P in Fig. 14-4. It is easy to see that we can draw a line (in fact, infinitely many lines) through P such that the whole set K lies *entirely on one side* of the line, i.e., entirely in one of the half spaces defined by the line. Analytically, we can find numbers s_1, s_2, and M such that K lies entirely in the half space $s_1x_1 + s_2x_2 \leq M$.[1] This means that if (x_1,x_2) is any point of K, we have $s_1x_1 + s_2x_2 \leq M$. Now go one step further. Take any point like Q or R or S which lies *on the boundary* of K. Then we can still find a line through Q or R or S such that all of K lies on the line or on one side of the line, i.e., still lies entirely in one of the half spaces defined by the line. The only difference now is that some of the points of K will lie squarely on the line. In the case of Q, only Q itself lies on the line. In the case of S, a whole piece of the boundary of K lies on the line. In the case of R, we have a choice. There are lines which contain only R among the points of K. But we could also choose either of the two linear portions of the boundary of K as our line, in which case infinitely many points of K are on the line. Note also that at Q or S there is only one line of this kind, but at R there are many.

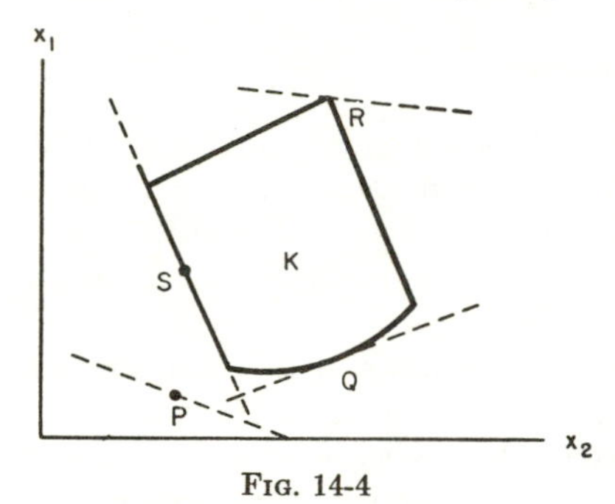

FIG. 14-4

Exactly the same thing holds in three or more dimensions. If we choose any point of the boundary of K, we can find one or more planes which go through the point in question, and such that the whole set K lies in one of the half spaces defined by the plane. Planes like this are

[1] We can always make the inequality $\leq$. For if we have a half space $t_1x_1 + t_2x_2 \geq M$, we can multiply both sides by -1, which changes the direction of the inequality so that it reads $(-t_1)x_1 + (-t_2)x_2 \leq -M$.

called *supporting planes* (or lines) of the convex set. Through every point of the boundary of a convex set there is at least one supporting plane. Another way of stating this is as follows. Let K be a convex set and Q, $(\bar{x}_1,\bar{x}_2, \ldots ,\bar{x}_n)$, be a point of its boundary. Then we can always find numbers $s_1, s_2, \ldots, s_n, M$ such that $s_1x_1 + s_2x_2 + \cdots + s_nx_n \leq M$ for every point of K, and $s_1\bar{x}_1 + s_2\bar{x}_2 + \cdots + s_n\bar{x}_n = M$ for Q itself.

This theorem is not true for sets which are not convex. Back in Fig. 14-3, it is easy to pick out boundary points of D, E, and F that do not have supporting lines. For convex sets the theorem of the supporting plane is geometrically obvious. One can feel the way a convex set *must* bend away back on itself and how this permits us to draw a supporting line or plane. The contact between a supporting line and a convex set is a generalization of the common notion of tangency (but note that support is much simpler and more elementary—no fussing with derivatives).

There is a crucial connection between supporting planes and maximization. This should be suggested by the occurrence (well known to economists) of tangency in maximum and minimum problems. We had a point $(\bar{x}_1,\bar{x}_2, \ldots ,\bar{x}_n)$ on the boundary of a convex set K. If we drew the supporting plane through this point, we could say that $s_1x_1 + \cdots + s_nx_n \leq M$ for every point of K. But in addition,

$$s_1\bar{x}_1 + \cdots + s_n\bar{x}_n = M$$

Comparing, we can see that

$$s_1x_1 + s_2x_2 + \cdots + s_nx_n \leq s_1\bar{x}_1 + s_2\bar{x}_2 + \cdots + s_n\bar{x}_n \quad (14\text{-}1)$$

for every point of K. We can translate this to say: *The linear function* $s_1x_1 + \cdots + s_nx_n$ *reaches its maximum in K at the point* $\bar{x}_1, \ldots, \bar{x}_n$. There may be ties, but the maximum is certainly also attained at the point in question. It may seem odd that a point like S in Fig. 14-4 can maximize anything, but a little reflection shows that if the coefficients of the supporting plane are all negative, maximizing the linear function will require making all the variables as small as possible.

That this apparatus is of importance to linear programming is clear from the fact that we are already talking of maximizing a linear function over a convex set (and we know that the feasible set in linear programming is always convex).

14-4. EFFICIENCY AND LINEAR PROGRAMMING

We are now equipped to go back and establish the fundamental interrelationships between the welfare concept of efficiency and the purely computational technique of linear programming. It was already suggested and made graphically plausible in Sec. 14-2 that there is such a connection. There is a certain kind of linear-programming maximum problem whose solution inevitably turns out to be an efficient point. And in addition, *every* efficient point turns up as a solution to at least one such maximum problem. This is interesting enough. But what is really important from the welfare-economics point of view is that although this whole setup is formulated in sheerly physical-technological terms, a notion of "price" rises inevitably out of the analysis. After all, a linear program maximizes a weighted sum of outputs. An economist can hardly resist thinking of the weights as prices and the sum as a value, especially when it turns out that at the maximum the "prices" are proportional to marginal rates of substitution. We want to make all this a little more rigorous.

To start with, we have a feasible set of outputs, namely, those which satisfy *all* the inequalities:

$$\begin{aligned} a_{11}x_1 + \cdots + a_{1n}x_n &\leq r_1 \\ \cdots\cdots\cdots&\cdots\cdots \\ a_{m1}x_1 + \cdots + a_{mn}x_n &\leq r_m \end{aligned} \tag{14-1a}$$

We are interested not in *all* feasible outputs, but only in the efficient ones. For example, suppose a bundle of outputs satisfies all the *strict* inequalities—then it is clearly inefficient. It would be possible to increase *all* outputs simultaneously (if ever so slightly) without violating any of the inequalities. Thus the first bundle can be dominated and is inefficient. Conversely, suppose that there is one of the inequalities in which all the a coefficients are positive. Then if we have a bundle which makes that inequality hold with an equals sign, the bundle is efficient. Any dominating bundle would violate the inequality.[1]

Now suppose we arbitrarily think of the resource limits as the constraints of a linear program and maximize a *positively weighted* objective function $p_1x_1 + \cdots + p_nx_n$, $p_i > 0$. Let $(\bar{x}_1,\bar{x}_2, \ldots ,\bar{x}_n)$ be any solution of the maximum problem (there may of course be a tie). We assert that $\bar{x} = (\bar{x}_1,\bar{x}_2, \ldots ,\bar{x}_n)$ is an efficient point.

Proof. Suppose $\bar{x}$ were not efficient. Then, by definition, there would have to be a feasible bundle $y = (y_1, \ldots ,y_n)$ satisfying (14-1a) and such that $y_i \geq \bar{x}_i$ for every i, and somewhere along the line, $y_j > \bar{x}_j$. Since every p_i is positive, $p_iy_i \geq p_i\bar{x}_i$ for every i, and somewhere along the line, $p_jy_j > p_j\bar{x}_j$. But then $p_1y_1 + \cdots + p_jy_j + \cdots + p_ny_n > p_1\bar{x}_1 + \cdots + p_j\bar{x}_j + \cdots + p_n\bar{x}_n$. This is a contradiction, because $\bar{x}$ was supposed to be a solution of the linear-programming problem and

[1] Why must every a coefficient be positive?

here we have a feasible y which gives a greater value to the objective function. Since the existence of a dominating y leads to a contradiction, no such y can exist and $\bar{x}$ must be efficient. Q.E.D.

Observe that the weights p_i must be strictly positive. Otherwise we can't conclude that $p_j y_j > p_j \bar{x}_j$ for some j, because it might be that although $y_j > \bar{x}_j$, $p_j = 0$. We are in the situation shown in Fig. 14-2b: a solution of the maximum problem can easily be inefficient with respect to any output carrying a zero weight.

Now we come to the more interesting half of the basic interrelationship. We start with any arbitrary efficient point $\bar{x}$ and show that there is at least one set of nonnegative weights $\bar{p} = (\bar{p}_1, \ldots, \bar{p}_n)$ such that $\bar{x}$ is a solution of the linear-programming maximum problem with $\bar{p}$ as weights. This will mean that if we solve all possible linear programs, we trace out the *whole* efficient set.

We know that the feasible outputs form a convex set and that the efficient program $\bar{x}$ is on the boundary of the feasible set [i.e., can't satisfy all strict inequalities in (14-1a)]. Here we use the results of Sec. 14-3. There must be a supporting plane through $\bar{x}$, perhaps more than one; i.e., there is at least one set of numbers $(\bar{p}_1, \ldots, \bar{p}_n)$ such that $\bar{p}_1 x_1 + \bar{p}_2 x_2 + \cdots + \bar{p}_n x_n \leq M$ for every feasible program $(x_1, \ldots, x_n)$ and $\bar{p}_1 \bar{x}_1 + \bar{p}_2 \bar{x}_2 + \cdots + \bar{p}_n \bar{x}_n = M$. As in the last section, we can say that $\bar{p}_1 x_1 + \cdots + \bar{p}_n x_n \leq \bar{p}_1 \bar{x}_1 + \cdots + \bar{p}_n \bar{x}_n$ for every feasible x, or that $(\bar{x}_1, \ldots, \bar{x}_n)$ maximizes $\bar{p}_1 x_1 + \cdots + \bar{p}_n x_n$ subject to (14-1a).

All that is left to prove is that the $\bar{p}$'s can be chosen nonnegative, so that it is legitimate to interpret them as "efficiency prices." Our proof is again by contradiction. We assume that, say, $\bar{p}_1$ is negative and show that this leads to nonsense; it follows then that $\bar{p}_1 \geq 0$, and similarly for every $\bar{p}_i$. But first we point out a self-evident property of the set of feasible outputs: if we start with any feasible program and reduce some outputs without increasing any outputs, we still get a feasible (if less desirable) program. In Fig. 14-2, any point southwest of a feasible point is also feasible, provided, of course, that it doesn't involve negative outputs. This property is not necessary for the theorem that we are about to prove, but it simplifies things. Afterward we can show how to get around it.

Now let us suppose that $\bar{p}_1$ is negative. Start with $\bar{p}_1 \bar{x}_1 + \bar{p}_2 \bar{x}_2 + \cdots + \bar{p}_n \bar{x}_n$ and *reduce* $\bar{x}_1$ a little. By what has just been said, we still get a feasible point (unless $\bar{x}_1 = 0$, a case which will be covered in a moment). But if $\bar{p}_1$ is negative, by *reducing* $\bar{x}_1$ to x_1 we *increase* $\bar{p}_1 \bar{x}_1$ to $\bar{p}_1 x_1$. Thus $(x_1, \bar{x}_2, \ldots, \bar{x}_n)$ is feasible, *and* $\bar{p}_1 x_1 + \bar{p}_2 \bar{x}_2 + \cdots + \bar{p}_n \bar{x}_n > \bar{p}_1 \bar{x}_1 + \bar{p}_2 \bar{x}_2 + \cdots + \bar{p}_n \bar{x}_n$, since the last $n - 1$ terms on each side are identical and $\bar{p}_1 x_1 > \bar{p}_1 \bar{x}_1$. Note that this violates the basic inequality of the supporting plane: we have found a feasible point in the half space of the supporting plane, which is impossible. So it can't be that $\bar{p}_1$ is negative, and a similar argument can be made for each $\bar{p}_i$ in turn.

We can now examine several minor subtleties. In the first place, we can dispense with the fact that if some outputs in a feasible program are reduced, the new program is still feasible. Actually, we only used the fact that this is true if the initial program is an efficient one. In Fig. 14-5, if the feasible set were *PQRS*, we could always work instead with an enlarged feasible set (such as *TPQRS* or even the unbounded set *UPQRV*) which has the *same* efficiency frontier as the original set and has the desired property to boot. For this enlarged set we can find a supporting plane with nonnegative coefficients as in the text. If we then simply throw away the added portions, this plane will still be a supporting plane of the original set.

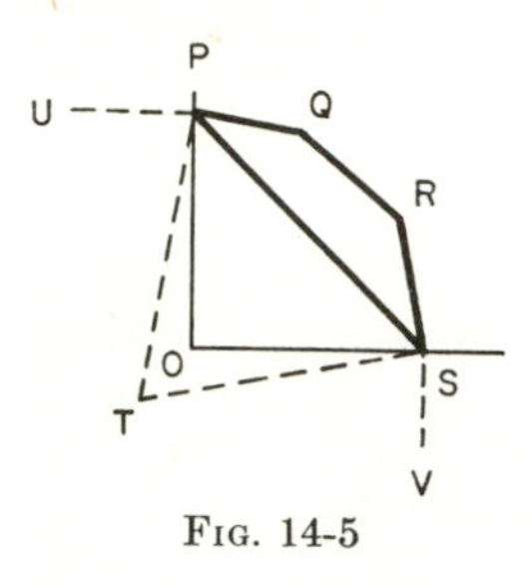

Fig. 14-5

In the same way we handle the situation in which the proof can't proceed as in the text because $\bar{x}_1$ is already zero and can't be reduced any more. The feasible set can be extended by horizontal and vertical lines to *UPQRSV* so that any variable can be reduced. A support plane for the extended set will also do for the original one.

Potential mathematicians among our readers will have spotted that we have only shown that the p's can't be negative—not that they must be positive. But a glance at point F in Fig. 14-2b will show that if there is a horizontal supporting line, there must also be one with negative slope (i.e., with $\bar{p}_1$ and $\bar{p}_2$ both positive). *GF* itself is such a line, and there are others. But in more general nonlinear cases, there may be only a horizontal supporting line at an efficient point. For example, if the feasible set is part of a circle as in Fig. 14-6, point A is efficient and the *only* tangent at A is horizontal.

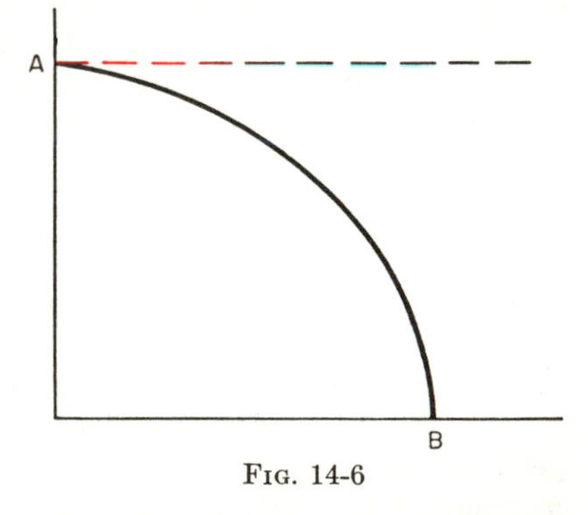

Fig. 14-6

Now where are we? Given any efficient point $\bar{x}$, we have found a supporting plane with positive (or nonnegative) coefficients. In other words, we have associated with each efficient point at least one set of weights such that the given efficient point is a solution of the linear-programming maximum problem with those weights in the objective function. We remember from Fig. 14-2 that the supporting plane (or budget line) has a sort of tangency to the feasible set at the efficient point. Hence, except at corners, the slope of the supporting plane (which is a ratio of p's) equals the slope of the efficiency frontier (which is a marginal rate at which one commodity can be efficiently substituted for another). It is irresistible to interpret the p's as a set of implicit efficiency prices associated with the given efficient point. At a corner or edge of the efficiency locus, the situation is slightly more complicated. No unique marginal rate of substitution exists—the slope of the efficiency frontier is different, depending on the direction we take. We can imagine these different slopes as putting limits to "the" MRS. We also know that at a corner or edge there will be more than one supporting plane, i.e., more than one associated set of p's. (At a really flat place on the boundary, the supporting plane is unique and literally coincides with the boundary.)

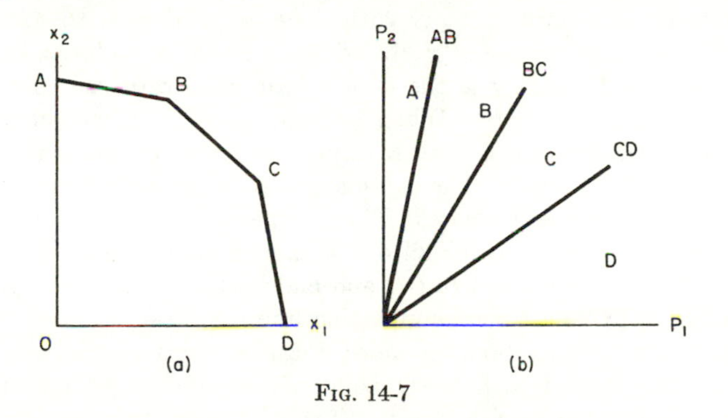

Fig. 14-7

But *all* these supporting planes will have slopes *within* the limits set for the MRS. So that even at corners the generalized correspondence between associated p ratios and MRS's persists.

To sum up briefly: We have found that any time you assign an arbitrary set of prices and solve the linear program of maximizing value of output you wind up at an efficient point; and that, conversely, *every* efficient point is the solution to at least one maximum-value problem with certain price weights. The efficiency-price concept rises out of the problem itself—it was not put there by institutional assumptions. But we now know that any institutional setup that results in the maximization of value sums will achieve efficient (but not necessarily "good") programs.

In the two-dimensional case we can easily dig out some more about the association between efficient points and price constellations. Let us draw in Fig. 14-7*a* a feasible set whose efficiency frontier *ABCD* has two corners. Now with each point of *ABCD* we associate one or more supporting (or tangent, or budget) lines and, more particularly, the coefficients of the supporting line. Only the ratios of the coefficients matter, for they alone determine slopes. The line $10x_1 + 2x_2 = 20$ is the same as the line $x_1 + \frac{1}{5}x_2 = 2$.

In Fig. 14-7*b* we are going to do the following. Each supporting line in Fig. 14-7*a* determines a *ray* in Fig. 14-7*b*, namely, the ray with slope p_2/p_1. This is the same thing as the ray through the point (p_1,p_2): it also goes through all points (kp_1,kp_2), among which we do not care to distinguish. Drawing the diagram is simplified by the following geometrical fact: the ray through (p_1,p_2) is always perpendicular (or "normal") to the lines $p_1x_1 + p_2x_2 =$ constant.

Now at A the supporting lines consist of the horizontal line $x_2 =$ constant (or $0 \cdot x_1 + p_2 \cdot x_2 =$ constant) *and* all lines intermediate in slope between the horizontal and AB, inclusive. Thus to A we make correspond in Fig. 14-7*b* the vertical ray [through $(0,p_2)$] and all intermediate rays up to and including the ray with slope perpendicular to AB. Thus we get the sector (or "cone") labeled A in 14-7*b*, boundaries inclusive. The right-hand boundary is the *only* ray or price ratio associated with the efficient points on AB. When we come to B itself we get not only the ray marked AB but also all rays intermediate between this and the ray perpendicular to BC. The ray marked BC does duty for every point of BC, including the end points. Then we get another zone belonging to the vertex C, a single ray for all of CD, and finally a cone for D which begins with the ray normal to CD and ends with the ray through $(p_1,0)$ corresponding to the vertical supporting line $p_1x_1 + 0 \cdot x_2 =$ constant.

In three dimensions (and in more, for that matter) the same thing happens. Associated with every point of the efficiency frontier is one or more rays or sets of price ratios. If there is more than one associated ray (which will be true at edges or corners of the efficiency locus), there is a whole cone of price rays. At a corner there is a genuine solid cone of price rays; at an edge there is a flat two-dimensional, fan-shaped cone such as the ones shown in Fig. 14-7*b*, except that this one is standing right up in three-dimensional space.

14-5. COMPETITIVE EQUILIBRIUM AND EFFICIENCY

It is now a relatively simple matter to broaden our sights to include the full Walras-Cassel model of competitive equilibrium as described in Chap. 13. We can show that there is a close relationship between the purely "economic" notion of competitive equilibrium and the purely "technological" notion of efficiency. Indeed we have the following theorem. Under the assumptions of Chap. 13, *if firms maximize their profits competitively with respect to any given set of positive prices for goods and factors, the resulting configuration of inputs and outputs will be efficient. Conversely, if we are given any efficient input-output situation, there is a set of nonnegative prices for which this situation is the competitive profit-maximizing equilibrium.*

Before proving this theorem we have to make some remarks about the definition of competitive equilibrium. In Chap. 13, a competitive equilibrium was a set of prices (p's), factor returns (v's), factor supplies (r's), and outputs (x's) that satisfied the equations and inequalities (13-1a), (13-2), (13-3a), and (13-4). In laying down the definitions we made no specific mention of profit maximizing on the part of firms. But as Secs. 13-4 and 13-5 showed, it is there all the same. We demonstrated (on p. 370 of Chap. 13) that any competitive equilibrium solution maximized $\sum_{i=1}^{n} p_i x_i$ and minimized $\sum_{i=1}^{m} r_i v_i$, subject to the dual constraints (13-1a) and (13-3a). Moreover, any solution of this pair of dual maximum and minimum problems that also satisfied the demand and supply equations would be a competitive equilibrium. Now instead of separately maximizing total revenue $\Sigma p_i x_i$ and minimizing total costs $\Sigma r_i v_i$, we might just as well speak of maximizing the difference between them, total profits, $\Sigma p_i x_i - \Sigma r_i v_i$. Duality theory tells us that the maximum value of total profits achieved at a competitive equilibrium is zero. Elsewhere it is negative. We need only add that in our linear, constant-returns-to-scale technology it does not matter whether we think of profits being maximized in the aggregate or by individual competitive firms, all facing *the same prices.* Thus, with some verbal emendations, the competitive equilibrium studied in Chap. 13 is identical with competitive equilibrium defined in terms of profit maximization. This is also evident from the discussion in Sec. 13-4 relating to marginal-productivity conditions.

Now to proceed. Suppose we have a competitive equilibrium, a set of prices, outputs, factor prices, and factor supplies such that profits are maximized. Under our assumptions of additivity and noninterference of production processes, we can imagine the economic system to be integrated into one giant firm. This firm, taking the p's and v's as given, maximizes aggregate profits. Total profits are represented as revenues minus costs, $\sum_{1}^{n} p_i x_i - \sum_{1}^{m} v_j r_j$. Technology tells us that we must have inequalities (13-1a) of Chap. 13 satisfied; i.e.,

$$
\begin{array}{c}
a_{11}x_1 + \cdots + a_{1n}x_n \leq r_1 \\
\cdots\cdots\cdots\cdots\cdots\cdots \\
a_{m1}x_1 + \cdots + a_{mn}x_n \leq r_m
\end{array}
$$

We have to prove that the input-output pattern is efficient. But suppose it weren't. Then we could make some change in the pattern which would only increase some x's and (perhaps) decrease some of the binding r's. (Decreasing nonbinding r's does not make a pattern inefficient; these are superfluous anyway!) But increasing x's has the effect of increasing revenues $\Sigma p_i x_i$, since the p's are positive. And decreasing scarce r's has the effect of decreasing costs $\Sigma r_j v_j$, since the prices of these factors are positive. Thus any such rearrangement would increase profits, $\Sigma p_i x_i - \Sigma r_j v_j$, and leave the constraints satisfied. This is impossible, since we started with a competitive equilibrium which, by definition, maximized total profits. No such rearrangement can exist. The competitive pattern is efficient, and the first half of the theorem is proved.

Now for the second half of the theorem. We have an arbitrary efficient bundle of outputs and inputs $(\bar{x}_1, \ldots, \bar{x}_n)$ and $(\bar{r}_1, \ldots, \bar{r}_m)$. We have to find a set of commodity and factor prices with respect to which this input-output situation forms a competitive profit-maximizing equilibrium.[1]

In the previous section we showed that if $\bar{x}$ is efficient, there certainly exists at least one price constellation $\bar{p}$ such that $\bar{p}_1 x_1 + \cdots \bar{p}_n x_n$ is maximized at $\bar{x}$ among all feasible outputs. Let us use these $(\bar{p}_1, \ldots, \bar{p}_n)$ as our commodity prices.

Now where are factor prices going to come from? The reader who remembers our earlier analysis of the Walras-Cassel system will see at once that the factor prices must satisfy the dual inequalities (13-3a), page 369:

$$
\begin{aligned}
a_{11}v_1 + \cdots + a_{m1}v_m &\geq p_1 \\
\cdots\cdots\cdots&\cdots\cdots \\
a_{1n}v_1 + \cdots + a_{mn}v_m &\geq p_n
\end{aligned}
$$

In fact we can go even further and take those v's which solve the dual linear-programming problem of minimizing $\bar{r}_1 v_1 + \cdots + \bar{r}_m v_m$ subject to the price-cost inequalities. Let us call them $(\bar{v}_1, \ldots, \bar{v}_n)$.

Our problem is solved: we have $\bar{p}$'s and $\bar{v}$'s satisfying the usual competitive price-cost relationships. All we have left to do is to verify that if this constellation of $\bar{p}$'s and $\bar{v}$'s is presented to a profit-maximizing firm, the firm will choose to produce the given outputs $\bar{x}$ with the given inputs $\bar{r}$. But this is easy. Total profits are $\Sigma \bar{p}_i \bar{x}_i - \Sigma \bar{v}_j \bar{r}_j$. Actually, the duality theorem tells us that profits are zero; we have to show that no higher profit is attainable at the given prices. The price-cost inequal-

[1] Naturally we can't mean by this that the demand and supply functions will be satisfied. For given supply and demand functions there may be only one full competitive equilibrium. All we can require of an efficient production pattern is that there exist some constellation of prices which would lead profit-maximizing firms to exactly that pattern.

ities say that for every production process profits per unit of output are nonpositive. Since we have additivity and constant returns to scale, this means that there is no additional profit to be gained by expanding any output. In addition, the duality relationships say that any process actually being used has zero unit profits, while only those not being used may have negative profits. Therefore no extra profit is to be gained by reducing any output which is actually being produced. Profits are being maximized, and every efficient production pattern is a competitive equilibrium for some set of prices.

14-6. COMPETITION AND LINEAR PROGRAMMING

In the last few sections we have forged a link between the solution of certain linear programs and the concept of efficiency and another link between efficiency and competitive equilibrium or the activity of profit maximizing. Two links make a chain, and the chain connects linear programming at one end and competitive profit maximizing at the other. Much of this terrain is familiar from Chap. 13. The difference is that here our objective has been the normative welfare-economic properties of competition. We have found a two-way correspondence: competitive equilibria are efficient, and the set of efficient points consists of nothing but all possible competitive equilibria. One way to show that a situation is not efficient is to show that it could never have come about by way of competition. Just this argument can be used to prove the distorting effects of excise taxes and the nondistorting character of lump-sum taxes and subsidies.

One important aspect of this connection between programming and competition was not emphasized in the preceding discussion. Linear programming is, so to speak, a centralized computational way of finding efficient patterns or even of exploring the whole efficient set. Competitive profit maximizing, on the other hand, is a decentralized, atomistic way of doing the same thing. To be sure, our proofs talked about aggregate profits, and the individual firms appeared nowhere. We can imagine them to be there, none the less. The point about additive constant-returns-to-scale production processes is precisely that the boundaries between profit-maximizing firms lose significance. Where one firm or industry stops and the next one starts is a matter of importance to the owner and his heirs, but to nobody else. The functioning of the system doesn't depend on it at all. It is a commonplace of the theory of the competitive industry that under constant returns to scale the allocation of output among firms is indeterminate. We can, if we like, imagine equal firms, all of some given small size.

In any case the profit maximizing that leads to efficiency, that solves the linear program, can be carried out by firms. Nobody needs to compute on a large scale. Each firm needs only to know the price of what it sells and the wages of the factors it uses. The market breaks up the big linear program into little profit decisions, and as long as everybody faces the same prices, the electronic computer and the higgling of the market place both lead to the efficiency frontier.

However, all this begs the dynamic question completely. Computers sometimes solve problems by iteration, by successive approximation. Someone must always make sure then that the iterations converge, that the approximations actually get better and better. Similarly it is one thing to say that a competitive equilibrium is efficient and quite another to suggest that over time the usual competitive process will get closer and closer to its equilibrium, to an efficient point. This question was discussed briefly way back in Chap. 2. It turns out that too "perfect" a model of competition may oscillate endlessly around its equilibrium, just as too "perfect" a pendulum may do. In competitive markets, as in pendulums, a little friction may be needed for stability. But to pursue this would take us too far afield.

14-7. THE BASIC THEOREM OF WELFARE ECONOMICS

To conclude this chapter we can show how the theorem on efficiency fits into the broader frame of welfare economics. As a preliminary step we have to look still further into our definition of competitive equilibrium, and also to introduce another important concept, that of a Pareto-optimum.

We have already remarked that our treatment of competitive equilibrium tended to blur the role of the individual firm, and we gave an excuse. Even more so does our earlier definition slight the individual consumer. Households don't appear in it explicitly; their behavior is summed up in the market commodity demand and factor supply functions. Since our main interest was in the production side of the model, this simplification was quite permissible. Even now, with welfare economics our explicit objective, it would take us too far afield and multiply the number of variables by too high a factor if we incorporated household behavior explicitly in our model. But for expository purposes, since this is familiar ground for most economists, we can afford a loose and intuitive treatment. So we shall only remark that instead of taking the supply and demand functions for granted, we could have proceeded differently. We could have started with a given distribution of the ownership of resources among individual households. For any set of factor and commodity prices we could suppose each household to maximize its standard utility function subject to its own budget constraint. Granted the usual assumptions (convexity of indifference curves, nonsatiation, etc.), this would determine for each household a unique offer of resources and demand for commodities. Adding up over all households, we would get a point on each of the demand and supply functions (13-2) and (13-4). By varying the prices and the factor returns, we could trace out the whole collection of demand and supply surfaces. All this justifies the assertion that we could have omitted (13-2) and (13-4) from the earlier definition of competitive equilibrium and substituted the requirement that each household adjust its demand and offer so as to maximize its satisfaction subject to its budget constraint. What is important here is that the budget constraint of each household involves *the same factor and commodity prices.*

Now we turn to another familiar notion, that of a Pareto-optimum. This is a standard tool of welfare economics; its importance comes from the fact that it provides a weak, universally acceptable, criterion of when one economic configuration is "better" in a welfare sense than another. We say that situation A is "better" than situation B if in A no household feels worse off than it does in B and at least one household feels better off. The reader can think in terms of Edgeworth box diagrams and the like. A Pareto-optimum is a configuration of household consumptions and household factor supplies that (1) *is feasible* in the sense that the summed-up consumption quantities (our x's) and the summed up factor supplies (r's) make up a possible input-output vector in the technology [i.e., satisfy the inequalities (14-1a)], and (2) *has the property that no other feasible configuration is "better."* Obviously we need only restrict ourselves to feasible configurations. There can be lots of Pareto-optima, corresponding to different distributions of real income. In a "pure exchange" Edgeworth box, the set of Pareto-optima consists of all the points on the "contract curve." The reason that we had to expand our definition of competitive equilibrium is precisely because of the way the welfare notion of Pareto-optimum is constructed around the households' utility functions. Although the explicit definition of a Pareto-optimum is thus concentrated on the demand side, we saw earlier that productive efficiency is a necessary (if not sufficient) condition for welfare optimality.

The main propositions of welfare economics are usually stated in terms of a long string of equivalences among marginal rates of substitution in consumption and marginal rates of transformation in production. More recently it has become common to sum up all these in one brief and easily understood theorem which contains everything of significance and provides the backbone of modern welfare economics. This fundamental theorem states: *Every competitive equilibrium is a Pareto-optimum; and every Pareto-optimum is a competitive equilibrium.*[1]

What this means is, first, that competition, defined either in terms of utility and profit maximization with unique prices or equivalently in terms of our systems of equations and inequalities of the previous chapter, brings about just those marginal equivalences which guarantee that the production and consumption configuration is Pareto-optimal; and, second, that any real configuration of production and consumption patterns which is Pareto-optimal (a statement entirely free of explicit price considerations) will be found to define a set of implicit prices in terms of which it is actually a competitive equilibrium. From this proposition, and particularly from its second half, flow all the classical statements about monopoly, tariffs, taxation, etc.

We have set out to prove that competitive equilibria are Pareto-optima, and Pareto-optima are competitive equilibria. We already know (1) that all Pareto-optima are efficient points, (2) that maximum-profit competitive equilibria are efficient, and (3) that efficient points are maximum-profit competitive equilibria. These three propositions sum up most of what we have to say about production. We must now show how the gaps can be filled in and the argument completed. The linear-programming nature of the technology will play no special role in the reasoning. Since the argument is just the classical one familiar to all students of economic theory, we shall simply sketch the steps briefly and refer once again to the literature.[2]

Let us start with a full competitive equilibrium and see why it must be a Pareto-optimum. The familiar argument goes like this: We know that production is efficient; we know more, that at this efficient point the price ratio between any pair of commodities measures the rate at which they can efficiently be transformed into each other (by transfer of

[1] This line of thought goes back to Pareto and Barone. See also P. A. Samuelson, *Foundations of Economic Analysis*, Harvard University Press, Cambridge, Mass., 1948. The best recent treatments are to be found in K. Arrow, "An Extension of the Basic Theorems of Classical Welfare Economics," in *Proceedings of the Second Berkeley Symposium on Mathematical Statistics and Probability*, pp. 507–532, University of California Press, Berkeley, Calif., 1951; and in G. Debreu, "The Coefficient of Resource Utilization," *Econometrica*, **19**:273–292 (July, 1951). In this chapter we are largely following the beautifully concise treatment by Debreu in an unpublished Cowles Commission Discussion Paper, "Linear Spaces and Classical Economics," dated Jan. 29, 1953.

[2] In addition to the items given in the above footnote 1, we might mention O. Lange, "The Foundations of Welfare Economics," *Econometrica*, **10**:215 (1942), and Abba Lerner, "The Economics of Control," p. 57, The Macmillan Company, New York, 1944.

resources).[1] Also, in the usual way, consumers have arranged their offers of factors and purchases of commodities so that *the same price ratio* measures the ratio of marginal utilities (the marginal rate of substitution) for the pair of commodities. Hence at the margin, for each consumer, commodities may be indifferently substituted at the same rate at which they may be interchanged in production. From this and the fact that indifference curves have a convexity opposed to that of the efficiency locus or transformation curve, we can deduce the Pareto-optimality of the competitive configuration. For example, the Edgeworth-box-diagram reasoning tells us that the competitive situation lies on the contract curve *for the total commodity amounts* actually produced. Thus the pure-exchange conditions for optimality are met. And if, for example, we shift resources to get a little more x_2 and a little less x_1, the convexity tells us that if we give up a unit of x_1 we get in return less x_2 than the amount that would just compensate *any* consumer for the loss of the x_1.

This argument can be made a little more rigorous in the following easy way: Every household has a budget constraint that it must meet. This constraint is of the form $\Sigma p_i x_i^{(s)} \leq \Sigma v_j r_j^{(s)} + t^{(s)}$, where the superscript refers to the household; it says that expenditures cannot exceed income from the sale of factor services plus some given income from external sources or transfers. For our purposes we can put all $t^{(s)} = 0$. In addition, since we rule out satiation, the equality will hold for all consumers. Another way of saying that each consumer maximizes his satisfaction subject to his budget constraint is to say that any consumption-factor-supply situation which is preferred to the one actually observed must violate the budget constraint at the observed prices. Otherwise it would displace the observed choice. With strongly convex indifference curves it is even true that situations *indifferent* to the observed situation will violate the budget constraint. Even if indifference curves have flat portions, it is at least true that no situation indifferent to the observed one can cost literally less.

Now suppose our competitive equilibrium $(\bar{x},\bar{r})$ were not a Pareto-optimum. Then there must be a feasible configuration (x,r) that is better in the sense that no household is worse off and at least one is better off. Then by the previous paragraph, for all households the net budget at $(\bar{x},\bar{r})$ is no more than at (x,r), and for one household at least it is smaller; i.e., for all households $\Sigma p_i x_i^{(s)} - \Sigma v_j r_j^{(s)} \geq \Sigma p_i \bar{x}_i^{(s)} - \Sigma v_j \bar{r}_j^{(s)}$, and for at least one household the inequality holds. We can add these budget

[1] At a vertex like C in Fig. 14-1*a* there are two marginal rates of transformation depending on which commodity is to be substituted for which. Here we can say that the price ratio is intermediate between these two critical rates.

inequalities up over all households, and we find that $\Sigma p_i x_i - \Sigma v_j r_j > \Sigma p_i \bar{x}_i - \Sigma v_j \bar{r}_j$, where symbols without superscripts are just the all-economy totals in our usual notation. But if we know anything, we know by now that a competitive equilibrium maximizes net profits or net value of output, $\Sigma p_i x_i - \Sigma v_j r_j$, over all feasible input-output configurations. (The maximum value is zero, but it is a maximum all the same.) Therefore no such "better" situation as (x,r) can exist, or else $(\bar{x},\bar{r})$ would not be maximizing net value of output. Thus the competitive equilibrium $(\bar{x},\bar{r})$ must be a Pareto-optimum.

The next item on our program is a "proof" of the reverse implication, that every Pareto-optimum is a competitive equilibrium. From the welfare-prescription point of view this is the more important half of the theorem. The first half tells us only that in a competitive equilibrium the lot of any household can be improved only by redistribution of real income from some other household. It says nothing about the goodness or badness of monopolistic elements in a noncompetitive situation. But the second half bites deeper, and in fact provides the fundamental welfare argument against monopoly, against indirect taxation, and against tariffs. Once we have proved that every Pareto-optimum is a competitive equilibrium, we can argue as follows: The existence of a monopolistic seller (marginal cost = marginal revenue < price) or of an excise tax or of a tariff means that somewhere in the system we can find a place where not all households and firms are maximizing with respect to the *same* set of prices (consumers take the post-tax price of the taxed or dutied commodity as given, sellers take the net price). But a unique set of price parameters for everyone is the hallmark of competitive equilibrium. A situation with monopoly, excise, or tariff can thus under no circumstances be a competitive equilibrium.[1] Then it can under no circumstances be a Pareto-optimum, for every Pareto-optimum *is* a competitive equilibrium. The remarkable thing about this reasoning is that it demonstrates the intrinsic, more than institutional, nature of the price system. Markets and market prices there need not be, but starting from a completely non-"capitalistic" definition of a welfare optimum we show that things must organize themselves or be organized *as if* there were a set of universal prices or exchange ratios. The latter can actually be calculated from the data of the problem.

The "proof" we shall give amounts to nothing more than reminding the reader of what every economic theorist knows. Naturally a rigorous and really quite simple proof can be given, but it involves one or two

[1] The fallacy (or one of the fallacies) of supposing that we need only price proportional to marginal cost, or a universal constant degree of monopoly, is evident in our model. Factor prices can't possibly follow this rule if they have any internal uses in households.

tricky points which have nothing to do with our particular problem of linear models of production, and so it will be omitted here. However, the reader interested in these points is referred to the papers of Arrow and Debreu listed earlier in this chapter. The tricky points have to do with the consumption side of the model. In the first place, one has to be quite precise about the continuity of individual preferences and about the exact nonsatiation assumptions to be made. More important is the fact that consumers cannot be "vertically integrated" in the way that firms can. Because firms maximize a linear sum of prices times quantities, and because of the simple technology we have assumed, it is easy to see that we need only worry about a maximum of aggregate profits over the set of feasible aggregate input-output points. The boundaries of firms are unimportant. Not so with households. There is no grandiose aggregate utility we can think of maximizing. And because of the way that a Pareto-optimum is defined, the boundaries of households are important. The way around this difficulty is to stay in the world of commodities and out of the world of utilities. For each household, think of the set of all commodity bundles preferred or indifferent to the bundle achieved in the given Pareto-optimum. Because of the assumed shape of indifference curves, this set of commodity bundles is convex. Now we can define an aggregate concept as the set of grandiose over-all commodity bundles having the property that there is *some* distribution of these commodity amounts among households that will leave each household in a position preferred or indifferent to the initial situation. This set can be shown to be convex. Naturally some of its members are not feasible. But at least one member is feasible, namely, the initial Pareto-optimal position. The Southwest boundary of this set now plays the role of a "community-indifference curve"; because of Pareto-optimality this boundary can do no more than touch the feasible set of outputs. Then we appeal to a slight generalization of the supporting-plane theorem of Sec. 14-3—also a visually "obvious" result. Roughly speaking, if we have two convex sets that have no points in common, we can find a plane which separates them, i.e., such that one of the convex sets lies entirely on one side of the plane, while the other lies entirely on the opposite side of the plane. Under some circumstances (namely, if at least one of the sets has "interior points"), the same theorem holds if the two sets are permitted to have boundary points in common. As in Sec. 14-6, the coefficients of the separating plane can be interpreted as prices. In fact, the separating plane is nothing but the supporting plane at the efficient (and now Pareto-optimal) input-output point, as described in the last section. What we have now shown is that this same plane can serve as a budget plane with respect to which utilities are maximized. And this is what is meant by a Pareto-optimum being a competitive equilibrium.

We now get on with our main task, that of outlining the proof that a Pareto-optimum is a competitive equilibrium. We are given outputs and factor inputs at a Pareto-optimum. We already know from earlier work that the input-output point must be efficient and that therefore on the production side it is representable as a competitive profit-maximizing equilibrium with respect to a certain set of positive prices. We have to make it plausible that on the consumption side the picture is also one that would be brought about by all households maximizing utility, subject to a budget constraint. It is essential that this budget constraint should involve the same prices for each household and that these prices be the *same* ones that we found on the production side.

Once the aggregate outputs and factor amounts are specified, as they are in our problem, the factor-supply-consumption side of the economic system reduces to a standard Edgeworth-box-diagram pure-exchange situation. The reader can draw such a diagram for himself, see how the size of the box corresponds to the given input-output data, and draw in the indifference curves (they will look a little odd in the one-factor, one-good case). If the given situation is Pareto-optimal, it must lie somewhere on the contract curve, the locus of points of tangency of the two properly convex indifference maps.[1] At each point on the contract curve, hence also at the given Pareto-optimum, marginal rates of substitution are equalized for all consumers, for every pair of goods, or pair of factors, or factor-good pair. This set of common marginal rates of substitution will be our implicit price ratios. If these ratios were to be embodied in market prices, households maximizing satisfaction within their budgets would find themselves in exactly the given situation.

The implicit price ratios are the same for each household. It only remains to show that these price ratios must match the ones found on the production side. But this follows from the standard classroom argument. Suppose that there were a price ratio that differed. On the production side, the price ratios were shown to measure marginal rates of efficient transformation. Suppose that 1 unit of A foregone will enable the output of B to be increased by 3 units, so the price ratio $P_A/P_B = 3$. Suppose on the consumption side, $P_A/P_B = 2$; 1 unit of A foregone will need to be compensated by 2 units of B, if utility is to be maintained at the previous level. (Note that *this rate of substitution is the same for all consumers.*) Then give up 1 unit of A, produce instead 3 units of B. Let each household give up its share of the A foregone and compensate each household with its proportional share of 2 units of B. All households are as well off as they were. But there is still 1 unit of B to be given away. Give it to any nonsatiated household. We have found a

[1] It must also satisfy the initial restrictions on resource ownership, and there might also have to be a system of lump-sum transfers.

situation better in the Pareto sense than the initial situation. But this contradicts the assumption that we started with a Pareto-optimum. Thus we must abandon the possibility that rates of transformation in production differ from rates of substitution in consumption. The implicit price ratios must be the same, whether they are deduced from the efficient nature of production or the contract-curve nature of consumption. Any Pareto-optimum can be thought of as the joint response of profit maximizers and utility maximizers to a certain set of prices, the same prices for everyone. A Pareto-optimum is a competitive equilibrium.

14-8. GENERALIZATIONS

It is worth mentioning, although we cannot pursue the subject here, that the welfare theorem we have just finished discussing can be proved under wider conditions than the ones we used. Readers interested in the subtleties (which sometimes lead to complicated mathematics) should turn to the papers of Arrow and Debreu already mentioned.

The main generalizations are in two directions. We worked within a linear technology which defines society's feasible input-output combinations by a system of linear inequalities. Thus if production is feasible at all,[1] the feasible set of input-output points is a convex polygon. The essential characteristic is the *convexity;* the linearity may be a convenience, but not a necessity. In the linear model the combination of constant returns to scale and the additivity of processes get us convexity. Universal *decreasing* returns to scale would result in even more convexity (no flat places on the efficiency frontier). So we could get along with a mixture of constant and decreasing returns to scale scattered through the separate processes available. All that is required is that society's ultimate feasible set should be convex. The efficiency frontier or transformation curve (or surface) must show nonincreasing marginal rates of transformation. Once this is assumed, the proofs go through much as we sketched them. The main mathematical tool, the existence of the supporting plane, is available in the more general case.

Another direction in which it is possible to generalize has to do with the precise assumptions to be made about consumers' tastes. We can insist on more or less continuity in consumer preferences; we can have more or less strict assumptions about nonsatiation; we can pay more or less attention to the case of commodities not consumed at all by some consumers. But all this is hardly to the point for expositional purposes.

Finally, it is possible, interesting, and important to extend the theory of welfare economics to cover production, consumption, saving, and capi-

[1] One can imagine, if not observe, a situation in which the inequalities are *inconsistent.* Production cannot take place at all within the given resource limitations.

tal formation over time. If everyone's horizon were finite, we could simply consider the same physical good as a different commodity in each period of time, and a cheap generalization of the standard results would be easily available. But the essence of the problem is that consistent valuation of goods and incomes over a finite period of time *requires* some assumptions about terminal values of stocks, capital, etc. These terminal values can only refer to the still more distant future. So the problem is inescapably infinite. We could assume the arrival of doomsday at the end of the period, but then the results thus deduced would be valid only for a world in which doomsday is expected *with certainty.*

"Cheap" generalizations of the kind just mentioned will get us no further on the road toward a combination of welfare theory over time and the theory of capital. It is necessary to begin from first principles with due attention to the peculiarities of capital. This we did in Chap. 12. In that chapter the attack was mainly from the production-efficiency side, fully exploiting the goods-made-from-previous-goods structure of the problem. The price implications emerged eventually. For an approach slightly more similar in spirit to the welfare economics of the present chapter, the reader is referred to an interesting paper by E. Malinvaud.[1]

[1] "Capital Accumulation and Efficient Allocation of Resourees," *Econometrica*, **21**:233–268 (April, 1953).

15

Elements of Game Theory

15-1. INTRODUCTION

The theory of games has been hailed as a landmark in the history of ideas. "Ten more such books," Jacob Marschak[1] once wrote, "and the progress of economics is assured." In the ten-odd years since its definitive presentation[2] game theory has had important applications to the science of military tactics and has contributed to a revolution in the theory of statistics. When it comes to economic problems, for which the theory was originally designed, the value of its contribution is more in doubt. Because of this question and because the theory has a close affinity with linear programming, game theory has a place in this book. We shall not pursue the subject very far, however, developing the concepts just enough to bring out their bearing on economic problems and discussing the technique just enough to demonstrate its relationship to linear programming.[3]

The underlying insight of game theory is that parlor games, economic markets, and military battles are all instances of social situations in which the participants pursue conflicting interests. Moreover the fate of each participant—the poker player's winnings, the firm's profits, the army's casualties—depends in part on the actions of the other participants. Let us call such a situation, in which the outcome is controlled jointly by a number of participants with incompatible objectives, a game. We ask: How should a rational person involved in a game act? A general answer to this question would throw light on special instances of conflict situations, including those in the sphere of economics.

[1] "Neumann's and Morgenstern's New Approach to Static Economics," *Journal of Political Economy*, 54:115 (1946).

[2] J. von Neumann and O. Morgenstern, *Theory of Games and Economic Behavior*, Princeton University Press, Princeton, N.J., 1944.

[3] To economists who wish a readable, nontechnical, and completely honest primer of game theory we recommend J. D. Williams, *The Compleat Strategyst*, McGraw-Hill Book Company, Inc., New York, 1954. Those prepared to brave some mathematics will find a well-rounded treatise in J. C. C. McKinsey, *Introduction to the Theory of Games*, McGraw-Hill Book Company, Inc., New York, 1952.

15-2. DUOPOLY: AN OPPOSITION OF INTERESTS

Following time-honored precedent we illustrate the problem and method of game theory by an example drawn from the area of duopoly, but not—for reasons that will soon emerge—by the classic duopoly example of Cournot and his successors. The essence of the duopoly problem, as of game theory, is that the returns earned by each competitor depend on what both of them do. Table 15-1 presents a simplified duopoly problem which displays this feature. Firm 1 can choose between offering 100 and 200 units of product; Firm 2 can do the same. The profits received by the two firms depend upon both of these choices and are shown in the body of the table.

TABLE 15-1. PROFITS OF DUOPOLISTS ($1,000)

Firm 1's offering, units	Firm 1's profit		Firm 2's profit	
	Firm 2's offering, units		Firm 2's offering, units	
	100	200	100	200
100	5	4	5	6
200	3	6	7	4

Let us now imagine ourselves in the situation of Firm 1. The most attractive opportunity that beckons is to offer 200 units in the hope that Firm 2 will also offer 200 units. But this is also a dangerous choice, the more so because Firm 2 could wish for nothing better than for us to offer 200 units so that he could offer 100 units, thereby reaping a profit of $7,000 and leaving us a meager $3,000. All in all, it may seem best to count on Firm 2's offering 100 units. Then we should be well advised to offer 100 units also and remain content with $5,000 profit. Once we reach this stage, though, it seems reasonable to expect Firm 2 to expect us to be this shrewd. Thus Firm 2 will expect us to offer 100 units, and on this expectation will offer 200 units, hoping for $6,000 profit. This being so, it seems as though the $6,000 profit, which we mentally relinquished at the very outset, can be realized after all; all we need do is offer 200 units. On the other hand . . . .

We could, of course, go on like this forever. We have become involved in a process of endless regress, and no solution can be obtained by continuing to try to form expectations about our opponent's expectations of our expectations. The only hope is to cut through this welter of possibilities by some bold stroke, and this is precisely what the theory of games purports to do.

15-3. BASIC DEFINITIONS AND CLASSIFICATIONS

By a game we shall mean a social situation in which there are several individuals, each pursuing his own interest and in which no single individual can determine the outcome. Parlor games, of course, fulfill this definition. So does the duopoly problem that we just discussed. So does much of warfare. And so, obviously, do many economic situations. The elements necessary to describe a game are a list of the individuals concerned, a specification of the choices open to each, and a specification of the way in which those choices determine the results realized by each participant.

The most fundamental classification of games is according to the number of participants. One-person games are excluded by our definition. The basic classes are two-person games and more-than-two-person games. The theories of these two sorts of games are very different, though closely related. We shall devote almost all our attention to two-person games (for reasons that will emerge presently), reserving a few remarks about more-than-two-person games until the very end of the chapter.

We now focus our attention on two-person games. Our duopoly example illustrates such a game, but the data assumed have one peculiarity to which we now draw attention. Notice that in every one of the four possible contingencies, the sum of the profits of the two firms is the same, namely, $10,000. For this reason our example would be called a constant-sum game, or, more usually and less correctly, a zero-sum game. The contrasting class is non-constant-sum games, and we have altered our data so as to produce one in Table 15-2. Using the data of Table

TABLE 15-2. PROFITS OF DUOPOLISTS, NON-ZERO-SUM VARIANT ($1,000)

Firm 1's offering, units	Firm 1's profit		Firm 2's profit	
	Firm 2's offering, units		Firm 2's offering, units	
	100	200	100	200
100	5	4	5	7
200	3	6	6	2

15-2, we see that joint profits can now vary from $8,000 to $11,000, depending on the choices made by the two firms. The change may seem minor, but the consequence is fundamental since von Neumann's principal method of analysis applies only to the constant-sum case. Now the interests of the two firms, though somewhat divergent, are no longer directly opposed. This is the situation that has occupied the attention of students of markets with few participants from Cournot to Fellner, and one of the main issues of the discussion has been whether the duopolists would maximize their joint profit.[1] The theory of two-person games does not purport to throw much light on this clouded discussion; it merely emphasizes the distinction between the two kinds of games. When the game is not of the constant-sum sort, the clear opposition of interest is contaminated and new complications enter.

From here on we narrow our attention to two-person, constant-sum games. This makes our theory disappointingly specialized, and, indeed, whether the theory so delimited is ever relevant in economics is an open question. Definitely, game theory does not crack the nut of duopoly theory. But it is worth examining the special constant-sum case to which the theory does provide a satisfying solution with some applications.

15-4. STRATEGIES AND THE PAY-OFF MATRIX

Most parlor games take the form of a sequence of moves, and so do many conflict situations in economics, warfare, and elsewhere. A description of a game in terms of its sequence of moves, including the rules controlling each option, is called, in the lingo of game theory, a description of the game in *extensive* form. The fact that games in practice occur in the form of a time sequence of moves, countermoves, and, perhaps, chance shocks is purely adventitious from the point of view of game theory. To convince yourself that the sequential aspect of games is inessential, you have only to conceive of two chess players who are also skiers. Shortly before their big championship match, we may suppose, one of them experiences the kind of accident that might be expected. But pluckily, as he lies in his bed of pain, he writes a letter to the referees, explaining his absence and including a (very bulky!) envelope in which he specifies *how he would move* in every possible contingency. We do not render this situation any less plausible if we now assume that the second skiing chess player is as unfortunate as the first and also as plucky. The referee, then, with the two envelopes in his hand, can play the match every bit as well as the two players face to face and with the identical result. If we call the contents of these two envelopes the strategies of the two players, we see that the play of a game, in spite of all its intricacy, amounts merely to a choice of strategy on the part of each opponent. For, once both opponents have determined what they will do in every possible contingency, the outcome of the game is determined.[2] It is most

[1] Cournot concluded that they would not maximize their joint profit; Fellner that they would (approximately). The opinions of the other authorities are similarly diverse.

[2] It is neither practice nor practical in most conflict situations (e.g., chess) to confront in advance all possible situations that may eventuate.

convenient to think of games in this way, i.e., as amounting to a choice of strategy, and to represent the field of choice as in Table 15-3.

TABLE 15-3. A PAY-OFF MATRIX

Strategy of Player 1	Strategy of Player 2				
	1	2	3	...	n
1	a_{11}	a_{12}	a_{13}	...	a_{1n}
2	a_{21}	a_{22}	a_{23}	...	a_{2n}
3	a_{31}	a_{32}	a_{33}	...	a_{3n}
.	...	...	...	...	...
.					
.					
m	a_{m1}	a_{m2}	a_{m3}	...	a_{mn}

Such a table is called a pay-off matrix. This table represents the possibilities in a game played by two players, 1 and 2. Player 1 is assumed to have m strategies, identified by numbers and listed vertically in rows. Similarly, Player 2 is assumed to have n strategies listed in columns. We shall assume both m and n to be finite, though in some games, like chess or bridge, they may be very large. The entries in the body of the table are the winnings of Player 1. For example, Table 15-3 tells us that if Player 1 uses his Strategy 2 and Player 2 uses his Strategy 3, then Player 1 will win a_{23}.

Because we are restricting ourselves to constant-sum games, there is no need for a similar table for Player 2. His winnings are simply some constant less Player 1's, and therefore, just as Player 1 is interested in reaching the largest entry possible in such a table, Player 2 is interested in reaching the smallest possible entry.[1] Going back to the duopolists of Table 15-1 we see that we can suppress the last two columns and express all the relevant information in the form of a pay-off matrix as in Table 15-4.

TABLE 15-4. PAY-OFF MATRIX FOR THE DUOPOLISTS, CONSTANT-SUM CASE

Strategy of Firm 1	Strategy of Firm 2 (pay-off in $1,000)		Row min
	100	200	
100	5	4	4
200	3	6	3
Column max.	5	6	

[1] Player 1 is called the maximizing player; Player 2 the minimizing player. It is a well established convention to list the strategies of the maximizing player vertically.

It should be noted that the entries in the pay-off matrix are the objective returns (e.g., sums of money) received by the maximizer and, by consequence, denied to the minimizer. This presupposes that the two players are actually contending over these measurable returns and not over prestige, safety, utility, or anything else. If the participants in an economic conflict situation are contending over something other than money, e.g., utility, then that something is what should be entered in the pay off matrix in order to have the matrix represent the relevant results of various struggles. Clearly, the unvoiced presupposition that the stakes can adequately be expressed as money (or some other objective measure of value) is not innocuous because the contrary is often the case. The assumption is essential because only if the stakes are something that can be shared or exchanged, like a sum of money, does it make sense to conceive of a constant-sum game or, indeed, of the sum of the returns to two players. We therefore assume that monetary pay-offs are an adequate measure of the results of the game.[1]

15-5. THE EVALUATION OF STRATEGIES AND THE WORTH OF A GAME

The fundamental difficulty in the analysis of conflict situations, we have already seen, is to decide what each opponent is to expect of the other. In game theory we visualize each player as acting on three presuppositions: (1) his opponent's interests are diametrically opposed to his own; (2) his opponent has all the information necessary to construct the pay-off matrix of the situation; (3) his opponent is shrewd enough so that if the opponent knew in advance what strategy the player had selected, the opponent would make a wise choice of his own strategy.[2]

Let us apply these considerations to Table 15-3. Suppose that Player 1 (the maximizing player, remember) is looking down the list of his strategies and comes to Strategy 3. We assume him to look across the row for Strategy 3 and note that, say, a_{32} is the lowest number occurring on that row. He then is assumed to reason: If I adopt my Strategy 3, then my opponent may, for all I can do about it, adopt his Strategy 2. Thus by adopting Strategy 3, I run the risk of winning only a_{32} (which may be a negative number). This number, which is the most that Player 1 can definitely count on winning if he plays Strategy 3, we shall call the worth of Strategy 3 to Player 1. In general, the worth of any

[1] Some element of approximation in practice is, of course, permissible. The problem of measuring the pay-offs is considered further in Appendix A.

[2] Assumption (3) may be weakened slightly. All that the argument requires is that each player fear that his opponent may make a wise choice, and wish to be protected against this possibility.

strategy to Player 1 is the minimum entry on the row of pay-offs corresponding to that strategy.

The situation as seen by Player 2 is a little different. He chooses columns, and his concern is to concede as little as possible to his opponent. Thus the worth to Player 2 of any of his strategies is the maximum entry in the column of pay-offs corresponding to that strategy because if he adopts that strategy he cannot prevent Player 1 from winning that much.

Let us apply these considerations to Table 15-4. If Firm 1 offers 100 units, there is nothing to prevent Firm 2 from offering 200 units, thereby keeping Firm 1's earnings down to $4,000. Hence the top entry in the "row min" column, and the other entries are derived similarly. In general, the worth of any strategy is a measure of the worst that can happen if that strategy is adopted.

The worths can now serve, tentatively, as a guide for decision. Firm 1 can, by offering 100 units, assure itself of earning at least $4,000, no matter what Firm 2 does. Similarly, Firm 2, by offering 100 units, can prevent Firm 1 from earning more than $5,000, no matter what.[1]

Let us generalize. The maximizing player can assure himself of winning an amount at least equal to the greatest of the row minima by proper choice of strategy and irrespective of what his opponent does. This number, called for short the "maxmin," is defined as the worth of the game to the maximizing player because his opponent cannot prevent him from realizing it.[2] Worth, then, means guaranteed rock-bottom worth. Of course, the maximizer can realize a smaller return than the worth of the game by selecting a strategy other than the one that yields the maxmin, but he does this on his own volition.

The game has a worth to the minimizing player, too. This is the quantity that he, by proper play, can prevent the maximizer from exceeding, and it is the smallest of the numbers in his worth row, i.e., the minimum of the column maxima. For short this is called the "minmax." It is easy to prove that the maxmin can never exceed the minmax.[3]

[1] It is not viciousness that makes this desirable from Firm 2's point of view, but the fact that in this conflict situation Firm 2 always earns the difference between $10,000 and Firm 1's profit.

[2] We have to split a hair here. Later on we shall define the "value" of the game. This is a more fundamental concept than "worth" as we have just defined it, and the two do not always agree. Our concept of "worth" is the same as von Neumann's concept of the "value of the minorant game."

[3] *Proof:* Suppose that in Table 15-3 entry a_{ij} equals the maxmin and entry a_{kl} equals the minmax. Then a_{ij}, being a row minimum, is less than or equal to every entry in its row and, in particular, to a_{il}. But a_{il} cannot exceed the maximum of its column, which is a_{kl}. Thus a_{ij} cannot exceed a_{il} which cannot exceed a_{kl}, and the statement is proved.

In the little example of Table 15-4, the worth of the game to Firm 1 is $4,000 (the maxmin) and the worth to Firm 2 is $5,000 (the minmax). In less technical words, Firm 1 can be sure of earning at least $4,000 and Firm 2 can be sure of preventing Firm 1 from earning more than $5,000. Please note the gap here, for it is this gap that caused the trouble in Sec. 15-2.

15-6. STRICTLY DETERMINED GAMES AND SOME APPLICATIONS

Since the gap causes trouble, we pause to consider another example that doesn't have one. Suppose, then, that the data were as shown in Table 15-5. In this table the maxmin (greatest row minimum) and the

TABLE 15-5. PAY-OFF MATRIX FOR THE DUOPOLISTS, STRICTLY DETERMINED CASE ($1,000)

Strategy of Firm 1	Strategy of Firm 2		Row min
	100	200	
100	5	6	5
200	3	4	3
Column max...........	5	6	

minmax (smallest column maximum) are both equal to $5,000. This means that Firm 1 can be sure of earning at least $5,000 (barring its own stupidity) and that Firm 2 can be sure of preventing Firm 1 from earning more than $5,000. It is not attributing undue insight or cautiousness to Firm 1 to assume that in such a situation they would realize that Firm 2 can prevent them from earning more than $5,000. Thus the sensible thing for them to do is to settle on the strategy that guarantees them $5,000, which is as much as they can hope to obtain. By similar reasoning we are led to expect that Firm 2 will choose the strategy that allows Firm 1 to earn $5,000 and no more. Thus the choices and the earnings of both players are determined. Oligopolistic indeterminacy has vanished.

A glance at the table makes it obvious why this is so. No matter what Firm 2 may do, Firm 1 is better off if it offers 100 units, and no matter what Firm 1 may do, Firm 2 is better off if it offers 100 units. The problem has become trivial. Yet we can learn something from it. Table 15-5 is an example of the concept of domination. If one of a player's strategies is better than some second strategy (from that player's point of view) against every single opposing strategy, then the first strategy is said to dominate the second. It is never to a player's advantage to use a dominated strategy. In the example, 200 units is a dominated strategy for both firms.[1]

The fact that each of the duopolists has only one undominated strategy makes the example of Table 15-5 trivial. But we shall see in a moment that a conflict situation may be easily solvable (in the sense of discovering rational policies for the opponents) even when this isn't so. The important criterion for the simple solvability of such a situation is the equality of the minmax and the maxmin. If the minmax equals the maxmin then, in effect, the maximizing player both is guaranteed a certain amount of winnings (their common value) and can be prevented from winning more. Thus he "will" adopt a strategy that will win him the maximum available amount, and the minimizer will adopt a strategy that will prevent the maximizer from winning more. The strategies of both players and the outcome of the situation are all clearly determinate. We have a *strictly determined* game.

The essential formal characteristic of a strictly determined game, we have said, is that the minmax and the maxmin are equal. If this equality holds, an obvious extension of the reasoning of footnote 3 on page 423 shows that there must be some element in the pay-off matrix (a_{il}, in the notation of the footnote) which is simultaneously the maximum of its column and the minimum of its row. Such an element, since it is the crest of a hill looking in one direction and the trough of a valley looking in the other, is called a "saddle-point." One way to state the situation is to say that a game is strictly determined if and only if it has a saddle-point.

We now suggest two simplified, but perhaps meaningful, economic models that lead to matrices with saddle-points, i.e., to strictly determined games.

Our first example concerns a firm whose overriding concern is its market position as measured by its share of the market. The instrumentality being considered for influencing the market share is the size of the advertising budget, which may be either small, medium, or large, and the consequences of the various possible decisions are shown in Table 15-6. The entries in the table are market shares; ours is clearly the leading firm in the industry.

The logical structure of this table is the same as that of Table 15-5. Large advertising expenditures dominate the other two strategies for both

[1] This definition of domination is a little stricter than it should be. An accurate definition is: Strategy i is said to dominate Strategy j if the pay-off to Strategy i is at least as favorable as the pay-off to Strategy j no matter what strategy the opponent may use and if the pay-off to Strategy i is more favorable than that to Strategy j for at least one opposing strategy.

TABLE 15-6. PAY-OFF MATRIX, ADVERTISING MODEL

Our advertising budget	Other firms' advertising budget			Row min
	Small	Medium	Large	
Small	75	65	60	60
Medium	85	75	70	70
Large	90	80	75	75
Column max	90	80	75	

the minimizing and the maximizing player, there is a saddle-point where the two dominating strategies intersect, the maxmin equals the minmax. Both participants advertise to the fullest extent permissible.

This application, trivial though it is, merits comment. In the first place we have assumed only three possible levels of advertising budget, but this assumption was clearly inconsequential. We might have conceived of five or ten or five hundred levels without affecting the result. What is consequential along these lines is the assumption that the choices are discrete, i.e., that they do not form a continuum. We shall adhere to this assumption throughout our discussion of game theory. We do so because games with a continuous range of choices do not always have solutions. But those games that do are not necessarily any more difficult or easy to handle than the discrete ones.

Second, we have been quite vague about who the second person in this game was. There might have been some major single competitor, or the opponent might have been the entire rest of the industry. It really doesn't matter much, if the one player is concerned with protecting himself against all contingencies. Two-person games can thus be used to a certain extent to deal with situations in which there are not literally two participants, and this is often a useful extension.

Our third, and crucial, comment concerns the objective, or pay-off function, that was adopted. By concentrating on market shares instead of gross sales or net profit or other more usual criteria, we guaranteed that we would have a constant-sum game. This is a limitation, to be sure, but sometimes market shares are important. What was more damaging was the fact that the cost of advertising was regarded as negligible. No wonder, then, that everyone advertised as much as permitted. It was practically free and had a positive marginal return. The neglect of cost is probably the most vitiating aspect of this example, and yet, let's face it, costs can rarely be introduced into the framework of the simple, constant-sum, two-person game. We shall have more comments about the objective function later, and they will tend to confirm this statement.[1]

The second example of strictly determined games is less trivial. In a famous paper Hotelling[2] considered the policies of a pair of duopolists who had to choose positions along some continuum which might be a street, a transcontinental railroad, or a scale of quality. The issue is best portrayed by a diagram, like Fig. 15-1. The entire market is

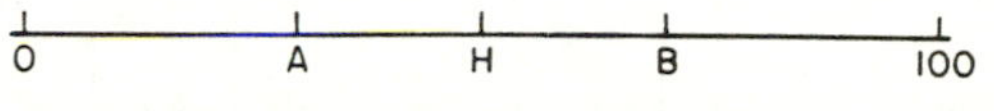

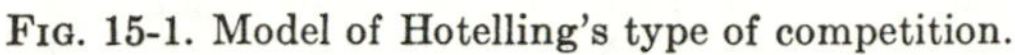

FIG. 15-1. Model of Hotelling's type of competition.

arranged along a scale from 0 to 100 and scaled off in such a way that the distance from 0 to any point, for example A, measures the proportion of the market to the left of that point. If two firms enter this market they must locate at some points, say A and B. Then we construct point H halfway between A and B and assume that each firm gets all the trade on its side of point H. Where should the two firms locate? We assume that total demand will be the same whatever the two firms do.

This model fits very neatly into the format of a two-person constant-sum game. Let us convert it to discrete form by assuming that the firms can locate only at points 0, 20, 40, 60, 80, and 100. We make the additional assumption that if the two firms choose the same location, they split the market 50-50. Then a little elementary arithmetic yields the pay-off matrix of Table 15-7, where Firm A is the maximizer.

Inspection of the table shows that the extreme strategies, 0 and 100, are dominated by 20 and 80, both for the maximizer and the minimizer. It never pays to locate at the end point. So the 6×6 game is reduced to a 4×4 game. The minmax and the maxmin both equal 50, and there are four saddle-points (in fact, multiple saddle-points will always imply a kind of saddle-plateau in the middle of the table). Each firm can secure itself half the market by locating as close to the center of the

[1] As an exercise, the reader can work out a closely similar case. Suppose that the demand curve for mineral water is of unit elasticity, viz.,

$$p = p(x_1 + x_2) = \frac{k}{x_1 + x_2}$$

Assuming zero variable costs and a discrete set of outputs, $x_i (i = 1, 2)$, for a pair of duopolists, draw up the game matrix and find the saddle-point solution. Note that three or more sellers can be handled equally well in this singular case. Why are unit elasticity of demand and the absence of variable costs necessary assumptions? What would happen if the duopolists could vary their outputs continuously?

[2] Harold Hotelling, "Stability in Competition," in G. J. Stigler and K. E. Boulding (eds.), *Readings in Price Theory*, pp. 467–484, Richard D. Irwin, Chicago, 1952.

TABLE 15-7. PAY-OFF MATRIX FOR HOTELLING'S MODEL

Location of Firm A	Location of Firm B						Row min
	0	20	40	60	80	100	
0	50	10	20	30	40	50	10
20	90	50	30	40	50	60	30
40	80	70	50	50	60	70	50
60	70	60	50	50	70	80	50
80	60	50	40	30	50	90	30
100	50	40	30	20	10	50	10
Column max......	90	70	50	50	70	90	

scale as permissible, and it doesn't matter which of the two central positions it chooses. This conclusion, that both firms should locate close together and toward the middle of the scale, is the same as the one that Hotelling arrived at. This argument from game theory is rather simple and direct, avoiding Hotelling's use of the calculus.[1]

15-7. CHANCE AND EXPECTED VALUES

It is usual in economics to describe alternative courses of action as if the consequences of each alternative could be predicted exactly. Ordinary demand curves and production functions, when taken literally, illustrate this supposition of certainty. A careful economist usually will remark somewhere in the course of his discussion that consequences are really not exactly predictable; that the results of his analysis are only "a first approximation"; and that allowances (presumably minor and, at any rate, usually not evaluated) have to be made for uncertainty and chance.

Our discussion of game theory has followed this precedent up to this point. No hint of uncertainty or randomness has entered our exposition or illustrations. But now—and perhaps this is one of the virtues of game

[1] This example gives rise to an easily solved continuous game. The method is to construct the Edgeworth box diagram showing the position of one firm along the horizontal axis and the position of the other along the vertical. If the reader constructs the diagram for this game, he will find, perhaps to his surprise: (1) there are not two sets of indifference curves, but only one; (2) each indifference contour is a straight line with a sharp discontinuity; (3) the two reaction curves almost coincide; and (4) the point at which the reaction curves cross (the Cournot point) is also both Stackelberg leadership points and the game theory saddle-point. Moreover, the Edgeworth contract curve is spread out into a region that covers the whole diagram. Aspect 1 is characteristic of all constant-sum continuous games; aspect 4 holds for all strictly determined constant-sum continuous games.

theory—we have to face up to this issue.[1] To see why chance fluctuations have to be taken into account, consider Table 15-8, which repre-

TABLE 15-8. PAY-OFF MATRIX FOR THE DUOPOLISTS IN THE CASE OF UNCERTAIN OUTCOMES

Strategy of Firm 1, units	Strategy of Firm 2, units	
	100	200
100	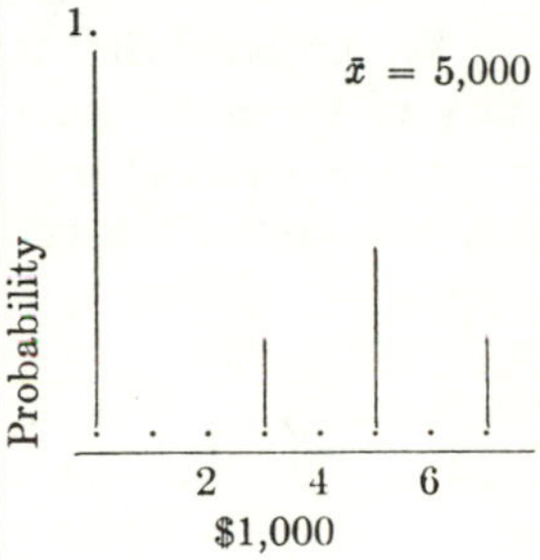 1. $\bar{x}$ = 5,000 Probability 2 4 6 \$1,000	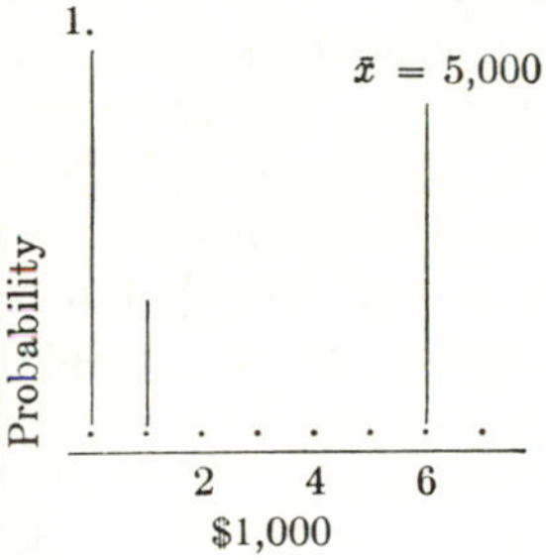 1. $\bar{x}$ = 5,000 Probability 2 4 6 \$1,000
200	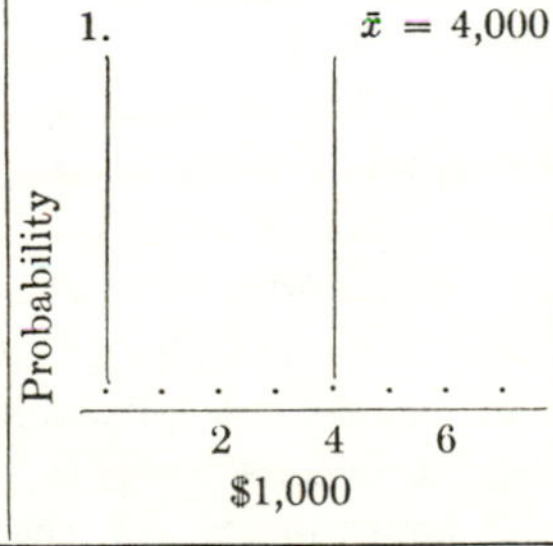 1. $\bar{x}$ = 4,000 Probability 2 4 6 \$1,000	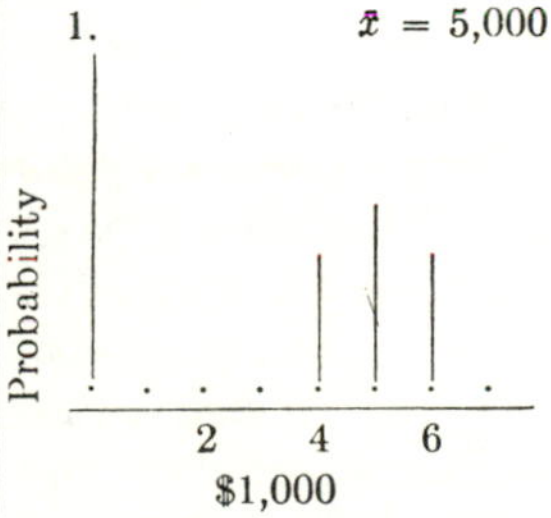 1. $\bar{x}$ = 5,000 Probability 2 4 6 \$1,000

sents our original pair of duopolists faced with some element of chance. In this table the univalued pay-offs of Tables 15-4 and 15-5 have been supplanted by probability distributions. The little graph in the upper left-hand corner, for example, signifies that if both duopolists offer 100 units, then Firm 1 has a 0.25 probability of earning \$3,000, a 0.50 probability of earning \$5,000, and a 0.25 probability of earning \$7,000.

When faced with this more complicated (and perhaps more usual) scheme of pay-offs it is no longer obvious which of the four outcomes Firm 1 would regard as the most desirable and which one Firm 2 (which gets whatever Firm 1 loses) would regard as most desirable from its point of view. The economist's standard technique for resolving issues of this kind is to construct a subjective preference ordering in which each of the

[1] We shall not really go the whole hog. In F. H. Knight's terminology we are going to introduce "risk" (i.e., the results of variations with definable probability distributions) but not "uncertainty."

firms would rank the four outcomes in 1, 2, 3, 4 order of desirability. Such an ordering would be insufficient for the purposes of game theory, however, for two reasons. First, it would sacrifice the constant-sum feature since two individuals' rankings cannot be added in any meaningful way. Second, it would not disclose the relationship between the pay-offs and probabilities of each probability distribution and the desirability of that distribution. An analysis of that relationship is, as we shall see, essential to the development of game theory.

A study of the connection between the desirabilities of the various pay-offs in a probability distribution, the probabilities attached to each of those pay-offs, and the desirability of the probability distribution as a whole is, therefore, an important part of game theory. In order not to interrupt the chain of exposition, we have sketched this analysis separately in Appendix A, but we can report the main results here. First, under reasonable assumptions a numerical (and, in a sense, summable) measure can be assigned to the desirability of either a definite pay-off or a probability distribution of pay-offs. This measure is called, naturally, "utility." Second, the utility of a probability distribution of pay-offs is equal to the mathematical expectation of the utilities of the individual pay-offs. In other words, the utility of a probability distribution of pay-offs is equal to the sum (or integral) constructed by multiplying each pay-off by its probability (or probability density) and adding (or integrating). Third, if the pay-offs are sums of money and if the utility of a sum of money is proportional to its amount, then the utility of a probability distribution of pay-offs can be taken as equal to the mathematical expectation of the distribution. In economic applications of game theory it is ordinarily assumed that the conditions for this kind of conclusion are fulfilled. The details of the construction of the von Neumann–Morgenstern measure of utility and the justification of these conclusions are given in Appendix A.

As applied to the game situation of Table 15-8, these conclusions, in particular the third, state that the outcome of $4,000 for certain associated with the strategy pair (200, 100) is less desirable to Firm 1 than the outcome associated with any other strategy pair and that the firm is indifferent among the other three outcomes, diverse though they are. Thus Table 15-9 includes everything in Table 15-8 that is relevant to decisions. Table 15-9, in turn, is solvable at a glance. If Firm 1 offers 100 units, it is assured of $5,000 (expected value, of course) whatever Firm 2 may do, while if it offers 200 units, it runs a risk of netting only $4,000 without any compensating gain. Similar considerations apply to Firm 2's offering 200 units. Question for the reader: Is there a saddle-point in Table 15-9?

Thus use of the postulate that if a strategy results in a probability distribution of outcomes the expected value of that distribution may be

TABLE 15-9. EXPECTED VALUES OF THE PAY-OFF MATRIX FOR THE DUOPOLISTS IN THE CASE OF UNCERTAIN OUTCOMES

Strategy of Firm 1, units	Strategy of Firm 2, units	
	100	200
100	\$5,000	\$5,000
200	4,000	5,000

regarded as *the* pay-off makes it possible to solve the situation represented in Table 15-8. Situations of this kind are clearly extremely common. In economic and military life it is probably quite exceptional to find a pay-off matrix that is free of probability distributions. Even in the severely simplified world of parlor games, the most interesting examples, e.g., bridge, poker, dice, include a probability element. We have seen that the postulate brings such games and situations within the compass of game theory, if their pay-off matrices have saddle-points.

Games that include probability elements, like the ones just mentioned, are known technically as "games with chance moves." This terminology stems from the conceptual device of analyzing such games by imagining a third participant in the game, called "chance," who makes certain moves from time to time, in accordance with the rules of the game, but does not participate in the pay-offs. Thus, in bridge it is chance that decides on the cards held by the actual players; in warfare it is chance that decides whether a particular message is received by friendly forces or is intercepted and decoded by the enemy; and in economic context it is chance that decides precisely how effective a given advertising campaign is. In all these cases chance makes its decisions in accordance with probability distributions prescribed in the rules of the game, and it is these chance moves which determine the probability distributions of pay-offs associated with each strategy adopted by the live participants.

But pay-offs that take the form of probability distributions play an even more central role in game theory than would be inferred from the fact that chance moves are so prevalent in games that have scientific and sporting interest. Such pay-off distributions provide the key to the solution of games without saddle-points. This is the subject of the next section.

15-8. MIXED STRATEGIES; GAMES WITHOUT SADDLE-POINTS

We have postponed it long, but now the time has come when we really must try to solve the special duopoly problem first stated. This problem, recall, boiled down to the pay-off matrix of Table 15-4. The difficulty was that there seemed to be no possible equilibrium position. There were only four possible pairs of strategies, but examining each of them in turn we found that in every case one of the duopolists or the other could increase his profit unilaterally by shifting his strategy. This is the economic significance of the absence of a saddle-point in the pay-off matrix.

Let us consider Table 15-4 again. According to that table, if Firm 1 offers 100 units, it can count on earning $4,000 (the row minimum), and if it offers 200 units, it can count on earning $3,000. Yet we also saw that $5,000 (the minmax) is the lowest ceiling that Firm 2 can place on Firm 1's earnings. The gap between $5,000 and $4,000 shows that by adopting either strategy Firm 1 is missing an opportunity. What opportunity? Evidently to keep Firm 2 guessing; to open possibilities that may lure Firm 2 into mistaken decisions.

To seize this opportunity Firm 1 will have to behave unpredictably; that is, it must not adopt either pure strategy as a policy but must hold open the possibility of using either. To be sure, if Firm 1 adopts an unpredictable strategy, it must be content with an unpredictable or random pay-off. But the postulate of the previous section entails that this will be worthwhile if the expected value of this random pay-off is larger than the earnings that could be obtained by following a more predictable policy.

Thus we are led to seek an unpredictable policy for Firm 1 which will yield the largest possible expected value of pay-off on the conservative assumption that Firm 2 follows the best possible counterstrategy. Suppose that Firm 1 leaves the ultimate decision up to some chance device that has a probability p of deciding in favor of offering 100 units and a probability $1 - p$ of deciding on 200 units. If Firm 2 offers 100 units, Firm 1's expected earnings are

$$E_1(p) = 5{,}000p + 3{,}000(1 - p)$$

and if Firm 2 offers 200 units, Firm 1's expected earnings are

$$E_2(p) = 4{,}000p + 6{,}000(1 - p)^{1}$$

If Firm 2 behaves correctly for any value of p selected by Firm 1, Firm 1 must expect the lesser of these two values. Thus the expected pay-off corresponding to a chance device with a probability p of offering 100 units is

$$E(p) = \min(E_1, E_2)$$

[1] If Firm 2 were sometimes to use 100, sometimes 200, Firm 1 would get a result somewhere between E_1 and E_2.

The situation is shown graphically in Fig. 15-2. In this figure the upward sloping diagonal shows E_1 as a function of p, the downward sloping diagonal shows E_2, and the heavily weighted broken line, made up of segments of the two diagonals, shows $E(p)$. Firm 1's problem is to find a random device that will correspond to the highest value of $E(p)$. This value occurs at the corner point, which is where $p = 0.75$ and $E(p) = \$4,500$. Thus if Firm 1 cuts an ordinary deck of cards and offers 200 units if it cuts to a spade and 100 units in any other case, it will assure itself of an expected pay-off of \$4,500 whatever quantity Firm 2 offers. This is the highest expected pay-off that Firm 1 can achieve irrespective of Firm 2's policy.

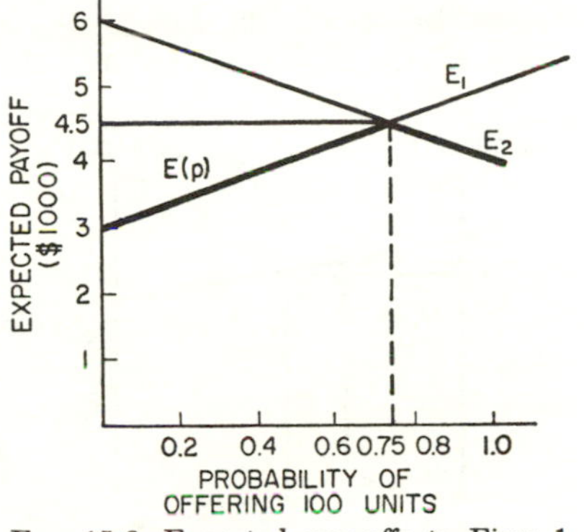

FIG. 15-2. Expected pay-offs to Firm 1, duopoly example.

Firm 2 has a similar problem which can be solved by a similar analysis. Its solution shows that Firm 2 should leave its decision to a toss of a coin, i.e., adopt $p = 0.5$. By so doing it can prevent Firm 1 from attaining an expected value of more than \$4,500. Note that the gap that disturbed us at the beginning of this section is now closed; Firm 1 can obtain an expected value of \$4,500, and Firm 2 can prevent Firm 1 from doing any better.

This is an example of the solution of games by the use of mixed strategies, the crucial advance achieved in the theory of games. Let us call any of the courses of action open to a participant in a conflict situation a "pure strategy." A mixed strategy is a probability distribution that assigns a definite probability to the choice of each pure strategy. Von Neumann and Morgenstern have shown, and we shall prove below, that in any game without a saddle-point[1] there exists a pair of mixed strategies that forms an equilibrium position. That is, if one participant adopts his member of the pair, the other participant can do no better than to adopt his member. In other words, these mixed strategies guarantee to each participant the most desirable expected pay-off that he can attain against competent opposition. This pair, then, represents an equilibrium position and a solution to the conflict situation.

[1] If a game is strictly determined with a saddle-point corresponding to a pair of pure strategies, this will obviously be a special and limiting case of the use of mixed strategies. All the probabilities are either 1 or 0.

15-9. GRAPHIC ANALYSIS OF SIMPLE GAMES

Additional insight can be gained from a somewhat different graphic representation of the duopolists' problem. Again we use the data of Table 15-4. Figure 15-3 shows the consequences to Firm 1 of offering 100 units. If Firm 2 also offers 100 units, Firm 1 will realize \$5,000 as indicated by the intersection of the diagonal with the axis labeled "Firm 2 offers 100 units." If Firm 2 offers 200 units, Firm 1 will realize \$4,000, as shown on the other vertical axis. The rest of the diagonal line shows the results if Firm 2 follows a mixed strategy. Thus if Firm 2 uses $p = 0.5$ of offering 200 units, the expected value to Firm 1 of offering 100 units is \$4,500. If Firm 2 uses $p = \frac{1}{3}$ of offering 200 units, the expected value of Firm 1's earnings will be \$4,667, etc.

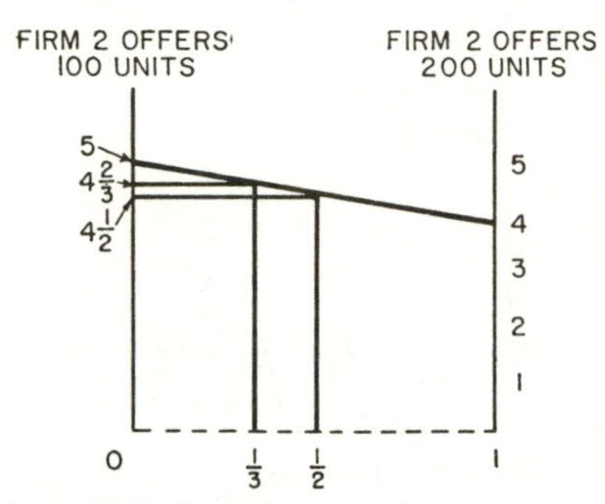

FIG. 15-3. Duopoly example—results to Firm 1 of offering 100 units. Pay-offs in \$1,000.

Figure 15-3 shows the results of one of Firm 1's pure strategies against all the pure and mixed strategies open to Firm 2. In Fig. 15-4 we add Firm 1's other pure strategy. In addition we note that any of Firm 1's mixed strategies will correspond to a line extending between the two vertical axes and entirely comprehended between the pure-strategy lines because the result of any mixed strategy is a weighted average of the results of the pure strategies involved in it. The dashed line in the figure represents one of Firm 1's mixed strategies. Since this dashed line bisects the vertical distance between the two pure strategy lines, it represents $p = 0.50$.

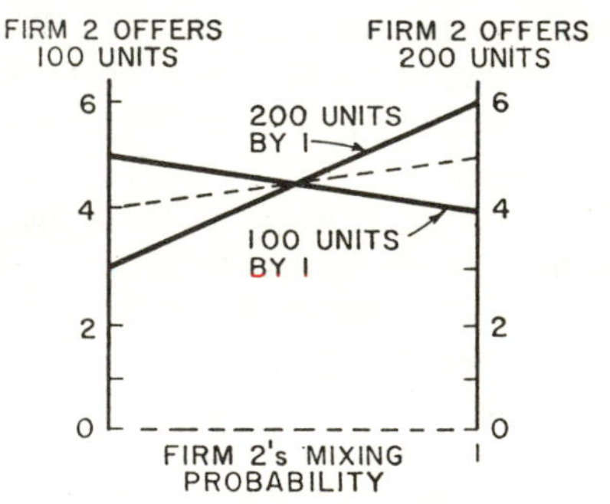

FIG. 15-4. Duopoly example—pay-offs to Firm 1's strategies, \$1,000.

In terms of this graph Firm 1's problem is to find the dashed line (mixed strategy line) whose lowest point is as high as possible. Clearly this will be the horizontal line through the intersection of the two pure-strategy lines. Algebraically we find the line from the condition that Firm 1's expected earnings must be the same when Firm 2 offers 100 units as when it offers 200 units. If we let p denote the probability that

Firm 1 will offer 100 units as a result of its chance device, we have

$$5{,}000p + 3{,}000(1 - p) = 4{,}000p + 6{,}000(1 - p)$$
$$p = 3/4$$

This is the result we found before.

This graphic approach can be applied to any game in which one of the participants has precisely two strategies. Williams[1] has provided an interesting illustration:

The Firm of Gunning and Kappler manufactures an amplifier having remarkable fidelity in the range above 10,000 cycles—it is exciting comment among dog whistlers in the carriage trade. Its performance depends critically on the characteristics of one small, inaccessible condenser. This normally costs Gunning and Kappler $1, but they are set back $10 if they have to replace a defective one.

Of course, there are some alternatives open to them: They know a test procedure which will catch a defect 3 times out of 4; unfortunately it costs $1 to apply the test. They know another which is surefire and of negligible direct cost, except that, 9 times out of 10, it results in breakage of the good condensers. It is possible for them to buy a superior-quality condenser, at $4, which is fully guaranteed; the manufacturer will make good the condenser and the costs incurred in changing it.

The problem facing Gunning and Kappler reduces to the two- by four-strategy game shown in Table 15-10 and Fig. 15-5. The entries in this

TABLE 15-10. PAY-OFF MATRIX FOR GUNNING AND KAPPLER

	Gunning and Kappler's strategy			
	1. No test	2. $1 test	3. Sure-fire test	4. Guaranteed condenser
1. Defect	10	4¼	1	4
2. No defect	1	2	10	4

pay-off matrix are the sum of the costs of inspection (if any), replacement (if needed), and purchase. Gunning and Kappler are minimizers; there is no active maximizer.[2] Since Gunning and Kappler are minimizers, their best mixed strategy corresponds to the line that (1) can be constructed as a weighted average of the lines corresponding to their four alternatives, and (2) has its highest point as low as possible. The first condition restricts the choice to lines entirely contained in the irregular

[1] *Op. cit.*, pp. 76–78.

[2] Thus this isn't really a game. Some of game theory's most fruitful applications are, as here, concerned with play against an imagined opponent.

polygon *ABCDEFG*. The second condition leads us to seek the lowest horizontal line that fulfills the first condition. This is clearly the horizontal through the intersection of pure-strategy lines 2 and 3.

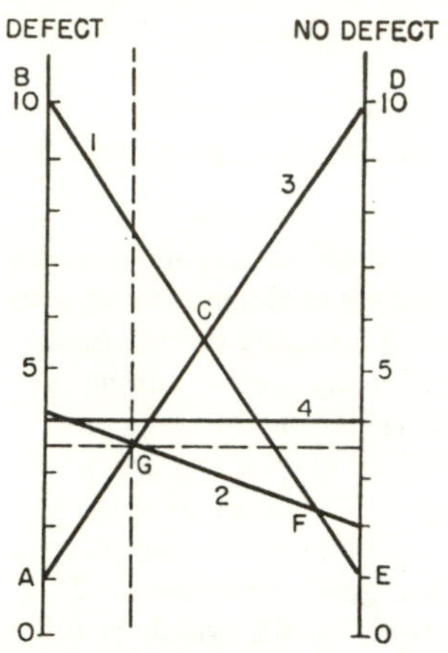

FIG. 15-5. Pay-off diagram for Gunning and Kappler.

The mixed strategy corresponding to this horizontal will be made up of Strategies 2 and 3 only. Let p be the probability in this mixture of using Strategy 2. Then p must satisfy

$$4.25p + (1 - p) = 2p + 10(1 - p)$$

since the expected cost must be the same, whichever strategy nature uses. This gives at once $p = 4/5$. Thus Gunning and Kappler should use a chance device which gives odds for the four strategies in the proportions 0:4:1:0. The expected cost will be \$3.60, and no other policy can assure them a lower expected cost.

Now suppose that "nature" is short for Natural Reproducers, Inc., a supply house that, all unknown to *G* and *K*, would like to buy them out eventually as cheaply as possible. Nature now becomes an active maximizer. What should it do? Nature's choice amounts to a choice of the probability that a defect will occur, which, diagrammatically, is a choice of a vertical line somewhere between the two axes. We have drawn in the best choice from nature's point of view (and the worst from *G* and *K*'s) because the minimum cost possible for any vertical line is the height at which the lowest diagonal crosses that line. The best vertical is the one that goes through the highest point on the lower boundary of *G* and *K*'s accessible region, i.e., the intersection of Strategies 2 and 3, or the same point we found before. This intersection corresponds to a proportion of about 71 per cent defectives. The reader should satisfy himself that an inimical nature could not harm *G* and *K* by supplying them with a higher proportion of defectives.

Once both participants have more than two strategies, the above graphic crutch fails us. One method of solution that is often effective in such cases is to convert the problem of finding an optimal mixed strategy into a linear-programming problem. We shall present one method for doing this now and discuss the matter in some detail in Chap. 16.

15-10. A GENERAL METHOD OF SOLUTION

Consider the general game whose pay-off matrix is shown in Table 15-3. By a mixed strategy for Player 1 we shall mean a set of probabilities x_1,

$x_2, \ldots, x_m$, one for each of the choices open to Player 1. It is clear that any mixed strategy has a determinate expected value when played against any of Player 2's strategies. For example, the expected value of the mixture $x_1, x_2, \ldots, x_m$ against Player 2's Strategy 1 is, using the first column of Table 15-3,

$$E_1 = x_1a_{11} + x_2a_{21} + \cdots + x_ma_{m1}$$

and, in general, the expected value of this mixture against Player 2's Strategy j is

$$E_j = x_1a_{1j} + x_2a_{2j} + \cdots + x_ma_{mj} \qquad j = 1, 2, \ldots, n \quad (15\text{-}1)$$

Let V denote the smallest of the expected values against Player 2's n strategies so that

$$V = \min (E_1, E_2, \ldots, E_n) \quad (15\text{-}2)$$

In these terms, Player 1's problem is to choose $x_1, x_2, \ldots, x_m$ so as to make V as large as possible. In words, Player 1 seeks the mixed strategy for which Player 2's most advantageous strategy gives him (Player 1) as large an expected pay-off as possible.

We shall think of Player 1 as having the choice variables $x_1, x_2, \ldots, x_m, V$. His objective is to make V as large as possible. In doing so he is restricted by Eqs. (15-1) and (15-2), which may be recapitulated as

$$x_1a_{1j} + x_2a_{2j} + \cdots + x_ma_{mj} \geq V \qquad j = 1, 2, \ldots, n \quad (15\text{-}3)$$

He is also restricted by the fact that the $x_1, \ldots, x_m$ must be probabilities. Thus they are nonnegative, and their sum must be unity. We also have to impose an additional restriction on the choice of V. We know in advance that V must be at least as great as the maxmin of the pay-off matrix because the maximizing player can assure himself of the maxmin by choosing the strategy that corresponds to it. Therefore, if the maxmin is nonnegative, so is V. Now suppose that the maxmin is negative. By adding a large enough positive constant to every entry in the pay-off matrix we can derive a new game in which the maxmin is nonnegative. And the optimal strategy for this new game will be the same as the one for the original game because adding a constant to all pay-offs amounts to giving a flat subsidy to the maximizer and does not affect the relative desirability of the various strategies. Thus either we can assure ourselves in advance that the V for any game is nonnegative or we can make this so by a trivial modification of the game. Therefore, without loss of generality, we may assume that V is nonnegative.

It is now clear that the maximizer's problem is the same as the following linear-programming problem: To choose numbers $x_1, x_2, \ldots, x_m$, V so as to maximize V subject to the following linear restraints:

$$\begin{aligned} x_1 \geq 0, x_2 \geq 0, \ldots, x_m \geq 0, V \geq 0 \\ a_{11}x_1 + a_{21}x_2 + \cdots + a_{m1}x_m - V \geq 0 \\ \cdots\cdots\cdots\cdots\cdots\cdots \\ a_{1n}x_1 + a_{2n}x_2 + \cdots + a_{mn}x_m - V \geq 0 \\ x_1 + x_2 + \cdots + x_m = 1 \end{aligned}$$

This is the kind of problem we already know how to solve by, for example, the simplex method.[1]

The situation is similar when considered from the minimizer's point of view. The minimizer's mixed strategy is a set of probabilities $y_1, y_2, \ldots, y_n$, one for each of the choices open to him; and he desires to choose these so that the greatest of the expected values

$$a_{i1}y_1 + a_{i2}y_2 + \cdots + a_{in}y_n \qquad i = 1, 2, \ldots, m$$

is as small as possible. If we denote the greatest of these expected values by W, an argument analogous to the one we have just gone through shows that the minimizer has to solve the following linear-programming problem: To choose numbers $y_1, y_2, \ldots, y_n$, W so as to minimize W subject to the following linear restraints:

$$\begin{aligned} y_1 \geq 0, y_2 \geq 0, \ldots, y_n \geq 0, W \geq 0 \\ a_{11}y_1 + a_{12}y_2 + \cdots + a_{1n}y_n - W \leq 0 \\ \cdots\cdots\cdots\cdots\cdots\cdots \\ a_{m1}y_1 + a_{m2}y_2 + \cdots + a_{mn}y_n - W \leq 0 \\ y_1 + y_2 + \cdots + y_n = 1 \end{aligned}$$

Comparison of this formulation with the maximizer's problem shows that the two are duals of each other. Thus when optimal strategies are selected, V and W will be equal. The common value of V and W, that is, the most that the maximizer can assure himself by proper choice of mixed strategies and the least that the minimizer must concede, is called the "value" of the game.

We illustrate with an example of ruthless industrial warfare. Firm A is fighting for its life against the determination of Firm B to drive it out of the industry. Firm A has the choice of raising its price, leaving it

[1] Inspection of the restraining equations shows that if all the a_{ij} are finite, this problem has a finite solution, i.e., that the restraints set an upper bound to the attainable values of V. By virtue of only the first and last restraints V cannot exceed the largest of the numbers $a_{11}, a_{21}, \ldots, a_{m1}$. Taking the other restraints into account, V cannot exceed $\min_j \max_i a_{ij}$. Even this, of course, is not the least upper bound to the range of V; it is just the lowest upper bound disclosed by superficial inspection of the restraints.

unchanged, or lowering it. Firm *B* has the same three options. Firm *A*'s gross sales in the event of each of the nine possible pairs of choices are shown in Table 15-11. It will be seen that at Firm *B*'s current prices,

TABLE 15-11. PAY-OFF MATRIX FOR INDUSTRIAL-WARFARE EXAMPLE (Gross Sales of Firm *A*)

Firm *A*'s price change	Firm *B*'s price change		
	Increase	None	Decrease
Increase	90	100	110
None	110	100	90
Decrease	120	100	80

Firm *A*'s demand curve has unit elasticity. If Firm *B* raises its prices, Firm *A*'s demand curve will shift up and become elastic. A price reduction by Firm *B* produces the opposite effect on Firm *A*'s demand curve.

What, now, should the two firms do? The linear-programming problem for Firm *A* is to choose x_1, x_2, x_3, V so as to maximize V subject to

$$\begin{aligned} x_1 \geq 0,\ x_2 \geq 0,\ x_3 \geq 0,\ V &\geq 0 \\ 90x_1 + 110x_2 + 120x_3 - V &\geq 0 \\ 100x_1 + 100x_2 + 100x_3 - V &\geq 0 \\ 110x_1 + 90x_2 + 80x_3 - V &\geq 0 \\ x_1 + x_2 + x_3 &= 1 \end{aligned}$$

This problem can be solved by standard linear-programming methods, and it turns out that the solution is not unique. The two alternative basic solutions are

(1) $x_1 = 2/3,\ x_2 = 0,\ x_3 = 1/3,\ V = 100$

(2) $x_1 = 1/2,\ x_2 = 1/2,\ x_3 = 0,\ V = 100$

Firm *B*'s problem is to minimize W subject to

$$\begin{aligned} y_1 \geq 0,\ y_2 \geq 0,\ y_3 \geq 0,\ W &\geq 0 \\ 90y_1 + 100y_2 + 100y_3 - W &\leq 0 \\ 110y_1 + 100y_2 + 90y_3 - W &\leq 0 \\ 120y_1 + 100y_2 + 80y_3 - W &\leq 0 \\ y_1 + y_2 + y_3 &= 1 \end{aligned}$$

There are two basic solutions to this problem also:

(1) $y_1 = 0,\ y_2 = 1,\ y_3 = 0,\ W = 100$

(2) $y_1 = 1/2,\ y_2 = 0,\ y_3 = 1/2,\ W = 100$

Thus Firm *B* might just as well leave its price unchanged, irrespective of Firm *A*'s intentions. Firm *A* should toss a coin between raising its prices and leaving them unchanged. In any event, Firm A can expect to maintain its current gross sales but not to increase them.

Williams[1] has an example with an interesting interpretation. The problem concerns an investor who has $100,000 to commit for 1 year and appeals to his broker for advice. The broker provides the investor with Table 15-12, giving the probable returns to three sorts of investment in the event of three possible political climates. The broker wisely declines to express any opinion about which climate is likely to prevail.

TABLE 15-12. RETURN TO $100,000 IN THREE CONTINGENCIES

Type of investment	In case there is:		
	War	Cold war	Peace
Bonds	2,900	3,000	3,200
War babies	18,000	6,000	−2,000
Mercantiles	2,000	7,000	12,000

How to invest? We regard the investor as a maximizer with three available strategies and formulate his problem as follows: To maximize V, the expected return in the most unfavorable eventuality, subject to

$$\begin{aligned} x_1 \geq 0,\ x_2 \geq 0,\ x_3 \geq 0,\ V &\geq 0 \\ 2{,}900x_1 + 18{,}000x_2 + 2{,}000x_3 - V &\geq 0 \\ 3{,}000x_1 + 6{,}000x_2 + 7{,}000x_3 - V &\geq 0 \\ 3{,}200x_1 - 2{,}000x_2 + 12{,}000x_3 - V &\geq 0 \\ x_1 + x_2 + x_3 &= 1 \end{aligned}$$

The solution is $x_1 = 0$, $x_2 = 0.294$, $x_3 = 0.706$, $V = \$6{,}710$. This solution can be interpreted in either of two ways. First, the investor should not buy bonds and should leave his choice between war babies and mercantiles up to a chance device with the odds indicated. Or, second, he should invest $29,400 in war babies and $70,600 in mercantiles. The justification for this second interpretation is, in Williams's words:[2]

> In this example . . . we have done something which appears at variance with our own rules; namely, we have countenanced an interpretation of the odds as a *physical mixture* of strategies on a single play of the game. That is, we have permitted the player to use a little of this strategy and a little of that strategy, instead of insisting that he use just one, basing his choice on a *suitable* chance device.
>
> Indeed, it is evident that we have violated some principle; otherwise the physical mixture would not be possible. Any possible set of actions should be

[1] *Op. cit.*, pp. 101–103.
[2] *Ibid.*, p. 103. Italics in the original.

represented by some pure strategy; so the possibility of using a physical mixture should not arise.

This anomalous situation is traceable to the fact that there is an infinite game which is closely related to the finite game stated above. In the infinite game the player could invest, in infinitely many ways, in mixtures of securities—we are ignoring the practical limitations concerning the divisibility of securities and money. Moreover, the partial payoff from each security is proportional to the amount purchased and the total payoff is just the sum of the partial ones. Such situations may be analyzed as infinite games, which turn out to have saddle-points, or as finite games, which turn out to require mixed strategies (usually). By interpreting the latter as a physical mixture, we arrive at a solution which is equivalent to the saddle-point of the associated infinite games.

We present one more example of a game situation without a saddle-point, partly as an exercise and partly as a vehicle for some final interpretive comments. Two automobile companies, A and B, foresee that they will have to merge in a year or two. The managers of both firms therefore perceive that their overriding short-run concern is to prepare for the negotiations by achieving as favorable an earnings-per-share rate as possible in comparison with the rate achieved by the other firm. Thus the managers of Firm A wish to maximize the quotient obtained by dividing Firm A's earnings per share by Firm B's earnings per share, and Firm B's managers wish to minimize that figure.

With this consideration very much in mind, each firm must decide on the horsepower of the new model. The consequences, in terms of this critical ratio, of various choices on the parts of the two firms are shown in Table 15-13. This table is based on the idea that as horsepower

TABLE 15-13. THE AUTOMOBILE MERGER: RATIO OF FIRM A'S PROFIT PER SHARE TO FIRM B'S PROFIT PER SHARE

A's horsepower	B's horsepower			
	250	275	300	325
250	0.6	0.4	0.3	0.8
275	0.9	0.6	0.4	0.7
300	0.8	0.7	0.6	0.4
325	0.5	0.4	0.5	0.6

increases sales increase too, but profit per unit falls sharply. Thus, in general, a firm loses out, on the one hand, if its horsepower is less than that of its opponent and, on the other hand, if its horsepower is very much greater.

Inspection of Table 15-13 shows that Firm B will surely not adopt a 250-horsepower design because this yields a less favorable result (i.e., a higher profit ratio for Firm A) than 275 horsepower against every one of Firm A's choices. In other words, 250 horsepower is a dominated strategy for Firm B. Eliminating this strategy leaves B with a choice of three horsepowers and A with a choice of four. We leave the rest of the solution, which is straightforward, as an exercise. The reader should find: For Firm A, $x_{275} = 1/13$, $x_{300} = 5/13$, $x_{325} = 7/13$; for Firm B, $y_{275} = 1/13$, $y_{300} = 7/13$, $y_{325} = 5/13$; value of the game $= 69/130 = 0.53$.

Two features of this illustration merit comment. First, the pay-off function selected is a peculiar one, although perhaps appropriate in some circumstances. Second, we have assumed throughout the problem that both firms adopt the same objective function and that both estimate the consequences of the various choices in the same way. For example, we have assumed that Firm A estimates that if they use a 250-horsepower engine and Firm B uses a 275-horsepower engine, then the earnings ratio will turn out to be 0.4, at least in an expected value sense, and we have assumed that Firm B's estimate of the consequences agrees exactly. Agreement on such tenuous forecasts is a strong assumption. Yet we were forced to make all these assumptions in order to have a constant-sum game.

15-11. NON-CONSTANT-SUM GAMES AND MANY-PERSON GAMES

It may appear that the approach just given should apply to non-constant-sum games like that of Table 15-2. To see what happens, let us use the data of Table 15-2 and solve two games: one in which Firm 1 is a maximizer against a passive opponent and one in which Firm 2 is a maximizer. The result of the first game is that Firm 1 should mix its strategies with a probability of $3/4$ of offering 100 units and a $1/4$ probability of offering 200 units. Similarly Firm 2 should use a $5/6$ probability of offering 100 units and a $1/6$ probability of offering 200 units. The trouble with this solution is that either firm can do better, provided the other firm uses the mixture just given. An easy computation shows that if Firm 1 uses the solution given it can expect a profit of \$4,500 whatever Firm 2 may do, while if Firm 1 simply offers 100 units while Firm 2 uses the mixed strategy given, Firm 1's expected profits are increased to \$4,833. Similarly, if Firm 2 uses the mixed strategy its expected profit is \$5,333, while if Firm 2 simply offers 200 units and Firm 1 uses the mixed strategy, Firm 2's expected earnings increase to \$5,750.

This contrasts with the situation found to be characteristic of constant-sum games, for in the non-constant-sum example there is a gain to offset the risk of deviating from a safe mixed strategy. Put differently, by cooperating, the players can increase the pool to be divided. In non-constant-sum games if one player uses a safe mixed strategy the other player can increase his expected earnings by not doing so. The situation reduces to the kind of jockeying for advantage that we encountered when we first considered the duopoly example, and the device of introducing mixed strategies does not help.

A sensible thing for the duopolists to do is collude, for this is the only way in which their joint profit can be maximized. This, indeed, is the solution given by von Neumann and Morgenstern. If collusion be ruled out, as it must be in many economic contexts, then we are up against the problem of the "conjectural reaction." What each participant does depends upon his surmises about his opponent's choice of strategy, and, as a matter of general theory, there are no compelling reasons in favor of any one particular surmise. We shall not discuss further the solutions that have been proposed for this problem; we agree with McKinsey that "despite the great importance of general games for the social sciences, there is not available so far any treatment of such games which can be regarded as even reasonably satisfactory."[1]

Games with more than two participants, whether constant-sum or not, are even more difficult than non-constant-sum two-person games. It is easy to see why if we consider the next stage of complexity, the three-person constant-sum game.

In a three-participant situation no pair of participants has directly opposing interests, for, in general, any pair of participants can benefit by forming a coalition against the third and thereby maximizing their joint return. So far the situation involving any pair is like the non-constant-sum duopoly problem just discussed. The additional complication enters when we recognize that in the duopoly case joint profits are maximized at the expense of a passive public, while in the three-person case the joint profits of any pair must be maximized at the cost of a third participant who need not remain passive. In the duopoly case there was only one possible coalition that could form; in the three-person case there are three possible coalitions, and once any of them has formed, the outsider is free to try to break it up.

[1] J. C. C. McKinsey, *Introduction to the Theory of Games*, p. 343, McGraw-Hill Book Company, Inc., New York, 1952. Our discussion of non-constant-sum games may seem to establish a little more than it actually does. A sufficient condition for the mixed-strategy approach to yield optimal strategies is for the game to be constant-sum. But this is not a necessary condition. The necessary condition is that the entries in the pay-off matrixes of the two participants be related by a linear formula. We relegate this weakening of the conditions to a footnote because the broadening it secures is mostly illusory. What the weaker condition says is that if one participant measures his profits in dollars and the other measures his in francs and if the two take into account different amounts of fixed cost, then the situation is still essentially of a constant-sum sort.

The theory of many-person games in the hands of von Neumann and Morgenstern is essentially a theory of coalitions, their formation and revision. The underlying idea is that two persons in such a situation cannot do worse by acting jointly than by acting severally, and may do better. Thus a many-person game tends to be reduced to a two-"person" game in which each "person" is a coalition. The problems then become: Which coalitions will form and how will the winnings be divided among the members of the coalition? To pursue the answers proposed for these questions would lead us into a specialized discussion, and since these answers are not very satisfactory we refrain.

The upshot of our cursory discussion of general games is that game theory provides convincing solutions of conflict situations only in the two-person constant-sum case. In more general cases the concepts and approach of game theory have provided a convenient and suggestive framework for analysis, but no really satisfactory basis for finding concrete solutions has yet been proposed.

15-12. GAME THEORY AS AN ECONOMIC TOOL

Our discussion has disclosed severe limitations on the applicability of game theory to economic problems. It appeared that only in the simplest case, that of two-person constant-sum games, did game theory lead to a complete solution, in the sense of determining the maximum gains that each player could expect to obtain and finding the strategies that yielded those maxima. To solve even this case we had to make two restrictive assumptions. First, we noticed that the very idea of having a constant-sum game implied that the stakes of interest were objectively measurable and transferable. Second, we had to assume that the players' subjective attitudes toward gains and losses were such that they regarded the expected value of a probability distribution of pay-offs as just as desirable as the whole chance configuration represented by the probability distribution.[1]

Most economic situations, we noted, take the form of games in which either the sum of the winnings is not constant or in which the number of players is greater than two. In these more complicated games the conclusions attained by game theory were much less definite. Our major conclusions were that coalitions would tend to form (if they promised any advantage) and that if the formation of coalitions converted the situation into a two-person game with the coalitions as players, then the principles governing the simple case applied.

[1] We were able to go beyond these restrictions slightly in the case of strictly determined games, where it was unnecessary to introduce mixed strategies.

It is not surprising, therefore, that the 13 years that have elapsed since the publication of *The Theory of Games* have seen no important applications of game theory to concrete economic problems. The theory of games has had a profound impact on statistics and on military science; in economics it is still merely a promising and suggestive approach.

What, in view of all these limitations, has game theory to contribute to economics? Oddly enough, since game theory is an attempt to determine optimal strategies explicitly, the contribution seems to be qualitative rather than quantitative. The conceptual framework developed in game theory provides a useful set of constructs for the qualitative discussion of problems of opposing interest in economics. For example, the concept of coalitions maintained by means of side payments among the members is an excellent theoretical counterpart to cartels and similar institutions. If this vocabulary had been in use at the time that Edgeworth and von Stackelberg wrote, they could hardly have overlooked, as they did, the possibility and importance of such coalitions.

Perhaps the most novel and fruitful concept of all those introduced by game theory is that of mixed strategies. This concept has been the key to the valuable applications in statistics and military science. Its applicability to economic problems has yet to be explored fully.

This upshot may be discouraging, but it would be unreasonable to expect anything more. Economic problems, and particularly the interactions of the economic objectives of diverse individuals, are enormously complex; one could hardly expect to reduce them to the level of parlor games, even theoretically. Game theory provides solutions to simple conflict situations and valuable hints for understanding more complicated ones. This is as much help as an economist can expect from a new branch of mathematics.

16

Interrelations between Linear Programming and Game Theory

16-1. INTRODUCTION

The economist is interested in the problem of game theory and in the problem of linear programming, each for its own sake. From his point of view, these are two separate subjects. But it turns out that, from the mathematician's viewpoint, these two subjects are closely related—which is in the nature of a happy coincidence, since any computing methods devised for one of these theories can be used to solve problems arising in the other. It turns out (1) that every two-person, zero-sum game problem can be computed by converting it into a related linear-programming problem; and (2) what is perhaps even more important, that every linear-programming problem can be artificially converted into a two-person, zero-sum game, so that we can, if we wish, compute the solution to the former by computing a solution to the latter.

One might have suspected this tie-up from the fact that both theories share John von Neumann as a progenitor (but extrapolation of such heuristic reasoning would bring much of mathematics into the linear-programming camp). Also a suspicious clue would be provided by the fact that saddle-points of certain simple linear functions—so-called "bilinear forms"—occur in both theories. However, historically the two theories did develop separately, until finally through the work of von Neumann, George Brown, Dantzig, and Gale, Kuhn, and Tucker, the complete tie-up was brought to light.[1]

16-2. CONVERSION OF A GAME INTO A LINEAR-PROGRAMMING PROBLEM

The easiest half of the tie-up between the two theories is the recognition that either player of a game can regard his problem of finding an

[1] See T. C. Koopmans (ed.), *Activity Analysis of Production and Allocation*, chaps. 19, 20, and 22, John Wiley & Sons, Inc., New York, 1951.

optimal mixed strategy as a standard problem of linear programming. The mathematical reasoning is quite simple, but the interpretation in common-sense words is a little complicated.[1]

Suppose I am Player 1 and you pay me a_{ij} if I play pure strategy i and you play pure strategy j of the game whose pay-off matrix is

$$\begin{bmatrix} a_{11}\,a_{12} & \cdots & a_{1j} & \cdots & a_{1n} \\ a_{i1}\,a_{i2} & \cdots & a_{ij} & \cdots & a_{in} \\ a_{m1}\,a_{m2} & \cdots & a_{mj} & \cdots & a_{mn} \end{bmatrix}$$

Of course I am interested in finding a best mixed strategy, i.e., a best set of probabilities with which I should play each of my m possible strategies. Call my unknown optimal mixed strategy $(x_1,x_2, \ldots ,x_m)$, or simply x, where it is understood that $\sum_1^m x_i = 1$ and $x_i \geq 0$. I want to maximize the expected pay-off to me, thereby keeping down the expected pay-off to you.

For the moment, imagine my optimal mixed strategy x to be given. You are then in a position to calculate your expected loss to me for each of *your* n pure strategies $(1, 2, \ldots, j, \ldots, n)$. Thus, if you play Strategy 1, my gain and your loss will be

$$E_1 = x_1a_{11} + x_2a_{21} + \cdots + x_ma_{m1} = E_1(x)$$

a linear function of my mixed strategy. Likewise if you were to play any pure strategy j, your loss would be

$$E_j = x_1a_{1j} + x_2a_{2j} + \cdots + x_ma_{mj} = E_j(x) \qquad j = 1, 2, \ldots, n$$

Some of these losses $E_1, \ldots, E_n$ will be bigger than others. If you are smart, you will of course shun any pure strategy that gives you a bigger loss; instead you will play only those of your strategies that give rise to minimum E's. The value of these minimum E's, it will be recalled, is defined as the "value of the game" to me, V; i.e.,

Value of game to Player 1 = minimum of Player 2's possible losses when he plays any pure strategy against Player 1's optimal mixed strategy

or $$V = \min\,(E_1, \ldots ,E_n) = \min\left(\sum_1^m x_ia_{i1}, \ldots, \sum_1^m x_ia_{in}\right)$$

and where Player 1 must have solved the problem of finding the best x's so as to maximize the above minimum value; i.e., Player 1 picks $x_1, \ldots, x_m$ so as to maximize the above minimum value.

[1] The development now to be given is a slightly more rigorous restatement of the argument of Sec. 15-10.

In short,

$$V = \text{maximum with respect to } x \text{ of } \min \left(\sum_{1}^{m} x_i a_{i1}, \ . \ . \ . \ , \sum_{1}^{m} x_i a_{in} \right) \tag{16-1}$$

By the above definition of the value of a game, we may write down the following inequalities for each pure strategy of Player 2 and for the optimal mixed strategy of Player 1:[1]

$$\begin{aligned} v_1 &= x_1 a_{11} + x_2 a_{21} + \cdots + x_m a_{m1} - V \geq 0 \\ &\cdots\cdots\cdots\cdots\cdots\cdots\cdots\cdots \\ v_n &= x_1 a_{1n} + x_2 a_{2n} + \cdots + x_m a_{mn} - V \geq 0 \\ & \quad\ x_1 \quad + x_2 \quad + \cdots + x_m \qquad\quad = 1 \\ & \qquad\qquad x_1 \geq 0, x_2 \geq 0, \ . \ . \ . \ , x_m \geq 0 \end{aligned} \tag{16-2}$$

We may note as a digression that, by taking weighted averages of these relations, we can derive a similar inequality for *any* mixed strategy $(y_1, \ . \ . \ . \ , y_n = 1 - y_1 - \cdots - y_{n-1})$ of Player 2. As an example, if Player 2 adopted the mixed strategy of playing each of his n pure strategies with equal probabilities $1/n$, we would have

$$\bar{v} = x_1 \bar{a}_{1\cdot} + x_2 \bar{a}_{2\cdot} + \cdots + x_m \bar{a}_{m\cdot} - V \geq 0 \tag{16-3}$$

where $$\bar{a}_{i\cdot} = \frac{a_{i1} + a_{i2} + \cdots + a_{in}}{n} = \frac{\sum_{j=1}^{n} a_{ij}}{n}$$

and $\bar{v}$ is defined as the nonnegative expression written in (16-3).

Player 1 is interested in picking his optimal x_i so as to maximize V, the value of the game to himself. In selecting his optimal x's, he must remember that the constraints of (16-2) will hold. Hence, his maximum problem can be written

$$\text{Subject to (16-2), maximize } 1 \cdot V \tag{16-4}$$

We might think of V as a new variable, called x_{m+1}, and if we were to write out this last maximum problem in terms of $(x_1, \ . \ . \ . \ , x_m, x_{m+1})$, we would find that it looks almost exactly like a standard linear-programming problem.

The qualification "almost exactly" is needed because we have as yet no reason to restrict x_{m+1} or V to being a nonnegative variable. Without such a restriction, our problem is not quite in the standard programming form.

[1] The nonnegative v's defined by Eqs. (16-2) could be interpreted as the "avoidable losses" of each of Player 2's pure strategies when used against Player 1's optimal strategy.

Obviously, V may in some cases turn out to be negative, depending upon the numerical values of the a_{ij}; for example, if every $a_{ij} < 0$, V is certainly negative. One simple way of ensuring that V is nonnegative is to add to every element of the original $[a_{ij}]$ matrix a positive constant E, sufficiently large to make the new a matrix have a nonnegative V. Adding such a constant to all elements will obviously not change the optimal x's at all and will simply change V by the added amount.

For the moment, let us suppose this has been done,[1] so that the a_{ij}'s we deal with are barred from yielding a negative V. Then the formulation (16-4) can have adjoined to it the condition $x_{m+1} \geq 0$, and it will then be recognizable as a standard problem in linear programming.

To summarize: Any game $[a_{ij}]$ that (by prior adjustment if necessary) is known to have a nonnegative value may be converted into the standard linear-programming maximum problem form:[2]

Subject to

$$\left.\begin{array}{l} -x_1a_{11} - x_2a_{21} - \cdots - x_ma_{m1} + x_{m+1}1 \leq 0 \\ \cdots\cdots\cdots\cdots\cdots\cdots\cdots\cdots \\ -x_1a_{1n} - x_2a_{2n} - \cdots - x_ma_{mn} + x_{m+1}1 \leq 0 \\ x_1 + x_2 + \cdots + x_m + x_{m+1}0 \leq 1 \\ x_1 \geq 0,\ x_2 \geq 0,\ \ldots,\ x_m \geq 0,\ x_{m+1} \geq 0 \end{array}\right\} \tag{16-5}$$

maximize

$$0x_1 + \cdots + 0x_m + 1x_{m+1} = Z$$

Thus, consider the 2×3 game matrix

$$\begin{bmatrix} 1 & -1 & 2 \\ -1 & 1 & 3 \end{bmatrix}$$

[1] If the least algebraic element of the game matrix is $\min_{i,j} a_{ij}$, we may safely set $E \geq |\min_{i,j} a_{ij}|$. Even this is too conservative a figure: $E \geq |\max_i \min_j a_{ij}|$ = maximum of row minima will do since the value of the game to Player 1 cannot be less than what he would earn by playing his best pure strategy. See J. von Neumann and Oskar Morgenstern, *Theory of Games and Economic Behavior*, 3d ed., p. 100, Princeton University Press, Princeton, N.J., 1953, for a discussion of the "majorant" and "minorant" games.

[2] It will be noted that two minor changes have been made in rewriting the constraints of (16-3). The direction of the inequalities has been reversed by changing an algebraic sign. This was done to conform to the usual convention for a maximum problem. More important, the sum of the probabilities has now been written as "less than or equal to" 1 rather than as "equal to." This too fits into the usual convention of linear programming. It is permissible to make such a change: as long as V is positive, Player 1 will never waste any of "the 100 per cent of probability granted to him," instead preferring to play whenever possible; if $V = 0$, he would be indifferent between $x_1 + \cdots + x_m = 1$ or $x_1 + \cdots + x_m < 1$, but the ratios of the x's would still be determinate and there will always be an optimal solution in which the equality is realized.

Its value can be shown to be zero, since Player 2 will never play his third strategy and what is left will be the "penny-matching" matrix of *Theory of Games.*[1] However, even if we had not been able to recognize that its V is nonnegative, we could, by noting that -1 is its least element, easily modify it by adding $E = 1$ to every element, giving us a new game matrix:

$$\begin{bmatrix} 2 & 0 & 3 \\ 0 & 2 & 4 \end{bmatrix}$$

The value Z of this new game can be found by solving the following linear-programming problem:

Subject to

$$\begin{aligned} -2x_1 + 0x_2 + x_3 &\leq 0 \\ 0x_1 - 2x_2 + x_3 &\leq 0 \\ -3x_1 - 4x_2 + x_3 &\leq 0 \\ x_1 + x_2 + 0x_3 &\leq 1 \\ x_1 \geq 0,\ x_2 \geq 0,\ x_3 &\geq 0 \end{aligned}$$

maximize

$$0x_1 + 0x_2 + 1 \cdot x_3 = Z = 1 + \text{value of original } [a_{ij}] \text{ game.}$$

Using any desired method, we can solve this and determine that

$$(x_1,x_2,x_3) = (0.5,\ 0.5,\ 1)$$

so that the value of our adjusted game is 1 and of our original game is zero.

Thus far we have ignored the optimizing behavior of Player 2 in seeking his best mixed strategy $(y_1, \ldots, y_n)$. He obviously wants to maximize $-V$ or, what is the same thing, to minimize V. But in putting ourselves into the shoes of Player 2, we could get a set of inequalities exactly like those of (16-2), except that now y takes the place of x; $-a_{ji}$ takes the place of a_{ij}; and $-V$ takes the place of V. The reader may write these down. By convention V was made nonnegative and $-V$ nonpositive. It is therefore convenient to define a new nonnegative variable $y_{n+1} = -(-V)$; and corresponding to Player 1's maximum problem formulation (16-5), we now have for Player 2 the standard linear-programming *minimum* problem:[2]

[1] Von Neumann and Morgenstern, *op. cit.*, p. 94.

[2] Note that whenever the value of the game to Player 2 is actually negative, he would rather not play at all and thereby earn zero; for this reason we require $y_1 + \cdots + y_n \geq 1$, so that he cannot fail to use all his probabilities. Of course, if $-V = 0$, the y's might exceed unity and lose their probability interpretation; their ratios would still be determinate, and we can always confine our attention to the optimal y's that add up to exactly 1.

Subject to

$$\begin{array}{rcl} -a_{11}y_1 - a_{12}y_2 - \cdots - a_{1n}y_n + 1 \cdot y_{n+1} & \geq & 0 \\ \cdots\cdots\cdots\cdots\cdots\cdots\cdots & & \\ -a_{m1}y_1 - a_{m2}y_2 - \cdots - a_{mn}y_n + 1 \cdot y_{n+1} & \geq & 0 \\ y_1 + \quad y_2 + \cdots + \quad y_n + \quad 0y_{n+1} & \geq & 1 \\ y_1 \geq 0,\ y_2 \geq 0,\ \ldots,\ y_n \geq 0,\ y_{n+1} & \geq & 0 \end{array} \tag{16-6}$$

minimize

$$0y_1 + \cdots + 0y_n + 1 \cdot y_{n+1} = Z^* \qquad \text{or} \qquad -V$$

It was earlier shown that every linear-programming problem has a dual problem. Suppose we ask for the meaning of the dual problem to (16-5), i.e., to Player 1's linear-programming problem. By writing out the dual explicitly, the reader can verify the following important fact: *The linear-programming dual to Player* 1*'s maximum problem is none other than the minimum problem of Player* 2 *expressed as a linear-programming problem; conversely, the dual to Player* 2*'s minimum problem is the original maximum problem of Player* 1.

This is a rather natural relationship. Any method of solving one of the linear-programming problems that yields information about its dual thereby automatically provides information about the other player's optimal strategies. Thus the "simplex method" applied to our previous numerical example would provide information about the solution to Player 2's problem, which is as follows:

Subject to

$$\begin{array}{rcl} -2y_1 - 0y_2 - 3y_3 + 1 \cdot y_4 & \geq & 0 \\ -0y_1 - 2y_2 - 4y_3 + 1 \cdot y_4 & \geq & 0 \\ y_1 + \quad y_2 + \quad y_3 + 0 \cdot y_4 & \geq & 1 \\ y_1 \geq 0,\ y_2 \geq 0,\ y_3 \geq 0,\ y_4 & \geq & 0 \end{array}$$

minimize

$$0y_1 + \cdots + 0y_3 + 1 \cdot y_4 = Z^*$$

It is easy to verify that (0.5, 0.5, 0, 1) is its solution.

16-3. ALTERNATIVE METHODS OF CONVERSION

This completes our discussion of the conversion of a game to a linear-programming problem. A few further remarks and references may be of interest to the curious. Dantzig gives a slightly different formulation of the same problem.[1] His Eqs. (3) follow from rewriting our relations (16-2). Noting that Dantzig's a_{ij} = our a_{ji}, we can derive Dantzig's formulation by substituting v_1, as defined in the first relation of our (16-2) into each of the following relations. Then rewriting V in terms of v_1 and

[1] In Koopmans (ed.), *Activity Analysis of Production and Allocation*, p. 331.

requiring that it be a maximum, we arrive at Dantzig's first formulation:

$$x_1a_{11} + x_2a_{21} + \cdots + x_ma_{m1} - v_1 = V$$

to be a maximum, subject to

$$\begin{array}{l}
x_1(a_{12} - a_{11}) + x_2(a_{22} - a_{21}) + \cdots + x_m(a_{m2} - a_{m1}) - v_2 + v_1 \geq 0 \\
\cdots\cdots\cdots\cdots\cdots\cdots\cdots\cdots\cdots\cdots \\
x_1(a_{1n} - a_{11}) + x_2(a_{2n} - a_{21}) + \cdots + x_m(a_{mn} - a_{m1}) - v_n + v_1 \geq 0 \\
x_1 + x_2 + \cdots + x_m = 1 \\
x_1 \geq 0, x_2 \geq 0, \ldots, x_m \geq 0, v_1 \geq 0, v_2 \geq 0, \ldots, v_n \geq 0
\end{array} \tag{16-7}$$

As Dantzig recognizes, this introduces a superficial asymmetry into the problem since the first relation of (16-2) is treated differently from the others. By using the $\bar{a}_{i.}$'s and $\bar{v}$ defined in (16-3), this superficial asymmetry may be avoided, and we have

$$x_1\bar{a}_{1.} + x_2\bar{a}_{2.} + \cdots + x_m\bar{a}_{m.} - \bar{v} = V$$

to be a maximum, subject to

$$\begin{array}{l}
-v_1 = -x_1(a_{11} - \bar{a}_{1.}) - x_2(a_{21} - \bar{a}_{2.}) - \cdots - x_m(a_{m1} - \bar{a}_{m.}) + \bar{v} \leq 0 \\
\cdots\cdots\cdots\cdots\cdots\cdots\cdots\cdots\cdots\cdots \\
-v_n = -x_1(a_{1n} - \bar{a}_{1.}) - x_2(a_{2n} - \bar{a}_{2.}) - \cdots - x_m(a_{mn} - \bar{a}_{m.}) + \bar{v} \leq 0 \\
x_1 + x_2 + \cdots + x_m = 1 \\
x_1 \geq 0, x_2 \geq 0, \ldots, x_m \geq 0, \bar{v} \geq 0
\end{array} \tag{16-8}$$

The interested reader can verify that (16-8) follows by subtracting (16-3) from each of the relations of (16-2) and rearranging terms.

Both (16-7) and (16-8) have the disadvantage that Player 1's dual linear-programming problem does not directly link up with Player 2's minimum problem. For this reason, Dantzig's second, symmetric formulation of the problem is of interest. So long as the value of the game to Player 1 is assured to be definitely positive, we can divide all the constraints of our (16-5) by x_{m+1}. This will lead to a linear-programming problem in the new variables

$$(\bar{x}_1, \ldots, \bar{x}_m, x_{m+1}^{-1}) = \left(\frac{x_1}{x_{m+1}}, \ldots, \frac{x_m}{x_{m+1}}, \frac{1}{x_{m+1}}\right)$$

The reader can verify that minimizing $1/x_{m+1}$ subject to these constraints is equivalent to solving our game problem and leads to the same relations as shown on the bottom of page 331 of *Activity Analysis*,[1] or to the following:

Subject to

$$\begin{aligned} \bar{x}_1 a_{11} + \cdots + \bar{x}_m a_{m1} &\geq 1 \\ \cdots\cdots\cdots&\cdots\cdots \\ \bar{x}_1 a_{1n} + \cdots + \bar{x}_m a_{mn} &\geq 1 \\ \bar{x}_1 \geq 0, \ \ldots, \ \bar{x}_m &\geq 0 \end{aligned} \tag{16-9}$$

minimize

$$\bar{x}_1 + \cdots + \bar{x}_m = \frac{1}{V}$$

Formulations (16-5) or (16-6) are subject to the inessential restriction that the value of the game is assumed to be nonnegative.[2] Formulations (16-7) and (16-8) are free of this restriction. Formulation (16-9) depends upon the value of the game being assumed to be nonzero as well as nonnegative. This means that the important class of skew-symmetric games must be adjusted so as to increase their V from zero; it also means that for small V, formulation (16-9) will be numerically sensitive and inaccurate.

A number of further formulations in terms of linear programming would be possible for the game problem, but may be left to the interested specialist.

16-4. CONVERTING A LINEAR-PROGRAMMING PROBLEM INTO A GAME

We can now turn to the other half of the relation between linear programming and games. Since a great deal of ingenuity has been devoted to solving two-person, zero-sum games, it is convenient to know that all the theorems in this field are at the disposal of anyone confronted with a linear-programming problem. For such a problem can *always* be converted into a game.

[1] Using this formulation, Dorfman (*Activity Analysis of Production and Allocation*, chap. 22) has worked by the simplex method a 5 × 6 numerical problem. On p. 335, he and H. Rubin have noted how information about the dual problem and about Player 2's optimal strategies can be gleaned from these computations. This is confirmed by the dual relation of (16-5) to (16-6) and by the theory of the simplex method.

[2] Instead of adding E to all a_{ij} so as to rule out $V < 0$, we can alternatively in (16-5) and (16-6) define $x_{m+1} = V + E = y_{n+1}$, where E is large enough to assure $V + E > 0$. This will put E rather than zeros on the right of the constraints of (16-5) and (16-6) and will require us to maximize $-E + x_{m+1} = -Ex_1 - \cdots - Ex_m + x_{m+1}$ in (16-5) and to minimize $-E + y_{n+1} = -Ey_1 - \cdots - Ey_n + y_{n+1}$ in (16-6).

There is more than one way of performing this conversion. However, the most convenient method seems to be that of converting our programming problem into a *skew-symmetric*, or "symmetric," game.[1] Though the symmetric game appears to have more rows and columns, it has the advantage of automatically yielding the solution to the dual linear-programming problem as well, and it contains so many zeros that its larger size is not much of a problem.

Consider the standard linear-programming problem, as follows:

Subject to

$$\begin{aligned} a_{11}x_1 + a_{12}x_2 + \cdots + a_{1n}x_n &\leq b_1 \\ \cdots\cdots\cdots\cdots\cdots\cdots\cdots\cdots\cdots\cdots\cdot \\ a_{m1}x_1 + a_{m2}x_2 + \cdots + a_{mn}x_n &\leq b_m \\ x_1 \geq 0, x_2 \geq 0, \ldots, x_n &\geq 0 \end{aligned} \tag{16-10}$$

maximize

$$c_1x_1 + c_2x_2 + \cdots + c_nx_n = Z$$

Recall that its dual can be written as follows:

Subject to

$$\begin{aligned} a_{11}y_1 + a_{21}y_2 + \cdots + a_{m1}y_m &\geq c_1 \\ \cdots\cdots\cdots\cdots\cdots\cdots\cdots\cdots\cdots\cdots\cdot \\ a_{1n}y_1 + a_{2n}y_2 + \cdots + a_{mn}y_m &\geq c_n \\ y_1 \geq 0, y_2 \geq 0, \ldots, y_m &\geq 0 \end{aligned} \tag{16-11}$$

minimize

$$b_1y_1 + b_2y_2 + \cdots + b_ny_m = Z^*$$

There is no reason why we shouldn't combine these two n-variable and m-variable problems into a third super $(m + n) \times (m + n)$ variable problem. Suppose we try to select x's *and* y's that maximize $Z - Z^*$ subject to *all* the inequalities of both problems. We have already seen[2] that the basic duality principle guarantees that *for optimal x's and y's* the Z of our original problem and the Z^* of its dual must be exactly equal. So we happen to know that the maximum $Z - Z^*$ of our new problem will eventually have to be zero for our final solution, even though it will be negative for all feasible x's and y's that are nonoptimal. Note that our super problem does contain both of the dual problems inside it,

[1] The specialist may be referred to *Activity Analysis of Production and Allocation*, p. 327, Theorem 6 of Gale, Kuhn, and Tucker, for a discussion of how an $m \times n$ programming problem can be transformed into an $(m + 1) \times (n + 1)$ game of *unknown* value. In recent years the advantage of the skew-symmetric game formulation, whose value is known to be zero, has become apparent. See p. 372, Theorem 7, and p. 334, Eqs. (14), of Dantzig.

[2] Chap. 4, pages 100–104.

since $Z - Z^*$ will be a maximum if, and only if, Z and $-Z^*$ are separately maximized. This is because the third problem can be decomposed into its two independent parts, since one set of constraints involves only x's, and the other set only y's.

We already know the maximum-value answer to our new super problem. It is zero. Our real point in formulating the problem, however, is to introduce a special kind of symmetry or skew symmetry into the problem; this skew symmetry results from adding or adjoining the dual problem to the original problem. We do not as yet know the optimal x's or y's and solving our super problem will get us both of them.

We can now write down explicitly the new super problem, which, it is to be emphasized, is a problem in linear programming; as yet, we have not made any transposition into the form of a game. Our super problem, in standard maximum form, looks as follows:

Subject to

$$\begin{array}{l} 0y_1 + \cdots + 0y_m + a_{11}x_1 + \cdots + a_{1n}x_n \leq b_1 \\ \cdots\cdots\cdots\cdots\cdots\cdots\cdots\cdots \\ 0y_1 + \cdots + 0y_m + a_{m1}x_1 + \cdots + a_{mn}x_n \leq b_m \\ -a_{11}y_1 - \cdots - a_{m1}y_m + 0x_1 + \cdots + 0x_n \leq -c_1 \\ \cdots\cdots\cdots\cdots\cdots\cdots\cdots\cdots \\ -a_{1n}y_1 - \cdots - a_{mn}y_m + 0x_1 + \cdots + 0x_n \leq -c_n \\ y_1 \geq 0, \ldots, y_m \geq 0, x_1 \geq 0, \ldots, x_n \geq 0 \end{array} \tag{16-12}$$

maximize

$$-b_1y_1 - \cdots - b_my_m + c_1x_1 + \cdots + c_nx_n = Z - Z^*$$

or what is the same thing, since we know the optimal $Z - Z^* = 0$, let us replace the maximum statement of the last line by the equivalent requirement,

$$-b_1y_1 - \cdots - b_my_m + c_1x_1 + \cdots + c_nx_n \geq 0 \tag{16-12a}$$

where the $>$ sign has been redundantly inserted in the full knowledge that no feasible solution will ever require its presence. The complete set of Eqs. (16-12), including (16-12a), define the optimum solution (x, y, $Z - Z^*$) of our original problem, of its dual, and of the new all-inclusive super problem.

As we look at this maximum problem, we instantly note its skew-symmetric (or perhaps antisymmetric would be the better word) form. The a_{ij}'s appear twice, once with a plus sign and once in transposed form a_{ji} with a minus sign. The b's and c's both appear twice, once in the vertical and once in the horizontal, and with opposite algebraic signs. The negative signs enter very naturally since we have, so to speak, converted the dual *minimum* problem into a maximum problem by changing signs; moreover, to keep the constraints in standard form, with the sign $\leq$ being used rather than $\geq$, it was necessary to introduce negative signs.

Our super problem is a standard maximum problem. Does *it* have a dual? Of course it must. We have simply to transpose its coefficients in the usual manner and formulate the related minimum problem. But in the case of our super problem, because of its special antisymmetry and zeros, what do we get? We get, as the reader can immediately verify, essentially the same super problem back again—the same a_{ij}'s and a_{ji}'s. *The dual to our super problem is itself.* The mathematician would say that this is a "self-adjoint," or "self-dual," case and would expect our skew-symmetric formulation to lead to this type of pattern.[1]

In a game all columns look alike, so it is natural to ask: How can we treat the right-hand b and c coefficients of the programming problem so that they will look like ordinary a's? To aid in this, we go on to ask: What happens to the solution to any linear-programming problem if we multiply the right-hand coefficients of its constraints by the positive constant z?. Reflection shows that we can then multiply any previous optimal solution by the same z to get a new optimal solution. And the quantity to be maximized, Z, will also be multiplied by the same constant z. All this is obvious once we reflect on the constant-returns-to-scale property of our linear relations: thus, it must cost twice as much to buy an optimal diet that requires twice the nutrients, and the relative proportions of the optimal diet will not be disturbed.

It is convenient to apply such a positive scale factor to our super problem. If the right-hand coefficients of the constraints of (16-12) are all multiplied by z, then the best $Z - Z^*$ will be multiplied by z, and because the quantity maximized, $Z - Z^*$, previously equaled zero, it will still be zero. However, any previously optimal solution $(y_1, \ldots, y_m, x_1, \ldots, x_n)$ will now be equal to $(zy_1, \ldots, zy_m, zx_1, \ldots, zx_n)$. If we like, we can now regard z as one of our variables just like the x's and y's: thirdly, we can transfer the right-hand coefficients of the constraints over to the left-hand side. The only advantage in doing this lies in the fact that a game is always written without any right-hand coefficients.

[1] We could, of course, construct a super-super problem by combining together our super problem *and* its identical-twin dual. But there is absolutely no point in doing so since it can be easily verified that this simply leads to a copying down *twice* of our x and y variables and of the a_{ij}'s and a_{ji}'s. A super-super-super problem would involve even more pointless copying.

How many new variables do we now have? Clearly we have $n + m + 1$ new variables, namely, $(zy_1, \ldots, zy_m, zx_1, \ldots, zx_n, z)$, and the full statement of our super problem, as given by (16-12) and (16-12a), can be written

$$\begin{array}{rcl} 0zy_1 + \cdots + 0zy_m + a_{11}zx_1 + \cdots + a_{1n}zx_n - b_1z &\leq& 0 \\ \cdots\cdots\cdots\cdots\cdots\cdots\cdots\cdots && \\ 0zy_1 + \ldots + 0zy_m + a_{m1}zx_1 + \cdots + a_{mn}zx_n - b_mz &\leq& 0 \\ -a_{11}zy_1 - \cdots - a_{m1}zy_m + 0zx_1 + \cdots + 0zx_n + c_1z &\leq& 0 \\ \cdots\cdots\cdots\cdots\cdots\cdots\cdots\cdots && \\ -a_{1n}zy_1 - \cdots - a_{mn}zy_m + 0zx_1 + \cdots + 0zx_n + c_nz &\leq& 0 \\ Z^* - Z = b_1zy_1 + \cdots + b_mzy_m - c_1zx_1 - \cdots - c_nzx_n + 0z &\leq& 0 \\ zy_1 \geq 0, \ldots, zy_m \geq 0, zx_1 \geq 0, \ldots, zx_n \geq 0, z &>& 0 \end{array} \tag{16-13}$$

and where only the ratios zy_i/z and zx_j/z are significant.

The connection of our super problem with a skew-symmetric game can now finally be made. If we look closely at the inequalities defining our super problem, what do we see? We see that these same inequalities would arise for Player 2's optimal mixed strategy if he were playing a game, which had a pay-off matrix to Player 1 involving the a_{ij}'s, $-a_{ji}$'s, b's, and c's of our original problem.

Thus, examine the skew-symmetric game with pay-off matrix from Player 2 to Player 1 of

$$\begin{array}{cccccccc} 0 & \ldots & 0 & a_{11} & \ldots & a_{1n} & -b_1 \\ \cdots & & & & & & \cdots \\ 0 & \ldots & 0 & a_{m1} & \ldots & a_{mn} & -b_m \\ -a_{11} & \ldots & -a_{m1} & 0 & \ldots & 0 & +c_1 \\ \cdots & & & & & & \cdots \\ -a_{n1} & \ldots & -a_{mn} & 0 & \ldots & 0 & +c_n \\ b_1 & \ldots & b_m & -c_1 & \ldots & -c_n & 0 \end{array} \tag{16-14}$$

and suppose that Player 2 must find his optimal mixed strategy for the probabilities of playing the different columns, $(Y_1, \ldots, Y_n; Y_{n+1}, \ldots, Y_{n+m}; Y_{n+m+1})$. Now we know from the skew symmetry or symmetry of the game that the value of the game to Player 1 is 0, since neither player has any advantage denied the other. Consequently, Player 2's optimal mixed strategy Y_j must be such as to result in Player 1's receiving not more than (the value of the game or) zero for any pure strategy that Player 1 plays against Player 2's optimal Y_j. If we write down this condition for each of Player 1's pure strategies, we have equations essentially like those of (16-1), but with $V = 0$ and Y's replacing the x's.

These conditions become

$$\begin{array}{l}
0\ Y_1 + \cdots + 0\ Y_m + a_{11} Y_{m+1} + \cdots + a_{1n} Y_{m+n} - b_1 Y_{m+n+1} \leq 0 \\
\cdots\cdots\cdots\cdots\cdots\cdots\cdots\cdots\cdots\cdots\cdots\cdots \\
0\ Y_1 + \cdots + 0\ Y_m + a_{m1} Y_{m+1} + \cdots + a_{mn} Y_{m+n} - b_m Y_{m+n+1} \leq 0 \\
-a_{11} Y_1 - \cdots - a_{m1} Y_m + 0\ Y_{m+1} + \cdots + 0\ Y_{m+n} + c_1 Y_{m+n+1} \leq 0 \\
\cdots\cdots\cdots\cdots\cdots\cdots\cdots\cdots\cdots\cdots\cdots\cdots \\
-a_{1n} Y_1 - \cdots - a_{mn} Y_m + 0\ Y_{m+1} + \cdots + 0\ Y_{m+n} + c_n Y_{m+n+1} \leq 0 \\
b_1\ Y_1 + \cdots + b_m\ Y_m - c_1\ Y_{m+1} - \cdots - c_n\ Y_{m+n} + 0 \leq 0 \\
Y_1 + \cdots + Y_m + Y_{m+1} + \cdots + Y_{m+n} + Y_{m+n+1} = 1 \\
Y_1 \geq 0, \ldots, Y_m \geq 0, Y_{m+1} \geq 0, \ldots, Y_{m+n+1} \geq 0
\end{array} \tag{16-14a}$$

Except for the inessential normalization condition

$$Y_1 + \cdots + Y_{m+n+1} = 1$$

and the condition $Y_{n+m+1} \geq 0$ rather than > 0, the reader can verify that the game inequalities of (16-14*a*) are identical with the super-linear-programming problem's inequalities of (16-13), provided we identify the Y_j's with our previous x's, y's, and z. Thus,

$$Y_1 = zy_1, \ldots, Y_m = zy_m, Y_{m+1} = zx_1, \ldots, Y_{m+n} = zx_n, Y_{m+n+1} = z$$

The reader should carefully verify this basic identity between the game and programming inequalities.

A recipe for solving our original linear-programming problem can now be given: (1) Using the a's, b's, and c's of the programming problem, set up the associated skew-symmetric game of the type (16-14). (2) By any known device, solve this game for an optimal mixed strategy $(Y_1, \ldots, Y_{m+n+1})$ or $(zy_1, \ldots, zx_1, \ldots, z)$, where $Y_{m+n+1} \neq 0$. If the original linear programming and its dual have optimal x's and y's that are all *finite*, then the final component $Y_{m+n+1} = z$ will *never* equal zero for any optimal strategy. But if the set of optimal x's and y's is unbounded, then there will always be an extreme optimal game strategy whose last component is zero. For example, consider a problem like $Z = x_1 - x_2$ to be a maximum subject to $x_1 - x_2 \leq -1$ and $x_1 \geq 0$, $x_2 \geq 0$; its optimum is given by the dual variable $y_1 = 1$ and any nonnegative x's such that $x_1 - x_2 = -1$. The associated super skew-symmetric game matrix is verified to be

$$\begin{bmatrix} 0 & 1 & -1 & 1 \\ -1 & 0 & 0 & 1 \\ 1 & 0 & 0 & -1 \\ -1 & -1 & 1 & 0 \end{bmatrix}$$

and this has an optimal mixed strategy $(zy_1, zx_1, zx_2, z) = (\frac{1}{3}, 0, \frac{1}{3}, \frac{1}{3})$. Hence, $y_1 = 1$, $x_1 = 0$, $x_2 = 1$ is an optimum solution. However, $(0, \frac{1}{2}, \frac{1}{2}, 0)$ is also an optimal mixed strategy for the game. But since its last component z vanishes, it is an extraneous mixed strategy and does *not* solve our programming problem. Of course, any positive weighted average of the above two strategies will also be optimal; hence $[\lambda/3, (1-\lambda)/2, (1-\lambda)/2 + \lambda/3, \lambda/3]$ is an optimal game strategy and $[y, x_1, x_2] = \{1, \frac{3}{2}[(1-\lambda)/\lambda], \frac{1}{2}[(3-\lambda)/\lambda]\}$ for any $0 < \lambda \leq 1$ are legitimate solutions to the programming problem. All the solutions of our original problem are generated by varying λ through its admissible range. Gale, Kuhn, and Tucker[1] have noted that for any "extraneous" optimal game strategy $(Y_1^0, \ldots, Y_m^0, Y_{m+1}^0 \ldots, Y_{n+m}^0, 0)$, it must follow that

$$b_1Y_1^0 + \cdots + b_nY_m^0 = 0 = c_1Y_{m+1}^0 + \cdots + c_mY_{m+n}^0$$

Hence, in problems such as the nutrition problem, in which the b's and c's are all one-signed, extraneous solutions will not arise. If the a's are all one-signed, the same can be shown to be true. It may be remarked that for any extraneous strategy like that just written, provided $Y_1^0 + \cdots + Y_m^0 > 0 < Y_{m+1}^0 + \cdots + Y_{m+n}^0$, then $[\lambda Y_1^0, \ldots, \lambda Y_m^0, \ldots, (1-\lambda)Y_{m+n}^0]$ is also an extraneous optimal strategy, $0 \leq \lambda \leq 1$. (There will always exist at least one such game solution if the programming problem has a solution.) (3) Then $Y_1/Y_{m+n+1} = y_1, \ldots, Y_m/Y_{m+n+1} = y_m, Y_{m+1}/Y_{m+n+1} = x_1, \ldots, Y_{m+n}/Y_{m+n+1} = x_n$ gives the optimum solutions to the linear-programming super problem and to the dual problems, and z can be computed from $c_1x_1 + \cdots + c_nx_n$ or $b_1y_1 + \cdots + b_my_m$.

16-5. FINAL REVIEW AND ELABORATION[2]

This completes the discussion of how to convert a linear-programming problem into a skew-symmetric game. A good review of this procedure, and of the earlier described reverse problem of converting a game into a linear-programming problem, is provided by the following: (1) start with any numerical game matrix; (2) convert it into a linear-programming problem, as described in the first part of this chapter; (3) then convert this linear-programming problem into a skew-symmetric game by the methods of the second half. Note that when this has been done, we not only end up with a review of both procedures—but in addition we have worked through one rigorous way of converting *any* game into a skew-symmetric game.

We begin with our earlier 2×3 matrix

$$\begin{bmatrix} 2 & 0 & 3 \\ 0 & 2 & 4 \end{bmatrix}$$

[1] *Activity Analysis of Production and Allocation*, p. 328.

[2] This section can be skipped.

known to have a positive value to Player 1. It is converted into the linear-programming problem with coefficients as shown in Sec. 16-2. The skew-symmetric game matrix (16-14) associated with the super problem that includes the dual has a matrix of coefficients, which (the reader should verify) is to be written down as follows:

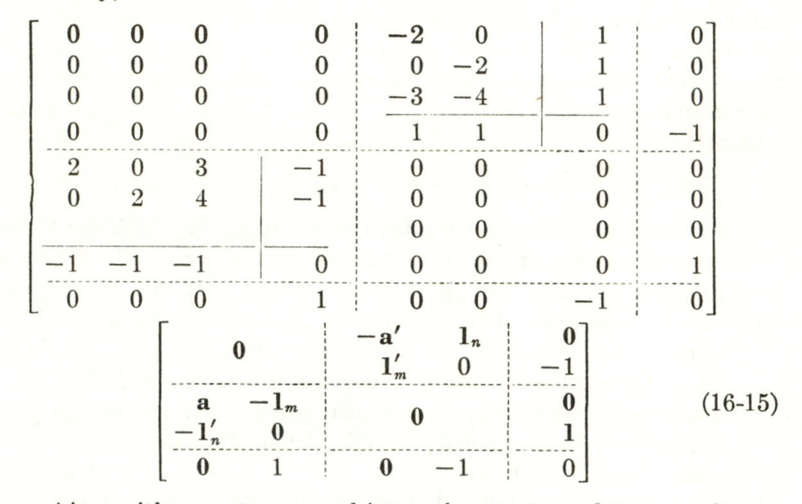

where $\mathbf{a}'$ is $\mathbf{a}$ with rows transposed into columns, $\mathbf{1}_n$ and $\mathbf{1}_m$ are columns of n and m 1's, $\mathbf{1}'_n$ and $\mathbf{1}'_m$ are rows of n and m 1's. If we note that $\mathbf{a}$ is bordered by -1's, the pattern is easy to remember. The optimal solution $(Y_1,Y_2,Y_3;Y_4;Y_{4+1},Y_{4+2};Y_{4+3};Y_{7+1})$ can be verified to be (0.1, 0.1, 0; 0.2; 0.1, 0.1; 0.2; 0.2); dividing through by the final element, we find our optimal strategies for Players 2 and 1, and the value of the game can all be written down as $(y_1,y_2,y_3;V;x_1,x_2;V;1) = (0.5, 0.5, 0; 1; 0.5, 0.5; 1; 1)$, which the reader can verify to be the correct solution by simply remembering the penny-matching origin of the problem.

In the general case, so long as $V > 0$, it can be shown that

$$[Y_1, \ldots, Y_n;Y_{n+1};Y_{(n+1)+1}, \ldots, Y_{(n+1)+m};Y_{(n+1)+(m+1)};Y_{(n+1)+(m+1)+1}] = (zy_1, \ldots, zy_n;zV;zx_1, \ldots, zx_m;zV;z)$$

where $z = 1/(3 + 2V)$. If $V = 0$, the same formula yields a solution; but the extraneous solution $(0.5y_1, \ldots, 0.5y_n; 0; 0, \ldots, 0; 0; 0.5)$ will also arise; and any positive weighted average of these two different solutions will also be a solution. If $V < 0$, only the above extraneous solution will occur and only the optimal y strategies will be computed, Player 1's x's being unobservable by this method.

Gale, Kuhn, and Tucker and George Brown and von Neumann[1] have discussed various other closely related ways of converting *any* game into

[1] D. Gale, H. Kuhn, and A. W. Tucker, "Linear Programming and the Theory of Games," in T. C. Koopmans (ed.), *Activity Analysis of Production and Allocation;*

skew-symmetric form, a form which is seen to involve no loss of generality and which has many expositional and computational advantages. One method, due to von Neumann, involves having each player simultaneously play the game $[a_{ij}]$ as it appeared to Player 1 and also simultaneously play the game $[-a_{ji}]$ as it originally appeared to Player 2. Thus, we may think of Player 1 as originally being called "white" (e.g., having the *first* move in a game of chess); similarly Player 2, who faces the pay-off matrix $[-a_{ji}]$ can be thought of as playing "black" (e.g., having the *second* move in chess). Von Neumann introduces symmetry into the problem by having each player simultaneously name a pair of strategies and play both sides of the game at once; thus, Player 1 will name strategy i for white and j for black; Player 2 will name strategy k for his black and q for his white; then the pay-off from Player 2 to Player 1 will be defined as $a_{ik} - a_{jq}$. Since each player has a choice of mn pairs of strategies, this von Neumann method converts an $m \times n$ game into a very much larger one, namely, $mn \times mn$, but in actual computation, many fewer variables can handle the situation.

These same authors discuss a second method for making a game skew-symmetric. It is essentially[1] that given at the beginning of this section. Both methods can be given a common-sense, but involved, heuristic interpretation. For the method given above, the reasoning is as follows:

I as Player 1 have the choice of picking a white strategy exactly as in a normal game; you as Player 2 have the usual choice of picking one of your black strategies. But in addition, I have the option of naming any

---

George Brown and J. von Neumann, "Solutions of Games by Differential Equations," in H. Kuhn and A. W. Tucker (eds.), *Contributions to the Theory of Games*, Annals of Mathematics Study 24, Princeton University Press, Princeton, N.J., 1950.

[1] After we have converted the game matrix a into the super-linear-programming problem, we can if we wish replace the normalization conditions $x_1 + \cdots + x_m \leq 1 \leq y_1 + \cdots + y_n$ by the single condition $-y_1 - \cdots - y_n + x_1 + \cdots + x_m \leq 0$. This will suggest a slight variant to (16-15)'s conversion of a into a skew-symmetric form, namely,

$$\begin{bmatrix} \mathbf{0} & -\mathbf{a}' & \mathbf{1}_n \\ \mathbf{a} & \mathbf{0} & -\mathbf{1}_m \\ -\mathbf{1}_n' & \mathbf{1}_m & 0 \end{bmatrix}$$

It can be verified that the optimal strategies for this new game (with the variables newly numbered) have the property

$$(Y_1, \ldots, Y_n; Y_{n+1}, \ldots, Y_{n+m}; Y_{n+m+1}) = (ty_1, \ldots, ty_n; tx_1, \ldots, tx_m; tV)$$

where $t = 1/(2 + V)$, provided that $V \geq 0$. Hence, $2Y_{n+m+1}/(1 - Y_{n+m+1}) = V$. This last formula is valid for the case $V = 0$, but in this case the full set of optimal strategies is given by

$$(Y_1, \ldots, Y_n; Y_{n+1}, \ldots, Y_{n+m}; Y_{n+m+1}) \\ = [\lambda y_1, \ldots, \lambda y_n; (1 - \lambda)x_1, \ldots, (1 - \lambda)x_1, \ldots, (1 - \lambda)x_m; 0]$$

where $\frac{1}{2} \leq \lambda \leq 1$. If $V < 0$, the optimal strategies for the new game will be $(y_1, \ldots, y_n; 0, \ldots, 0; 0)$, which yields only the optimal set of black strategies.

black strategy; and you can name any white strategy if you wish. However, if the value of the game is assumed nonnegative to white, what is there to keep both of us from always naming white strategies, thereby leading always to no-contest and zero pay-off? Nothing, until we add the following feature. Two new strategies are available to me: (1) I can make the guess that you will name a white strategy rather than a black; or (2) I can make the guess that you will name a black strategy rather than a white. If I make such a guess and it is right, you pay me $1; if my guess is wrong, I pay you $1. If I guess that you will play a black or white strategy and you name neither, then there will be zero pay-off. But now what is there to keep both of us from always making such guesses and avoiding ever naming a black or white strategy? To penalize such behavior, I am permitted one last option: (3) I can make the prediction that you will make the guess that I will play white. If you do make that guess, my prediction is verified and I receive $1. If you make the opposite guess—that I will play black—my prediction costs me $1; if you play a black or white strategy or indulge in a similar (3) prediction about my guesses, there is zero pay-off.

We have now defined a symmetrical set of strategies for each player. What then must happen? It turns out that I will play a black and a white strategy with equal frequency; moreover, I must play the different white and black strategies with relative frequencies as dictated by their optimal mixed strategies. It further turns out that as often as I play black or white, I must make a final (3) prediction about your guesses. The remainder of the time I will divide equally between guessing (1) that you will play white rather than black, or (2) that you will play black rather than white. The frequency with which I make such guesses will be directly proportionate to V, the value of the game to white playing against black: when the values of the game equal $1, the amount wagered on guesses 1 and 2 and predictions 3, the five different modes of behavior will be equally likely.

All the above is very complicated. However, it is logically equivalent to the mathematical principles involved in converting a game to symmetric form and provides a concrete interpretation of the resultant mixed strategies.

16-6. CONVERTING SPECIAL PROGRAMMING PROBLEMS INTO NONSYMMETRIC GAMES

Often in linear programming we encounter problems with special properties: (1) all the right-hand coefficients $[b_i]$ in the constraints are of the same algebraic sign; (2) all the $[c_j]$ coefficients in the linear expression to be maximized (or minimized) are of the same algebraic sign; and (3) all the $[a_{ij}]$ coefficients are of the same algebraic sign.

The simple comparative-advantage problem of international trade is of this form when we assume positive amounts of all limited resources and all international prices positive. The simple minimum-diet problem is of this special type, since all the minimum requirements of nutrients are positive and all food prices are positive.[1] Still other examples could be given.

The only problems of any interest with the above special properties can be put into the minimum form as follows:

Subject to

$$y_1a_{11} + \cdots + y_ma_{m1} \geq c_1$$
$$\cdots\cdots\cdots\cdots\cdots$$
$$y_1a_{1n} + \cdots + y_ma_{mn} \geq c_n \qquad (16\text{-}16)$$

minimize

$$y_1b_1 + \cdots + y_mb_m$$
$$y_1 \geq 0, \ldots, y_m \geq 0$$

where $c_i \geq 0$, $b_j \geq 0$, $a_{ij} \geq 0$.

Clearly, if any $c_i = 0$, we can remove the ith row, ignoring it as a constraint because any nonnegative x's will automatically satisfy it. Similarly if any $b_j = 0$, we can look in the jth column for any rows with nonzero a_{ij}. Each such row can be ignored since by setting y_j = maximum for such rows of $[c_i/a_{ij}]$, we can satisfy those constraints at zero cost, and having satisfied them can proceed to ignore them in computing the optimal remaining y's.

It follows that we can imagine our problem as having only b's and c's that are definitely positive, with all the rows and columns corresponding to zero $[c_i]$ or $[b_j]$ having been removed from the problem matrix.[2]

For the remainder of this section, therefore, we consider problems of the form (16-16) but with $b_i > 0$, $c_i > 0$, and no restriction on the a_{ij} except that we must have a feasible problem with finite $Z = Z^*$.

[1] The case of zero requirements or zero prices can be ignored in the simple diet problem because of the fact that the a_{ij} loadings of nutrient per unit of food are assumed to be nonnegative. This means that any food of zero price can be assumed to have been bought until the nutrient requirements have been fulfilled for any nutrient that it positively contains. And any nutrient for which we have zero further requirements can always be ignored. We should note, however, that changing the simple-diet problem to one in which the prescribed requirements are *both* minimum and maximum will lead to a problem in which the b's or c's do *not* agree in sign.

[2] The dual variable x_i corresponding to any $c_i = 0$, must of course be zero since the constraint in question does not bind. This contrasts with the dual variable y_j corresponding to any $b_j = 0$ being equal to max $(c_1/a_{j1}, c_2/a_{j2}, \ldots, c_n/a_{jn})$, it being understood that no a_{ji} in the parenthesis can be zero.

It may be noted that if we may permit some a_{ij}'s to be negative, then we cannot so easily rule out the case of zero b's and c's. So if we do permit some a_{ij}'s to be negative, we had better assume from the beginning that the b's and c's are strictly positive. We had also better be sure that the negative a_{ij}'s are not such as to lead to an infinite Z and an infeasible dual.

We rewrite (16-16) dividing each row i by c_i and replace each y_j by new variables $u_j = b_j y_j$ to get the following:

Subject to

$$\left.\begin{array}{ll} & u_1 A_{11} + \cdots + u_m A_{m1} \geq 1 \\ & \cdots\cdots\cdots\cdots\cdots \\ & u_1 A_{1n} + \cdots + u_m A_{mn} \geq 1 \\ \text{minimize} & \\ & Z = u_1 + \cdots + u_m \\ \text{or maximize} & \\ & V = \dfrac{1}{u_1 + \cdots + u_m} \\ & u_1 \geq 0, \ldots, u_m \geq 0 \\ \text{where} & A_{ij} = \dfrac{a_{ij}}{c_i b_j} \end{array}\right\} \qquad (16\text{-}17)$$

But note that, except for change of notation, (16-17) is precisely of the form (16-9) which Dantzig and others had used in converting a positive-valued game into a programming problem. It follows that we can solve the game

$$\begin{bmatrix} A_{11} & \ldots & A_{1n} \\ \cdots & \cdots & \cdots \\ A_{m1} & \ldots & A_{mn} \end{bmatrix} = \left[\frac{a_{ij}}{c_i b_j}\right]$$

for Player 1's best strategies $[U_1, \ldots, U_m]$ and value $V > 0$. Then (16-17) has for its solution $(u_1, \ldots, u_m, Z) = (U_1 V, \ldots, U_m V, V^{-1})$. To solve the dual of our original problem, we set

$$(x_1, \ldots, x_n, Z^*) = (W_1 V, \ldots, W_n V, V^{-1})$$

where the W's are Player 2's optimal strategies.[1]

The above method will not work if the b and c coefficients do not satisfy special conditions. Occasionally, but very rarely, we may have a priori knowledge that will permit us to add terms to both sides of the constraints and enable us to define new b and c coefficients with the desired properties. But such artifices cannot, in general, be prescribed.

Of course, by chance, we might encounter still other linear-programming problems that can be converted directly into an $m \times n$ game. Thus, suppose the program happened to have the special zeros of the matrix in (16-5) or (16-5a). Then, reversing the discussion of (16-5), we could reduce these programming problems to a game. But such lucky encounters will be rare.

[1] G. Morton, "Notes on Linear Programming," *Economica*, pp. 408–409, November, 1951, gives a "Man-versus-nature" interpretation to this reduction of the diet problem to a game. Von Neumann has also given this reduction of an $m \times n$ programming problem with positive b's and c's to an $m \times n$ game with positive values.

Appendix A

Chance, Utility, and Game Theory

In Chap. 14 we saw that strategies that lead to probability distributions rather than to definite numerical pay-offs are essential to the theory of games. Such strategies are fundamental in the solution of nonstrictly determined games, and they turn up elsewhere as well. It is thus necessary for players to be able to evaluate the desirabilities of probability distributions just as they evaluate the desirabilities of definite pay-offs. For this purpose game theory adopts the following postulate: In evaluating the desirabilities of probability distributions of pay-offs players should pay attention exclusively to the expected values of those probability distributions. We do not have to labor the point that the solution of nonstrictly determined games by the use of mixed strategies, the major achievement of game theory, rests squarely on this postulate.

At the same time, this assumption goes against the grain. Everyone knows that there is a lot more to a probability distribution than its expected value, and it seems irresistible intuitively that the reasonable player should take into account other characteristics such as the dispersion, skewness, etc. The purpose of this appendix is to sketch a justification of the postulate.

We can get to the heart of the issue by considering it in its baldest form. Suppose a choice is to be made between these two outcomes: (1) A 50-50 chance of winning or losing \$10; (2) a 50-50 chance of winning or losing \$10,000. It would be presumptuous for von Neumann and Morgenstern or anyone else to maintain that because these two outcomes have the same monetary expected value a reasonable man should be indifferent between them, and von Neumann and Morgenstern do *not* maintain this. They would agree that a man may quite reasonably prefer outcome 1 if he is more fearful of losing so great a sum as \$10,000 than he is desirous of gaining it. What they do maintain is that a reasonable man need pay attention only to the *expected value of the utilities* associated with the various outcomes, the so-called "moral expectation." To bring out the contrast, let $U(x)$ denote the "utility" of winning x dollars. Then, in our example, the moral expectation of outcome 1 is

$\frac{1}{2}U(-10) + \frac{1}{2}U(10)$, and the moral expectation of outcome 2 is $\frac{1}{2}U(-10{,}000) + \frac{1}{2}U(10{,}000)$. For comparison, the actuarial expectations of the two outcomes are

$$\tfrac{1}{2}(-10) + \tfrac{1}{2}(10) = \tfrac{1}{2}(-10{,}000) + \tfrac{1}{2}(10{,}000) = 0$$

The equality of the actuarial expectations does not preclude the possibility that

$$\tfrac{1}{2}U(-10) + \tfrac{1}{2}U(10) > \tfrac{1}{2}U(-10{,}000) + \tfrac{1}{2}U(10{,}000)$$

in which case outcome 1 should, on the game-theory postulate, be preferred.

These considerations immediately raise numerous questions concerning the numerical measurability of utility and other matters, and we shall have to consider some of them later. For the sake of the argument, though, let us postpone these questions and concede for the moment that meaningful numbers, such as $U(x)$, can be assigned. It still is not immediately evident that the reasonable man should take account of only the expected value of his moral expectation and should disregard its variance, its skewness, and all the rest. We now demonstrate the reasonableness of this postulate by starting from more immediately appealing assumptions and showing that the game-theory postulate follows from them.

For this argument let us define a "gamble" to be any situation in which two outcomes are possible and in which the probabilities of the two outcomes are known. We shall denote a gamble by $[x_1, x_2; p]$, meaning that the two possible outcomes are x_1 and x_2, that the probability of x_1 is p and the probability of x_2 is $1 - p$.

As our starting point we take the "strong independence axiom," one version of which is: If x_1 is indifferent to y_1 and x_2 is indifferent to y_2, then the gamble $[x_1, x_2; p]$ is indifferent to the gamble $[y_1, y_2; p]$ irrespective of what the outcomes are and of the value of p. This axiom says merely that if the consequences of two gambles are pairwise indifferent and their probabilities are the same, then there is no reason to prefer one over the other.

Let $U[\ \ ]$ denote the utility or desirability of any situation, be it a definite payment or a gamble. Our task is to show that

$$U[x_1, x_2; p] = pU[x_1] + (1 - p)U[x_2] \tag{A-1}$$

whatever x_1, x_2, and p may be, i.e., that the utility of a gamble is equal to the expected value of the utilities of the outcomes. To do this we conceive of a very desirable outcome called M and a very undesirable one called N and consider gambles in which M and N are the stakes. Such gambles, which have the form $[M, N; p]$, will be called "standard gambles" and will be used to evaluate the desirability of other gambles and of fixed outcomes. We assume (this is really a second axiom) that it is possible to find a standard gamble that is just as desirable as outcome x_1 (i.e., the standard gamble and x_1 are indifferent alternatives). Let this gamble be $[M, N; p_1]$. Clearly, the more desirable x_1, the greater will be p_1, the probability of the favorable outcome in the standard gamble indifferent to x_1.

Similarly let $[M, N; p_2]$ be the standard gamble indifferent to x_2 and $[M, N; p_3]$ be the standard gamble indifferent to the original gamble $[x_1, x_2; p]$. Then, by the strong-independence axiom, the compound gamble

$$\{[M, N; p_1], [M, N; p_2]; p\}$$

is indifferent to the original gamble $[x_1, x_2; p]$. The compound gamble $\{[M, N; p_1], [M, N; p_2]; p\}$ is a two-stage uncertain event, the outcomes of whose first stage are the ordinary gambles indicated. The classic example of such a compound gamble is the Irish sweepstakes. In the case at hand, p is the probability that the outcome of the initial chance event will be $[M, N; p_1]$, and p_1 is the probability that M will result from the second stage. Thus the probability of winning M via $[M, N; p_1]$ is pp_1. Similarly, the probability of winning M via $[M, N; p_2]$ is $(1 - p)p_2$, and therefore the total probability of winning M from the compound gamble is $pp_1 + (1 - p)p_2$. Now notice that the compound gamble has the same ultimate outcomes as $[M, N; p_3]$, and since they are both indifferent to $[x_1, x_2; p]$, they are indifferent to each other. Therefore the probability of winning M must be the same in these two gambles, and

$$p_3 = pp_1 + (1 - p)p_2 \tag{A-2}$$

Since the probabilities occurring in standard gambles indicate relative desirability, they can be used as Paretian indexes of ophelimity, or utility. Thus the utility of any situation, say S, is measured by the probability, say p_s, of winning M in the standard gamble indifferent to S; that is, we may write $U[S] = p_s$, where p_s satisfies $U[S] = U[M, N; p_s]$. Then $p_1 = U[x_1]$, $p_2 = U[x_2]$, $p_3 = U[x_1, x_2; p]$. Substituting these values in Eq. (A-2), Eq. (A-1) results and is proved. This argument generalizes readily to uncertain events with more than two outcomes. The essential assumption on which our argument is founded is worth restating. It is that we can reduce a compound gamble to a simple one via the rules of probability without changing its desirability.

In the course of this argument we answered the question of whether it is possible to assign definite numbers to the desirabilities of different situations. Our answer was that one way of doing this, and of course not the only way, is to measure the utility of any situation by the probability of the favorable outcome in the standard gamble indifferent to it. This procedure will assign higher numbers to more desirable situations and thus fulfill all the requirements of an ordinal utility indicator of the sort conceived by Pareto. Of course, any monotonic function of p_s could also serve as an ordinal utility indicator, but only the one we have chosen will have the very handy property expressed in Eq. (A-1) and employed in the theory of games. Thus for purposes of game theory it must be assumed that utility is measured in this way. This is, by the way, an empirically observable measure of utility, arbitrary only up to scale and origin constants as determined by the choice of M and N. A more complete proof would show (1) that the choice of M and N can be arbitrary without really affecting anything, and (2) that the utility indicator can be extended to outcomes worse than M or better than N.

Let us now make an additional and restrictive assumption. Let us assume that utility can be measured adequately in monetary terms. Then the utility of any sum of money, say x, is proportional to its amount. Since the scale of measurement of utility is arbitrary, we may as well take the factor of proportionality to be unity and write $U[x] = x$. Substituting this relation into Eq. (A-1), we find

$$U[x_1, x_2; p] = px_1 + (1 - p)x_2$$

or the utility of the gamble is indeed equal to its actuarial expected value. But notice that a powerful particularizing assumption—in effect, that of constant marginal utility of money—was needed in order to produce this result.

This analysis has significant consequences for game theory for it calls attention to the distinction between the utilities derived by the players from the outcome of the game and the objective pay-offs that they receive or make. The essential stakes, with which the players concern themselves, are, clearly, the utilities and not the numerical pay-offs. Economists have long since become reconciled to the conclusion that the utilities derived by two or more persons from a social situation are not summable. What, then, becomes of the concept of a constant-sum game? Answer: It must go by the board.

But all is not lost. If the utilities derived by the two participants from any outcome are connected by a linear formula, then although the concept of a total value of the game is meaningless, the strategic recommendations derived from an analysis of constant-sum games will still be valid. Thus if the utilities derived by both players are proportional to their monetary pay-offs (positive or negative), we are on safe ground. If their utilities are proportional to the market shares they obtain, we are on safe ground again. We have used both of these situations in our examples. In general, if there is any external measure of returns that is proportional to the utility scales of both players, and if the game is constant-sum in terms of this measure, then game-theory analysis is applicable. On the other hand, if the utility derived by either player depends in any more complicated manner on the objective results of the game, then the situation is essentially that of a non-constant-sum game, with the unfortunate consequences noted in the text.

Appendix B

The Algebra of Matrices

B-1. INTRODUCTION

As with other branches of mathematics, the use of matrices confers two considerable advantages. The first is analytic: mathematicians have proved a lot of theorems about matrices. When an economic problem can be formulated in matrix terms all those theorems are at our disposal and can occasionally lead to interesting conclusions that might otherwise be exceedingly difficult to achieve. The second advantage is purely notational: this is not to be underrated. The choice of a neat, natural, and suggestive notation is sometimes half the battle in grasping a complicated setup; it frees the mind for more useful activity than simply keeping things straight. In especially fortunate cases—and matrix algebra is one of these—the two advantages interact. The notation, apart from being a convenient shorthand, also suggests operations, maneuvers, and theorems, some of which turn out to be valid.

At various places in this book the matrix notation has been too convenient to be resisted. In most cases, fortunately, it turns out that the notation itself and a few elementary theorems are all that is needed to enable us to reap the blessing. To make the exposition essentially self-contained we include this brief appendix. It is intended to carry the reader just far enough for the purposes of the text and to give him some feeling for the way in which linear systems can be handled. No proofs are given, and we make no attempt at generality or rigor. For those who would like to go further and deeper, there are now available several excellent introductory textbooks on linear algebra.[1]

B-2. VECTORS AND VECTOR SPACES

Nearly everyone is familiar with the fact that any point in the plane can be represented by a pair of numbers, its *coordinates*. If we draw a

[1] For example, S. Perlis, *Theory of Matrices*, Addison-Wesley Publishing Company, Cambridge, Mass., 1952; G. Birkhoff and S. MacLane, *A Survey of Modern Algebra*, The Macmillan Company, New York, 1948.

horizontal axis, and label it the x_1 axis, and a vertical, or x_2 axis, then we can represent any point uniquely by giving its x_1 coordinate and its x_2 coordinate *in order*. The point (a,b) is the point that lies directly above the point marked a on the x_1 axis and level with the point marked b on the x_2 axis. Naturally a or b or both can be negative. If b is negative, the word "above" in the last sentence is replaced by "below." Note that the *order* of the coordinates is important. The points represented by (a,b) and (b,a) are *not* the same (unless $a = b$). Conversely, to find the coordinates of any point we simply drop perpendiculars to the x_1 and x_2 axes and read off the coordinates from the bases of the perpendiculars.[1]

We have a complete correspondence between geometrical objects, namely, points of the plane, and purely algebraic objects, namely, ordered pairs of numbers. These algebraic objects that correspond so exactly to points have a name of their own. They are called *vectors*. Thus a vector is any pair of numbers (x_1,x_2). To be exact we should call these pairs two-dimensional vectors, since we must now notice that we could do for ordinary three-dimensional space, the space we live in, exactly what we have just done for the plane. Take any corner of the room as an origin and notice that there are three mutually perpendicular lines meeting there. Label these the x_1, x_2, and x_3 axes. Then any point in the room is in unique correspondence with an *ordered triple* of numbers, namely (x_1,x_2,x_3), its x_1, x_2, and x_3 coordinates, and vice versa.[2] Any ordered list of three numbers can be called a three-dimensional vector.

The whole point of this new vocabulary is that geometrically defined objects can be translated into algebraic concepts via the vector notion. For example, the circle in the plane with center at (a,b) and radius r becomes the collection of all vectors (x_1,x_2) such that

$$(x_1 - a)^2 + (x_2 - b)^2 = r^2$$

[1] (1) There is nothing holy about our original choice of axes. *Any* pair of mutually perpendicular lines would have served us just as well. (2) In fact the axes don't even have to be mutually perpendicular. Any pair of nonparallel lines would do. Only instead of moving perpendicularly, we move always parallel to the *other* axis. With these axes, to find the point (a,b) we move out to the point a on the x_1 axis and draw a parallel to the x_2 axis. Then find the point b on the x_2 axis and draw a parallel to the x_1 axis. Where the parallels intersect is our sought-for point. (3) There are wholly different ways of representing points in the plane. We could, for example, fix an origin and a base line in the plane and represent any point by two numbers: r = its *distance* from the origin, and θ = the *angle* a ray from the origin makes with the base line. These are called "polar coordinates."

[2] The interested reader can reread the last footnote and ask himself what corresponding qualifications need to be made in three dimensions. *Hint:* Any three lines can serve as axes as long as they meet in some common point and don't all lie in the same plane.

The reader should make up some questions to ask himself on this point. For example: What corresponds in three-dimensional geometry to the set of vectors with $x_1 = 0$? To the vectors with $x_2 = 10$? To the vectors with $x_1 = 2x_2$? To the vectors with $x_1 = x_2 = x_3$?

With three dimensions our geometrical analogy comes to an end. But algebra is subject to no such visual limitation. There is no reason at all why we can't go on to define four-, five-, and n-dimensional vectors as ordered quadruples, quintuples, or n-tuples of numbers. And so we shall. By definition, then, an n-dimensional vector is simply an ordered list of n numbers, $(x_1, x_2, \ldots, x_i, \ldots, x_n)$. The number x_i is called the ith coordinate, or component, of the vector. And by extension we can sometimes think of an n-dimensional vector as a point in n-dimensional space with the given coordinates.

We have an algebraic counterpart to a point in two-, or three-, or n-dimensional space. Just as the totality of *all* points in the plane is called "the plane," or "two-dimensional space," we shall call the totality of *all* two-dimensional vectors "two-dimensional vector space" and nickname it R^2; and we shall treat similarly higher-dimensional vector spaces R^3, R^4, R^n.

So far we haven't learned to *do* anything with vectors other than to identify them with points. We need some operations. Here we can begin formally and then check to see what our formal definitions amount to geometrically. Consider two n-dimensional vectors

$$\mathbf{x} = [x_1, \ldots, x_j, \ldots, x_n] \qquad \text{and} \qquad \mathbf{y} = [y_1, \ldots, y_j, \ldots, y_n]$$

By the "sum" of these vectors we shall mean the n-dimensional vector $\mathbf{x} + \mathbf{y} = [x_1 + y_1, \ldots, x_j + y_j, \ldots, x_n + y_n]$; i.e., the sum of two vectors is a vector of the same dimension as the summands, each of whose components is the sum of the corresponding components of the summands. Thus the sum of $(-1, 0, 3, 1.2)$ and $(10, -2, 1.6, -1.5)$ is $(9, -2, 4.6, -0.3)$.

Two vectors are said to be "equal" if they are of the same dimension and if their corresponding components are equal; that is, $\mathbf{x} = \mathbf{y}$ if and only if $x_1 = y_1, x_2 = y_2, \ldots, x_j = y_j, \ldots, x_n = y_n$. It will be seen that two vectors are equal if and only if they represent the *same* geometrical point. In combination with the preceding paragraph we can observe that $\mathbf{z} = \mathbf{x} + \mathbf{y}$ means that $\mathbf{z} = [z_1, \ldots, z_j, \ldots, z_n]$ and $z_1 = x_1 + y_1, \ldots, z_j = x_j + y_j, \ldots, z_n = x_n + y_n$. As a consequence of what we know about the addition of ordinary numbers we can say that $\mathbf{x} + \mathbf{y} = \mathbf{y} + \mathbf{x}$ (because $x_j + y_j = y_j + x_j$) and

$$(\mathbf{x} + \mathbf{y}) + \mathbf{z} = \mathbf{x} + (\mathbf{y} + \mathbf{z}) = \mathbf{x} + \mathbf{y} + \mathbf{z}$$

[because $(x_j + y_j) + z_j = x_j + (y_j + z_j) = x_j + y_j + z_j$].[1]

[1] We shall simply leave it at this. Some readers may know that there *are* mathe-

It is convenient to have a special symbol for the null vector, i.e., the vector all of whose components are zero. We shall call it ϕ. Clearly for any vector $\mathbf{x}$, $\mathbf{x} + \phi = \mathbf{x}$.

Now suppose that $\mathbf{x}$ is a vector and c is an ordinary number (sometimes called a "scalar"). Then we shall define the *product* of c and $\mathbf{x}$ as the vector each of whose components is simply the corresponding component of $\mathbf{x}$ multiplied by c. If $\mathbf{x} = [x_1, \ldots, x_j, \ldots, x_n]$, then

$$c\mathbf{x} = [cx_1, \ldots, cx_j, \ldots, cx_n]$$

Put a little differently, $\mathbf{y} = c\mathbf{x}$ means $y_1 = cx_1, \ldots, y_n = cx_n$. Again the rules of ordinary arithmetic show us that $(c + d)\mathbf{x} = c\mathbf{x} + d\mathbf{x}$ [since $(c + d)x_j = cx_j + dx_j$] and $c(d\mathbf{x}) = (cd)\mathbf{x} = dc\,\mathbf{x} = d(c\mathbf{x})$ [since $c(dx_j) = cd\,x_j = dc\,x_j = d(cx_j)$]. Also $0\mathbf{x} = \phi$, for any x, and equally obviously $c\mathbf{x} = \mathbf{x}c$.

By $-\mathbf{x}$ we mean the vector $(-1)\mathbf{x} = [-x_1, \ldots, -x_j, \ldots, -x_n]$, and thus the difference of two vectors, say $\mathbf{y} - \mathbf{x}$, is the vector $\mathbf{y} + (-1)\mathbf{x}$, that is, $\mathbf{y} - \mathbf{x} = [y_1 - x_1, \ldots, y_j - x_j, \ldots, y_n - x_n]$. We are now possessed of such concepts as the linear combination of vectors, for example,

$$a\mathbf{x} + b\mathbf{y} + c\mathbf{z} = [ax_1 + by_1 + cz_1, \ldots, ax_j + by_j + cz_j, \ldots, ax_n + by_n + cz_n]$$

We can solve some simple equations involving vectors (remember that two vectors are equal only if all their components agree). Thus $c\mathbf{x} + \mathbf{y} = \phi$ has the solution $\mathbf{x} = -(1/c)\mathbf{y}$ (provided $c \neq 0$). What if $c = 0$? The equation $3\mathbf{x} + 4(1,2,3,4) = (0, -0.4, 100, -16.7)$ has the solution $\mathbf{x} = (-4/3, -2.8, 88/3, -10.9)$.

To conclude this section, let us return to the geometry of two dimensions to see how our algebraic definitions look in this light. In Fig. B-1 we have drawn the points $\mathbf{x}$ and $\mathbf{y}$ corresponding to (x_1,x_2) and (y_1,y_2). The point corresponding to $\mathbf{z} = \mathbf{x} + \mathbf{y}$ is the point each of whose coordinates is the sum of the corresponding coordinates of $\mathbf{x}$ and $\mathbf{y}$. The reader can see for himself that $\mathbf{z}$ can be found by the familiar process of completing the parallelogram which has the origin $\mathbf{x}$, and $\mathbf{y}$ as three of its four vertices. The point $\mathbf{z}$ will be the fourth.

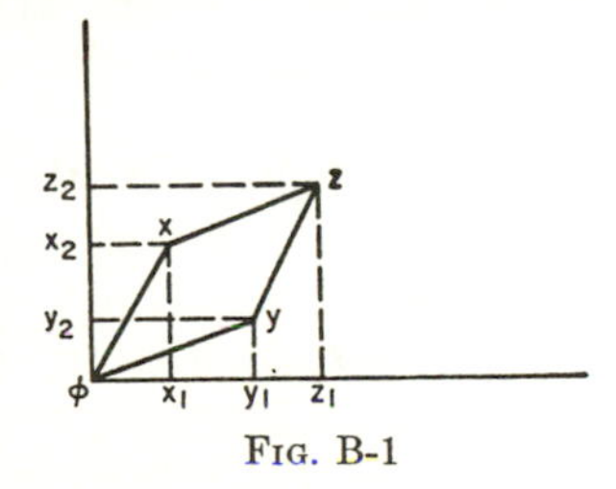

Fig. B-1

matical objects which don't obey these simple so-called commutative and associative rules. In fact, we have reversed the mathematicians' way of doing things. They would *characterize* n-dimensional vector space by these and similar properties and then prove that we can think of vectors as ordered n-tuples of numbers. But for economists these fine points are irrelevant.

Now how about multiplying a vector by a scalar? Figure B-2 will help us understand the meaning of multiplication. This operation leaves the ratio of the coordinates the same. Thus if we multiply **x** by a scalar c, we get another point on the ray from the origin through **x**. If $c > 1$, we get a point farther out; if $0 < c < 1$, we get a point between **x** and the origin. If c is negative, we simply go back through the origin to the opposite quadrant, and out the same ray a distance c times the distance from the origin to **x**. By combining the two processes of scalar multiplication along a ray and addition by completing the parallelogram, the reader should be able to start with any pair of points **x** and **y** and find, say, the point $(2\mathbf{x} - 0.5\mathbf{y})$. Any linear combination of **x** and **y** goes the same way. To find $a\mathbf{x} + b\mathbf{y} + c\mathbf{z}$ we proceed by finding $a\mathbf{x} + b\mathbf{y}$ and then adding this to $c\mathbf{z}$. Our algebra tells us that we could alternatively find $b\mathbf{y} + c\mathbf{z}$ and add $a\mathbf{x}$, to get the same result.

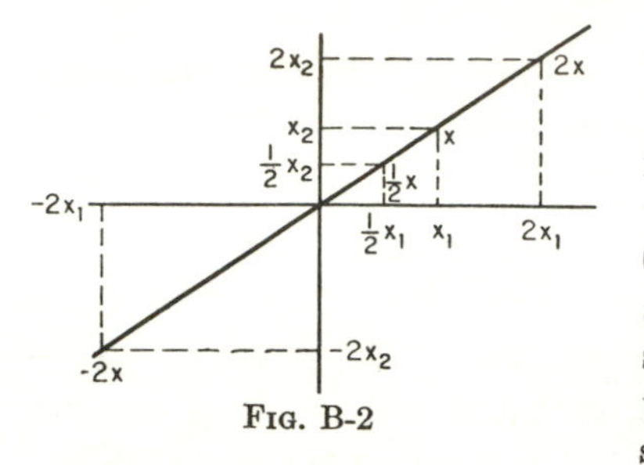

FIG. B-2

The notion of a vector fits very neatly into the linear-programming scheme. In the diet problem, for instance, if there are m nutrients, we can represent the nutrient content of a unit of a particular food by an m-dimensional vector, say $\mathbf{u} = [u_1, \ldots, u_m]$, $\mathbf{v} = [v_1, \ldots, v_m]$, **w**, **x**, **y**, **z**, etc., for as many foods as are available. If we consume a (a number) units of the first food, the nutrient intake will be the vector $a\mathbf{u}$. The nutrient content of a diet consisting of a, b, c, d, e, f units, respectively, of each of the six foods will have a nutrient content given by the vector $a\mathbf{u} + b\mathbf{v} + c\mathbf{w} + d\mathbf{x} + e\mathbf{y} + f\mathbf{z}$. This has to be compared with the vector of requirements $\mathbf{R} = [R_1, \ldots, R_m]$. The case of the flow of commodities through the competitive firm goes similarly.

B-3. LINEAR INDEPENDENCE—BASES

The reader may have noticed that in the last paragraph we have come very close to talking about simultaneous linear equations. Suppose we ask whether there is any diet that *exactly* fulfills the dietary requirements; i.e., is there any set of numbers a, b, c, d, e, f such that the vector equation $a\mathbf{u} + b\mathbf{v} + c\mathbf{w} + d\mathbf{x} + e\mathbf{y} + f\mathbf{z} = \mathbf{R}$ holds. Since equality between vectors means component-by-component equality, we are really asking whether the following *system* of m linear equations in six unknowns has a solution:

$$
\begin{aligned}
au_1 + bv_1 + cw_1 + dx_1 + ey_1 + fz_1 &= R_1 \\
au_2 + bv_2 + cw_2 + dx_2 + ey_2 + fz_2 &= R_2 \\
\cdots\cdots\cdots\cdots\cdots\cdots\cdots\cdots \\
au_m + bv_m + cw_m + dx_m + ey_m + fz_m &= R_m
\end{aligned}
\qquad \text{(B-1)}
$$

In fact, one of the notational bonuses of the vector technique is that it enables us to compress our discussion of systems of linear equations.

Still another way of looking at Eqs. (B-1) is to ask whether the vector $\mathbf{R}$ can be expressed as a linear combination of the vectors $\mathbf{u}, \mathbf{v}, \ldots, \mathbf{z}$. This leads us to the important concept of *linear dependence*. A set $\mathbf{u}, \mathbf{v}, \ldots, \mathbf{z}$ of vectors all belonging to the same vector space R^n is said to be linearly dependent if (1) one of them is the null vector, or (2) if one of them can be expressed as a linear combination of the preceding ones. If neither of these things is true, the vectors are linearly independent. The more usual definition is to say that the vectors are linearly independent if the only linear combination of them which is equal to the null vector is the trivial combination $0 \cdot \mathbf{u} + 0 \cdot \mathbf{v} + \cdots + 0 \cdot \mathbf{z}$, that is, if $a\mathbf{u} + b\mathbf{v} + c\mathbf{w} + d\mathbf{x} + e\mathbf{y} + f\mathbf{z} = \phi$ implies $a = b = c = \cdots = f = 0$. But these two definitions are easily shown to be equivalent.

In R^2 linear dependence has a simple graphical meaning. Consider two non-null vectors $\mathbf{u} = [u_1,u_2]$ and $\mathbf{v} = [v_1,v_2]$. The set $\mathbf{u}, \mathbf{v}$ is linearly dependent provided $\mathbf{v}$ is a linear combination of $\mathbf{u}$, that is, if $\mathbf{v} = a\mathbf{u}$ for some scalar a. But we saw in the previous section that $a\mathbf{u}$ is a vector or point on the ray through $\mathbf{u}$. Thus $\mathbf{u}$ and $\mathbf{v}$ in R^2 are linearly dependent if they represent points on the same ray. Now consider three non-null vectors from R^2, $\mathbf{u}, \mathbf{v}, \mathbf{w}$. They are linearly dependent if $\mathbf{v} = a\mathbf{u}$ or if $\mathbf{w}$ is a linear combination of $\mathbf{u}$ and $\mathbf{v}$, that is, if $\mathbf{w} = b\mathbf{u} + c\mathbf{v}$ for some scalars b and c. This latter condition states

$$
\begin{aligned}
bu_1 + cv_1 &= w_1 \\
bu_2 + cv_2 &= w_2
\end{aligned}
$$

By substitution we can solve this pair of linear equations in the unknowns b and c to find

$$
b = \frac{w_1v_2 - w_2v_1}{u_1v_2 - u_2v_1} \qquad c = \frac{u_1w_2 - u_2w_1}{u_1v_2 - u_2v_1}
$$

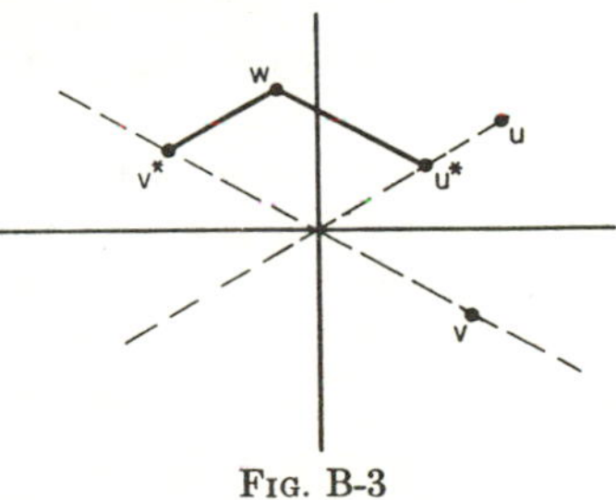

FIG. B-3

That is, $\mathbf{w}$ can *always* be expressed as a linear combination of $\mathbf{u}$ and $\mathbf{v}$—always, that is, unless the denominators in the above solution should be zero. But in that case $u_1v_2 = u_2v_1$, or $v_1/u_1 = v_2/u_2$ and if we call this common ratio a, we see that $\mathbf{u}$ and $\mathbf{v}$ are themselves dependent. We have shown that in either case the vectors $\mathbf{u}, \mathbf{v}, \mathbf{w}$ are linearly dependent; i.e., one can't find *three* linearly independent vectors in R^2. Geometrically this is easy to see. In Fig. B-3, if $\mathbf{u}$ and $\mathbf{v}$ are any pair of independent vectors (i.e., do not lie on the same ray), then any third point $\mathbf{w}$ can be expressed as a linear combination of $\mathbf{u}$ and $\mathbf{v}$ by the two processes of expansion along a ray and formation of a parallelogram. In the diagram

$\mathbf{w} = \mathbf{u}^* + \mathbf{v}^*$ where $\mathbf{u}^*$ and $\mathbf{v}^*$ are multiples of $\mathbf{u}$ and $\mathbf{v}$. What we have in effect done is to choose the rays through $\mathbf{u}$ and $\mathbf{v}$ as new axes and express $\mathbf{w}$ in this new coordinate system.

So much for R^2. How about R^3? In three-dimensional vector space, two vectors are linearly dependent if they lie on the same ray. Three vectors are linearly dependent if they lie in a *plane* which also passes through the origin. Three linearly independent vectors can be found, namely, any three which do not lie in such a plane (e.g., one vector along each axis). But any set of *four* vectors is linearly dependent. In effect, if the first three are independent, we can choose them as a new set of axes and always express the fourth in terms of the new coordinate system.

The further generalization is obvious. One can find n linearly independent n-dimensional vectors,[1] but not $n + 1$. If we are given any n linearly independent vectors of R^n we can express every vector of R^n as a linear combination of them.

This is really a theorem in simultaneous equations. Suppose in R^6 we try to express an arbitrary vector $\mathbf{c}$ as a linear combination of $\mathbf{u}$, $\mathbf{v}$, $\mathbf{w}$, $\mathbf{x}$, $\mathbf{y}$, $\mathbf{z}$. Then we must have scalars $a, b, \ldots, f$ such that

$$\begin{aligned} au_1 + bv_1 + \cdots + fz_1 &= c_1 \\ au_2 + bv_2 + \cdots + fz_2 &= c_2 \\ \cdots\cdots\cdots&\cdots\cdots \\ au_6 + bv_6 + \cdots + fz_6 &= c_6 \end{aligned} \qquad \text{(B-1}a\text{)}$$

We now know that these six equations in six unknowns have a solution if the vectors $\mathbf{u}, \mathbf{v}, \ldots, \mathbf{z}$ are linearly independent (i.e., if the *columns* of coefficients form a linearly independent set of vectors).[2]

A set of linearly independent vectors, equal in number to the number of dimensions, is called a "basis" of R^n. A basis of R^n has the important property that any vector of R^n can be expressed as a linear combination of the basis vectors. (Recall that in the m-nutrient diet problem a feasible diet using just m foods is called a "basic" solution. If the m foods have linearly independent nutrient vectors, they form a basis of R^m. Any food's nutrient vector can then be expressed as a linear combination of the m foods. This linear combination is exactly the "equivalent combination" appearing in the simplex method.)

[1] A mathematician would use the maximal number of independent vectors to *define* the notion of "dimension," and he would do it without even talking about such things as coordinates or components. But he then proves that any n-dimensional vector space is really our R^n, or space of n-tuples, perhaps thinly disguised. We *start* with R^n because it's the space in which we are fundamentally interested.

[2] Even if the columns are linearly dependent, there *might* be a solution, e.g., if $\mathbf{c}$ should happen to be proportional to $\mathbf{u}$, or a linear combination of $\mathbf{u}$, $\mathbf{w}$, and $\mathbf{z}$, etc. But we know there will be some $\mathbf{c}$'s for which no solution exists if the columns are linearly dependent.

Actually, to express a given vector in terms of a particular basis may be no easy task. It amounts to solving a set of simultaneous equations like (B-1a). For some special bases the job is much simplified. Most especially, consider the n vectors

$$\mathbf{e}_1 = [1, 0, 0, \ldots, 0] \qquad \mathbf{e}_3 = [0, 0, 1, \ldots, 0],$$
$$\mathbf{e}_2 = [0, 1, 0, \ldots, 0] \qquad \mathbf{e}_n = [0, 0, 0, \ldots, 1]$$

(Note that $\mathbf{e}_i$ is here a *vector*, not a component of a vector.) In the first place they are linearly independent (try to express any one as a linear combination of the others). In the second place it is trivially easy to express any vector of R^n in terms of these. If $\mathbf{x} = [x_1,x_2,x_3, \ldots ,x_n]$, then $\mathbf{x} = [x_1\mathbf{e}_1 + x_2\mathbf{e}_2 + x_3\mathbf{e}_3 + \cdots + x_n\mathbf{e}_n]$. Write out the equations corresponding to (B-1a) and see how easy they become. The vectors $\mathbf{e}_i$ are called the "unit vectors" of R^n, and obviously they will come in very handy. Geometrically the unit vectors are vectors along the axes themselves. To express $\mathbf{x}$ in terms of the $\mathbf{e}_i$ is to find the coordinates of $\mathbf{x}$ with respect to the original axes themselves. Hence the components x_i of $\mathbf{x}$ are themselves the sought-for coefficients.

B-4. MAPPINGS

Every economist has some acquaintance with the notion of a function of one variable or even of several variables. To say that $y = f(x)$ means that to each[1] value of x we make correspond a particular value of y, namely, $f(x)$. Similarly for $y = g(x_1,x_2, \ldots ,x_n)$. Can we generalize this notion to the case in which $\mathbf{y}$ and $\mathbf{x}$ are vectors? Can we write a vector $\mathbf{y}$ as some kind of function of a vector $\mathbf{x}$? Why not? In fact, there is even no need for $\mathbf{x}$ and $\mathbf{y}$ to belong to the same vector space, i.e., to have the same number of components. Suppose we consider on the one hand R^n and on the other R^m. Suppose we have a rule which associates with each point or vector of R^n a point or vector of R^m. Then again we can write $\mathbf{y} = f(\mathbf{x})$, only now $\mathbf{x}$ stands for an n-dimensional point or vector and $\mathbf{y}$ for an m-dimensional one. We say that the function f is a *mapping* from R^n to R^m.

Let us understand what this means. A vector $\mathbf{x} = [x_1,x_2, \ldots ,x_n]$ determines, via the rule or mapping, a particular vector

$$\mathbf{y} = [y_1,y_2, \ldots ,y_n]$$

That is to say, knowledge of the components of $\mathbf{x}$ determines all the components of $\mathbf{y}$. We could express this by saying that each component y_i

[1] The function f need be defined only for some special values of x, for example, for nonnegative x, or for x equal to a whole number, or for x between 0 and 1.

is a function of all the components of $\mathbf{x}$. In short, the notation $\mathbf{y} = f(\mathbf{x})$ is simply shorthand for m different functions each of n variables, to wit:

$$\begin{aligned} y_1 &= f_1(x_1,x_2, \ldots ,x_n) \\ y_2 &= f_2(x_1,x_2, \ldots ,x_n) \\ &\cdots\cdots\cdots\cdots \\ y_m &= f_m(x_1,x_2, \ldots ,x_n) \end{aligned}$$

There is, of course, no reason why the m functions f_i should be the same, or even anything like each other.

In the important special case in which $m = n$, $\mathbf{y} = f(\mathbf{x})$ maps R^n into itself, via n functions each of the n variables $(x_1, \ldots ,x_n)$.

Here are some simple examples of mappings of R^2 into itself, which the reader should be able to visualize graphically:

$$\begin{aligned} y_1 &= x_1 \\ y_2 &= 0 \end{aligned} \tag{B-2}$$

This is called a projection on the x_1 axis.

$$\begin{aligned} y_1 &= x_2 \\ y_2 &= x_1 \end{aligned} \tag{B-3}$$

$$\begin{aligned} y_1 &= \frac{x_1}{x_1 + x_2} \\ y_2 &= \frac{x_2}{x_1 + x_2} \end{aligned} \tag{B-4}$$

Here we must exclude cases in which $x_1 + x_2 = 0$. Note that $y_2/y_1 = x_2/x_1$.

$$\begin{aligned} y_1 &= \frac{x_1}{(x_1^2 + x_2^2)^{1/2}} \\ y_2 &= \frac{x_2}{(x_1^2 + x_2^2)^{1/2}} \qquad x_1 \text{ and } x_2 \text{ not both zero} \end{aligned} \tag{B-5}$$

$$\begin{aligned} y_1 &= x_1^2 \\ y_2 &= x_2^2 \end{aligned} \tag{B-6}$$

$$\begin{aligned} y_1 &= a_{11}x_1 + a_{12}x_2 \\ y_2 &= a_{21}x_1 + a_{22}x_2 \end{aligned} \tag{B-7}$$

B-5. LINEAR TRANSFORMATIONS

The idea and notation of a vector is especially suited to the study of a special but exceedingly important class of mappings called *linear transformations*. The mapping $T(\mathbf{x})$ of R^n into R^m is a linear transformation if it has the two following properties:

$$\begin{aligned} T(a\mathbf{x}) &= aT(\mathbf{x}) \\ T(\mathbf{u} + \mathbf{v}) &= T(\mathbf{u}) + T(\mathbf{v}) \end{aligned}$$

for every scalar a and all vectors $\mathbf{u}$, $\mathbf{v}$, and $\mathbf{x}$ of R^n. These are obviously properties which deserve the adjective "linear." They say that if the vector $\mathbf{x}$ is multiplied by a scalar, then the "image" $T(\mathbf{x}) = \mathbf{y}$ is multiplied by the same scalar; and the image of the sum of two vectors is the sum of the two images (which are also vectors). It follows from these properties that if a and b are scalars and $\mathbf{x}$ and $\mathbf{y}$ are vectors of R^n, then $T(a\mathbf{x} + b\mathbf{y}) = aT(\mathbf{x}) + bT(\mathbf{y})$, and in fact

$$T(a\mathbf{u} + b\mathbf{v} + \cdots + f\mathbf{z}) = aT(\mathbf{u}) + bT(\mathbf{v}) + \cdots + fT(\mathbf{z})$$

Once again the case $m = n$ is of special interest: we have then a linear transformation of R^n.

The reader should check his understanding by deciding which of the examples of mappings at the end of the previous section are linear.[1]

Some properties of linear transformations are evident from the definition itself. For example, $T(\phi) = \phi$; that is, a linear transformation maps the null vector into the null vector. *Proof:* $T(\phi) = T(0\mathbf{x}) = 0T(\mathbf{x}) = \phi$ for any vector $\mathbf{x}$.

Here is a rather more important example. A linear transformation (or any mapping, for that matter) is called 1:1 if no vector $\mathbf{y}$ is the image of more than one vector $\mathbf{x}$; that is, if $T(\mathbf{u}) = T(\mathbf{v})$ implies $\mathbf{u} = \mathbf{v}$. The 1:1 linear transformations are very important; their importance resides in the fact that if $\mathbf{y} = T(\mathbf{x})$ and T is 1:1, then if we are presented with $\mathbf{y}$ we can locate the $\mathbf{x}$ from which it came. More on this later.[2] We can easily show that a linear transformation T is 1:1 if and only if the only vector it maps into the null vector is the null vector itself. We already know that in any case $T(\phi) = \phi$; we now state that T is 1:1 if and only if it maps no other vector into the null vector. *Proof:* (1) We already know that $T(\phi) = \phi$; hence, trivially, if T is 1:1, there can be no other $\mathbf{x}$ such that $T(\mathbf{x}) = \phi$. (2) Suppose that $T(\mathbf{x}) = \phi$ implies $\mathbf{x} = \phi$; if $T(\mathbf{x}) = T(\mathbf{y})$, then $T(\mathbf{x}) - T(\mathbf{y}) = \phi$, but $T(\mathbf{x}) - T(\mathbf{y}) = T(\mathbf{x} - \mathbf{y})$, and hence $\mathbf{x} - \mathbf{y} = \phi$ and $\mathbf{x} = \mathbf{y}$. Q.E.D. A linear transformation of R^n which is 1:1 is also described as "nonsingular."

B-6. MATRICES

So far we have Hamlet without the Prince of Denmark: the notion of a matrix has not yet appeared in this appendix on matrices. Now the time has come. Nor have all the preliminaries been wasted, for in this section we shall show that a linear transformation and a matrix are "really" one and the same thing. But first some definitions.

[1] *Answer:* 2, 3, and 7.

[2] In anticipation, there is an *inverse* mapping $\mathbf{x} = T^{-1}(\mathbf{y})$ which associates with each $\mathbf{y}$ its "parent" $\mathbf{x}$. If T is not 1:1, $\mathbf{y}$ may have many "parents."

A matrix is just a rectangular array of numbers, e.g.,

$$\begin{bmatrix} 1 & 2 & 3 & 4 \\ 5 & 6 & 7 & 8 \end{bmatrix} \quad \text{or} \quad \begin{bmatrix} 0 & 5 \\ -3 & 1 \end{bmatrix} \quad \text{or} \quad \begin{bmatrix} a_{11} & a_{12} & \cdots & a_{1n} \\ a_{21} & a_{22} & \cdots & a_{2n} \\ \cdots & \cdots & \cdots & \cdots \\ a_{m1} & a_{m2} & \cdots & a_{mn} \end{bmatrix}$$

$$\text{or} \quad [0 \quad 0 \quad a \quad b] \quad \text{or} \quad \begin{bmatrix} c_1 \\ c_2 \\ c_3 \end{bmatrix}$$

Each number that appears in a matrix is called an "element." When we have no particular numerical values in mind, a general matrix of m rows and n columns (described as m by n, or $m \times n$) is usually written in double-subscript notation as in the third example: the first subscript identifies the row, the second the column. A matrix with one row or one column is usually called a "row vector," or a "column vector"; we have tacitly been writing vectors as rows, for typographical convenience, but this is not essential. Matrices are frequently symbolized by a single letter or by giving the typical element inside brackets. Thus the third example might be indicated by $\mathbf{A}$ or by $[a_{ij}]$. A matrix with as many rows as columns is said to be "square."

Now let us return to linear transformations from R^n to R^m. Because of the linearity properties, if we know what a particular linear mapping does to the vectors comprising a basis for R^n, we can *compute* what it does to *any* vector of R^n. In other words, if we know the image of each vector of a basis, we know essentially *all* about the transformation. Here is where the special basis consisting of the n unit vectors $\mathbf{e}_1, \mathbf{e}_2, \ldots, \mathbf{e}_n$ comes in particularly handy.

Suppose we know $T(\mathbf{e}_1), T(\mathbf{e}_2), \ldots, T(\mathbf{e}_n)$ and would like to know $T(\mathbf{x})$, where $\mathbf{x}$ is an arbitrary vector of R^n. But $\mathbf{x}$ can be expressed as a linear combination of the $\mathbf{e}$'s, namely, $\mathbf{x} = [x_1\mathbf{e}_1 + x_2\mathbf{e}_2 + \cdots + x_n\mathbf{e}_n]$. Then by linearity

$$T(\mathbf{x}) = T(x_1\mathbf{e}_1 + x_2\mathbf{e}_2 + \cdots + x_n\mathbf{e}_n) = x_1T(\mathbf{e}_1) + x_2T(\mathbf{e}_2) + \cdots + x_nT(\mathbf{e}_n)$$

and $T(\mathbf{x})$ is easily calculable.

Now let us write our vectors of R^n (and of R^m) as column vectors.

$$\mathbf{x} = \begin{bmatrix} x_1 \\ x_2 \\ \cdot \\ \cdot \\ \cdot \\ x_n \end{bmatrix} \quad \text{and} \quad \mathbf{y} = T(\mathbf{x}) = \begin{bmatrix} y_1 \\ y_2 \\ \cdot \\ \cdot \\ \cdot \\ y_m \end{bmatrix}$$

Each $T(\mathbf{e}_i)$ is a vector of R_m, say,

$$T(\mathbf{e}_1) = \begin{bmatrix} a_{11} \\ a_{21} \\ \cdot \\ \cdot \\ \cdot \\ a_{m1} \end{bmatrix}, T(\mathbf{e}_2) = \begin{bmatrix} a_{12} \\ a_{22} \\ \cdot \\ \cdot \\ \cdot \\ a_{m2} \end{bmatrix}, \ldots, T(\mathbf{e}_j) = \begin{bmatrix} a_{1j} \\ a_{2j} \\ \cdot \\ \cdot \\ \cdot \\ a_{mj} \end{bmatrix}, \text{ etc.}$$

Now we can write

$$\mathbf{y} = \begin{bmatrix} y_1 \\ y_2 \\ \cdot \\ \cdot \\ \cdot \\ y_m \end{bmatrix} = T(\mathbf{x}) = \sum_{i=1}^{n} x_i T(\mathbf{e}_i) =$$

$$\begin{bmatrix} a_{11}x_1 + a_{12}x_2 + \cdots + a_{1j}x_j + \cdots + a_{1n}x_n \\ a_{21}x_1 + a_{22}x_2 + \cdots + a_{2j}x_j + \cdots + a_{2n}x_n \\ \cdots\cdots\cdots\cdots\cdots\cdots\cdots \\ a_{i1}x_1 + a_{i2}x_2 + \cdots + a_{ij}x_j + \cdots + a_{in}x_n \\ \cdots\cdots\cdots\cdots\cdots\cdots\cdots \\ a_{m1}x_1 + a_{m2}x_2 + \cdots + a_{mj}x_j + \cdots + a_{mn}x_n \end{bmatrix} \quad \text{(B-8}$$

or

$$y_i = \sum_{j=1}^{n} a_{ij}x_j$$

In the last expression we could separate out the matrix of m rows and n columns, each of whose columns is simply the image $T(\mathbf{e}_i)$ of a unit vector. Call it $\mathbf{A}$. Thus,

$$\mathbf{A} = \begin{bmatrix} a_{11} & a_{12} & \cdots & a_{1n} \\ a_{21} & a_{22} & \cdots & a_{2n} \\ \cdots & \cdots & \cdots & \cdots \\ a_{m1} & a_{m2} & \cdots & a_{mn} \end{bmatrix}$$

In this way every *linear transformation* of R^n to R^m is associated with an $m \times n$ *matrix*, and every $m \times n$ matrix represents a linear transformation from R^n to R^m.

It suggests itself immediately (and if it doesn't we now suggest it) that we conventionally *define* the last vector in (B-8) as the *product* of the matrix $\mathbf{A}$ and the vector $\mathbf{x}$. Thus,

$$\mathbf{y} = T(\mathbf{x}) = \mathbf{Ax} \quad \text{(B-8}a\text{)}$$

Since the transformation T operates only on vectors from R^n, we see that the product $\mathbf{Ax}$ makes sense only if $\mathbf{x}$ has as many components (or *rows*) as $\mathbf{A}$ has *columns*.

Looking back at (B-8) we now have a rule for multiplying any $m \times n$ matrix by any $n \times 1$ column vector, as follows. The product is an $m \times 1$ column vector whose first component is the sum of the cross products of the elements of the first row of $\mathbf{A}$ with the corresponding component of $\mathbf{x}$; the second component is the sum of the cross products of the elements of the second row of $\mathbf{A}$ with the components of $\mathbf{x}$, etc. For example,

$$\begin{bmatrix} 1 & 2 & 3 & 4 \\ 5 & 6 & 7 & 8 \end{bmatrix} \begin{bmatrix} 9 \\ 10 \\ 11 \\ 12 \end{bmatrix} = \begin{bmatrix} 9 + 20 + 33 + 48 \\ 45 + 60 + 77 + 96 \end{bmatrix} = \begin{bmatrix} 110 \\ 278 \end{bmatrix}$$

At the end of Sec. B-4 we gave six examples of mappings of R^2, three of which were linear. Verify that the matrices associated with (B-2), (B-3), and (B-7) are, respectively,

$$\begin{bmatrix} 1 & 0 \\ 0 & 0 \end{bmatrix}, \begin{bmatrix} 0 & 1 \\ 1 & 0 \end{bmatrix}, \begin{bmatrix} a_{11} & a_{12} \\ a_{21} & a_{22} \end{bmatrix}$$

Note also that the matrix associated with a linear transformation of R^n is a square $n \times n$ matrix.

B-7. FURTHER OPERATIONS WITH MATRICES

Take any linear transformation from R^n to R^m, say T, and let c be a scalar. Then it is natural to mean by the linear transformation cT the transformation T_c which maps $\mathbf{x}$ into c times $T(\mathbf{x})$. Formally, $T_c(\mathbf{x}) = cT(\mathbf{x})$. If T corresponds to the matrix $\mathbf{A}$, then the reader should verify that cT corresponds to a matrix which is derived from $\mathbf{A}$ by multiplying *every element* by c and which we shall call $c\mathbf{A}$. Thus, by *definition,* to multiply a matrix by a scalar, multiply every element of the matrix by the scalar. For example,

$$2\begin{bmatrix} 0 & 1 \\ 0.5 & -2 \end{bmatrix} = \begin{bmatrix} 0 & 2 \\ 1 & -4 \end{bmatrix}$$

If the scalar should happen to be -1, we see that the negative of a matrix is the original matrix with the sign of every element changed.

Now consider two linear mappings S and T, each of which operates on vectors of R^n and maps them into vectors of R^m. Then we can define a new linear transformation $S + T$ as the transformation which maps any vector $\mathbf{x}$ of R^n into the vector $S(\mathbf{x}) + T(\mathbf{x})$ of R^m.[1] Suppose $\mathbf{A}$ is the $m \times n$ matrix which describes T and $\mathbf{B}$ is the $m \times n$ matrix identified with S. What matrix is associated with $S + T$? A simple example will give us the answer: Let

$$\mathbf{A} = \begin{bmatrix} a_{11} & a_{12} \\ a_{21} & a_{22} \end{bmatrix} \qquad \mathbf{B} = \begin{bmatrix} b_{11} & b_{12} \\ b_{21} & b_{22} \end{bmatrix}$$

Then

$$\begin{aligned}(S + T)(\mathbf{x}) = S(\mathbf{x}) + T(\mathbf{x}) &= \begin{bmatrix} b_{11} & b_{12} \\ b_{21} & b_{22} \end{bmatrix}\begin{bmatrix} x_1 \\ x_2 \end{bmatrix} + \begin{bmatrix} a_{11} & a_{12} \\ a_{21} & a_{22} \end{bmatrix}\begin{bmatrix} x_1 \\ x_2 \end{bmatrix} \\ &= \begin{bmatrix} (b_{11} + a_{11})x_1 + (b_{12} + a_{12})x_2 \\ (b_{21} + a_{21})x_1 + (b_{22} + a_{22})x_2 \end{bmatrix} = \begin{bmatrix} b_{11} + a_{11} & b_{12} + a_{12} \\ b_{21} + a_{21} & b_{22} + a_{22} \end{bmatrix}\begin{bmatrix} x_1 \\ x_2 \end{bmatrix}\end{aligned}$$

The matrix of $[S + T]$, which it is natural to call $[\mathbf{B} + \mathbf{A}]$, is a matrix obtained by adding the corresponding elements of $\mathbf{B}$ and $\mathbf{A}$. Briefly $\mathbf{B} + \mathbf{A} = \mathbf{A} + \mathbf{B} = [a_{ij} + b_{ij}]$. Here is our definition of the *sum* of two matrices. Obviously two matrices can be added only if they have the same number of rows and the same number of columns. By $\mathbf{A} - \mathbf{B}$, we mean $\mathbf{A} + (-1)\mathbf{B}$. For example,

$$\begin{bmatrix} 1 & 6 \\ 2 & 5 \\ 3 & 4 \end{bmatrix} + \begin{bmatrix} 3 & -2 \\ 2 & -1 \\ 1 & 0 \end{bmatrix} = \begin{bmatrix} 4 & 4 \\ 4 & 4 \\ 4 & 4 \end{bmatrix}$$

$$\begin{bmatrix} 1 & 6 \\ 2 & 5 \\ 3 & 4 \end{bmatrix} - \begin{bmatrix} 3 & -2 \\ 2 & -1 \\ 1 & 0 \end{bmatrix} = \begin{bmatrix} -2 & 8 \\ 0 & 6 \\ 2 & 4 \end{bmatrix}$$

Now we must define the *product* of two matrices. Imagine two linear transformations S and T, and ask what happens if first we apply T to a vector $\mathbf{x}$ and then apply S to $T(\mathbf{x})$, that is, if we compute $S[T(\mathbf{x})]$. First off, if T is a mapping from R^n to R^m, $T(\mathbf{x})$ is a vector of R^m, say y. If $S[T(\mathbf{x})] = S(\mathbf{y})$ is to have any meaning at all, S must be a mapping from R^m to some vector space, say R^k. Then $S[T(\mathbf{x})]$ is a vector of R^k. The combined linear[2] transformation $S[T(\mathbf{x})]$ which we may call $ST(\mathbf{x})$, the product of S and T, is a mapping from R^n to R^k. $\mathbf{A}$, the matrix of T, is $m \times n$; $\mathbf{B}$, the matrix of S, must be $k \times m$; and the matrix of ST must be $k \times n$. Let us call this matrix $\mathbf{BA}$ the "product" of the two matrices and see if we can compute what it is. We know first that

[1] If you enter into the spirit of the mathematical game you will ask: How do you know the mapping thus defined is *linear? Proof:*

$$\begin{aligned}(S + T)(a\mathbf{u} + b\mathbf{v}) &= S(a\mathbf{u} + b\mathbf{v}) + T(a\mathbf{u} + b\mathbf{v}) = aS(\mathbf{u}) + bS(\mathbf{v}) + aT(\mathbf{u}) + bT(\mathbf{v}) \\ &= a[S(\mathbf{u}) + T(\mathbf{u})] + b[S(\mathbf{v}) + T(\mathbf{v})] = a(S + T)(\mathbf{u}) + b(S + T)(\mathbf{v})\end{aligned}$$

[2] Prove that ST is linear!

$$\mathbf{y} = T(\mathbf{x}) = \begin{bmatrix} a_{11}a_{12} & \cdots & a_{1n} \\ a_{21}a_{22} & \cdots & a_{2n} \\ \cdots & \cdots & \cdots \\ a_{m1}a_{m2} & \cdots & a_{mn} \end{bmatrix} \begin{bmatrix} x_1 \\ x_2 \\ \cdot \\ x_n \end{bmatrix}$$

$$= \begin{bmatrix} a_{11}x_1 + a_{12}x_2 + \cdots + a_{1n}x_n \\ a_{21}x_1 + a_{22}x_2 + \cdots + a_{2n}x_n \\ \cdots\cdots\cdots\cdots \\ a_{m1}x_1 + a_{m2}x_2 + \cdots + a_{mn}x_n \end{bmatrix}$$

and

$$\mathbf{z} = ST(\mathbf{x}) = \begin{bmatrix} b_{11}b_{12} & \cdots & b_{1m} \\ b_{21}b_{22} & \cdots & b_{2m} \\ \cdots & \cdots & \cdots \\ b_{k1}b_{k2} & \cdots & b_{km} \end{bmatrix} \begin{bmatrix} \Sigma a_{1j}x_j \\ \Sigma a_{2j}x_j \\ \cdots \\ \Sigma a_{mj}x_j \end{bmatrix}$$

$$= \begin{bmatrix} b_{11}\Sigma a_{1j}x_j + b_{12}\Sigma a_{2j}x_j + \cdots + b_{1m}\Sigma a_{mj}x_j \\ b_{21}\Sigma a_{1j}x_j + b_{22}\Sigma a_{2j}x_j + \cdots + b_{2m}\Sigma a_{mj}x_j \\ \cdots\cdots\cdots\cdots \\ b_{k1}\Sigma a_{1j}x_j + b_{k2}\Sigma a_{2j}x_j + \cdots + b_{km}\Sigma a_{mj}x_j \end{bmatrix}$$

In the last monstrous term, sort out the coefficients of $x_1, x_2, \ldots, x_n$ in each row. For example, extracting the x_1's in the first row, we find that the coefficient is $b_{11}a_{11} + b_{12}a_{21} + \cdots + b_{1m}a_{m1}$. The coefficient of x_2 in the first row is $b_{11}a_{12} + b_{12}a_{22} + \cdots + b_{1m}a_{m2}$. The coefficient of x_3 in the second row is $b_{21}a_{13} + b_{22}a_{23} + \cdots + b_{2m}a_{m3}$, etc. After all this we recognize that the coefficient of x_k in the ith row is $\sum_{j=1}^{m} b_{ij}a_{jk}$. Thus summarizing, finally, we see that

$$\mathbf{z} = ST(\mathbf{x}) = \begin{bmatrix} \Sigma b_{1j}a_{j1}x_1 + \Sigma b_{1j}a_{j2}x_2 + \cdots + \Sigma b_{1j}a_{jn}x_n \\ \Sigma b_{2j}a_{j1}x_1 + \Sigma b_{2j}a_{j2}x_2 + \cdots + \Sigma b_{2j}a_{jn}x_n \\ \cdots\cdots\cdots\cdots \\ \Sigma b_{kj}a_{j1}x_1 + \Sigma b_{kj}a_{j2}x_2 + \cdots + \Sigma b_{kj}a_{jn}x_n \end{bmatrix}$$

$$= \begin{bmatrix} \Sigma b_{1j}a_{j1} & \Sigma b_{1j}a_{j2} & \cdots & \Sigma b_{1j}a_{jn} \\ \Sigma b_{2j}a_{j1} & \Sigma b_{2j}a_{j2} & \cdots & \Sigma b_{2j}a_{jn} \\ \cdots & \cdots & \cdots & \cdots \\ \Sigma b_{kj}a_{j1} & \Sigma b_{kj}a_{j2} & \cdots & \Sigma b_{kj}a_{jn} \end{bmatrix} \begin{bmatrix} x_1 \\ x_2 \\ \cdot \\ x_n \end{bmatrix} = \mathbf{BAx}$$

All the sums run on j from 1 to m, which just fits because **B** has m columns (the second subscript) and **A** has m rows (the first subscript). The product matrix **BA** has k rows and n columns, as it must. In words: The product of a $k \times m$ matrix **B** and an $m \times n$ matrix **A** is a $k \times n$ matrix. To find the element of **BA** in the first row and first column, take the sum of the cross products of the elements in the first *row* of **B** and the corresponding elements in the first *column* of **A**. To find the element in the second row and third column of **BA**, take the sum of cross products of corresponding elements in the second row of **B** and the third column of **A**. To find the (i,j) element of **BA**, sum the products of corresponding elements in the ith row of **B** and the jth column of **A**.

Example:

$$\begin{bmatrix} 0 & 1 \\ 1 & 3 \\ 3 & 0 \end{bmatrix} \begin{bmatrix} 4 & 2 \\ 1 & 3 \end{bmatrix} = \begin{bmatrix} 1 & 3 \\ 7 & 11 \\ 12 & 6 \end{bmatrix}$$

Sample computation:

$$(0 \times 4) + (1 \times 1) = 1$$
$$(1 \times 4) + (3 \times 1) = 7 \qquad \text{etc.}$$

Note that as distinct from everyday multiplication of numbers, the *order* in which the factors are written is vital in matrix multiplication. **BA** and **AB** are different. In fact, since to multiply two matrices the first factor must have as many columns as the second factor has rows, **BA** may be perfectly well defined while **AB** has no meaning at all. (Remember we are really talking about linear transformations from one vector space to another.) In the numerical example, the first matrix factor is 3×2, the second, 2×2. In reverse order, no multiplication is possible. (*Easy exercise:* When is multiplication in both orders possible?) Even if **BA** and **AB** are both defined, there is no reason why **BA** and **AB** should be the same.

Example:

$$\begin{bmatrix} a & b & c \\ d & e & f \\ g & h & i \end{bmatrix} \begin{bmatrix} 0 & 0 & 1 \\ 0 & 1 & 0 \\ 1 & 0 & 0 \end{bmatrix} = \begin{bmatrix} c & b & a \\ f & e & d \\ i & h & g \end{bmatrix} \qquad \begin{bmatrix} 0 & 0 & 1 \\ 0 & 1 & 0 \\ 1 & 0 & 0 \end{bmatrix} \begin{bmatrix} a & b & c \\ d & e & f \\ g & h & i \end{bmatrix} = \begin{bmatrix} g & h & i \\ d & e & f \\ a & b & c \end{bmatrix}$$

B-8. IDENTITY MATRIX—INVERSES

There is no such operation as dividing one matrix by another, but there is a similar concept which we must now describe.

Perhaps the simplest linear transformation of R^n into itself is the one which maps any $\mathbf{x}$ exactly into itself; that is, $T(\mathbf{x}) = \mathbf{x}$. It is usually called the *identity* transformation. It is easily seen that the matrix corresponding to this simple transformation is the matrix

$$\begin{bmatrix} 1 & 0 & 0 & \dots & 0 \\ 0 & 1 & 0 & \dots & 0 \\ 0 & 0 & 1 & \dots & 0 \\ \dots & \dots & \dots & \dots & \dots \\ 0 & 0 & 0 & \dots & 1 \end{bmatrix}$$

called the *identity matrix* and usually designated as **I** or $\mathbf{I}_n$ (or sometimes as **E**, for Einheit). It is a square matrix, since it maps a vector space into itself. What happens if we multiply another $n \times n$ matrix, say **A**, by **I**, either before or after; i.e., what is **IA** or **AI**? We can answer this question without actually doing any matrix multiplication—by just thinking about transformations. For if we represent the identity mapping by I and the mapping corresponding to **A** by T, we have $I(\mathbf{x}) = \mathbf{x}$; hence $TI(\mathbf{x}) = T(\mathbf{x})$ and $IT(\mathbf{x}) = T(\mathbf{x})$. In other words $IT = TI = T$, as linear transformations. Since matrices merely represent transformations, it must follow that $\mathbf{AI} = \mathbf{IA} = \mathbf{A}$; that is, multiplying a matrix by **I** just gives us back the original matrix. And since consistency, though the hobgoblin of little minds, is the main virtue of mathematics, it indeed works that way:

$$\begin{bmatrix} 1 & 0 \\ 0 & 1 \end{bmatrix}\begin{bmatrix} a & b \\ c & d \end{bmatrix} = \begin{bmatrix} a & b \\ c & d \end{bmatrix}\begin{bmatrix} 1 & 0 \\ 0 & 1 \end{bmatrix} = \begin{bmatrix} a & b \\ c & d \end{bmatrix}$$

The identity matrix plays the same role in matrix multiplication that the number 1 does in multiplying numbers.

Now dividing 6 by 3 is the same thing as multiplying 6 by $\frac{1}{3}$, the reciprocal of 3. How do we know that $\frac{1}{3}$ is the reciprocal of 3? Because $3 \times \frac{1}{3} = \frac{1}{3} \times 3 = 1$. Take any square matrix **A**. There *may* be another square matrix **B** with the nice property that $\mathbf{BA} = \mathbf{AB} = \mathbf{I}$. If such is the case, **B** is called the "inverse" of **A**, usually written $\mathbf{A}^{-1}$. Thus $\mathbf{AA}^{-1} = \mathbf{A}^{-1}\mathbf{A} = \mathbf{I}$. By virtue of the same definition, **A** is the inverse of $\mathbf{A}^{-1}$. Inverse matrices play much the same role as reciprocals of numbers. For example, if

$$\mathbf{A} = \begin{bmatrix} 3 & 4 \\ 4 & 6 \end{bmatrix}$$

then

$$\mathbf{A}^{-1} = \begin{bmatrix} 3 & -2 \\ -2 & 1.5 \end{bmatrix}$$

because

$$\begin{bmatrix} 3 & 4 \\ 4 & 6 \end{bmatrix}\begin{bmatrix} 3 & -2 \\ -2 & 1.5 \end{bmatrix} = \begin{bmatrix} 1 & 0 \\ 0 & 1 \end{bmatrix}$$

Actually, *finding* the inverse of a given matrix or even discovering whether a given matrix *has* an inverse may be a formidable computational job.

Let us view the inverse matrix from the linear-mapping point of view. Suppose T maps R^n into itself: what **x** is mapped into **y** by T? If, and only if, T is 1:1, this question has a unique answer. This answer sets up a correspondence between y and x which we can call the *inverse* mapping to T and designate T^{-1}. By definition, then, if $\mathbf{y} = T(\mathbf{x})$, $\mathbf{x} = T^{-1}(\mathbf{y})$. It can be shown that the inverse of a linear mapping is linear. Now what happens if we apply T to a vector **x** and then apply T^{-1} to $T(\mathbf{x})$; that is, what about the mapping $T^{-1}T$? Obviously, $T^{-1}T(\mathbf{x}) = \mathbf{x}$, for if $\mathbf{y} = T(\mathbf{x})$, $T^{-1}T(\mathbf{x}) = T^{-1}(\mathbf{y}) = \mathbf{x}$: all we are doing is going from $\mathbf{x}$ to $\mathbf{y}$ via T and then asking where did $\mathbf{y}$ come from, and clearly it came from $\mathbf{x}$. So $T^{-1}T$ is the identity mapping $T^{-1}T = I$. But if $\mathbf{A}$ is the matrix of T, and $\mathbf{B}$ is the matrix of T^{-1}, we know that $\mathbf{BA}$ is the matrix of $T^{-1}T$, and so $\mathbf{BA} = \mathbf{I}$, $\mathbf{B} = \mathbf{A}^{-1}$. The inverse of a matrix is the matrix corresponding to the inverse mapping of the original matrix. And now we know which matrices have inverses and which do not. If a mapping is 1:1, there is an inverse mapping, otherwise not. So square matrices belonging to a 1:1 mapping have inverses. Such matrices, like the mappings they describe, are called "nonsingular."

Finally, consider two 1:1 transformations S and T and apply them in succession to $\mathbf{x}$ to get $\mathbf{z} = ST(\mathbf{x})$; that is, $\mathbf{y} = T(\mathbf{x})$ and

$$\mathbf{z} = S(\mathbf{y}) = S[T(\mathbf{x})] = ST(\mathbf{x})$$

Since S and T are both 1:1, we can get back uniquely from $\mathbf{z}$ to its unique "parent" $\mathbf{y}$ and from $\mathbf{y}$ to $\mathbf{z}$'s unique "grandparent" $\mathbf{x}$. Hence the product transformation ST is nonsingular and has an inverse. Moreover, we have practically worked out what that inverse mapping is. To get from $\mathbf{z}$ back to $\mathbf{y}$ we have $\mathbf{y} = S^{-1}(\mathbf{z})$ and to get from $\mathbf{y}$ to $\mathbf{x}$ we have $\mathbf{x} = T^{-1}(\mathbf{y})$. Hence $\mathbf{x} = T^{-1}[S^{-1}(\mathbf{z})] = T^{-1}S^{-1}(\mathbf{z})$. This says that $(ST)^{-1} = T^{-1}S^{-1}$: the inverse of a product is the product of the inverses in reverse order. As usual, we can deduce immediately a theorem about matrices. If $\mathbf{A}$ and $\mathbf{B}$ are, respectively, the nonsingular square matrices belonging to T and S, then we know that $\mathbf{BA}$ is the matrix of ST. And $(\mathbf{BA})^{-1}$ is the matrix of $(ST)^{-1}$. But $(ST)^{-1} = T^{-1}S^{-1}$, and the matrix of this transformation is $\mathbf{A}^{-1}\mathbf{B}^{-1}$. Hence the theorem $(\mathbf{BA})^{-1} = \mathbf{A}^{-1}\mathbf{B}^{-1}$: *The inverse matrix of a product is the product of the inverses in reverse order.* Remember, we have assumed $\mathbf{A}$ and $\mathbf{B}$ to be square and to possess inverses themselves.

B-9. INNER PRODUCT—TRANSPOSED MATRICES

In economics we often come across expressions of the form $\sum_{i=1}^{n} p_i x_i$; the maximand in a linear program is always written in this way. Such expressions also appear in statistics, under the name of "covariance," $\Sigma(x_i - \bar{x})(y_i - \bar{y})$. It is possible to think of such sum-of-cross-products expressions as generated by two vectors, namely, $\mathbf{p} = [p_1, \ldots, p_n]$ and $\mathbf{x} = [x_1, \ldots, x_n]$. Accordingly, we define $\Sigma p_i x_i$ as the *inner product* of the vectors $\mathbf{p}$ and $\mathbf{x}$, and we symbolize it by $(\mathbf{p},\mathbf{x})$. Obviously the order here makes no difference: $(\mathbf{p},\mathbf{x}) = (\mathbf{x},\mathbf{p})$. Other common notations are $\mathbf{p}'\mathbf{x}$ and $\mathbf{p}\cdot\mathbf{x}$.

There is still another way of looking at the inner product. If we write the vector **p** as a *row vector* and **x** as a *column vector*, **p** is $1 \times n$ and **x** is $n \times 1$ and we can compute the matrix product **px**, namely,

$$[p_1,p_2, \ldots ,p_n]\begin{bmatrix} x_1 \\ x_2 \\ \cdot \\ \cdot \\ \cdot \\ x_n \end{bmatrix} = [\mathbf{p},\mathbf{x}]$$

The product is a 1×1 matrix, i.e., an ordinary scalar, and turns out to be just the inner product $(\mathbf{p},\mathbf{x})$. The inner-product notion gives us a brief way of describing the technique of matrix multiplication. The product **BA**, where **B** is $k \times m$ and **A** is $m \times n$, is a $k \times n$ matrix whose (i,j) element is the inner product of the ith row of **B** and the jth column of **A**.

The "transpose" of an $m \times n$ matrix is an $n \times m$ matrix whose ith row is just the ith column of the original matrix. Thus if **A** is the matrix,

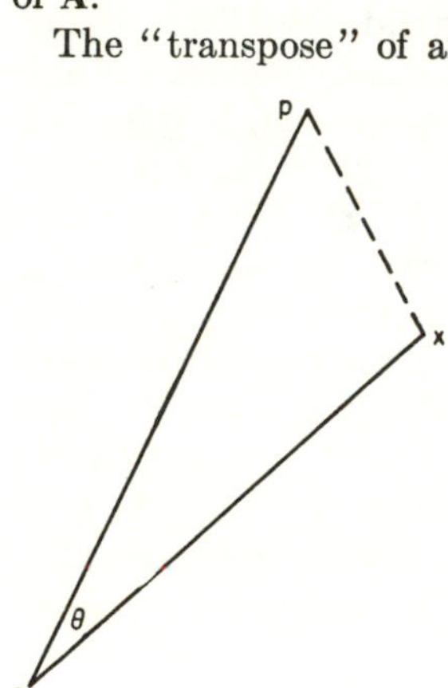

Fig. B-4

$$\mathbf{A} = \begin{bmatrix} a_{11} & a_{12} & a_{13} & a_{14} \\ a_{21} & a_{22} & a_{23} & a_{24} \\ a_{31} & a_{32} & a_{33} & a_{34} \end{bmatrix}$$

the transpose of **A**, usually written **A**′ or $\mathbf{A}^T$, is

$$\mathbf{A}' = \begin{bmatrix} a_{11} & a_{21} & a_{31} \\ a_{12} & a_{22} & a_{32} \\ a_{13} & a_{23} & a_{33} \\ a_{14} & a_{24} & a_{34} \end{bmatrix}$$

This notation is the origin of the expression $\mathbf{p}'\mathbf{x}$ for the inner product. If we think of all vectors as column vectors, then $\mathbf{p}'$, the transpose of a column, will be a row and $\mathbf{p}'\mathbf{x}$ is just the matrix product which yields the inner product.

Inner products have an important geometric significance. Figure B-4 shows two vectors **p** and **x** of n elements each, emanating from the origin 0. The angle between the vectors is designated θ, and the triangle formed by the two vectors has been completed by connecting their end points. By elementary trigonometry

$$|\mathbf{p} - \mathbf{x}|^2 = |\mathbf{p}|^2 + |\mathbf{x}|^2 - 2|\mathbf{p}|\,|\mathbf{x}| \cos \theta$$

where $|\mathbf{p}|$ and $|\mathbf{x}|$ denote the lengths of $\mathbf{p}$ and $\mathbf{x}$, respectively, and $|\mathbf{p} - \mathbf{x}|$ denotes the distance between their end points. By the Pythagorean theorem of geometry

$$|\mathbf{p}| = \sqrt{\Sigma_i p_i^2} \qquad |\mathbf{x}| = \sqrt{\Sigma_i x_i^2}$$

and $$|\mathbf{p} - \mathbf{x}| = \sqrt{\Sigma_i (p_i - x_i)^2} = \sqrt{\Sigma_i p_i^2 + \Sigma_i x_i^2 - 2\Sigma_i p_i x_i}$$

Substituting these in the formula connecting the lengths and $\cos \theta$ and remembering that $\Sigma p_i x_i$ is the inner product $(\mathbf{p},\mathbf{x})$ we find

$$-2(\mathbf{p},\mathbf{x}) = -2|\mathbf{p}|\,|\mathbf{x}| \cos \theta$$

or $$\cos \theta = \frac{(\mathbf{p},\mathbf{x})}{|\mathbf{p}|\,|\mathbf{x}|}$$

Since the denominator of this fraction is surely positive, this leads to the conclusion that the angle between two vectors is acute, right, or obtuse, according as their inner product is positive, zero, or negative.

For most purposes we need to make only slight use of the transpose idea. One theorem on transposes that occasionally turns up states that $(\mathbf{BA})' = \mathbf{A}'\mathbf{B}'$: the transpose of a product is the product of the transposes, in reverse order. A dull but useful exercise for the reader would be to compute both sides of the above equation for some matrices to verify the theorem, or better, prove the theorem from the definitions of a product and a transpose.

The mathematical significance of transposes is the following: Suppose we look at the inner product of a vector $\mathbf{y}$ not with $\mathbf{x}$ itself but with $\mathbf{Ax}$, namely, $(\mathbf{y},\mathbf{Ax})$. $\mathbf{Ax}$ is of course a vector with the same number of components as $\mathbf{y}$; it is simply the product of $\mathbf{x}$ and the matrix $\mathbf{A}$. Now we may ask: Is there a matrix $\mathbf{C}$ such that $(\mathbf{y},\mathbf{Ax}) = (\mathbf{x},\mathbf{Cy})$, that is, such that the inner product of $\mathbf{x}$ and $\mathbf{Cy}$ is the same as the inner product of $\mathbf{y}$ and $\mathbf{Ax}$? Straightforward computation shows that indeed there is such a $\mathbf{C}$, and it is none other than $\mathbf{A}'$; that is, $(\mathbf{y},\mathbf{Ax}) = (\mathbf{x},\mathbf{A}'\mathbf{y})$. Using this fact and recalling that $(\mathbf{x},\mathbf{A}'\mathbf{y}) = (\mathbf{A}'\mathbf{y},\mathbf{x})$, since order is irrelevant in an inner product, we can give a proof of the theorem that $(\mathbf{BA})' = \mathbf{A}'\mathbf{B}'$, namely,

$$(\mathbf{y},\mathbf{BAx}) = [\mathbf{y},\mathbf{B}(\mathbf{Ax})] = (\mathbf{Ax},\mathbf{B}'\mathbf{y}) = (\mathbf{B}'\mathbf{y},\mathbf{Ax}) = (\mathbf{x},\mathbf{A}'\mathbf{B}'\mathbf{y}) = (\mathbf{A}'\mathbf{B}'\mathbf{y},\mathbf{x})$$

Putting the ends together, $(\mathbf{y},\mathbf{BAx}) = (\mathbf{A}'\mathbf{B}'\mathbf{y},\mathbf{x})$. But $(\mathbf{BA})'$ is the matrix such that $(\mathbf{y},\mathbf{BAx}) = [(\mathbf{BA})'\mathbf{y},\mathbf{x}]$. Hence $(\mathbf{BA})' = \mathbf{A}'\mathbf{B}'$.

B-10. SIMULTANEOUS EQUATIONS

The equipment now at hand applies very neatly to the study of simultaneous linear equations. Since many of the equation systems that come up in this book are systems with as many equations as unknowns, we shall limit ourselves mainly to that case. A set of n linear equations in n unknowns can be written

$$\begin{aligned} a_{11}x_1 + a_{12}x_2 + \cdots + a_{1n}x_n &= b_1 \\ a_{21}x_1 + a_{22}x_2 + \cdots + a_{2n}x_n &= b_2 \\ \cdots\cdots\cdots\cdots\cdots\cdots\cdots& \\ a_{n1}x_1 + a_{n2}x_2 + \cdots + a_{nn}x_n &= b_n \end{aligned}$$

The a's and b's are given numbers; the problem is to find a set of numerical x's which will make all the equations true.

First off we can compress the notation considerably. If we let $\mathbf{A}$ be the matrix $[a_{ij}]$, $\mathbf{x}$ be the column vector

$$\begin{bmatrix} x_1 \\ \cdot \\ \cdot \\ \cdot \\ x_n \end{bmatrix}$$

and $\mathbf{b}$ be the column vector

$$\begin{bmatrix} b_1 \\ \cdot \\ \cdot \\ \cdot \\ b_n \end{bmatrix}$$

we have simply $\mathbf{Ax} = \mathbf{b}$.

Now actually solving linear equations is literally child's play. The old-fashioned high-school method of simply eliminating one unknown at a time and thus reducing the problem to fewer and fewer unknowns is still a good one and is still the backbone of most machine methods. It will always work; i.e., either it will produce a solution or it will break down in an obvious contradiction, which tells us that the equations have no solution, or are *inconsistent*. But we are not at the moment interested in computation problems, but rather in theory.

The system of equations $\mathbf{Ax} = \mathbf{b}$ can be thought of in the following way: The matrix $\mathbf{A}$ represents a linear transformation T. The equations state that $\mathbf{x}$ is a vector which is mapped or carried by T into the vector $\mathbf{b}$. We are asked what vector $\mathbf{x}$ is mapped by T into $\mathbf{b}$. There are three possible answers to this question: There may be no such vector $\mathbf{x}$; there may be exactly one such vector $\mathbf{x}$; there may be many such $\mathbf{x}$'s. The equations may have no solution, a unique solution, or many solutions. It is mainly the middle case that interests us here.

If the transformation T is 1:1, that is, if T and $\mathbf{A}$ are nonsingular, then b has indeed one "parent" which can be identified by the inverse mapping as $T^{-1}(\mathbf{b})$. Indeed $\mathbf{x} = T^{-1}(\mathbf{b})$ does satisfy the equation $T[T^{-1}(\mathbf{b})] = \mathbf{b}$. In matrix terms, $\mathbf{x} = \mathbf{A}^{-1}\mathbf{b}$. If we insert this in the given equations, we verify that it is indeed the solution

$$\mathbf{A}\mathbf{A}^{-1}\mathbf{b} = \mathbf{I}\mathbf{b} = \mathbf{b}$$

We could have seen this directly from the equations $\mathbf{A}\mathbf{x} = \mathbf{b}$. Multiply both sides on the left by $\mathbf{A}^{-1}$ to get

$$\mathbf{A}^{-1}\mathbf{A}\mathbf{x} = \mathbf{I}\mathbf{x} = \mathbf{x} = \mathbf{A}^{-1}\mathbf{b}$$

which is the desired solution.

We saw in Sec. B-8 that the inverse of

$$\begin{bmatrix} 3 & 4 \\ 4 & 6 \end{bmatrix}$$

is

$$\begin{bmatrix} 3 & -2 \\ -2 & 1.5 \end{bmatrix}$$

Hence the solution of the equations

$$\begin{aligned} 3x_1 + 4x_2 &= 5 \\ 4x_1 + 6x_2 &= 8 \end{aligned}$$

should be

$$\begin{bmatrix} x_1 \\ x_2 \end{bmatrix} = \begin{bmatrix} 3 & -2 \\ -2 & 1.5 \end{bmatrix} \begin{bmatrix} 5 \\ 8 \end{bmatrix} = \begin{bmatrix} -1 \\ 2 \end{bmatrix}$$

And a check shows that this is the case.

If we *know* the matrix $\mathbf{A}^{-1}$, then it is a trivial matter to solve any system of equations $\mathbf{A}\mathbf{x} = \mathbf{b}$, any system whose matrix of coefficients is $\mathbf{A}$. What this means is that we can solve the system easily for varying $\mathbf{b}$'s. But generally (especially if the number of rows and columns is large) we would not know $\mathbf{A}^{-1}$. The fact of the matter is, practically speaking, you don't solve equations by use of the inverse. Quite the contrary, you compute an inverse if you want it by solving linear equations. For example, suppose we want the inverse of

$$\begin{bmatrix} a_{11} & a_{12} \\ a_{21} & a_{22} \end{bmatrix}$$

i.e., we want a matrix

$$\begin{bmatrix} c_{11} & c_{12} \\ c_{21} & c_{22} \end{bmatrix}$$

such that

$$\begin{bmatrix} a_{11} & a_{12} \\ a_{21} & a_{22} \end{bmatrix} \begin{bmatrix} c_{11} & c_{12} \\ c_{21} & c_{22} \end{bmatrix} = \begin{bmatrix} 1 & 0 \\ 0 & 1 \end{bmatrix}$$

Element by element this says that

$$a_{11}c_{11} + a_{12}c_{21} = 1, \; a_{11}c_{12} + a_{12}c_{22} = 0$$
$$a_{21}c_{11} + a_{22}c_{21} = 0, \; a_{21}c_{12} + a_{22}c_{22} = 1$$

Here we have two sets of linear equations. In the first pair the unknowns are c_{11} and c_{21}, in the second the unknowns are c_{12} and c_{22}. By solving them we compute the elements c_{ij} of $\mathbf{A}^{-1}$. A sample computation: Eliminate c_{11} by $c_{11} = -(a_{22}/a_{21})c_{21}$ (supposing $a_{21} \neq 0$). Then substitute to get $-(a_{11}a_{22}/a_{21})c_{21} + a_{12}c_{21} = 1$, or $c_{21} = -a_{21}/(a_{11}a_{22} - a_{12}a_{21})$, provided the denominator isn't zero. (If it is, the matrix **A** isn't nonsingular and there will be no solution or many!)

If matrices aren't a computational convenience, why drag them into the discussion of simultaneous equations at all? In the first place, there is the sheer notational simplicity of writing $\mathbf{Ax} = \mathbf{b}$ instead of something full of eyesores such as double subscripts. Then the notation $\mathbf{x} = \mathbf{A}^{-1}\mathbf{b}$ tells us at once the important fact that each component of the solution $[x_1, x_2, \ldots, x_n]$ is a linear combination of the **b**'s. Moreover, there are important theorems on matrices, some of which are, naturally, too advanced to be discussed here, which make the matrix idea a great analytical convenience in the discussion of linear equations.

B-11. SIMULTANEOUS EQUATIONS AND RANK OF A MATRIX

So far we have discussed only the case in which the matrix **A** is nonsingular. We are not even through with that case yet. How can we tell when a square matrix is nonsingular (i.e., when it is 1:1, or possesses an inverse)? Early in this appendix we saw that solving linear equations can also be thought of as expressing one given vector as a linear combination of other given vectors. The equations $\mathbf{Ax} = \mathbf{b}$ can profitably be viewed in this light. If we symbolize the *columns* of **A** by $\mathbf{A}_1, \mathbf{A}_2, \ldots, \mathbf{A}_n$, then we can rewrite the equations as

$$\mathbf{A}_1x_1 + \mathbf{A}_2x_2 + \cdots + \mathbf{A}_nx_n = \mathbf{b} \tag{B-9}$$

expressing **b** as a linear combination of the columns of **A**. When can we guarantee that this is possible, whatever **b**? There are n columns $\mathbf{A}_i$, each with n components. If they are *linearly independent*, they form what we earlier called a basis of R^n (i.e., a legitimate set of axes) and *any* vector of R^n can be expressed *uniquely* as a linear combination of the columns (a point has one, and only one, set of coordinates with respect to a set of axes). Thus *if the columns of* **A** *are linearly independent, the equa-*

tions $\mathbf{Ax} = \mathbf{b}$ *possess a unique solution.* Hence we have the further theorem. *The* $n \times n$ *matrix* $\mathbf{A}$ *is nonsingular* (*represents a* 1:1 *transformation, or possesses an inverse*) *if and only if its columns are linearly independent.* Remember that the columns of $\mathbf{A}$ are nothing but the images of the unit vectors under the transformation (in other words, $\mathbf{A}_i = \mathbf{Ae}_i$; try it!). So still another way of describing a nonsingular matrix is to say that it maps the unit vectors into a basis for R^n.

What if the columns of $\mathbf{A}$ are linearly *dependent?* Then not every vector of R^n can be expressed as in (B-9). Two cases present themselves: either $\mathbf{b}$ is one of the vectors of R^n that *can* be expressed as a linear combination of the $\mathbf{A}_i$, or it is one of those that *can't*.[1] In the latter case, clearly, there is no solution to the equations. In the former case, just as clearly, there is a solution, and indeed there are infinitely many solutions. It is not hard to see why. Suppose $\mathbf{A}_n$ can be expressed as a linear combination of the other columns. Then if $\mathbf{b}$ is a linear combination of all the $\mathbf{A}_i$, it can also be represented as a combination of $\mathbf{A}_1, \ldots, \mathbf{A}_{n-1}$. As in the simplex method we can find an "equivalent combination" for $\mathbf{A}_n$ and introduce varying amounts of it to give many different solutions. More formally, if the columns are linearly dependent, then $\mathbf{Ay} = \boldsymbol{\phi}$ has a nontrivial solution. If, then, $\mathbf{A\bar{x}} = \mathbf{b}$,

$$\mathbf{A}(\mathbf{\bar{x}} + k\mathbf{y}) = \mathbf{A\bar{x}} + k\mathbf{Ay} = \mathbf{A\bar{x}} = \mathbf{b}$$

so along with $\mathbf{\bar{x}}$, $\mathbf{\bar{x}} + k\mathbf{y}$ is a solution for *any* scalar k.

For any matrix, square or not, the number of linearly independent columns is called the "rank" of the matrix.[2] Thus an $n \times n$ matrix is nonsingular if and only if its rank is n. The next section will show that a test of nonsingularity and a method of finding a rank is provided by the concept of a "determinant."

In terms of the above concepts the reader might like to investigate for himself the situation with systems of linear equations not equal in number to the number of unknowns. The question always reduces to one of whether the vector on the right-hand side can be expressed as a linear combination of the columns of the matrix on the left-hand side. If $\mathbf{A}$ is $m \times n$ and $m < n$ (more unknowns than equations), the greatest possible rank is m. In this case (i.e., if the rank of $\mathbf{A}$ is actually m), $\mathbf{b}$ is always expressible in terms of the columns of $\mathbf{A}$, but never uniquely. If the rank of $\mathbf{A}$ is less than m, there may again be no solution or many. If $m > n$ (more equations than unknowns), we can never be sure that a solution exists at all. Since there are fewer than m columns, they can-

[1] The set of vectors that can be so expressed is called the *subspace spanned* by the columns of $\mathbf{A}$. Hence the two cases occur according as $\mathbf{b}$ is or is not in that subspace.

[2] It is a theorem that the number of independent rows equals the number of independent columns. Hence (1) the rank of a matrix can never exceed the smaller of the number of rows or columns, and (2) a matrix and its transpose have the same rank.

not form a basis of R^m. Hence some **b**'s will not be in the subspace spanned by the columns, and for such **b**'s there will be no solution.

Let us return for a moment to the square $n \times n$ case. If the right-hand vector **b** is the null vector, the system $\mathbf{Ax} = \phi$ is described as *homogeneous*. If **A** is nonsingular, we know that there is a unique solution to these equations, and indeed that solution is $\mathbf{x} = \phi$ (recall that we long ago proved that a 1:1 transformation maps only the null vector into the null vector; here we have that proposition coming up again). When homogeneous equations come up in applications we usually want to know when they have a solution other than the "trivial" one, $\mathbf{x} = \phi$. The answer is: Homogeneous equations have a nonzero solution only when the rank of **A** is less than n, i.e., when **A** is singular or its columns are linearly dependent. For if **A** is singular, one of its columns can be expressed as a linear combination of the others, which is the same as saying that a nontrivial combination of the columns can be made equal to the null vector, or $\mathbf{A\bar{x}} = \phi$ for $\mathbf{\bar{x}} \neq \phi$, and hence $\mathbf{\bar{x}}$ is the desired nontrivial solution.[1] If $\mathbf{\bar{x}}$ is a solution, so is $k\mathbf{\bar{x}}$, for any scalar k.

Thus if the rank of **A** is $n - 1$, we get a whole family of solutions of the homogeneous equations, namely, $\mathbf{\bar{x}}$ and all its scalar multiples. If the rank of **A** is $n - 2$, there are *two* such families of solutions (and they are themselves linearly independent) and in addition any linear combination of solutions is a solution. If $\mathbf{A\bar{x}} = \phi$ and $\mathbf{Ay} = \phi$, then

$$\mathbf{A}(s\mathbf{\bar{x}} + t\mathbf{y}) = s\mathbf{A\bar{x}} + t\mathbf{A\bar{y}} = \phi + \phi = \phi$$

If the rank of **A** is $n - k$, then there are k linearly independent families of solutions, and again we can use all their linear combinations.

An excellent application of all this is the Walrasian system studied in Chap. 13 of the text. The factor market equations can be written $\mathbf{Ax} = \mathbf{r}$; the price-cost equations can be written $\mathbf{A'v} = \mathbf{p}$. **A** is $m \times n$ since there are m factors and n commodities. The *possibility* of satisfying these equations exactly, given the resource endowment, depends on the rank of **A**. Only here we have the added difficulty that only nonnegative **x**'s and **v**'s make sense, so something a bit more complicated is involved. We can think of the linear transformation **A** as mapping *nonnegative* vectors **x** (commodity outputs) into nonnegative factor-use vectors.

B-12. DETERMINANTS

Determinants are an almost unmitigated nuisance. You can "use" them to solve linear equations, but they are a very inefficient computa-

[1] Or else one of the columns of **A** (say the first) *is* ϕ, and then [1,0,0, . . . ,0] is a nontrivial solution.

tional device, and if they are to be used only analytically, we can do better with the concept of an inverse matrix. Still they do provide a "different" way of solving equations, and a direct way of computing an inverse matrix and of finding the rank of a given matrix, and so we will give them a once-over-lightly.

Suppose we solve the 2×2 system

$$\begin{bmatrix} a_{11} & a_{12} \\ a_{21} & a_{22} \end{bmatrix} \begin{bmatrix} x_1 \\ x_2 \end{bmatrix} = \begin{bmatrix} b_1 \\ b_2 \end{bmatrix}$$

by elimination. We write

$$\begin{aligned} a_{11}x_1 + a_{12}x_2 &= b_1 \\ a_{21}x_1 + a_{22}x_2 &= b_2 \end{aligned}$$

Multiply the first equation by a_{22} and the second by a_{12} and subtract. This eliminates x_2 and leaves

$$(a_{11}a_{22} - a_{12}a_{21})x_1 = a_{22}b_1 - a_{12}b_2$$

If the quantity in parentheses is not zero, we can divide it out and find a solution x_1. But if the quantity in parentheses is zero, and if the right-hand side is not zero, there is no solution x_1; and if the right-hand side is zero, any value of x_1 will do as a solution. This suggests that the expression $a_{11}a_{22} - a_{12}a_{21}$ has a lot to do with the *rank* of the matrix **A**. And indeed it has.

This number $a_{11}a_{22} - a_{12}a_{21}$ is called the "determinant" of the matrix **A** and is written $\det \mathbf{A} = \begin{vmatrix} a_{11} & a_{12} \\ a_{21} & a_{22} \end{vmatrix}$; that is, it is written in the same way as the matrix, but with vertical bars instead of brackets. Note that det **A** is a *number*, not a conventional object like a matrix.[1] To compute *any* 2×2 determinant (determinants are always square) simply multiply the northwest and southeast elements and subtract the product of the northeast and southwest elements. Thus

$$\begin{vmatrix} 3 & 4 \\ 4 & 6 \end{vmatrix} = 18 - 16 = 2; \begin{vmatrix} a & b \\ c & d \end{vmatrix} = ad - bc; \begin{vmatrix} 0 & 3 \\ -2 & 1 \end{vmatrix} = 0 - (-6) = 6$$

Later we shall define the value of a 3×3 or $n \times n$ determinant.

Now suppose that $\begin{vmatrix} a_{11} & a_{12} \\ a_{21} & a_{22} \end{vmatrix} = a_{11}a_{22} - a_{21}a_{12} = 0$. Then

$$\frac{a_{11}}{a_{12}} = \frac{a_{21}}{a_{22}} = c$$

say (if neither a_{12} nor $a_{22} = 0$, and these cases can be handled separately

[1] A 1×1 determinant *is* simply its single element. Thus $|12| = 12$.

as an exercise by the reader), and the columns of **A** are linearly dependent, since $\begin{bmatrix} a_{11} \\ a_{21} \end{bmatrix} = c \begin{bmatrix} a_{12} \\ a_{22} \end{bmatrix}$. What's more, if $a_{22}b_1 - a_{12}b_2 = 0$, then $b_1/a_{12} = b_2/a_{22} = d$, say, so that $\begin{bmatrix} b_1 \\ b_2 \end{bmatrix} = d \begin{bmatrix} a_{12} \\ a_{22} \end{bmatrix}$ and b *is* in the subspace spanned by $\begin{bmatrix} a_{12} \\ a_{22} \end{bmatrix}$, and we get solutions. Note, by the way, that $a_{22}b_1 - a_{12}b_1 = \begin{vmatrix} b_1 & a_{12} \\ b_2 & a_{22} \end{vmatrix}$.

On the other hand, if $\begin{vmatrix} a_{11} & a_{12} \\ a_{21} & a_{22} \end{vmatrix} \neq 0$, the columns of **A** *are* linearly independent by the same reasoning. For if $\begin{bmatrix} a_{11} \\ a_{21} \end{bmatrix} = c \begin{bmatrix} a_{12} \\ a_{22} \end{bmatrix}$, then $a_{11}/a_{12} = a_{21}/a_{22}$, and $a_{11}a_{22} - a_{12}a_{21} = 0$, which contradicts the assumption that det **A** is nonzero. Hence the vanishing of the determinant is a criterion of linear dependence. If the determinant is zero, the columns (and rows) are linearly dependent; if the determinant is not zero, the columns (and rows) are linearly independent.

One more thing. If det **A** $\neq 0$, so that a unique solution to the equations exists, we have all but computed that solution, namely,

$$x_1 = \frac{a_{11}b_1 - a_{12}b_2}{a_{22}a_{11} - a_{12}a_{21}} = \frac{\begin{vmatrix} b_1 & a_{12} \\ b_2 & a_{22} \end{vmatrix}}{\begin{vmatrix} a_{11} & a_{12} \\ a_{21} & a_{22} \end{vmatrix}}$$

and a little further calculation shows that

$$x_2 = \frac{\begin{vmatrix} a_{11} & b_1 \\ a_{21} & b_2 \end{vmatrix}}{\begin{vmatrix} a_{11} & a_{12} \\ a_{21} & a_{22} \end{vmatrix}}$$

For 2×2 systems we can state that if the determinant of the matrix doesn't vanish, the matrix is nonsingular, and the unique solution of the equations can be described thus: Each x_1 (or x_2) is a ratio of determinants—the denominator being the determinant of the system and the numerator the determinant we get from it by replacing the first (or second) column by the column of constants b. This rule, suitably extended to $n \times n$ systems, is called Cramer's rule.

We must now go on to define 3×3 and higher-order determinants. There are many ways to do this; we choose an inductive method. Consider the 3×3 determinant:

$$\begin{vmatrix} a_{11} & a_{12} & a_{13} \\ a_{21} & a_{22} & a_{23} \\ a_{31} & a_{32} & a_{33} \end{vmatrix}$$

If we cross out any row and any column, we are left with a 2×2 determinant, which we know how to evaluate. If we cross out the first row and first column, we get a determinant which is called the *minor* of a_{11}. If we attach the sign $(-1)^{1+1}$ to the minor, we get the *cofactor* of a_{11}, which we may call $\mathbf{A}_{11}$. Similarly, if we cross out the ith row and jth column, we get the minor of a_{ij}, and if we prefix the sign of $(-1)^{i+j}$, we get the cofactor of a_{ij}, namely, $\mathbf{A}_{ij}$. Thus the cofactor of any element is the determinant left if we delete the row and column in which that element lies, with an attached sign which is positive if the numbers of the row and column add up to an even number, negative if they add to an odd number. Thus in

$$\begin{vmatrix} 1 & 2 & 3 \\ 4 & 5 & 6 \\ 7 & 8 & 9 \end{vmatrix} \qquad \begin{aligned} \mathbf{A}_{11} &= +\begin{vmatrix} 5 & 6 \\ 8 & 9 \end{vmatrix} = -3, \ \mathbf{A}_{23} = -\begin{vmatrix} 1 & 2 \\ 7 & 8 \end{vmatrix} = 6, \\ \mathbf{A}_{32} &= -\begin{vmatrix} 1 & 3 \\ 4 & 6 \end{vmatrix} = 6,^{1} \text{ etc.} \end{aligned}$$

Now to define and evaluate a 3×3 determinant. Pick out a row or a column and compute the *cofactor* (sign and all) of each element in the row (or column). Multiply each element of the row (or column) by its own cofactor and add the products. The result is the determinant. It is a theorem that whichever row or column one chooses, the end result is always the same. For example, we can "expand" the above determinant first by the first row, then by the third column.[2] We get

$$1 \cdot \mathbf{A}_{11} + 2 \cdot \mathbf{A}_{12} + 3\mathbf{A}_{13} = 1 \cdot \begin{vmatrix} 5 & 6 \\ 8 & 9 \end{vmatrix} - 2 \cdot \begin{vmatrix} 4 & 6 \\ 7 & 9 \end{vmatrix} + 3 \cdot \begin{vmatrix} 4 & 5 \\ 7 & 8 \end{vmatrix}$$
$$= -3 + 12 - 9 = 0$$
$$3 \cdot \mathbf{A}_{13} + 6 \cdot \mathbf{A}_{23} + 9\mathbf{A}_{33} = 3 \cdot \begin{vmatrix} 4 & 5 \\ 7 & 8 \end{vmatrix} - 6 \cdot \begin{vmatrix} 1 & 2 \\ 7 & 8 \end{vmatrix} + 9 \cdot \begin{vmatrix} 1 & 2 \\ 4 & 5 \end{vmatrix}$$
$$= -9 + 36 - 27 = 0$$

We can now compute any 3×3 determinant. Naturally, practically speaking, if a row or column has a lot of zeros in it, that's the one along which to expand.

In the same way, the cofactor of any element of a 4×4 determinant is the 3×3 determinant obtained by deleting the row and column in

[1] Purely by accident this happens to have the same numerical value as $\mathbf{A}_{23}$.

[2] This theorem (which we shall not prove here) is of course a vital one. The proof is not hard, but it is lengthy, and it would involve us in the more usual definition of a determinant as the sum of $n!$ different terms, with appropriate signs, each term being a product of n elements, one from each row and one from each column. From this definition the desired result can be shown to follow. But the development is inevitably long and detailed.

which the given element lies (same convention as to the prefixed sign). A 4×4 determinant can be evaluated by expanding along any row or column and computing the sum of the products of each element in the row and its cofactor. This yields a number of 3×3 determinants, each of which can be reduced to 2×2 determinants, and evaluated. Hence we have a way of computing a determinant of *any* order.

It is a fact that if the determinant of a matrix vanishes, the matrix is singular, and its columns are linearly dependent and all the usual consequences follow. We found above that

$$\begin{vmatrix} 1 & 2 & 3 \\ 4 & 5 & 6 \\ 7 & 8 & 9 \end{vmatrix} = 0$$

It can be verified that

$$-1 \begin{bmatrix} 1 \\ 4 \\ 7 \end{bmatrix} + 2 \begin{bmatrix} 2 \\ 5 \\ 8 \end{bmatrix} = \begin{bmatrix} 3 \\ 6 \\ 9 \end{bmatrix}$$

so that the matrix is indeed singular. We have that

$$\begin{bmatrix} 1 & 2 & 3 \\ 4 & 5 & 6 \\ 7 & 8 & 9 \end{bmatrix} \begin{bmatrix} -1 \\ 2 \\ -1 \end{bmatrix} = \begin{bmatrix} 0 \\ 0 \\ 0 \end{bmatrix}$$

Cramer's rule extends to n equations in n unknowns, $\mathbf{Ax} = \mathbf{b}$. If det $\mathbf{A} \neq 0$, there is a unique solution; each x_i is a ratio of determinants. In the denominator is det $\mathbf{A}$. In the numerator is the determinant obtained from det $\mathbf{A}$ by replacing the ith column by $\mathbf{b}$. We repeat, however, that however nice it is to be able to state this, Cramer's rule is a very inefficient computational procedure, far inferior to high-school elimination for large systems.

Determinants also give us a foolproof, if foolish, way of finding the rank of a matrix, square or not. In an arbitrary matrix find the largest (in the sense of number of rows and columns) *nonzero* determinant that can be made by deleting rows and columns. If there is a nonzero $r \times r$ determinant and no nonzero determinant of more than r rows, the rank of the matrix is r.

There are various legal ways of manipulating determinants which the reader can find in any algebra text. For example, if the elements in a single row (or column) are all multiplied by a constant c, the value of the determinant is multiplied by c.

Proof:

$$\begin{vmatrix} ca_{11} & a_{12} & a_{13} \\ ca_{21} & a_{22} & a_{23} \\ ca_{31} & a_{23} & a_{33} \end{vmatrix} = ca_{11}\mathbf{A}_{11} + ca_{21}\mathbf{A}_{21} + ca_{31}\mathbf{A}_{31}$$
$$= c(a_{11}\mathbf{A}_{11} + a_{21}\mathbf{A}_{21} + a_{31}\mathbf{A}_{31})$$
$$= c \cdot \begin{vmatrix} a_{11} & a_{12} & a_{13} \\ a_{21} & a_{22} & a_{23} \\ a_{31} & a_{32} & a_{33} \end{vmatrix}$$

Hence a determinant with a row of zeros is zero. A determinant two of whose rows (or columns) are identical is zero (rows or columns linearly dependent).

Also,

$$\begin{vmatrix} a_{11} + d_1 & a_{12} & a_{13} \\ a_{21} + d_2 & a_{22} & a_{23} \\ a_{31} + d_3 & a_{32} & a_{33} \end{vmatrix} = \begin{vmatrix} a_{11} & a_{12} & a_{13} \\ a_{21} & a_{22} & a_{23} \\ a_{31} & a_{32} & a_{33} \end{vmatrix} + \begin{vmatrix} d_1 & a_{12} & a_{13} \\ d_2 & a_{22} & a_{23} \\ d_3 & a_{32} & a_{33} \end{vmatrix}$$

Proof: Expand along the first column. It now follows that we can add a multiple of a row (or column) to any other row (or column) without changing the value of the determinant. *Proof:* In the above equation, let $d_1 = ka_{12}$, $d_2 = ka_{22}$, $d_3 = ka_{23}$; then we get the original determinant plus k times a determinant with two identical columns. These properties can often be used to simplify the evaluation of a determinant.

There are many other theorems on determinants. We mention only a few odds and ends. In all evaluations of determinants, we may work with rows or with columns. Hence det $\mathbf{A}$ = det $\mathbf{A}'$. If $\mathbf{A}$ and $\mathbf{B}$ are square matrixes of the same size, det $(\mathbf{AB})$ = det $\mathbf{A} \cdot$ det $\mathbf{B}$.

Finally, determinants also offer a way to compute the inverse of a nonsingular square matrix. If $[b_{ij}] = [a_{ij}]^{-1}$, then $b_{ij} = \mathbf{A}_{ji}/\det \mathbf{A}$: the (i,j) element of the inverse is the cofactor of the (j,i) element divided by the determinant. Also det $\mathbf{A}^{-1} = 1/\det \mathbf{A}$.

B-13. THE LEONTIEF INVERSE

The static input-output model provides a very simple application of elementary matrix algebra. As shown in the text, the assumption of fixed input coefficients leads to the balance equations for total output

$$\mathbf{x} = \mathbf{Ax} + \mathbf{c}$$

where $\mathbf{A} = [a_{ij}]$ is the square $n \times n$ matrix of input coefficients, and $\mathbf{x}$ and $\mathbf{c}$ are n-dimensional column vectors of gross output and final demand, respectively. By their nature, all the a_{ij}, the c's, and the x's must be nonnegative.

We can rewrite the Leontief equations as

$$(\mathbf{I} - \mathbf{A})\mathbf{x} = \mathbf{c}$$

As usual, there are two ways of looking at this. We can think of $\mathbf{I} - \mathbf{A}$ as a linear transformation of R^n which carries gross outputs into final demands. The linearity reflects the basic superposition principle: if $(\mathbf{I} - \mathbf{A})\mathbf{x} = \mathbf{c}$ and $(\mathbf{I} - \mathbf{A})\mathbf{x}^* = \mathbf{c}^*$, then $(\mathbf{I} - \mathbf{A})(\mathbf{x} + \mathbf{x}^*) = \mathbf{c} + \mathbf{c}^*$. To produce the combined final demands, simply add the respective gross outputs. In this view, we are given the vector of final demands, and the problem is to find the gross outputs required.

If $\mathbf{I} - \mathbf{A}$ is a nonsingular mapping, the solution is $\mathbf{x} = (\mathbf{I} - \mathbf{A})^{-1}\mathbf{c}$. Only now this is not quite enough: to make economic sense, all the components of $\mathbf{x}$ must be ≥ 0. Not only must $\mathbf{I} - \mathbf{A}$ have an inverse mapping, but the inverse mapping must have the following property: it must map every vector of nonnegative components into a vector of nonnegative components. For this to be the case, the matrix $[\mathbf{I} - \mathbf{A}]^{-1}$ must have *all nonnegative elements*. If so much as one negative element appears in $[\mathbf{I} - \mathbf{A}]^{-1}$, then there is at least one final demand vector $\mathbf{c}$ which will lead to meaningless negative outputs. Suppose the (h,k) element of $[\mathbf{I} - \mathbf{A}]^{-1}$ is negative; then if $\mathbf{c}$ is a vector with very small components except for a large c_k, it will turn out that x_h is negative. The Hawkins-Simon conditions discussed in the text are necessary and sufficient conditions on $\mathbf{A}$ that ensure that $[\mathbf{I} - \mathbf{A}]^{-1}$ has nonnegative elements.

An alternative view of the linear equations is more akin to linear programming generally. We can think of the columns of $\mathbf{I} - \mathbf{A}$ as describing a "process" or activity for each industry, with one output and many inputs. A linear combination (with nonnegative coefficients) of the columns yields the net flow of commodities through the system if the industrial processes are operated at the levels indicated by the $\mathbf{x}$'s. The problem then is to find a nonnegative linear combination of columns that will yield the given vector of final demands or net commodity flows.

Because of the special form of the Leontief matrix, i.e., $\mathbf{I}$ minus a matrix of nonnegative elements, it is usually especially well behaved. The details can be found in the literature; here we merely indicate a lead. Multiplying out will show that

$$(\mathbf{I} - \mathbf{A})(\mathbf{I} + \mathbf{A} + \mathbf{A}^2 + \cdots + \mathbf{A}^n) = \mathbf{I} - \mathbf{A}^{n+1}$$

Here $\mathbf{A}^n$ means the nth power of $\mathbf{A}$, the matrix $\mathbf{A}$ multiplied by itself n times. Now let $n \to \infty$, and suppose that every element of $\mathbf{A}^{n+1}$ tends to zero.[1] Then in the limit we can write

$$(\mathbf{I} - \mathbf{A})(\mathbf{I} + \mathbf{A} + \mathbf{A}^2 + \cdots + \mathbf{A}^n + \cdots) = \mathbf{I}$$

[1] The elements of $\mathbf{A}^{n+1}$ are $(n + 1)$st-"round" input coefficients. In a productive system one would expect these to dwindle away to zero.

and this says that the infinite geometric series of matrices $\mathbf{I} + \mathbf{A} + \mathbf{A}^2 + \cdots$ is $(\mathbf{I} - \mathbf{A})^{-1}$. This is highly reminiscent of the ordinary geometric series in which $1 + a + a^2 + \cdots = 1/(1 - a) = (1 - a)^{-1}$, provided $-1 < a < 1$. The condition that every element of $\mathbf{A}^{n+1}$ tends to zero is naturally more complicated, and we can't discuss it here. But one fact we can learn: if this condition is satisfied so that

$$(\mathbf{I} + \mathbf{A} + \mathbf{A}^2 + \cdots) = [\mathbf{I} - \mathbf{A}]^{-1}$$

then clearly $[\mathbf{I} - \mathbf{A}]^{-1}$ consists of nonnegative elements only. For $\mathbf{I}$ and $\mathbf{A}$ have nonnegative elements, and so have $\mathbf{A}^2$, $\mathbf{A}^3$, etc., since their elements come from multiplying and adding the elements of $\mathbf{A}$, and in this way no negative numbers are ever generated. All we have shown is that for $\mathbf{A}^{n+1}$ to tend to zero is a sufficient condition for the nonnegativity of $[\mathbf{I} - \mathbf{A}]^{-1}$. It happens also to be necessary.

B-14. INEQUALITIES AND CONES

So far we have discussed linear equations exclusively, while much of the text has been concerned with linear inequalities. The reason for this is straightforward: equations are rather easier to deal with, both theoretically and computationally. In the simplex method, for instance, the very first step is to transmute the constraining inequalities into equations. The other kind of inequality with which we have had to deal in the text is the restriction on many variables to be nonnegative.

First a bit of notation. An inequality $\mathbf{x} < \mathbf{y}$ between *vectors* means that the inequality holds component by component (we compare only vectors of the same number of dimensions). We write $\mathbf{x} \leqq \mathbf{y}$ to mean $x_i \leq y_i$ for each i. It is sometimes convenient to write $\mathbf{x} \leq \mathbf{y}$ to mean $x_i \leq y_i$ for each i and $x_i < y_i$ for at least one i. Graphically (as in Fig. B-5), the shaded area consists of all vectors $\mathbf{x} \leqq \mathbf{y}$. The shaded area, not counting $\mathbf{y}$ itself, represents all vectors $\mathbf{x} \leq \mathbf{y}$. The shaded area less the whole horizontal and vertical parts of its boundary represents all vectors $\mathbf{x} < \mathbf{y}$. The vectors with all nonnegative components form the first quadrant in the plane and what is called the "nonnegative orthant" in R^n.

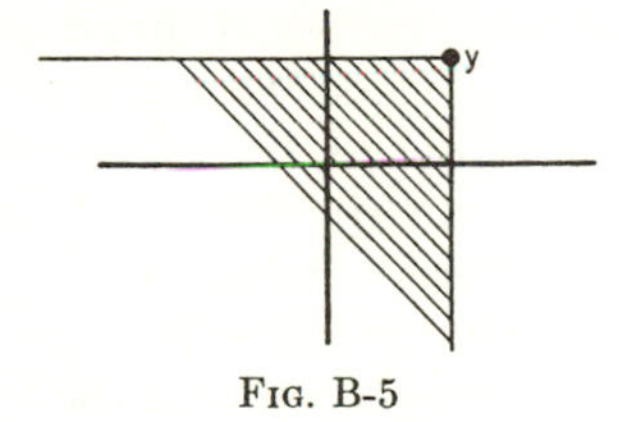

Fig. B-5

In the Walrasian system when we write $\mathbf{Ax} \leq \mathbf{r}$, where $\mathbf{A}$ is $m \times n$, we mean that the vector of outputs in R^n must be mapped into something like the shaded region in the diagram in R^m. Since all $a_{ij} \geq 0$ and since in addition $\mathbf{x} \geq 0$, $\mathbf{Ax}$ will also be in the nonnegative orthant of R^m.

Alternatively, we can think of $\mathbf{Ax}$ as a linear combination of the columns of $\mathbf{A}$, and the inequalities state that only those outputs are feasible which give a linear combination $\leq \mathbf{r}$. Earlier we learned to speak of the set of all linear combinations of the columns of $\mathbf{A}$ as the subspace spanned by the columns of $\mathbf{A}$. Here we have not quite that: we can't speak of *all* linear combinations, but only those with nonnegative coefficients. This restricted set of linear combinations is called the "cone" spanned by the columns of $\mathbf{A}$. The difference is seen in Fig. B-6. The two vectors $\mathbf{A}_1$ and $\mathbf{A}_2$ are linearly independent: they span the whole plane. But if we use only nonnegative linear combinations, we get the shaded area which is indeed a cone. (Multiply $\mathbf{A}_1$ and $\mathbf{A}_2$ by positive scalars and form a parallelogram. All vectors thus obtained will be in the cone.)

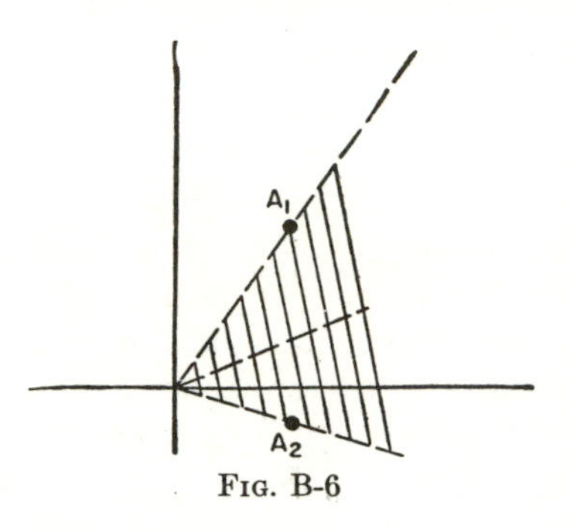

FIG. B-6

In the Walrasian case the columns of $\mathbf{A}$ consist of nonnegative elements. As vectors, they will all lie in the nonnegative orthant of R^m. The cone they span will also lie in the nonnegative orthant, but it will not be the whole nonnegative orthant unless among the columns of $\mathbf{A}$ are all the unit vectors of R^m (i.e., unless each factor is the sole input into at least one commodity). Now vectors in the cone spanned by the columns represent factor endowments that can be exactly used up by at least one nonnegative collection of outputs. Vectors not in the cone represent resource endowments that simply can't be used up, given the technological structure. This is why the Walras-Cassel system may fail if we try to write $\mathbf{Ax} = \mathbf{r}$. As shown in Fig. B-7, $\mathbf{r}$ lies outside the cone spanned by $\mathbf{A}_1$ and $\mathbf{A}_2$. There is no nonnegative vector $\mathbf{x}$ such that $\mathbf{Ax} = \mathbf{r}$. But if we only require $\mathbf{Ax} \leq \mathbf{r}$, then we allow any resource-use vector in the shaded rectangle. The common part of this rectangle and the cone represent the resource-use vectors which are feasible both from the supply and from the demand points of view.

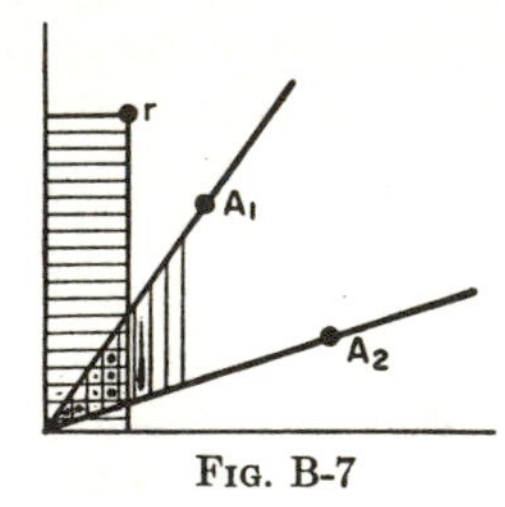

FIG. B-7

B-15. FARKAS'S THEOREM

The theory of linear inequalities and cones is clearly at the bottom of most of the discussion in this book, although we have mentioned this theory explicitly at only a few places. Farkas's theorem typifies the kind of relationship dealt with in this branch of mathematics and is particularly important for programming (cf. Chap. 8). We state and prove it here both because of its own importance and because it illustrates a whole family of theorems and a useful method of proof.

Farkas's Theorem. Let **A** be a matrix of n rows and k columns and let **c** be a vector of k elements. Suppose that for all k-element vectors **x** such that $\mathbf{Ax} \geqq 0$ it is true that the inner product $(\mathbf{c},\mathbf{x}) \geqq 0$. Then **c** must be a nonnegative combination of the rows of **A**; that is, there must exist a nonnegative, n-element vector **v** such that $\mathbf{A}'\mathbf{v} = \mathbf{c}$.

A simple graph will make this theorem intuitively obvious. Figure B-8 represents a k-dimensional space in which the rows of **A**, **c**, and **x** can all be plotted. $\mathbf{A}_1$, $\mathbf{A}_2$, $\mathbf{A}_3$ represent the rows of **A**. $\mathbf{B}_1$ is drawn at right angles to $\mathbf{A}_1$; $\mathbf{B}_2$ is drawn at right angles to $\mathbf{A}_3$. Now consider any vector **x** (not shown) that lies inside the cone $\mathbf{B}_1 0 \mathbf{B}_2$. It will make an acute angle with each of the vectors $\mathbf{A}_i$ (or at most a right angle), and therefore the inner products $(\mathbf{A}_i,\mathbf{x})$ will all be nonnegative. On the other hand, any vector **x** lying outside the cone $\mathbf{B}_1 0 \mathbf{B}_2$ will make an obtuse angle with at least one of the $\mathbf{A}_i$, and therefore, for such a vector, at least one of the inner products $(\mathbf{A}_i,\mathbf{x})$ will be negative. Therefore the cone $\mathbf{B}_1 0 \mathbf{B}_2$ is coextensive with the vectors **x** that have nonnegative inner products with all the vectors $\mathbf{A}_i$.

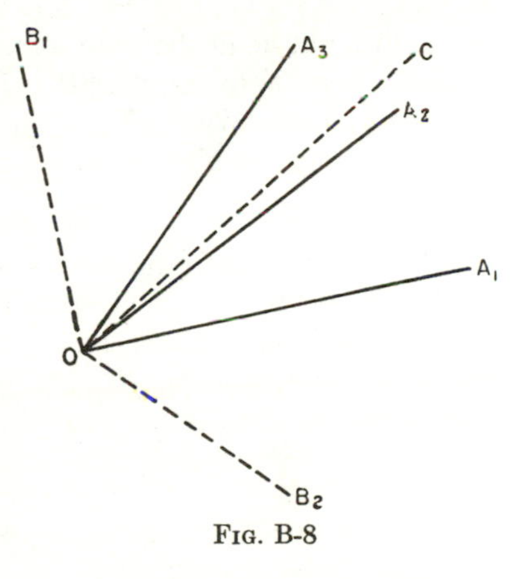

FIG. B-8

Now suppose that $(\mathbf{c},\mathbf{x}) \geqq 0$ for all vectors **x** in the cone $\mathbf{B}_1 0 \mathbf{B}_2$. Where must **c** lie? If it lay to the left of $\mathbf{A}_3$, any vector **x** lying on or sufficiently close to $\mathbf{B}_2$ (but still in the cone $\mathbf{B}_1 0 \mathbf{B}_2$) would make an obtuse angle with it, implying $(\mathbf{c},\mathbf{x}) < 0$. Similarly, if **c** lay to the right of $\mathbf{A}_1$, there would be vectors **x** in the cone, close to $\mathbf{B}_1$, that made an obtuse angle with **c**, implying $(\mathbf{c},\mathbf{x}) < 0$. It follows that if $(\mathbf{c},\mathbf{x}) \geqq 0$ for all vectors **x** in the cone $\mathbf{B}_1 0 \mathbf{B}_2$, **c** must lie somewhere in the cone $\mathbf{A}_1 0 \mathbf{A}_3$, as drawn. In short, any vector in the cone $\mathbf{A}_1 0 \mathbf{A}_3$, spanned by the vectors of the rows of **A**, makes a nonobtuse angle with all vectors in the cone $\mathbf{B}_1 0 \mathbf{B}_2$; no vector outside the cone spanned by the row vectors has this property.

It is clear, then, that if **c** has the assumed property, it lies within the cone framed by the vectors of the rows of **A**. In other words, it is an average of those vectors using nonnegative weights, as the theorem asserts. Thus the theorem is obvious; now we have to prove it.

Instead of proving Farkas's theorem directly, it will be easier to prove the following slight variant formulated by A. W. Tucker.

Tucker's Form of Farkas's Theorem. Let $\mathbf{A}$ be a matrix of n rows and k columns and let $\mathbf{c}$ be a k-element vector. Then either

1. There exists an n-element, nonnegative vector $\mathbf{v}$ such that $\mathbf{A}'\mathbf{v} = \mathbf{c}$ (that is, $\mathbf{c}$ is a nonnegative combination of the rows of $\mathbf{A}$), or

2. There exists a k-element vector $\mathbf{u}$ such that $\mathbf{Au} \geqq 0$ and $(\mathbf{c},\mathbf{u}) < 0$.

Tucker's form is obviously equivalent to Farkas's theorem. If Case 1 obtains, $\mathbf{c}$ is in the cone spanned by the rows of $\mathbf{A}$. If Case 2 obtains, there is some vector $\mathbf{u}$ that makes a nonobtuse angle with every row of $\mathbf{A}$ but makes an obtuse angle with $\mathbf{c}$, which, as we have already seen, means that $\mathbf{c}$ does not lie in the cone spanned by the rows of $\mathbf{A}$. Thus Farkas's theorem amounts to asserting that if Tucker's Case 2 does not apply, then his Case 1 does. We shall now prove Tucker's version of the theorem.

Proof. Consider all vectors $\mathbf{x}$ that can be expressed as

$$\mathbf{x} = \mathbf{A}'\mathbf{v}$$

where $\mathbf{v}$ is understood (and throughout this discussion) to be nonnegative. Each such vector $\mathbf{x}$ is a nonnegative combination of the rows of $\mathbf{A}$, and all such nonnegative combinations qualify as vectors of this type. Graphically these vectors form a kind of pyramid (which we, following accepted, illogical usage, shall call a "cone") with its apex at the origin and with the vectors whose coordinates are the rows of $\mathbf{A}$ along its edges. For example, let

$$\mathbf{A}_i = (a_{i1}, a_{i2}, \ldots, a_{in})$$

be the ith row of $\mathbf{A}$. Then $\mathbf{A}_i$ is in the cone and so is $v_i\mathbf{A}_i$, where v_i is any nonnegative number. Similarly, if $\mathbf{A}_j$ be defined in the same way as $\mathbf{A}_i$, then $v_i\mathbf{A}_i + v_j\mathbf{A}_j$ is a point in the cone, where v_i and v_j are any nonnegative numbers. The same sort of statement is true for combinations of three or more vectors. Note that the cone is convex. That is, if $\mathbf{x}^1$ and $\mathbf{x}^2$ are two points in the cone, then $w\mathbf{x}^1 + (1 - w)\mathbf{x}^2$ is also in the cone, where $0 \leq w \leq 1$.

Now assume that Tucker's Case 1 does not apply, i.e., that there is no nonnegative vector $\mathbf{v}$ such that $\mathbf{A}'\mathbf{v} = \mathbf{c}$. Then $\mathbf{c}$ is not in the cone. But there will be some vector, say $\mathbf{x}^0$, in the cone which lies at minimum possible distance from $\mathbf{c}$.[1] We shall show that the vector $\mathbf{u}$ defined by

$$\mathbf{u} = \mathbf{x}^0 - \mathbf{c}$$

[1] At this point we sweep a technical difficulty under the rug. How do we know that there is a vector in the cone at minimum distance from $\mathbf{c}$? The assertion sounds plausible and is also true; but, since the proof is rather detailed, we leave the statement with an appeal as to its plausibility.

is the vector whose existence is asserted in Case 2 of Tucker's form of the theorem. First note that $\mathbf{x}^0$ is at least as close to $\mathbf{c}$ as any point of the form $w\mathbf{x}^0$ (where w is an ordinary, nonnegative number), since all such points are in the cone. Thus $\mathbf{x}^0$ is the foot of the perpendicular from $\mathbf{c}$ to the line drawn from the origin through $\mathbf{x}^0$ (see Fig. B-9). Therefore,

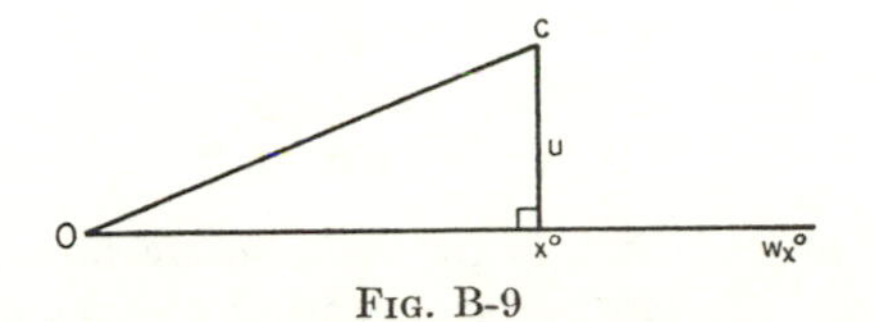

Fig. B-9

in the triangle formed by the origin (denoted by $\mathbf{0}$), $\mathbf{x}^0$, and $\mathbf{c}$ the angle at $\mathbf{x}^0$ is a right angle. Since the two sides that meet at $\mathbf{x}^0$ are $\mathbf{x}^0 - \mathbf{0} = \mathbf{x}^0$ and $\mathbf{x}^0 - \mathbf{c} = \mathbf{u}$, the inner product of these two vectors is zero, and we have

$$(\mathbf{u},\mathbf{x}^0) = 0$$

Consequently the angle at $\mathbf{c}$ must be acute, so

$$(\mathbf{x}^0 - \mathbf{c}, \mathbf{0} - \mathbf{c}) > 0$$

or

$$(\mathbf{u},-\mathbf{c}) > 0$$

or

$$(\mathbf{u},\mathbf{c}) < 0$$

Thus $\mathbf{u}$ satisfies one of the two requirements of Case 2. We now turn to the other, which requires us to show $(\mathbf{u},\mathbf{A}_i) \geq 0$ for all row vectors $\mathbf{A}_i$.

Let $\mathbf{x}$ be an arbitrary point of the cone distinct from $\mathbf{x}^0$ and consider the triangle whose vertices are $\mathbf{x}$, $\mathbf{x}^0$, and $\mathbf{c}$. This triangle is illustrated in Fig. B-10. Note that $\mathbf{u}$ is the side $\mathbf{cx}^0$ of the triangle. The essential fact

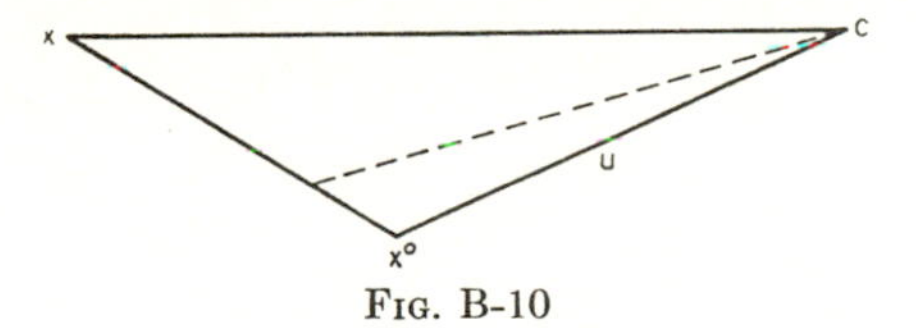

Fig. B-10

about this triangle is that since no point in the cone is closer to $\mathbf{c}$ than $\mathbf{x}^0$ is, side $\mathbf{cx}^0$ ($= \mathbf{u}$) is no longer than side $\mathbf{cx}$ (and, indeed, is no longer than any line that can be drawn from $\mathbf{c}$ to the side $\mathbf{x}^0\mathbf{x}$). This implies that the angle at $\mathbf{x}^0$ is not acute. For, since $\mathbf{cx}$ is at least as long as $\mathbf{cx}^0$, the angle opposite it in the triangle (the angle at $\mathbf{x}^0$) must be at least as great as the angle opposite $\mathbf{cx}^0$ (the angle at $\mathbf{x}$). Therefore if the angle at $\mathbf{x}^0$ were acute, the angle at $\mathbf{x}$ would be acute also and the perpendicular from $\mathbf{c}$ to the side $\mathbf{x}^0\mathbf{x}$ would reach that side at some interior point, thus determining a point in the segment $\mathbf{x}^0\mathbf{x}$ distinct from $\mathbf{x}^0$ and closer to $\mathbf{c}$ than $\mathbf{x}^0$. But since all points in the segment $\mathbf{x}^0\mathbf{x}$ are in the cone, this conclusion is inconsistent with the fact that $\mathbf{x}^0$ is a minimum-distance point of the cone from $\mathbf{c}$. Thus the assumption that the angle at $\mathbf{x}^0$ was acute leads to a contradiction, and we have established that this angle must be either right or obtuse; so

$$(\mathbf{x} - \mathbf{x}^0, \mathbf{c} - \mathbf{x}^0) \leq 0$$

Then

$$(\mathbf{x}^0 - \mathbf{x}, \mathbf{u}) \leq 0$$

and

$$(\mathbf{x}^0,\mathbf{u}) - (\mathbf{x},\mathbf{u}) \leq 0$$

We have already proved that $(\mathbf{x}^0,\mathbf{u}) = 0$, so we find $(\mathbf{x},\mathbf{u}) \geq 0$ for all points $\mathbf{x}$ in the cone. In particular, choosing $\mathbf{x} = \mathbf{A}_i$, we have

$$(\mathbf{A}_i,\mathbf{u}) \geq 0$$

for all i, completing the proof of the theorem.

By the way, it is easily seen that if Case 1 holds, Case 2 does not. For if $\mathbf{A}'\mathbf{v} = \mathbf{c}$, then $(\mathbf{c},\mathbf{x}) = (\mathbf{A}'\mathbf{v},\mathbf{x}) = (\mathbf{v},\mathbf{A}\mathbf{x})$, and if $\mathbf{A}\mathbf{x} \geq 0$ and $\mathbf{v} \geq 0$, the last inner product cannot be negative.

Bibliography

The literature relevant to this volume extends over a very wide range. At one extreme, the properties of convex polyhedral cones in finite dimensional vector spaces constitute the mathematical foundation of the methods of analysis treated herein. At the other extreme, the practical significance of these methods depends on the statistical evidence bearing on the temporal stability of capital-input coefficients and on similar empirical considerations. These matters and all those in the spectrum between them have accumulated extensive literatures in the past decade or more. As a result, any comprehensive bibliography would be so large and would include such a diversity of titles as to provide very little guidance to the reader who wished to pursue these topics further. Besides, there are now available a number of excellent comprehensive bibliographies of the major subdivisions of the topics covered by this book.[1]

Instead, therefore, of presenting an inclusive bibliography we list below a selection of basic works that cover fairly broad divisions of the general field, together with a few specialized contributions that seem necessary for conveying a reasonably rounded picture of the work being done.

LINEAR PROGRAMMING

Antosiewicz, H. A. (ed.), *Proceedings of the Second Symposium in Linear Programming*, 2 vols., National Bureau of Standards, Office of Scientific Publications, Washington, 1955.

Some thirty contributions emphasizing progress in the application of linear programming in the period 1952 to 1955. In addition to papers on practical applications, includes some theoretical papers on the relation of linear programming to economic theory and some contributions to the mathematical theory of linear programming. Participants included, *inter alios*, R. Bellman, A. Charnes, W. W. Cooper, G. B. Dantzig, J. Marschak, L. W. McKenzie, A. Orden, G. Tintner, A. W. Tucker. The papers vary considerably in mathematical difficulty.

[1] We hasten to express our obligation to the bibliographers of linear programming, input-output, game theory, and related topics. We have made liberal use of their work, particularly that of Vera Riley and Robert Loring Allen and of Lily Atiyah and James H. Griesmer.

Arrow, Kenneth J., and Gerard Debreu, "Existence of an Equilibrium for a Competitive Economy," *Econometrica*, **22**:265–290 (1954).

A fundamental paper applying modern analytic techniques to the economic theory of general equilibrium. Mathematically advanced.

Bellman, Richard, *Dynamic Programming*. Princeton University Press, Princeton, N.J. (To be published.)

Problems of optimal resource allocation over time are treated in a novel way which handles linear and nonlinear problems. Closely related to our Chaps. 11 and 12, economists will find that Bellman's theory throws additional light on the theory of capital. Mathematically advanced.

Charnes, A., and W. W. Cooper, "The Stepping-stone Method of Explaining Linear Programming Calculations in Transportation Problems," *Management Science*, **1**:49–69 (1954).

An expository article on the transportation problem.

———, ———, and A. Henderson, *An Introduction to Linear Programming*, John Wiley & Sons, Inc., New York, 1953.

Perhaps the most popular textbook on linear programming. The first half is an elementary exposition of the simplex method of solving linear-programming problems. The second half, which presupposes a moderate amount of mathematical training, develops the main theorems of linear programming by geometric arguments.

Fox, Karl A., "A Spatial Equilibrium Model of the Livestock-feed Economy in the United States," *Econometrica*, **21**:547–566 (1953).

Uses estimates of supply and demand functions and the theory of the transportation problem to determine an optimal distribution of livestock-feed production in 1949 to 1950, which is then compared with observed conditions.

Henderson, A., and R. Schlaifer, "Mathematical Programming," *Harvard Business Review*, **32**:73–100 (May–June, 1954).

A nontechnical survey of the business uses of linear programming.

Henderson, James M., "A Short-run Model for the Coal Industry," *Review of Economics and Statistics*, **37**:336–346 (1955). "Efficiency and Pricing in the Coal Industry," *ibid.*, **38**:50–60 (1956).

Uses the existing patterns of coal consumption, coal production, and costs of production and transportation to determine a least-cost pattern of coal shipments and the associated set of dual royalties. These are compared with observed conditions in the United States for 1947 to 1951.

Hoffman, A. J., M. Mannos, D. Sokolowsky, and N. Wiegmann, "Computational Experience in Solving Linear Programs," *Journal of the Society for Industrial and Applied Mathematics*, **1**:17–33 (1953).

A report on a test comparison of alternative methods of solving linear-programming problems.

Koopmans, T. C. (ed.), *Activity Analysis of Production and Allocation*, John Wiley & Sons, Inc., New York, 1951.

The fundamental collection of papers on the theory and application of linear programming and activity analysis. Includes, among others, Dantzig's basic memoir on the simplex method and Koopmans's major paper on resource allocation. Most of the papers are quite advanced mathematically. Excellent bibliography.

Kuhn, H. W., and A. W. Tucker (eds.), *Contributions to the Theory of Games*, I. Annals of Mathematics Studies no. 24. II. Annals of Mathematics Studies no. 28, Princeton University Press, Princeton, N.J., 1953. *Linear Inequalities and Related Systems*, Annals of Mathematics Studies no. 38, Princeton University Press, Princeton, N.J., 1956.

Three important collections of papers on the mathematical background of linear programming, game theory, and the analysis of linear inequalities. Most of the papers are advanced mathematically. Each volume includes an extensive bibliography.

——— and ———, "Nonlinear Programming," in J. Neyman (ed.), *Second Berkeley Symposium on Mathematical Statistics and Probability*, pp. 481–492, University of California Press, Berkeley, Calif., 1951.

The basic paper on nonlinear programming. Mathematically advanced.

McKenzie, L. W., "On Equilibrium in Graham's Model of World Trade and Other Competitive Systems," *Econometrica*, **22**:147–161 (1954).

An important application of the theory of linear inequalities to a problem in the theory of international trade. Mathematically advanced.

Manne, Alan S., *Scheduling of Petroleum Refinery Operations*, Harvard University Press, Cambridge, Mass., 1956.

Applies linear programming in conjunction with more conventional methods of cost accounting and economic analysis. Includes some consideration of the interpretation of dual variables.

Orden, A., and L. Goldstein (eds.), *Symposium on Linear Inequalities and Programming*, U.S. Air Force and National Bureau of Standards, Washington, 1952.

Twelve papers plus abstracts of a number of others dealing with methods of solving systems of linear inequalities and with applications of linear programming to managerial and business problems. Contributors include A. Charnes, W. W. Cooper, and B. Mellon, G. B. Dantzig, T. S. Motzkin, A. Orden, H. Raiffa, G. L. Thompson, and R. M. Thrall, M. A. Woodbury. Mathematical difficulty varies.

Riley, Vera, *Programming for Policy Decision. I. Mathematical Techniques*, BRS-2, vol. 1, The Johns Hopkins University, Operations Research Office, Chevy Chase, Md., 1954.

A very complete bibliography of linear programming and related branches of mathematics.

Solow, Robert, "On the Structure of Linear Models," *Econometrica*, **20**:29–46 (1952).

An analysis of some quantitative and qualitative necessary conditions for an economy describable by a system of linear equations to possess an economically meaningful solution. Provides an introduction to certain related properties of Leontief models and Metzler-Goodwin-Chipman multisector income-generation models. Mathematically advanced.

Symonds, Gifford H., *Linear Programming: The Solution of Refinery Problems,* Esso Standard Oil Company, New York, 1955.

A privately printed volume containing worked examples of eight typical oil-refinery problems which can be solved by linear-programming methods.

Tucker, A. W., "Linear Programming," *Proceedings, Annual Convention of the American Society for Quality Control, May 1953,* pp. 651–657.

An expository article leaning heavily on geometrical methods.

———, *Game Theory and Linear Programming,* Oklahoma Agricultural and Mechanical College, Stillwater, Okla., 1955.

An excellent textbook of the mathematics of linear programming and game theory. Presupposes a knowledge of elementary college algebra.

Vajda, S., *Theory of Games and Linear Programming,* Methuen & Co., Ltd., London, 1956.

A textbook of the mathematical aspects of game theory and linear programming. Assumes a fair degree of mathematical maturity.

Wald, Abraham, "On Some Systems of Equations of Mathematical Economics," *Econometrica,* **19**:368–403 (1951).

A translation of a pioneering paper in the use of modern algebraic techniques in the analysis of general equilibrium. The original paper by Wald, about 1935, is a precursor of recent developments in the analysis of economic problems by systems of linear inequalities. Addressed to nonmathematical readers.

INPUT-OUTPUT ANALYSIS

Barna, Tibor (ed.), *The Structural Interdependence of the Economy,* John Wiley & Sons, Inc., New York, 1955. A. Giuffre, Milan, 1955.

A symposium volume containing 21 papers on varied aspects of the theory and application of input-output analysis. Among the contributors are H. B. Chenery, W. D. Evans, W. W. Leontief, H. Markowitz, G. Morton, R. N. Stone.

Conference on Research in Income and Wealth, *Input-Output Analysis: An Appraisal,* Studies in Income and Wealth, vol. 18, Princeton University Press, Princeton, N.J., 1955.

Eight papers on the state of input-output analysis as of 1952. Contributions by C. F. Christ, W. D. Evans, and M. Hoffenberg, and W. W. Leontief are particularly valuable.

Evans, W. D., and M. Hoffenberg, "The Interindustry Relations Study for 1947," *The Review of Economics and Statistics,* **34**:97–142 (1952).

Presents the most detailed input-output table yet published, for the United States, 1947. Includes a brief and very lucid exposition of the input-output technique.

Leontief, Wassily W., *The Structure of American Economy, 1919–1939,* 2d ed., Oxford University Press, New York, 1951.

Leontief's basic monograph on his input-output approach. In addition to a statement of the theory, the volume includes input-output tables for the United States for 1919, 1929, and 1939 and three applications of the technique to current economic problems.

——— and Others, *Studies in the Structure of the American Economy*, Oxford University Press, New York, 1953.

Reports by members of the staff of the Harvard Economic Research Project on a number of aspects of input-output analysis. Among the topics treated are the stability of input-output coefficients over time, a dynamic input-output model, and detailed research into the production functions of a number of industries.

Morgenstern, Oskar (ed.), *Economic Activity Analysis*, John Wiley & Sons, Inc., New York, 1954.

Papers by members of the Economics Research Project, Princeton University. Includes a history of input-output analysis and its predecessors, a number of papers on the economic interpretation of the model and on the problems of using it empirically, and some useful (but rather advanced) papers on the mathematical properties of input-output matrices.

Netherlands Economic Institute, "Input-Output Relations," *Proceedings of a Conference on Inter-industrial Relations Held at Driebergen, Holland, 1953*, H. E. Stenfort Kroese N.V., Leyden, 1953.

Proceedings of an international symposium on the economic theory and applications of input-output analysis. Participants included T. Barna, W. D. Evans, R. M. Goodwin, T. C. Koopmans, W. W. Leontief, and R. N. Stone.

Riley, Vera, and R. L. Allen, *Interindustry Economic Studies*, Johns Hopkins Press, Baltimore, 1955.

A very complete bibliography of writings on input-output analysis, including references to the theory of input-output, mathematical aspects, and empirical studies using the technique.

In addition, T. C. Koopmans (ed.), *Activity Analysis of Production and Allocation*, cited above, includes some valuable papers on the theory of input-output analysis.

GAME THEORY

Hurwicz, Leonid, "The Theory of Economic Behavior," *American Economic Review*, **35**:919–925 (1945).

A brief introductory exposition of game theory.

McKinsey, J. C. C., *Introduction to the Theory of Games*, McGraw-Hill Book Company, Inc., New York, 1952.

Probably the best and most complete text on the mathematics of game theory. Presupposes about two years of college mathematics or its equivalent.

Marschak, Jacob, "Neumann's and Morgenstern's New Approach to Static Economics," *Journal of Political Economy*, **54**:97–115 (1946).

An elementary exposition of game theory.

von Neumann, John, and Oskar Morgenstern, *Theory of Games and Economic Behavior*, 3d ed., Princeton University Press, Princeton, N.J., 1953.

The authoritative presentation of the theory of games. Notoriously hard reading. McKinsey, *op. cit.*, supersedes it as far as the mathematical argument goes, but of course no other author can supplant von Neumann and Morgenstern's discussions of the basic philosophic and psychological assumptions and implications of their theory.

Stone, Richard, "The Theory of Games," *Economic Journal*, **58**:185–201 (1948).

An exposition and critique of game theory.

Williams, J. D., *The Compleat Strategyst*, McGraw-Hill Book Company, Inc., New York, 1954.

Easily the most remarkable literary tour de force in the literature of linear programming and game theory. A thoroughly delightful elementary (but accurate, and occasionally profound) presentation of game theory. Presupposes no previous study of mathematics or game theory.

JOURNALS

Much of the current work in linear programming and game theory appears in three journals, all of recent origin:

Management Science
Naval Research Logistics Quarterly
Operations Research (formerly *Journal of the Operations Research Society of America*)

An annual volume of any of these journals will include a half dozen papers or more dealing with specific instances in which linear programming and game theory have been applied to managerial and business problems, or with recent advances in the mathematical theory or methods of computation.

Econometrica frequently publishes papers dealing with economic aspects and implications of linear programming, game theory, and input-output analysis; and the major journals of economics often contain relevant papers.

Index

A

Acceleration principle, nonlinear, 286, 298
Activity (*see* Process)
Activity-analysis production model, 346, 358
  and general equilibrium system, 351
"Acyclicity," Frobenius concept of, 257
Additivity, assumption of, in Leontief statical system, 225
  in von Neumann general-equilibrium model, 300, 302, 304, 384
  in Walrasian system, 348, 405, 407, 415
Aggregate concept defined, 413
Aggregation of industries, 240–245
Aitken, Alexander Craig, 264
Algebra, applied to linear-programming problems, 64–78, 80–104
  matrix, applied to Leontief system, 253–254
    applied to linear programming problems, 470–506
Allocation, optimal, of budget, 27–28
  corresponds to competitive equilibrium, 117, 127
  as dual problem to valuation, 5, 39–63, 101, 174–183, 198
  as frequent economic problem, 1
  of resources, 281–282, 340–341, 391–393
Arrow, Kenneth Joseph, 63*n*.1, 250*n*.2, 353*n*.1, 381*n*.1, 410*n*.1, 413, 415
Atomistic competition, 33, 63, 319, 407
Austrian economists' structure of production, 205, 234, 235, 255
Automobile problem (hypothetical), determination of resource vâlues for, 171–172
  as example of linear programming for the firm, 133–138
  Kuhn-Tucker conditions applied to, 196–198
Automobile problem (hypothetical), in which marginal utilities are variable, 186–188
Autonomous multiplicand terms, 247–248

B

Barone, Enrico, 410*n*.1
Basic program, 149–150
Basic solutions (*see* Solutions, basic)
Basic variables, 81
Basing-point pricing systems, 126
Basis, 480, 492–493
  defined, 79–80, 476–477
  feasible, 80
Basis shifting in simplex method, 85, 90, 110*n*.1
  and degeneracy, 93
  and simultaneous equations, 92
Bilinear forms, 446
Block-triangular systems, 339
Böhm Bawerk, Eugen, 309, 325
Bounded set, 372, 372*n*.2
Brown, George, 446, 460
Budget, optimal allocation of, 27–28
Budget lines, 24–26, 304, 320, 330, 337, 340, 394–396, 402
Business-cycle theories, 286

C

Capital, accumulation of, 267, 309–345
  balanced growth of, 326–335, 344
  decumulation of, 341
  disinvestment of, 286, 325
  fixed, stocks of, 266
  formation of, negative, 323
    zero net, 322
  geometric growth of, 270, 275, 278–279, 326
  price and valuation aspects of, 295
  redundant, 266, 295, 296, 296*n*.2, 305

Capital, steady growth of, 326–335, 343–344
stock, 268, 274–275, 284, 289–296, 302–303
structure of, 285–286
theory of, 265–266
(*See also* Investment)
Capital-output ratios, fixed, 266
Carlson, Sune, 201, 202, 203*n*.1
Cassel, Gustav, 329, 345, 349, 352*n*.1, 353, 365
(*See also* Walras-Cassel model)
Champernowne, David Gawen, 381*n*.2
Chance and game theory, 428–431, 465–469
Chemical problem (hypothetical), 141–154
determination of resource values for, 172–174
dualism of pricing and allocation in, 178–183
Chipman, John Somerset, 254*n*.1, 357*n*.1
Choice variables (*see* Variables, choice)
Clark, John Bates, 204, 249
Closed set, 372, 372*n*.1
Coalitions in game theory, 443–445
Cobweb cycle, 274
Coefficients, detached, table of, 87
direct capital, 292*n*.1, 293–294
of fabrication, 348
flow, 286, 335
input, of Walras-Leontief type, 351
linearly dependent, 72–73
of a process, 30
of production, fixed, assumed by Leontief, 209, 230–231, 236, 247–250, 266, 286–288, 335, 349, 382–383, 499
assumed by Walras, 348–349, 358, 365–366
variable, 365–366
von Neumann input and output, 283*n*.1
Column vectors, 143, 148, 151*n*.2, 480–482, 488, 490
Community-indifference curve, 413
Comparative advantage, theory of, (*See* International trade problems)
Comparative cost, Ricardian theory of, 346
Competition and linear programming, 407–408
Competitive equilibrium, 127, 227, 305
allocation corresponding to, is optimal, 117
defined, 405, 409
Competitive equilibrium, and efficiency, 404–407
and linear programming, 407–408
as Pareto optimum, 409–410
in Walras-Cassel model, 352
and welfare economics, 390
(*See also* Equilibrium; General equilibrium)
Competitive markets, 43–44, 235, 286, 317–322, 340
Complete-description method of solving linear-programming problems, 67, 93–100
basic algebra of, 94–98
compared to simplex method, 100
example of, 98–100
Cones, 502–506
Conservation of discounted value of capital, law of, 322
Consolidation of industries, 240–245
Constant-cost assumptions, 36, 62, 138, 280, 346–347
Constant returns to scale, assumption of, in international-trade example, 41*n*.1
in Leontief statical system, 209, 220, 223, 225, 230, 237, 250, 251
in linear-programming theory, 10, 132, 456
in neoclassical model, 310, 334, 375, 376, 378, 379
in Ramsey model, 267, 271, 276
in Ricardian model, 347
and transformation locus, 281, 282
in von Neumann technology, 283*n*.1, 300, 304, 384
in Walrasian system, 348, 353, 356, 405, 407, 415
Constant-sum games (zero-sum), 101, 419–444, 446, 453, 468–469
Constraints (*see* Restraints)
Consumption, as instantaneous rate of flow, 283*n*.1
in Leontief model, 207, 211, 234, 234*n*.1, 237–238
levels of, maintainable, 322–325, 342–343
theory of, comparison of concepts with linear-programming methods in solving diet problem, 24-31
zero, 28, 280, 296, 322, 322*n*.1, 325, 326*n*.2, 334, 343
Consumption-possibility schedule (frontier), 219–224, 226–227, 322
(*See also* Production-possibility frontier

Consumptionless systems, closed, 325–326
original version of Leontief input-output analysis, 3
Contract curve, 363, 409, 411, 414–415
Edgeworth, 428*n*.1
Convergent multipliers in Leontief system, 253–254, 262–264
Convex curve, 27, 27*n*.1
Convex hull, 226
Convex sets, 372, 372*n*.1, 398, 401, 413
supporting planes of, 399, 401–404, 413
Convexity, 372*n*.2, 396–399, 411, 415
Coordinates, 470–471, 473–474, 476, 476*n*.1
polar, 471*n*.1
Corner solutions, 96, 99
Cornfield, Jerome, 3, 9*n*.1, 254*n*.1
Cornfield-Leontief multiplier process, 253–254, 262–264
Cost accounting, linear programming as aid to, 184
Cost and price relations in statical Leontief system, 234–237
Cost imputation, 294–295
Courant, Richard, 372*n*.1
Cournot, Antoine Augustin, 8, 418, 420, 420*n*.1
Cournot duopoly-reaction path, 274
Cournot point, 428*n*.1
Covariance in statistics, 487
Cramer's rule, 496, 498
Crout, Prescott D., 264

D

Dantzig, George Bernard, 3, 9*n*.1, 14*n*.1, 15*n*.2, 67, 83, 83*n*.1, 110*n*.1, 128, 245*n*.1, 339, 446, 451–452, 454*n*.1, 464
Debreu, Gerard, 63*n*.1, 353*n*.1, 381*n*.1, 410*n*.1, 413, 415
Decentralization, 59–63, 184, 286
Decomposable systems of industries, 255, 257–260
Degeneracy, 92–93
handling of, in transportation problem illustrated, 117–121
(*See also* Nondegeneracy assumption)
Desk calculator, use of, in computing for Leontief system, 264, 264*n*.1
Determinants, algebraic concept of, 70–74, 215, 288*n*.1, 494–499
Diet problem (hypothetical), 9–31, 40
concepts and generalizations on, 28–31
Diet problem (hypothetical), conversion of, to game, 463, 463*n*.1, 464*n*.1
dual of, 21, 45–59
economic price data for, 12–13
as first economic problem solved by linear programming, 9
graphic solution of, 21–24
health standards for, 10
and notion of vector, 474, 476
numerical example for, 13–15
nutritional composition of foods in, 10–12
solution of, by elimination, 15–21
and theory of consumption, 24–28
Diminishing returns, 271–275, 282, 306, 310, 324, 333, 380, 415
Discount rates, 265, 266
Discrete vs. continuous periods of time, 283*n*.1, 310
Disinvestment, 286, 325
Disposal activities, 164, 174
role of, in simplex method, 153–154
Disposal variables (*see* Variables, slack)
Domar, Evsey, D., 326
(*See also* Harrod-Domar type of linear models)
Doolittle, Myrick Hascall, 264, 264*n*.1
Dorfman, Robert, 453*n*.1
Activity Analysis of Production and Allocation, 453, 454*n*.1
"Dosing," concept of, 141
Dual problems, 39–63
in game theory, 101, 451, 454–456
maximizer's problem as dual of minimizer's problem, 438
of Leontief dynamic intertemporal model, 339–340
mathematically described, 39–41, 370
and relationship to direct problems, 40–41, 44, 101–102, 123–124
self-dual case in, 456
in von Neumann's model, 305
Dualism, 100–104
in dynamic system, 270–271
illustrated by chemical problem, 178–183
of pricing and allocation, 174–178
of quantities and prices in Leontief system, 236
Duality theorem, 228*n*.1, 368, 405
Duopoly, example for game theory drawn from, 418–444
Dwyer, Paul Sumner, 89*n*.1, 264
Dynamic efficiency, and competitive markets, 318–322
equations of, nonlinear, 316

E

Economic problems, applicability of game theory to, 2, 417, 419, 430, 431, 444–445
applicability of linear programming to, 8–9
Economic rent, 50, 56, 184, 294–295, 305
Edgeworth, Francis Ysidro, 445
Edgeworth box diagram, 363, 409, 411, 414, 428*n*.1
Efficiency, in capital accumulation, 309–345
and competitive equilibrium, 404–407
and linear programming, 399–404
productive, in welfare economics, 391–408
Efficiency envelope, 312–316, 320, 325, 330, 335–337, 343
Efficiency frontier, 33–36, 310–312, 403–404, 408, 411, 415
Efficiency locus (*see* Transformation locus)
Efficiency prices, 401, 403
Efficient points, 391–395, 400–404, 410
Electronic calculating machines, 25*n*.1, 408
typically required in linear-programming problems, 106
use of, in calculations for Leontief system, 264, 264*n*.1
in computations by simplex method, 84, 89
Elimination method of solution, in case of degenerate problems, 92–93
illustrated by diet problem, 15–21
Engel's curves, 243
Engineer's approach to economic analysis, 131–132
Equilibrium, in closed Leontief system, 3
defined by von Neumann, 329*n*.1, 382, 385
in duopoly, 432–433
(*See also* Competitive equilibrium; General equilibrium)
Equilibrium prices, 60, 207, 227, 235, 345
Equivalent combination, concept of, 163–165, 167, 173, 176
in simplex method, 157, 476, 493
in test for optimality of program, 152, 179–182
Euler's theorem on homogenous functions, 237, 322*n*.1, 328, 378
Evans, Wilmoth Duane, 3, 208*n*.1, 245*n*.1, 309
Excluded activities, 164
gain from, 173–174
Excluded variables, 81, 83, 88–90
Extreme solutions, 95, 97, 99–100

F

Fabrication, coefficients of, 348
Farkas, Julius, 191
Farkas' theorem, 191, 502–506
Feasible program, 30, 149–150, 162–164
Feasible solutions (*see* Solutions, feasible)
Fellner, William John, 420, 420*n*.1
Firm, competitive, 6, 404, 407–408
in economic example of game theory, 418–419, 423–425, 429–435, 442
general formulation of linear-programming problem for, in terms of matrix notation, 160–165
Hotelling's model of, 427–428
hypothetical examples concerning, advertising problem, 425–426
automobile-merger example, 441–442
automobile problem, 133–138
chemical problem, 141–154
increasing-cost example, 138–141
industrial-warfare example, 438–440
suit problem, 154–160
and linear programming, 154, 202
linear programming concept of, 130–133
nonlinear programming vs. conventional methods for, 189
and notion of vector, 474
valuation and duality applied to, 166–185
vertically integrated, 141, 413
Fisher, Irving, 317*n*.2, 325
Fixed-point theorem, 371–372
of Kakutani, 369, 371, 374
of von Neumann, 386
Flow prices, 307, 317–318
Flows, 236, 237, 284–290, 300–301, 307
considered by Carlson, 201
input-output tables of, 204–210
Frobenius, Ferdinand Georg, 254, 255, 257
Frobenius matrix, 297*n*.1

G

Gain, in allocation problem, 183
from excluding activity, 173–174
as value of scarce resource, 174
Gaitskell, Hugh Todd Naylor, 234*n*.1

Gaitskell multiplier chains, 253–254, 262–264
Gale, David, 386*n*.2, 446, 454*n*.1, 459, 460
Gamble, compound, 467
 defined, 466
 standard, 466–468
Game theory, and chance, 428–431, 465–469
 coalitions in, 443–445
 conversion, of linear-programming problems into problems of, 453–459
  of problems of, into linear-programming problems, 446–453, 459
 defined, 2
 and duality, 101, 451, 454–456
 as economic tool, 1, 444–445
 elements of, 417–445
 implications of, for economic, military, and statistical theory, 2, 417, 431, 445
 and input-output analysis, 4–5
 and linear programming, 4–5, 417, 436–442, 446–464
 presuppositions of player in, 422
 and utility, 465–469
Games, with chance moves, 428–431
 constant-sum, 101, 419–444, 446, 453, 468–469
 continuous, 426, 428*n*.1
 defined, 417, 419
 description of, in extensive form, 420
 discrete, 426, 427
 many-person, 419, 442–444
 minorant, 423*n*.2
 non-constant-sum, 442–444, 469
 nonsymmetric, conversion of special problems into, 462–464
 without saddle-points, 431–433, 441
 skew-symmetric, 453–462
 strictly determined, 424–428, 433*n*.1, 444*n*.1, 465
 two-person, 101, 419–444, 446, 453, 468–469
 value of, 438, 447–449, 452
  compared with worth of, 423*n*.2
 worth of, 423
Gauss, Karl Friedrich, 264, 264*n*.1
Gaussian "normal" curve, 200
General equilibrium, and efficiency, 392–393
 and input-output theory, 204
 and linear programming, 7, 346–389
 possibility of substitution in, 249
 stability of, 351*n*.2
General equilibrium, in von Neumann models, 283*n*.1, 300–305, 346, 381–388
 (*See also* Competitive equilibrium; Equilibrium)
Geometric growth of capital, 270, 275, 278–279, 326
Georgescu-Roegen, Nicholas, 381*n*.2, 387*n*.1
Global production function of firm, 6
 possibility of, 4, 4*n*.1, 381
Goodwin, Richard Murphey, 254*n*.1, 286
Graham, Frank Dunstone, 347, 374*n*.1
Gresham, Sir Thomas, 32*n*.1

H

Half spaces, 396
Harrod, Roy Forbes, 326, 329, 345
Harrod-Domar type of linear models, 275
Harvard Mark I computer, 264*n*.1
Hawkins, David, 257
Hawkins-Simon conditions, 215–218, 221, 288, 291, 292*n*.1, 500
Hicks, John Richard, 28*n*.1, 286, 351*n*.2
 Value and Capital, 28*n*.1
Hicksian model of multiple markets, 257
 affinity of Leontief's system to, 254
Hitchcock, Frank Lauren, 106
Hoffenberg, Marvin, 3, 208*n*.1, 245*n*.1
Homogenous equations, 494
Hotelling, Harold, 427, 428

I

Identity transformation of matrices, 485–487
Images, 479–481
Imperfect competition, 27
 (*See also* Perfect competition)
Imputed values, of activity, 194
 in Kuhn-Tucker conditions, 199–200
 and Lagrange multipliers, 197
 test for optimal, 179–183
Included activities, 164
Included variables, 89–91
Income analysis, basic identity of, 229
Income-multiplier analysis, 246
Income stream, formula for present discounted value of, 268
Increasing-cost example of firm (hypothetical), 138–141
Increasing returns to scale, 271, 273–274
Indecomposability, concept of, 254–257, 288*n*.1

Indifference maps, 24–25, 27
Indirect costs, in international-trade problem, 118–120
  in transportation problem, 112–113, 115–116
Industries, consolidation and aggregation of, 240–245
  indecomposable and decomposable groups of, 254–264, 288*n*.1
Inner product of vectors, 146, 146*n*.1, 148, 487–489
Input-output analysis, consolidation and aggregation in, 240–245
  cost and price relations in, 234–237
  described, 2–3, 204
  empirical-algebraic properties of, 253–254
  flow tables for, 204–210
  and game theory, 4–5
  in Leontief dynamic model, 283–300
  in Leontief statical model, 204–264, 266
  indecomposable and decomposable groups of industries in, 254–260
  and linear programming, 4, 204, 210–215, 228
  matrix algebra applied to, 499–501
  money flows in, 204, 237
  as new method of economic analysis, 1, 346
  numerical example of, 260–264
  prices in, 227–264
  production-possibility schedule for, 219–224
  real or nonprice relations in, 230–233
  solving of problems in, 215–219
  substitutability in, 204, 224–227, 248–252
  (*See also* Leontief production models)
Interest rates, 266, 268, 278, 317*n*.2, 334–355, 387–388
Intermediate goods, 326
  in Leontief input-output model, 2–3, 234, 347*n*.1, 355
  in neoclassical model, 375
  in Ricardian model, 347
  in von Neumann general-equilibrium model, 382–383
  in Walras-Cassel model, 348, 351*n*.3, 352, 355–356, 392
International-trade problems (hypothetical), problem of optimal clothing and food production in England and Portugal, 31–38, 40, 346–347
  conversion of, to game, 463
  dual of, 41–45, 59–63
International-trade problems (hypothetical), problem of optimal crop allocation among United Kingdom, France, and Spain, 117–121
  dual of, 127
Intertemporal efficiency conditions, 308
  in Leontief models, 335–345
  in smooth neoclassical models, 310–335
Inventories, accumulation of, 145–146
  in Leontief dynamic system, 6, 266
  in von Neumann general-equilibrium model, 383
Investment, 267
  and disinvestment, 286, 325
  and dynamic economic models, 265–266, 308
  and invisible-hand principle, 319–321
  rate of, in Paretian view, 302
  Williams' game-theory example of, 440–441
Invisible-hand principle, 286, 319–321
Irish sweepstakes as classic example of compound gamble, 467
Irreversibility and maximum rate of disinvestment, 286
Isoprofit contours, 136, 187
Isoquant surfaces, 8, 209, 287, 301–302, 311, 363, 376

J

Jacobs, Eugene Howard, 339
Jevons, William Stanley, 309
Joint production, 223, 227, 230, 252, 267, 278–280, 282, 283, 284*n*.1, 286–288, 300, 335, 347*n*.2, 355, 356, 383
Joint profit in duopoly, 419–420, 443

K

Kakutani fixed-point theorem, 369, 371–374
Keynes, John Maynard, 266*n*.2, 317*n*.2
Keynes-Kahn multiplier chain, 263
Keynesian effective demand, 366*n*.3
Keynesian multicountry income models, 257
  affinities of Leontief's system to, 254
Keynesian unemployment, 365*n*.1
Knight, Frank Hyneman, 429*n*.1
Koopmans, Tjalling C., 4, 128, 245*n*.1, 250*n*.2, 357*n*.1, 390*n*.1
  Activity Analysis of Production and Allocation, 250
Kuhn, Harold William, 189*n*.2, 198, 200, 446, 454*n*.1, 459, 460
Kuhn-Tucker optimality conditions, 189–201

L

Labor, allocation of, in three-crop international-trade example, 117–121
as commodity, 282
in general-equilibrium theory, 347, 358–366
in Leontief system, 205–207, 211–213, 219–221, 224–225, 227–230, 232, 234–240, 249, 253, 260, 322, 326, 347*n*.1
redundancy of, 359
in Ricardo's theory, 60
units of, as measure of value in dual of international-trade example, 127
in von Neumann's model, 382
Laderman, Jack, 9*n*.1, 15*n*.2
Lagrangean multipliers, 15, 196–197, 306, 307, 307*n*.1, 315
Lange, Oskar, 309, 351*n*.2, 410*n*.2
Laplace, Pierre Simon, 350
Leontief, Wassily W., 2, 3, 6, 204, 205, 209, 227, 230, 231*n*.1, 237–240, 244, 245, 245*n*.1, 246, 247*n*.3, 256, 264*n*.1, 266, 283*n*.1, 284, 286, 287, 289, 289*n*.3, 298, 300, 300*n*.1, 323, 325, 340, 341, 355, 385
(*See also* Cornfield-Leontief multiplier process; Walras-Leontief input coefficient)
Leontief production models, dynamic, 6, 283–300
balanced growth in, 295–297
causal indeterminacy of, 297–300
compared with von Neumann model, 300–305
intertemporal efficiency in, 335–345
no-substitution model, 304, 310
unlocked, 266, 384*n*.1
statical, 6, 204–264, 288, 291
closed-end system, 3, 245–248
connection with dynamic models, 266, 283–284, 300
consolidation and aggregation in, 240–245
cost and price relations in, 234–237
empirical algebraic properties of, 253–254
fixed-coefficient assumption in, 209, 230–231, 236, 247–250, 266, 286–288, 335, 349, 382–383, 499
indecomposable and decomposable groups of industries in, 254–260
inverse in, 499–501
labor in, 205–207, 211–213, 219–221, 224–225, 227–230, 232, 234–240, 249, 253, 260, 322, 326, 347*n*.1
linear-programming interpretation of, 210–215
locked, 266, 300
numerical example of, 260–264
open-end system, 236, 238, 245, 260, 288*n*.1, 289
as predictive device, 240
prices in, 227–264
production-possibility schedule for, 219–224
quantitative measurement of, 237–240
real or nonprice relations in, 230–233
substitutability in, 224–227, 248–252
(*See also* Input-output analysis)
Leontief trajectories, 298–300, 340–341, 344
Lerner, Abba Ptachya, 317*n*.2, 410*n*.2
Levels of processes (activities), 132–133, 167, 387*n*.1
and commodity outputs, 356–357
defined, 161–162
indicated by choice variable, 29
and marginal utility, 186
optimal, determined by linear programming, 108, 133, 141
use of input-output analysis to determine, 3
Linear dependence, 72–73, 474–477, 492–494, 496, 498
Linear economics, three branches of, 1–5
Linear programming, algebraic methods in, 64–78, 80–104
and commodities in budget, 28
and competition, 407–408
complete-description method for, 93–100
defined, 3–4, 8, 130
dual problems in, 39–63
and efficiency, 399–404
and firm, 130–165, 202
formal characteristics of problems in, 28–31
formulation of efficient capital program by, 337–339
and game theory, 4–5, 417, 436–442, 446–464
and general equilibrium, 7, 346–389
geometry used in, 78–83
implications of, for economic theory, 4, 8
and input-output analysis, 4, 204, 210–215, 228
and marginal analysis, 141, 183–184
as new method of economic analysis, 1
nondegenerate problems in, 74–78
and nonlinear programming, 198

Linear programming, simplex method for, 67–69, 74–78, 80–93
solution of problems in, 15–16, 66–67
theorems of, stated, 14, 162–165, 179
used by U.S. Air Force, 3–4
value implications of, 166–185
and varying marginal-cost conditions, 140
and vectors, 474
and welfare economics, 7, 390–416
Linear transformations, 478–487, 490–491, 493–494, 500

M

McKenzie, Lionel Wilfred, 374*n*.1
McKinsey, John Charles Chenoweth, 417*n*.3, 443
Macroaggregates in input-output analysis, 204
Macroanalysis vs. microanalysis, 243–245
Macroeconomic application of duality principles illustrated by national-income accounting, 236–240
Majorant and minorant games, 449*n*.1
Malinvaud, Edmond, 416
Malthus, Thomas Robert, 245, 247*n*.2, 281, 296, 325, 326, 326*n*.2, 329, 345
Malthusian economics, 238, 245, 246, 247*n*.2
Malthusian theory of population, 5, 281
Managerial planning, application of linear programming to, 4
Many-person games, 419, 442–444
Mappings, 477–487, 490, 493–494, 500
Marginal analysis, and diminishing marginal utilities, 188
and linear programming, 28, 28*n*.1, 133, 133*n*.1, 141, 165, 183–184
and nonlinear programming, 201–203
Marginal costs, 15, 35–36, 59, 140
in automobile problem, 138
in mixed-food case of diet problem, 50–52
Marginal productivity, 42, 166–169, 177, 250, 250*n*.2, 282, 283*n*.1, 285, 305–306, 308, 310, 375, 377, 405
Marginal rate of substitution, 25, 27, 268–269, 271–277, 308, 310, 312, 318–319, 335, 400, 403, 409, 411, 414–415
diminishing, 186, 306, 333
in Leontief system, 226–228
Marginal utility, 18, 350*n*.1, 411
of any activity defined, 186
diminishing, 27, 186–188
Marginalist theory of production, 6
Market-price ratio, 36
Markoff, Andrei Andreevich, 254
Marschak, Jakob, 417
Marshall, Alfred, 317*n*.2, 367
Marx, Karl, 325
Matrices, algebra of, 470, 479–506
definition of, 143, 143*n*.1, 479–480
determinants of, 494–499
elements of, 480, 482–485
identity, 485–487
input, 217*n*.1
inverse, 20, 217*n*.1, 485–487, 491–492, 495
Leontief system in terminology of, 253, 499–501
as linear transformation, 479, 481
nonsingular, 487, 492–499
notation, 480, 492
advantages of, 470
pay-off, 420–422, 424–426, 428, 429, 431, 435, 439, 443*n*.1, 457
penny-matching, 450, 460
product of, 483–485
and column vectors, 148
and scalars, 482
rank of, 492–495, 498
square, 480, 486, 487, 492–499
sum of, 483
transposed, 487–489
Max-min, 423–427, 437
Maximization and minimization, 28, 65
Maximizing player in game theory, 421–423, 427, 437–438
Menger, Karl, 381*n*.2
Metzler, Lloyd Appleton, 257, 351*n*.2
(*See also* Robertson-Metzler dynamical Keynesian systems)
Military implications of game theory, 2, 417, 431, 445
Mill, John Stuart, 385
Min-max, 423–427
Minimizing player in game theory, 421–423, 427, 438
Minimum-of-subsistence theory of wages, 245
Minkowski, Hermann, 254, 257
Mobilization planning aided by input-output model, 3
Money flows in input-output model, 204, 237
Moral expectation, 465–466
Morgenstern, Oskar, 245*n*.1, 430, 433, 443, 444, 449*n*.1, 465
Morton, G., 464*n*.1
Mosak, Jacob Louis, 257
Motzkin, Theodore Samuel, 67

N

National income, 208
accounting, 236–240
implications of input-output theory for, 204
statistics of, 322–323
National Research Council, 10, 12*n*.1, 59
Neisser, Hans, 360, 365
Neoclassical economics, compared with general-equilibrium system, 375–381
international trade theory of, 267
intertemporal efficiency in smooth case of, 310–335
no-corner transformation locus of, 304
production model for, 305–308
smooth marginal productivities of, 283*n*.1, 285
theories of Walras and Clark in, 204
Neoclassical marginalism, 8
smooth, 283*n*.1, 285
Net National Product, discounted, law of conservation of, 322
Net revenue, defined, 146
of a firm, 166–169
marginal, in Kuhn-Tucker conditions, 199–200
per unit of activity, 172
"Netting out" intrafirm and intra-industry transactions, 231*n*.1, 240, 256
Non-constant-sum games, 442–444, 469
Nondegeneracy assumption, 74–78
Nonlinear dynamic efficiency equations, 316
Nonlinear programming, 186–203
applied to automobile problem, 196–198
compared, with conventional theory of production, 201–203
with linear programming, 198
and dynamic production process, 266, 271–277
general formulations of, 198–201
Kuhn-Tucker optimality conditions in, 189–196
in point-rationing case, 27
problem of, stated, 186–189
Nonnegative orthant, 501
Null vectors, 473, 475, 479, 494

O

Objective function, 90, 104, 186, 198, 395
defined, 28–30
of dual, 42
in game theory, 426–427
nonlinear, 193–194, 198
Oligopolistic indeterminacy, 2, 350, 424
Optimal program (*see* Program, optimal)
Optimal solutions (*see* Solutions, optimal)
Ordered triple of numbers, 471
"Own" inputs, 211, 212*n*.1
Own-rates of interest, 281, 310, 313, 317–320, 325, 334–335

P

Parallelogram of forces, 5*n*.1, 79
Parametric programming, 38
Paretian indexes of ophelimity, 467
Pareto, Vilfredo, 410*n*.1, 468
Pareto-optimality, 302, 391, 408–415
Pay-off function, 426, 429–431, 442, 444
Pay-off matrix, 420–422, 424–426, 429, 431, 435, 439, 443*n*.1, 457
for Hotelling's model, 428
Penny-matching matrix, 450, 460
Perfect competition, 6, 166, 186, 235, 251, 350–351, 408
(*See also* Imperfect competition)
Personnel assignment as illustration of transportation problem, 121
Phase diagram, 333
Point-input, point-output model of Jevons, 309
Point-rationing case, 26–27
Present discounted value of income stream, formula for, 268
Price and programming, dualism of, 174–178, 198
Price ratio, competitive 236, 276, 278, 280, 319–322
Price-unit-cost relations, 357, 369, 378, 392
Prices, absolute and relative, 354
correspondence of, with efficient allocation of resources, 183
with optimal output, 38
efficiency, 401, 403
and equilibrium, 385
flow, 307, 317–318
and interest in steady growth, 334–335
in Leontief system, 224–225, 227–229, 233–237, 265, 270–271
in Walras-Cassel model, 352–354, 367–369
in welfare economics, 400–403
zero, 358–360, 365, 375, 379, 384, 463*n*.1
Probability distribution of pay-offs in game theory, 430–431, 465
Process (activity), alternative, 283*n*.1, 348, 355–356, 381–383
coefficients as essential characteristics of, 30

Process (activity), defined, 29, 132–133, 202–203
described in matrix algebra, 500
disposal, 164, 174
role of, in simplex method, 153–154
distinguished from commodity, 356–357
dynamic, in statical framework, 269
excluded, 164
gain from, 173–174
fictitious, 29, 29*n*.1
included, 164
in Leontief system, 224, 226–227, 301
in linear-programming problems, 29–30
marginal utility of, 186
net and gross revenues per unit of, concept of, 172
in optimal program, 166
unit imputed cost of, 194
unit-level operation of, 211
value of, 30
Process vector, defined, 30
Production function, 130–131, 141, 276, 349, 349*n*.1, 358, 375–378, 428
conventional, 201–203
global, 4, 4*n*.1, 6, 381
in Leontief system, 208–209, 230, 249–250, 274, 281, 286–287, 355
in Ramsey model, 267
smooth neoclassical, 305–308, 310
Production-possibility frontier (curve, schedule), 31, 219–224, 232–233, 267, 282, 289, 349, 373*n*.2
(*See also* Consumption-possibility schedule; Transformation locus)
Productive efficiency in welfare economics, 391–408
Program, 30
basic, 149–150
feasible, 30, 149–150, 162–164
optimal, 108, 149–150
feasible, 30, 162–165
for firm in perfect competition, 166–167
of investment and capital development, 267, 309–345
of nonlinear-programming problems, 189
and Kuhn-Tucker conditions, 194–196
and resource values, 173
test for, 150–154
(*See also* Solutions)
Proportionality, deviations from, 132
Punch cards, 235
use of, in computations by simplex method, 84
Pure variables, 20
Pythagorean theorem of geometry, 489

Q

Quantities and prices, duality between, in Leontief systems, 236, 270–271

R

Raiffa, Howard, 67
Ramsey, Frank, 266*n*.2, 283*n*.1
Ramsey model, 266–282
Recycling, 93
Resources, allocation of, 281–282, 340–341, 391–393
employment of, 358–366
fixed, 184
imputed values of, test for, 179–182
redundancy of, 359, 363, 365
valuation of, 169–174, 177
Restraints, applied to competitive firm in linear-programming problems, 161
as complicating feature of linear-programming problems, 15–16, 26
in dual problems, 42
elimination of dependent variables, with single, 19–20
with two or more, 20–21
as formal characteristic of linear-programming problems, 29
not necessarily satisfied exactly, 16
redundant, 348
Ricardian theory of comparative cost, 346–347
Ricardo, David, 31, 35–37, 60, 385
Robbins, Herbert Ellis, 372*n*.1
Robertson-Metzler dynamical Keynesian systems, 254
Rounds, 216–218, 253–254, 262–263, 500*n*.1
Row vectors, 480, 488
Rubin, Herman, 453*n*.1

S

Saddle-points, 386*n*.1, 388*n*.1, 425–427, 428*n*.1, 431, 433*n*.1, 441, 446
games without, 431–433, 441
Samuelson, Paul Anthony, 59*n*.1, 297*n*.1, 351*n*.2
Say's law, 246
Scalars, 473–474, 479, 482
Self-dual case, 456
Shadow prices, 15, 33, 33*n*.1, 103, 271, 276, 295, 317, 339–340, 344
duality relations of, 266

Shadow prices, Lagrange multipliers as, 307
ratio of, 36
in solution of dual, of diet problem, 47–50, 52–59
of international-trade problem, 61–62
in von Neumann model, 305
Simon, Herbert A., 257
(*See also* Hawkins-Simon conditions)
Simplex criterion, 82, 83, 90, 92, 93, 103, 109
Simplex method of solving linear-programming problems, 493*n*.1
applied to chemical example, 148–154
compared with complete-description method, 100
computation by, 84–85, 501
and degeneracy, 92–93
equivalent combination in, 157, 476, 493
example of, 85–92
finding optimum by, 80–84
fundamental theorems of, 74–78
and game theory, 438, 451
general argument of, 67–69
and simultaneous equations, 69, 89–92
(*See also* Basis shifting)
Simultaneous equations, 69–74, 235, 474, 476, 477, 489–494
existence of direct formulas for solving, 66–67
solution of, in simplex method illustrated, 89–92
Skew symmetry, 454–462, 465, 466
Slack variables, 29*n*.1, 67, 99, 102
"Slough-off" activities, 164, 174
in simplex solution of linear-programming problems, 153–154
Smith, Adam, 8
Smith, Harlan M., 247*n*.3
Social transformation curve, 223
Solow, Robert M., 254*n*.1, 257, 297*n*.1
Solutions, back, 264
basic, attention in solving linear-programming problems restricted to, 78
and basis shifting, 85
in complete-description method, 95, 97, 99–100
in diet problem, 476
and equivalent combination, 163
to nondegenerate linear-programming problems, 78
in simplex method, 69, 82
in transportation problem, 108–109
corner, 96, 99
Solutions, equilibrium, in Walrasian systems, existence of, 349–351, 354–355, 357–375
existence of, 66
extreme, 95, 97, 99–100
feasible, basic, 69, 75, 77–78, 82, 90, 94–95
and equivalent combination, 163
existence of, 104
in graphic approach, 80
in simplex method, 68–69, 75, 77, 83
for linear-programming problems, 66–67
optimal, 30
in simplex method, 68–69, 80–83, 102–104
vertex, 96
(*See also* Program)
Sraffa, Piero, 317*n*.2
Stackelberg, Heinrich von, 358*n*.1, 365, 445
Stackelberg leadership points, 428*n*.1
Stationary state, 325, 385
and maintainable consumption, 342–343
Statistical implications of game theory, 2, 417, 445
Stigler, George Joseph, 9*n*.1
Stock rents, 317–318
Strategies in game theory, dominated, 424–426
evaluation of, 422–424
extraneous, 459
mixed, 431–445, 447–448, 449*n*.1, 457–459, 462, 465
and pay-off matrix, 420–422
pure, 432–434, 441, 447–448, 449*n*.1, 457
symmetrical set of, 461–462
Strong independence axiom, 466, 467
Substitutability, 295, 349
in Leontief systems, 224–227, 248–252 304
Substitution, diminishing rates of, 282
elimination by, 17–19
Suit-manufacturing problem (hypothetical), 154–160
Supporting planes of convex set, 399, 401–404, 413
"Switching rules" of Leontief, 341

T

Table d'hôte rule, 55
Taylor, Fred Manville, 283*n*.1
Taylor series, 332
Thompson, Gerald Luther, 67, 386*n*.2

Thornton, Henry, 317*n*.2
Thrall, Robert McDowell, 67
Transformation curve, 223, 272–273, 276, 320, 323, 347–348, 349*n*.1, 411
Transformation function, 267, 310, 322, 358
Transformation locus (efficiency locus), 281–283, 283*n*.1, 288–294, 302–310, 312, 314, 327, 335–337, 341, 342, 411
(*See also* Production-possibility frontier)
Transportation problem (hypothetical), 106–117
dual of, 122–127
method of, applied to international-trade problem, 117–121, 127
other interpretations of, 121–122
simplicity of solution to, 127–128
Triangular capital model of Böhm Bawerk, 309
"Triangular" system of equations, 128
Triangularity, 205*n*.1, 255, 264
Tucker, Albert William, 189*n*.2, 198, 200, 446, 454*n*.1, 459, 460
form of Farkas' theorem according to, 504–506
(*See also* Kuhn-Tucker optimality conditions)
Two-person games, 101, 419–444, 446, 453, 468–469

U

Unit vectors, 477, 481
United Nations Commission on Living Standards, 33
U.S. Air Force, 3–4, 8, 255
U.S. Bureau of Labor Statistics, 12, 254*n*.1, 255, 262
Utility, numerical measurability of, 465–469
in pay-off matrix, 422, 430

V

Valk, Willem Lodewijk, 365
Valuation, as dual of allocation, 5, 39–63, 101, 174–183, 198
opportunity-cost type of, 172
problem of, for firm formulated, 175
of resources, 166–177
test for imputed, 179–182
(*See also* Imputed values)
Value-of-marginal-product equations, 318, 319*n*.1, 378
Variables, basic, 81
choice, in engineer's approach to economic analysis, 131
as formal characteristic of linear-programming problems, 29–30
in solution of linear-programming problems, 44
in dual problems, 44
excluded (nonbasic), 81, 83, 88–90
included, 89–91
pure, 20
slack, 29*n*.1, 67, 99, 102
Vector spaces, 470–474, 483, 485–486
Vectors, 78–80, 142–143, 470–479, 501–502
column, 143, 148, 151*n*.2, 480–482, 488, 490
inequality between, 501
inner product of, 146, 146*n*.1, 148, 487–489
linearly dependent, 475–477
multiplication of, by scalar, 474, 479
null, 473, 475, 479, 494
process, 30
program, 162
row, 480, 488
unit, 477, 481
Vertex solutions, 96
Vertically integrated firm, 141, 413
von Neumann, John, 1–2, 56, 246, 267, 296, 300, 325, 326*n*.2, 329–332, 344–346, 381–382, 383*n*.1, 385–388, 419, 423*n*.2, 430, 433, 443–444, 446, 449*n*.1, 460–461, 464*n*.1, 465
Theory of Games and Economic Behavior, 2, 445, 450
von Neumann general-equilibrium model, 267, 270, 283*n*.1, 284*n*.1, 300–305, 381–388

W

Wald, Abraham, 352*n*.1, 366–369
Walras, Léon, 7, 28*n*.1, 204, 249, 348, 348*n*.3, 349, 350*n*.1, 351*n*.2, 354, 355*n*.2
Elements of Pure Economics, 283*n*.1, 348
Walras-Cassel model, 346, 348, 351–375, 386, 392, 397, 404, 406
application of matrix algebra to, 494, 501–502
compared with neoclassical model, 375–381
question of existence of solution for, 349–351, 354–355, 357–375
Walras' law, 353–354, 379

Walras-Leontief input coefficient, 351
Welfare economics, 1, 207, 267, 321, 351*n*.1, 373*n*.2, 380
basic theorem of, 408–416
extended over time, 415–416
and linear programming, 7, 390–416
Whirlpools of interdependence in Leontief models, 205, 234, 234*n*.1, 264, 375
Wicksell, Knut, 317*n*.2
Wicksteed, Philip Henry, 249
Wilbur-Kelvin analog computer at Massachusetts Institute of Technology, 264*n*.1
Williams, John Davis, 417*n*.3, 435, 440–441

Z

Zero consumption, 28, 280, 296, 322, 322*n*.1, 325, 326*n*.2, 334, 343
Zero prices, 358–360, 365, 375, 379, 384, 463*n*.1
Zero-profit condition, 49, 53–55, 61–62, 247*n*.1, 276, 305, 385, 387–388, 406–407
Zero-sum games (*see* Constant-sum games)
Zeuthen, Frederik, 360

A CATALOG OF SELECTED
DOVER BOOKS
IN SCIENCE AND MATHEMATICS

CHALLENGING MATHEMATICAL PROBLEMS WITH ELEMENTARY SOLUTIONS, A.M. Yaglom and I.M. Yaglom. Over 170 challenging problems on probability theory, combinatorial analysis, points and lines, topology, convex polygons, many other topics. Solutions. Total of 445pp. 5⅜ × 8½. Two-vol. set.
Vol. I 65536-9 Pa. $5.95
Vol. II 65537-7 Pa. $5.95

FIFTY CHALLENGING PROBLEMS IN PROBABILITY WITH SOLUTIONS, Frederick Mosteller. Remarkable puzzlers, graded in difficulty, illustrate elementary and advanced aspects of probability. Detailed solutions. 88pp. 5⅜ × 8½.
65355-2 Pa. $3.95

EXPERIMENTS IN TOPOLOGY, Stephen Barr. Classic, lively explanation of one of the byways of mathematics. Klein bottles, Moebius strips, projective planes, map coloring, problem of the Koenigsberg bridges, much more, described with clarity and wit. 43 figures. 210pp. 5⅜ × 8½. 25933-1 Pa. $4.95

RELATIVITY IN ILLUSTRATIONS, Jacob T. Schwartz. Clear non-technical treatment makes relativity more accessible than ever before. Over 60 drawings illustrate concepts more clearly than text alone. Only high school geometry needed. Bibliography. 128pp. 6⅛ × 9¼. 25965-X Pa. $5.95

AN INTRODUCTION TO ORDINARY DIFFERENTIAL EQUATIONS, Earl A. Coddington. A thorough and systematic first course in elementary differential equations for undergraduates in mathematics and science, with many exercises and problems (with answers). Index. 304pp. 5⅜ × 8¼. 65942-9 Pa. $7.95

FOURIER SERIES AND ORTHOGONAL FUNCTIONS, Harry F. Davis. An incisive text combining theory and practical example to introduce Fourier series, orthogonal functions and applications of the Fourier method to boundary-value problems. 570 exercises. Answers and notes. 416pp. 5⅜ × 8½. 65973-9 Pa. $8.95

THE THOERY OF BRANCHING PROCESSES, Theodore E. Harris. First systematic, comprehensive treatment of branching (i.e. multiplicative) processes and their applications. Galton-Watson model, Markov branching processes, electron-photon cascade, many other topics. Rigorous proofs. Bibliography. 240pp. 5⅜ × 8½. 65952-6 Pa. $6.95

AN INTRODUCTION TO ALGEBRAIC STRUCTURES, Joseph Landin. Superb self-contained text covers "abstract algebra": sets and numbers, theory of groups, theory of rings, much more. Numerous well-chosen examples, exercises. 247pp. 5⅜ × 8½. 65940-2 Pa. $6.95

GAMES AND DECISIONS: Introduction and Critical Survey, R. Duncan Luce and Howard Raiffa. Superb non-technical introduction to game theory, primarily applied to social sciences. Utility theory, zero-sum games, n-person games, decision-making, much more. Bibliography. 509pp. 5⅜ × 8½. 65943-7 Pa. $10.95

Prices subject to change without notice.

Available at your book dealer or write for free Mathematics and Science Catalog to Dept. GI, Dover Publications, Inc., 31 East 2nd St., Mineola, N.Y. 11501. Dover publishes more than 175 books each year on science, elementary and advanced mathematics, biology, music, art, literary history, social sciences and other areas.